Advanced Linear Algebra

SECOND EDITION

The second edition of this engaging textbook for advanced undergraduate students and first-year graduate students covers all the core subjects in linear algebra.
  Key features include:

- An emphasis on ideas from analysis, in addition to pure algebra, and a balance of abstraction, practicality, and applications.
- Four chapters tackling some of the most important contemporary applications, including quantum mechanics, machine learning, data science, and quantum information and computation.
- A rich collection of exercises designed for students from diverse backgrounds, including a wealth of newly added ones in this edition.
- A full solution manual available to instructors at www.cambridge.org/Yang2e.
- Selected solutions provided at the back of the book, making it suitable for self-study as well as for use as a course text.

YISONG YANG is Professor at the Courant Institute of Mathematical Sciences, New York University. He is a fellow of the American Mathematical Society and the author of *Solitons in Field Theory and Nonlinear Analysis* (2001) and *Mathematical Physics with Differential Equations* (2023).

Advanced Linear Algebra

A Concise Text with Contemporary Applications

SECOND EDITION

YISONG YANG
New York University

CAMBRIDGE
UNIVERSITY PRESS

Shaftesbury Road, Cambridge CB2 8EA, United Kingdom

One Liberty Plaza, 20th Floor, New York, NY 10006, USA

477 Williamstown Road, Port Melbourne, VIC 3207, Australia

314–321, 3rd Floor, Plot 3, Splendor Forum, Jasola District Centre,
New Delhi – 110025, India

103 Penang Road, #05–06/07, Visioncrest Commercial, Singapore 238467

Cambridge University Press is part of Cambridge University Press & Assessment,
a department of the University of Cambridge.

We share the University's mission to contribute to society through the pursuit of
education, learning and research at the highest international levels of excellence.

www.cambridge.org
Information on this title: www.cambridge.org/9781009589840

DOI: 10.1017/9781009589826

First published 2015
Second edition 2025

Cover image: djgunner/iStock/Getty Images Plus

A catalogue record for this publication is available from the British Library

Library of Congress Cataloging-in-Publication Data
Names: Yang, Yisong, author.
Title: Advanced linear algebra : a concise text with contemporary
applications / Yisong Yang, New York University.
Description: Second edition. | Cambridge, United Kingdom ; New York, NY,
USA : Cambridge University Press, 2025. | Includes bibliographical
references and index.
Identifiers: LCCN 2024054710 | ISBN 9781009589840 (paperback) | ISBN
9781009589826 (ebook)
Subjects: LCSH: Algebras, Linear – Textbooks. | Algebras, Linear – Study and
teaching (Higher) | Algebras, Linear – Study and teaching (Graduate)
Classification: LCC QA184.2 .Y36 2025 | DDC 512/.5–dc23/eng/20250111
LC record available at https://lccn.loc.gov/2024054710

ISBN 978-1-009-58984-0 Paperback

Additional resources for this publication at www.cambridge.org/Yang2e.

For Sheng,
Peter, Anna, and Julia

Contents

Preface to the Second Edition

In this edition, a significant expansion is made to meet the interests of applying linear algebra to the development of some emerging areas of application of contemporary research, including machine learning, data analysis, and quantum information and quantum computation. For this purpose, four more chapters, in the form of subject excursions, as in Chapter 9, are added.

In Chapter 10, we present some fundamental problems in supervised machine learning based on methods of regression, data separation and classification, and optimization. We emphasize how the problems are formulated, approached, and eventually solved.

In Chapter 11, we present two important methods in data analysis: the low-rank data approximation and principal component analysis. These methods are essential for data compression and modeling. We emphasize understanding the underlying structures and effectiveness of the ideas and methods developed.

In Chapter 12, we study multilinear algebra. This subject is a natural continuation of the study of linear algebra and plays an indispensable role in many areas of mathematics, physics, and engineering. We focus on the foundations of the subject and consider a few important applications as illustrations.

In Chapter 13, we present an introduction to quantum information, quantum computation, and quantum entanglement. First, we consider quantum bits and their representation and processing by quantum gates. Next, we discuss untangled and entangled states. Then, we present the Bell inequality and Bell's theorem and discuss their profound quantum-mechanical implications. This chapter is also an area of application of the subjects of Chapter 12.

The original text (Chapters 1–9) has remained mostly intact, except for minor wording improvements in a few places. Taking advantage of this opportunity, exercise problems have further been modified and expanded, based on past experience using the first edition of the book in class. Solutions to all exercises are available to instructors via www.cambridge.org/Yang2e.

xi

The author would like to thank John Peloquin for valuable input given to him on several parts of the book, especially some of the contents of Chapter 12, and Roger Astley and Holly Paveling of Cambridge University Press for advice and suggestions that made his experience in preparing this edition rewarding.

Preface to the First Edition

This book is concisely written to provide comprehensive core materials for a year-long course in linear algebra for senior undergraduate and beginning graduate students in mathematics, science, and engineering. Students who gain a profound understanding and grasp of the concepts and methods of this course will acquire an essential knowledge foundation to excel in their future academic endeavors.

Throughout the book, the methods and ideas of analysis are greatly emphasized and used, along with those of algebra, wherever appropriate, and a delicate balance is established between the abstract formulation and practical origins of various subject matters.

The book is divided into nine chapters. The first seven chapters embody a traditional course curriculum. An outline of the contents of these chapters is sketched as follows.

In Chapter 1, we cover basic facts and properties of vector spaces. These include definitions of vector spaces and subspaces, concepts of linear dependence, bases, coordinates, dimensionality, dual spaces and dual bases, quotient spaces, normed spaces, and the equivalence of the norms of a finite-dimensional normed space.

In Chapter 2, we cover linear mappings between vector spaces. We start from the definition of linear mappings and discuss how linear mappings may be concretely represented by matrices with respect to given bases. We then introduce the notion of adjoint mappings and quotient mappings. Linear mappings from a vector space into itself comprise a special but important family of mappings and are given a separate treatment later in this chapter. Topics studied there include invariance and reducibility, eigenvalues and eigenvectors, projections, nilpotent mappings, and polynomials of linear mappings. We end the chapter with a discussion of the concept of the norms of linear mappings and use it to show that being invertible is a generic property of a linear mapping and then to show how the exponential of a linear mapping may be constructed and understood.

In Chapter 3, we cover determinants. As a nontraditional but highly motivating example, we show that the calculation of the topological degree of a differentiable map from a closed curve into the unit circle in $\mathbb{R}^2$ involves computing a 2×2 determinant, and the knowledge gained allows us to prove the Fundamental Theorem of Algebra. We then formulate the definition of a general determinant inductively, without resorting to the notion of permutations, and establish all its properties. We end the chapter by establishing the Cayley–Hamilton theorem. Two independent proofs of this important theorem are given. The first proof is analytic and consists of two steps. In the first step, we show that the theorem is valid for a matrix of distinct eigenvalues. In the second step, we show that any matrix may be regarded as a limiting point of a sequence of matrices of distinct eigenvalues. Hence the theorem follows again by taking limit. The second proof, on the other hand, is purely algebraic.

In Chapter 4, we discuss vector spaces with scalar products. We start from the most general notion of scalar products without requiring either nondegeneracy or positive definiteness. We then carry out detailed studies on nondegenerate and positive definite scalar products, respectively, and elaborate on adjoint mappings in terms of scalar products. We end the chapter by discussing isometric mappings in both real and complex space settings and noting their subtle differences.

In Chapter 5, we focus on real vector spaces with positive definite scalar products and quadratic forms. We first establish the main spectral theorem for self-adjoint mappings. We will not take the traditional path of first using the Fundamental Theorem of Algebra to assert that there is an eigenvalue and then applying the self-adjointness to show that the eigenvalue must be real. Instead, we will formulate an optimization problem and use calculus to prove directly that a self-adjoint mapping must have a real eigenvalue. We then present a series of characteristic conditions for a symmetric bilinear form, a symmetric matrix, or a self-adjoint mapping, to be positive definite. We end the chapter with a discussion of the commutativity of self-adjoint mappings and the usefulness of self-adjoint mappings for the investigation of linear mappings between different spaces.

In Chapter 6, we study complex vector spaces with Hermitian scalar products and related notions. Much of the theory here is parallel to that of the real-space situation with the exception that normal mappings can only be fully understood and appreciated within a complex space formalism.

In Chapter 7, we establish the Jordan decomposition theorem. We start with a discussion of some basic facts regarding polynomials. We next show how to reduce a linear mapping over its generalized eigenspaces via the Cayley–Hamilton theorem and the prime factorization of the characteristic polynomial of the mapping. We then prove the Jordan decomposition theorem. The key and often the most difficult step in this construction is a full understanding of how a nilpotent mapping is reduced canonically. We approach this problem inductively with the degree of a nilpotent mapping and show that it is crucial to tackle a mapping of degree 2. Such a treatment eases the subtlety of the subject considerably.

In Chapter 8, we present four selected topics that may be used as materials for some optional extra-curricular study when time and interest permit. In Section 8.1, we present the Schur decomposition theorem that may be viewed as a complement to the Jordan decomposition theorem. In Section 8.2, we give a classification of skew-symmetric bilinear forms. In Section 8.3, we state and prove the Perron–Frobenius theorem regarding the principal eigenvalues of positive matrices. In Section 8.4, we establish some basic properties of the Markov matrices.

In Chapter 9, we present yet another selected topic for the purpose of optional extra-curricular study: a short excursion into quantum mechanics using gadgets purely from linear algebra. Specifically, we will use $\mathbb{C}^n$ as the state space and Hermitian matrices as quantum-mechanical observables to formulate the over-simplified quantum-mechanical postulates, including Bohr's statistical interpretation of quantum mechanics and the Schrödinger equation governing the time evolution of a state. We next establish Heisenberg's uncertainty principle. Then, we prove the equivalence of the Schrödinger description via the Schrödinger equation and the Heisenberg description via the Heisenberg equation, of quantum mechanics.

Also provided in the book is a rich collection of mostly proof-oriented exercises to supplement and consolidate the main course materials. The diversity and elasticity of these exercises aim to satisfy the needs and interests of students from a wide variety of backgrounds.

At the end of the book, solutions to some selected exercises are presented. These exercises and solutions provide additional illustrative examples, extend the main course materials, and help the reader to master the subjects and methods covered in a broader range.

Finally some bibliographic notes conclude the book.

This text may be curtailed to meet the time constraints of a semester-long course. Here is a suggested list of selected sections for such a plan: Sections 1.1–1.5, 2.1–2.3, 2.5, 3.1.2, 3.2, and 3.3 (present the concept of adjugate matrices only); Section 3.4 (give the second proof of the Cayley–Hamilton theorem only, based on an adjugate matrix expansion); Sections 4.3, 4.4, 5.1, and 5.2

(omit the analytic proof that a self-adjoint mapping must have an eigenvalue but resort to Problem 5.2.1 instead); and Sections 5.3, 6.1, 6.2, 6.3.1, and 7.1–7.4. Depending on the pace of lectures and time available, the instructor may decide in the later stage of the course to what extent the topics in Sections 7.1–7.4 (the Jordan decomposition) can be presented productively.

The author would like to thank Patrick Lin, Thomas Otway, and Robert Sibner for constructive comments and suggestions, and Roger Astley of Cambridge University Press for valuable editorial advice, which helped improve the presentation of this book.

Notations and Conventions

We use $\mathbb{N}$ to denote the set of all natural numbers,

$$\mathbb{N} = \{0, 1, 2, \dots\},$$

and $\mathbb{Z}$ to denote the set of all integers,

$$\mathbb{Z} = \{\dots, -2, -1, 0, 1, 2, \dots\}.$$

We use i to denote the imaginary unit $\sqrt{-1}$. For a complex number $c = a + ib$, where a, b are two real numbers, we use

$$\bar{c} = a - ib$$

to denote the complex conjugate of c. We use $\Re\{c\}$ and $\Im\{c\}$ to denote the real and imaginary parts of the complex number $c = a + ib$. That is,

$$\Re\{c\} = a, \quad \Im\{c\} = b.$$

We use i, j, k, l, m, n to denote integer-valued indices or space dimension numbers; a, b, c to scalars; u, v, w, x, y, z to vectors; A, B, C, D to matrices; P, R, S, T to mappings; and U, V, W, X, Y, Z to vector spaces, unless otherwise stated.

We use t to denote the variable in a polynomial or a function or the transpose operation on a vector or a matrix.

When X or Y is given, we use $X \equiv Y$ to denote that Y, or X, is defined to be X, or Y, respectively.

Occasionally, we use the symbol $\forall$ to express "for all."

Let X be a set and Y, Z subsets of X. We use $Y \backslash Z$ to denote the subset of elements in Y that are not in Z.

Let S be a finite set. We use the notation $\#S$ to denote the number of elements in S.

We use the symbol $\hat{\ }$ to indicate an item that is left out at the place of an expression.

For convenience and conciseness, trivial situations are usually excluded from our general statements of results, unless otherwise specified.

1

Vector Spaces

In this chapter, we study vector spaces and their basic properties and structures. We start by stating the definition and discussing examples of vector spaces. Next, we introduce the notions of subspaces, linear dependence, bases, coordinates, and dimensionality. Then, we consider dual spaces, direct sums, and quotient spaces. Finally, we cover normed vector spaces.

1.1 Vector spaces

A *vector space* is a non-empty set consisting of elements called *vectors* that can be added and multiplied by some quantities called *scalars*. We start this section with the study of vector spaces.

1.1.1 Fields

The scalars to operate on vectors in a vector space are required to form a *field*, which may be denoted by $\mathbb{F}$, where two operations, usually called addition, denoted by "$+$," and multiplication, denoted by "$\cdot$" or omitted, over $\mathbb{F}$ are performed between scalars, such that the following axioms are satisfied:

(1) (Closure) If $a, b \in \mathbb{F}$, then $a + b \in \mathbb{F}$ and $ab \in \mathbb{F}$.

(2) (Commutativity) For $a, b \in \mathbb{F}$, there hold $a + b = b + a$ and $ab = ba$.

(3) (Associativity) For $a, b, c \in \mathbb{F}$, there hold $(a + b) + c = a + (b + c)$ and $a(bc) = (ab)c$.

(4) (Distributivity) For $a, b, c \in \mathbb{F}$, there holds $a(b + c) = ab + ac$.

(5) (Existence of zero) There is a scalar, called zero, denoted by 0, such that $a + 0 = a$ for any $a \in \mathbb{F}$.

(6) (Existence of unity) There is a scalar different from zero, called one, denoted by 1, such that $1a = a$ for any $a \in \mathbb{F}$.

1

(7) (Existence of additive inverse) For any $a \in \mathbb{F}$, there is a scalar denoted by $-a$ or $(-a)$, such that $a + (-a) = 0$.

(8) (Existence of multiplicative inverse) For any $a \in \mathbb{F} \setminus \{0\}$, there is a scalar denoted by a^{-1}, such that $aa^{-1} = 1$.

It is easily seen that zero, unity, and additive and multiplicative inverses are all unique. Besides, a field consists of at least two elements.

With the usual addition and multiplication, the sets of rational numbers, real numbers, and complex numbers denoted by $\mathbb{Q}$, $\mathbb{R}$, and $\mathbb{C}$, respectively, are all fields. These are infinite fields. However, the set of integers, $\mathbb{Z}$, is not a field because there is a lack of multiplicative inverses for its non-unit elements.

Let p be a prime $(p = 2, 3, 5, \ldots)$ and set $p\mathbb{Z} = \{n \in \mathbb{Z} \mid n = kp, k \in \mathbb{Z}\}$. Classify $\mathbb{Z}$ into the so-called cosets modulo $p\mathbb{Z}$, that is, some non-overlapping subsets of $\mathbb{Z}$ represented as $[i]$ $(i \in \mathbb{Z})$, such that

$$[i] = \{j \in \mathbb{Z} \mid i - j \in p\mathbb{Z}\}. \tag{1.1.1}$$

It is clear that $\mathbb{Z}$ is divided into exactly p cosets, $[0], [1], \ldots, [p-1]$. Use $\mathbb{Z}_p$ to denote the set of these cosets and pass the additive and multiplicative operations in $\mathbb{Z}$ over naturally to the elements in $\mathbb{Z}_p$ so that

$$[i] + [j] = [i+j], \quad [i][j] = [ij]. \tag{1.1.2}$$

It can be verified that, with these operations, $\mathbb{Z}_p$ becomes a field with its obvious zero and unit elements, $[0]$ and $[1]$. Of course, $p[1] = [1] + \cdots + [1]$ (p terms)$= [p] = [0]$. In fact, p is the smallest positive integer whose multiplication with the unit element results in the zero element. A number of such a property is called the *characteristic of the field*. Thus, $\mathbb{Z}_p$ is a *field of characteristic p*. For $\mathbb{Q}, \mathbb{R}$, and $\mathbb{C}$, since no such integer exists, we say that these fields are of *characteristic 0*. Thus, if we use $\mathrm{ch}(\mathbb{F})$ to denote the characteristic of a field $\mathbb{F}$, then $\mathrm{ch}(\mathbb{Z}_p) = p$ and $\mathrm{ch}(\mathbb{Q}) = 0$.

1.1.2 Vector spaces

Let $\mathbb{F}$ be a field. Consider the set of n-tuples denoted by $\mathbb{F}^n$, with the elements called vectors arranged in a row or column, such as

$$\begin{pmatrix} a_1 \\ \vdots \\ a_n \end{pmatrix} \quad \text{or} \quad (a_1, \ldots, a_n), \quad \text{where } a_1, \ldots, a_n \in \mathbb{F}. \tag{1.1.3}$$

Furthermore, we can define the addition of two vectors and the scalar multiplication of a vector by a scalar following the rules such as

$$\begin{pmatrix} a_1 \\ \vdots \\ a_n \end{pmatrix} + \begin{pmatrix} b_1 \\ \vdots \\ b_n \end{pmatrix} = \begin{pmatrix} a_1 + b_1 \\ \vdots \\ a_n + b_n \end{pmatrix}, \tag{1.1.4}$$

$$\alpha \begin{pmatrix} a_1 \\ \vdots \\ a_n \end{pmatrix} = \begin{pmatrix} \alpha a_1 \\ \vdots \\ \alpha a_n \end{pmatrix}, \text{ where } \alpha \in \mathbb{F}. \tag{1.1.5}$$

The set $\mathbb{F}^n$, modeled over the field $\mathbb{F}$ and equipped with the above operations, is a prototype example of a vector space.

More generally, we say that a set U is a *vector space over a field* $\mathbb{F}$ if U is non-empty, and there is an operation called *addition*, denoted by "+," between the elements of U, called *vectors*, and another operation called *scalar multiplication* between elements in $\mathbb{F}$, called *scalars*, and vectors, such that the following axioms hold:

(1) (Closure) For $u, v \in U$, we have $u + v \in U$. For $u \in U$ and $a \in \mathbb{F}$, we have $au \in U$.
(2) (Commutativity) For $u, v \in U$, we have $u + v = v + u$.
(3) (Associativity of addition) For $u, v, w \in U$, we have $u + (v + w) = (u + v) + w$.
(4) (Existence of zero vector) There is a vector, called zero and denoted by 0, such that $u + 0 = u$ for any $u \in U$.
(5) (Existence of additive inverse) For any $u \in U$, there is a vector denoted by $(-u)$, such that $u + (-u) = 0$.
(6) (Associativity of scalar multiplication) For any $a, b \in \mathbb{F}$ and $u \in U$, we have $a(bu) = (ab)u$.
(7) (Property of unit scalar) For any $u \in U$, we have $1u = u$.
(8) (Distributivity) For any $a, b \in \mathbb{F}$ and $u, v \in U$, we have $(a + b)u = au + bu$ and $a(u + v) = au + av$.

As in the case of the definition of a field, we see that it readily follows from the definition that zero vector and additive inverse vectors are all unique in a vector space. Besides, any vector multiplied by zero scalar results in zero vector. That is, $0u = 0$ for any $u \in U$.

Other examples of vector spaces (with obviously defined vector addition and scalar multiplication) include the following:

(1) The set of all polynomials with coefficients in $\mathbb{F}$ defined by

$$\mathcal{P} = \{a_0 + a_1 t + \cdots + a_n t^n \mid a_0, a_1, \ldots, a_n \in \mathbb{F}, n \in \mathbb{N}\}, \tag{1.1.6}$$

where t is a variable parameter.

(2) The set of all real-valued continuous functions over the interval $[a,b]$ for $a,b \in \mathbb{R}$ and $a < b$ usually denoted by $C[a,b]$.

(3) The set of real-valued solutions to the differential equation

$$a_n \frac{d^n x}{dt^n} + \cdots + a_1 \frac{dx}{dt} + a_0 x = 0, \quad a_0, a_1, \ldots, a_n \in \mathbb{R}. \tag{1.1.7}$$

(4) In addition, we can also consider the set of arrays of scalars in $\mathbb{F}$ consisting of m rows of vectors in $\mathbb{F}^n$ or n columns of vectors in $\mathbb{F}^m$ of the form

$$(a_{ij}) = \begin{pmatrix} a_{11} & a_{12} & \cdots & a_{1n} \\ a_{21} & a_{22} & \cdots & a_{2n} \\ \cdots & \cdots & \cdots & \cdots \\ a_{m1} & a_{m2} & \cdots & a_{mn} \end{pmatrix}, \tag{1.1.8}$$

where $a_{ij} \in \mathbb{F}, i = 1, \ldots, m, j = 1, \ldots, n$, called an $m \times n$ *matrix* and each a_{ij} is called an *entry* or *component* of the matrix. The set of all $m \times n$ matrices with entries in $\mathbb{F}$ may be denoted by $\mathbb{F}(m,n)$. In particular, $\mathbb{F}(m,1)$ or $\mathbb{F}(1,n)$ is simply $\mathbb{F}^m$ or $\mathbb{F}^n$. Elements in $\mathbb{F}(n,n)$ are also called *square matrices*.

1.1.3 Matrices

Here we consider some of the simplest manipulations on and properties of matrices.

Let A be the matrix given in (1.1.8). Then A^t, called the *transpose* of A, is defined to be

$$A^t = \begin{pmatrix} a_{11} & a_{21} & \cdots & a_{m1} \\ a_{12} & a_{22} & \cdots & a_{m2} \\ \cdots & \cdots & \cdots & \cdots \\ a_{1n} & a_{2n} & \cdots & a_{mn} \end{pmatrix}. \tag{1.1.9}$$

Of course, $A^t \in \mathbb{F}(n,m)$. Simply put, A^t is a matrix obtained from taking the row (column) vectors of A to be its corresponding column (row) vectors.

For $A \in \mathbb{F}(n,n)$, we say that A is *symmetric* if $A = A^t$, or *skew-symmetric* or *anti-symmetric* if $A^t = -A$. The sets of symmetric and anti-symmetric matrices are denoted by $\mathbb{F}_S(n,n)$ and $\mathbb{F}_A(n,n)$, respectively.

It is clear that $(A^t)^t = A$.

It will now be useful to introduce the notion of the *dot product*. For any two vectors $u = (a_1, \ldots, a_n)$ and $v = (b_1, \ldots, b_n)$ in $\mathbb{F}^n$, their dot product $u \cdot v \in \mathbb{F}$ is defined to be

$$u \cdot v = a_1 b_1 + \cdots + a_n b_n. \tag{1.1.10}$$

The following properties of the dot product can be directly examined:

(1) (Commutativity) $u \cdot v = v \cdot u$ for any $u, v \in \mathbb{F}^n$.
(2) (Associativity and homogeneity) $u \cdot (av + bw) = a(u \cdot v) + b(u \cdot w)$ for any $u, v, w \in \mathbb{F}^n$ and $a, b \in \mathbb{F}$.

With the notion of dot product, we can define the *product of two matrices* $A \in \mathbb{F}(m, k)$ and $B \in \mathbb{F}(k, n)$ by

$$C = (c_{ij}) = AB, \quad i = 1, \ldots, m, \quad j = 1, \ldots, n, \quad (1.1.11)$$

where c_{ij} is the dot product of the ith row of A and the jth column of B. Thus, $AB \in \mathbb{F}(m, n)$.

Alternatively, if we use u, v to denote two column vectors in $\mathbb{F}^n$, then

$$u \cdot v = u^t v. \quad (1.1.12)$$

That is, the dot product of u and v may be viewed as a matrix product of the $1 \times n$ matrix u^t and $n \times 1$ matrix v as well.

The matrix product (or *matrix multiplication*) enjoys the following properties:

(1) (Associativity of scalar multiplication) $a(AB) = (aA)B = A(aB)$ for any $a \in \mathbb{F}$ and any $A \in \mathbb{F}(m, k)$, $B \in \mathbb{F}(k, n)$.
(2) (Distributivity) $A(B + C) = AB + AC$ for any $A \in \mathbb{F}(m, k)$ and $B, C \in \mathbb{F}(k, n)$; $(A + B)C = AC + BC$ for any $A, B \in \mathbb{F}(m, k)$ and $C \in \mathbb{F}(k, n)$.
(3) (Associativity) $A(BC) = (AB)C$ for any $A \in \mathbb{F}(m, k)$, $B \in \mathbb{F}(k, l)$, $C \in \mathbb{F}(l, n)$.

Alternatively, if we express $A \in \mathbb{F}(m, k)$ and $B \in \mathbb{F}(k, n)$ as made of m row vectors and n column vectors, respectively, rewritten as

$$A = \begin{pmatrix} A_1 \\ \vdots \\ A_m \end{pmatrix}, \quad B = (B_1, \ldots, B_n), \quad (1.1.13)$$

then, formally, we have

$$AB = \begin{pmatrix} A_1 \cdot B_1 & A_1 \cdot B_2 & \cdots & A_1 \cdot B_n \\ A_2 \cdot B_1 & A_2 \cdot B_2 & \cdots & A_2 \cdot B_n \\ \cdots & \cdots & \cdots & \cdots \\ A_m \cdot B_1 & A_m \cdot B_2 & \cdots & A_m \cdot B_n \end{pmatrix}$$

$$= \begin{pmatrix} A_1 \\ \vdots \\ A_m \end{pmatrix} (B_1, \ldots, B_n)$$

$$= \begin{pmatrix} A_1B_1 & A_1B_2 & \cdots & A_1B_n \\ A_2B_1 & A_2B_2 & \cdots & A_2B_n \\ \cdots & \cdots & \cdots & \cdots \\ A_mB_1 & A_mB_2 & \cdots & A_mB_n \end{pmatrix}, \tag{1.1.14}$$

which suggests that matrix multiplication may be carried out with legitimate multiplications executed over appropriate matrix blocks.

If $A \in \mathbb{F}(m,k)$ and $B \in \mathbb{F}(k,n)$, then $A^t \in \mathbb{F}(k,m)$ and $B^t \in \mathbb{F}(n,k)$ so that $B^t A^t \in \mathbb{F}(n,m)$. Regarding how AB and $B^t A^t$ are related, here is the conclusion.

Theorem 1.1 *For $A \in \mathbb{F}(m,k)$ and $B \in \mathbb{F}(k,n)$, there holds*

$$(AB)^t = B^t A^t. \tag{1.1.15}$$

The proof of this basic fact is assigned as an exercise.

Other matrices in $\mathbb{F}(n,n)$ with interesting properties include the following:

(1) *Diagonal matrices* are of the form $A = (a_{ij})$ with $a_{ij} = 0$ whenever $i \neq j$. The set of diagonal matrices is denoted as $\mathbb{F}_D(n,n)$.
(2) *Lower triangular matrices* are of the form $A = (a_{ij})$ with $a_{ij} = 0$ whenever $j > i$. The set of lower triangular matrices is denoted as $\mathbb{F}_L(n,n)$.
(3) *Upper triangular matrices* are of the form $A = (a_{ij})$ with $a_{ij} = 0$ whenever $i > j$. The set of upper triangular matrices is denoted as $\mathbb{F}_U(n,n)$.

There is a special element in $\mathbb{F}(n,n)$, called the *identity matrix*, or *unit matrix*, and denoted by I_n, or simply I, which is a diagonal matrix whose diagonal entries are all 1 (unit scalar) and off-diagonal entries are all 0. For any $A \in \mathbb{F}(n,n)$, we have $AI = IA = A$.

Definition 1.2 A matrix $A \in \mathbb{F}(n,n)$ is called *invertible* or *non-singular* if there is some $B \in \mathbb{F}(n,n)$, such that

$$AB = BA = I. \tag{1.1.16}$$

In this situation, B is called the *inverse* of A and denoted by A^{-1}.

If $A, B \in \mathbb{F}(n,n)$ are such that $AB = I$, then we say that A is a *left inverse* of B and B a *right inverse* of A. It can be shown that a left or right inverse is simply the inverse. In other words, if A is a left inverse of B, then both A and B are invertible and the inverses of each other.

If $A \in \mathbb{R}(n,n)$ enjoys the property $AA^t = A^t A = I$, then A is called an *orthogonal matrix*. For $A = (a_{ij}) \in \mathbb{C}(m,n)$, we adopt the notation $\overline{A} = (\overline{a}_{ij})$ for taking the complex conjugate of A and $A^\dagger$ for taking the complex conjugate of the transpose of A, $A^\dagger = \overline{A}^t$, which is also commonly referred to as the *Hermitian conjugate* of A. If $A \in \mathbb{C}(n,n)$, we say that A is *Hermitian symmetric*, or simply *Hermitian*, if $A^\dagger = A$, and *skew-Hermitian* or *anti-Hermitian*, if $A^\dagger = -A$. If $A \in \mathbb{C}(n,n)$ enjoys the property $AA^\dagger = A^\dagger A = I$, then A is called a *unitary matrix*. We will see the importance of these notions throughout our study.

Exercises

1.1.1 Show that it follows from the definition of a field that zero, unit, additive, and multiplicative inverse scalars are all unique.

1.1.2 Let $p \in \mathbb{N}$ be a prime and $[n] \in \mathbb{Z}_p$. Find $-[n]$ and prove the existence of $[n]^{-1}$ when $[n] \neq [0]$. In $\mathbb{Z}_5$, find $-[4]$ and $[4]^{-1}$.

1.1.3 Show that it follows from the definition of a vector space that both zero and additive inverse vectors are unique.

1.1.4 Prove the associativity of matrix multiplication by showing that $A(BC) = (AB)C$ for any $A \in \mathbb{F}(m,k)$, $B \in \mathbb{F}(k,l)$, and $C \in \mathbb{F}(l,n)$.

1.1.5 Prove Theorem 1.1.

1.1.6 Let U be a vector space over a field $\mathbb{F}$. Prove that, if $u + v = u$ where u and $v \in U$, then $v = 0$.

1.1.7 Solve for the unknown x in the equation $[4]x + [3] = [6]$ over $\mathbb{Z}_7$.

1.1.8 Let $A \in \mathbb{F}(n,n)$ $(n \geq 2)$ and rewrite A as

$$A = \begin{pmatrix} A_1 & A_2 \\ A_3 & A_4 \end{pmatrix}, \tag{1.1.17}$$

where $A_1 \in \mathbb{F}(k,k), A_2 \in \mathbb{F}(k,l), A_3 \in \mathbb{F}(l,k), A_4 \in \mathbb{F}(l,l)$, $k, l \geq 1$, and $k + l = n$. Show that

$$A^t = \begin{pmatrix} A_1^t & A_3^t \\ A_2^t & A_4^t \end{pmatrix}. \tag{1.1.18}$$

1.1.9 Prove that the inverse of an invertible matrix is unique by showing the fact that if A, B, and $C \in \mathbb{F}(n,n)$ satisfy $AB = I$ and $CA = I$, then $B = C$.

1.1.10 Let $A \in \mathbb{C}(n,n)$. Show that A is Hermitian if and only if iA is anti-Hermitian.

1.1.11 Let $\mathbb{F} = \mathbb{Z}_3$. Compute $\#\mathbb{F}(3,4)$ and extend your result to obtain $\#\mathbb{F}(m,n)$ $(m,n$ are positive integers). (Recall that, if S is a finite set, then $\#S$ denotes the number of elements in S.)

1.1.12 Let $\mathbb{F}$ be a field and $n \in \mathbb{N}$. Then set

$$P_n = \{a_0 + a_1 t + \cdots + a_n t^n \mid a_0, a_1, \ldots, a_n \in \mathbb{F}\} \tag{1.1.19}$$

to be the set of polynomials of degrees up to n over $\mathbb{F}$. For $\mathbb{F} = \mathbb{Z}_3$, compute $\#P_n$.

1.1.13 Solve for the unknown x for the quadratic equations over the field $\mathbb{Z}_7$:

$$x^2 + x - a = 0 \tag{1.1.20}$$

when $a = [6]$ and $a = -[1]$, respectively.

1.1.14 Let $\mathbb{F}$ be a field and recall the binomial formula for $a, b \in \mathbb{F}$:

$$(a+b)^n = \sum_{k=0}^{n} \binom{n}{k} a^k b^{n-k}, \quad \binom{n}{k} = \frac{n!}{k!(n-k)!}, \quad k = 0, 1, \ldots, n. \tag{1.1.21}$$

If $\mathrm{ch}(\mathbb{F}) = p$, where p is a prime, prove the "Formula of Freshman's Dream":

$$(a+b)^p = a^p + b^p, \quad \forall a, b \in \mathbb{F}. \tag{1.1.22}$$

1.1.15 In calculus, it is known that, if the derivative of a function over $\mathbb{R}$ is zero, then the function is a constant. Show that this is not true for a function over a finite field by considering the function $f : \mathbb{Z}_p \to \mathbb{Z}_p$ (p being a prime), such that for $f(x) = x^k$, we *define* formally

$$f'(x) = kx^{k-1}, \quad k = 0, 1, 2, \ldots, \tag{1.1.23}$$

where x is a variable.

1.2 Subspaces, span, and linear dependence

Let U be a vector space over a field $\mathbb{F}$ and $V \subset U$ a non-empty subset of U. We say that V is a *subspace* of U if V is a vector space over $\mathbb{F}$ with the inherited addition and scalar multiplication from U. It is worth noting that, when checking whether a subset V of a vector space U becomes a subspace, one only needs to verify the closure axiom (1) in the definition of a vector space since the rest of the axioms follow automatically as a consequence of (1).

The two trivial subspaces of U are those consisting only of zero vector, $\{0\}$, and U itself. A nontrivial subspace is also called a *proper subspace*.

Consider the subset $\mathcal{P}_n$ ($n \in \mathbb{N}$) of $\mathcal{P}$ defined by

$$\mathcal{P}_n = \{a_0 + a_1 t + \cdots + a_n t^n \mid a_0, a_1, \ldots, a_n \in \mathbb{F}\}. \tag{1.2.1}$$

It is clear that $\mathcal{P}_n$ is a subspace of $\mathcal{P}$ and $\mathcal{P}_m$ is subspace of $\mathcal{P}_n$ when $m \leq n$.

Consider the set S_a of all vectors $(x_1, \ldots, x_n)$ in $\mathbb{F}^n$ satisfying the equation

$$x_1 + \cdots + x_n = a, \tag{1.2.2}$$

where $a \in \mathbb{F}$. Then S_a is a subspace of $\mathbb{F}^n$ if and only if $a = 0$.

Let $u_1,\ldots,u_k$ be vectors in U. The *linear span* of $\{u_1,\ldots,u_k\}$, denoted by $\mathrm{Span}\{u_1,\ldots,u_k\}$, is the subspace of U defined by

$$\mathrm{Span}\{u_1,\ldots,u_k\} = \{u \in U \mid u = a_1u_1 + \cdots + a_ku_k, a_1,\ldots,a_k \in \mathbb{F}\}. \quad (1.2.3)$$

Thus, if $u \in \mathrm{Span}\{u_1,\ldots,u_k\}$, then there are $a_1,\ldots,a_k \in \mathbb{F}$ such that

$$u = a_1u_1 + \cdots + a_ku_k. \quad (1.2.4)$$

We also say that u is *linearly spanned* by $u_1,\ldots,u_k$ or *linearly dependent* on $u_1,\ldots,u_k$. Therefore, the zero vector, 0, is linearly dependent on any finite set of vectors.

If $U = \mathrm{Span}\{u_1,\ldots,u_k\}$, we also say that U is *generated* by the vectors $u_1,\ldots,u_k$.

For $\mathcal{P}_n$ defined in (1.2.1), we have $\mathcal{P}_n = \mathrm{Span}\{1,t,\ldots,t^n\}$. Thus $\mathcal{P}_n$ is generated by $1,t,\ldots,t^n$. Naturally, for two elements p and q in $\mathcal{P}_n$, say $p(t) = a_0 + a_1t + \cdots + a_nt^n$, $q(t) = b_0 + b_1t + \cdots + b_nt^n$, we identify p and q if and only if all the coefficients of p and q of like powers of t coincide in $\mathbb{F}$, or, $a_i = b_i$ for all $i = 0, 1, \ldots, n$.

In $\mathbb{F}^n$, define

$$e_1 = (1,0,0,\ldots,0), \quad e_2 = (0,1,0,\ldots,0), \quad e_n = (0,0,\ldots,0,1). \quad (1.2.5)$$

Then $\mathbb{F}^n = \mathrm{Span}\{e_1,e_2,\ldots,e_n\}$ and $\mathbb{F}^n$ is generated by $e_1,e_2,\ldots,e_n$.

Thus, for S_0 defined in (1.2.2), we have

$$\begin{aligned}(x_1,x_2,\ldots,x_n) &= -(x_2 + \cdots + x_n)e_1 + x_2e_2 + \cdots + x_ne_n \\ &= x_2(e_2 - e_1) + \cdots + x_n(e_n - e_1),\end{aligned} \quad (1.2.6)$$

where $x_2,\ldots,x_n$ are arbitrarily taken from $\mathbb{F}$. Therefore,

$$S_0 = \mathrm{Span}\{e_2 - e_1,\ldots,e_n - e_1\}. \quad (1.2.7)$$

For $\mathbb{F}(m,n)$, we define $M_{ij} \in \mathbb{F}(m,n)$ to be the vector such that all its entries vanish except that its entry at the position (i,j) (at the ith row and jth column) is 1, $i = 1,\ldots,m, j = 1,\ldots,n$. We have

$$\mathbb{F}(m,n) = \mathrm{Span}\{M_{ij} \mid i = 1,\ldots,m, j = 1,\ldots,n\}. \quad (1.2.8)$$

The notion of spans can be extended to cover some useful situations. Let U be a vector space and S be a (finite or infinite) subset of U. Then define

$$\mathrm{Span}(S) = \text{The set of linear combinations}$$

$$\text{of all possible finite subsets of } S. \quad (1.2.9)$$

It is obvious that $\mathrm{Span}(S)$ is a subspace of U. If $U = \mathrm{Span}(S)$, we say that U is spanned or generated by the set of vectors S.

As an example, we have

$$\mathcal{P} = \text{Span}\{1, t, \ldots, t^n, \ldots\}. \tag{1.2.10}$$

Alternatively, we can also express $\mathcal{P}$ as

$$\mathcal{P} = \cup_{n=0}^{\infty} \mathcal{P}_n. \tag{1.2.11}$$

The above discussion motivates the following formal definition.

Definition 1.3 Let $u_1, \ldots, u_m$ be m vectors in the vector space U over a field $\mathbb{F}$. We say that these vectors are *linearly dependent* if one of them may be written as a linear span of the rest of them or linearly dependent on the rest of them. Or equivalently, $u_1, \ldots, u_m$ are linearly dependent if there are scalars $a_1, \ldots, a_m \in \mathbb{F}$ where $(a_1, \ldots, a_m) \neq (0, \ldots, 0)$ such that

$$a_1 u_1 + \cdots + a_m u_m = 0. \tag{1.2.12}$$

Otherwise $u_1, \ldots, u_m$ are called *linearly independent*. In this latter case, the only possible vector $(a_1, \ldots, a_m) \in \mathbb{F}^n$ to make (1.2.12) fulfilled is the zero vector, $(0, \ldots, 0)$, such that a single vector is linearly independent if and only if it is a nonzero vector.

To proceed further, we need to consider the following system of linear equations:

$$\begin{cases} a_{11}x_1 + \cdots + a_{1n}x_n & = \quad 0, \\ \cdots\cdots\cdots\cdots\cdots\cdots\cdots & \cdots \quad \cdots \\ a_{m1}x_1 + \cdots + a_{mn}x_n & = \quad 0, \end{cases} \tag{1.2.13}$$

over $\mathbb{F}$ with unknowns $x_1, \ldots, x_n$.

Theorem 1.4 *In the system (1.2.13), if $m < n$, then the system has a nontrivial solution $(x_1, \ldots, x_n) \neq (0, \ldots, 0)$.*

Proof   We prove the theorem by using induction on $m + n$.

The beginning situation is $m + n = 3$ when $m = 1$ and $n = 2$. It is clear that we always have a nontrivial solution.

Assume that the statement of the theorem is true when $m + n \leq k$, where $k \geq 3$.

Let $m + n = k + 1$. If $k = 3$, the condition $m < n$ implies $m = 1, n = 3$, and the existence of a nontrivial solution is obvious. Assume then $k \geq 4$. If all the coefficients of the variable x_1 in (1.2.13) are zero, that is, $a_{11} = \cdots = a_{m1} = 0$, then $x_1 = 1, x_2 = \cdots = x_n = 0$ is a nontrivial solution. So we may assume one of the coefficients of x_1 is nonzero. Without loss of generality, we assume $a_{11} \neq 0$. If $m = 1$, there is again nothing to show. Assume $m \geq 2$. Dividing the first equation in (1.2.13) by a_{11}, if necessary, we can further assume $a_{11} = 1$. Then, adding the $(-a_{i1})$-multiple of the first equation to the ith equation, in (1.2.13), for $i = 2,\ldots,m$, we arrive at

$$\begin{cases} x_1 + & a_{12}x_2 + \cdots + a_{1n}x_n & = & 0, \\ & b_{22}x_2 + \cdots + b_{2n}x_n & = & 0, \\ & \cdots\cdots\cdots\cdots\cdots & \cdots & \cdots \\ & b_{m2}x_2 + \cdots + b_{mn}x_n & = & 0. \end{cases} \tag{1.2.14}$$

The system below the first equation in (1.2.14) contains $m - 1$ equations and $n - 1$ unknowns $x_2,\ldots,x_n$. Of course, $m - 1 < n - 1$. So, in view of the inductive assumption, it has a nontrivial solution. Substituting this nontrivial solution into the first equation in (1.2.14) to determine the remaining unknown x_1, we see that the existence of a nontrivial solution to the original system (1.2.13) follows. $\qquad\square$

The importance of Theorem 1.4 is seen in the following.

Theorem 1.5 *Any set of more than m vectors in* $\mathrm{Span}\{u_1,\ldots,u_m\}$ *must be linearly dependent.*

Proof Let $v_1,\ldots,v_n \in \mathrm{Span}\{u_1,\ldots,u_m\}$ be n vectors where $n > m$. Consider the possible linear dependence relation

$$x_1v_1 + \cdots + x_nv_n = 0, \tag{1.2.15}$$

for some $x_1,\ldots,x_n \in \mathbb{F}$.

Since each $v_j \in \mathrm{Span}\{u_1,\ldots,u_m\}$, $j = 1,\ldots,n$, there are scalars $a_{ij} \in \mathbb{F}$ ($i = 1,\ldots,m, j = 1,\ldots,n$) such that

$$v_j = \sum_{i=1}^{m} a_{ij}u_i, \quad j = 1,\ldots,n. \tag{1.2.16}$$

Substituting (1.2.16) into (1.2.15), we have

$$\sum_{j=1}^{n} x_j \left(\sum_{i=1}^{m} a_{ij}u_i \right) = 0. \tag{1.2.17}$$

Regrouping the terms in (1.2.17), we arrive at

$$\sum_{i=1}^{m} \left(\sum_{j=1}^{n} a_{ij}x_j \right) u_i = 0, \tag{1.2.18}$$

which may be fulfilled by setting

$$\sum_{j=1}^{n} a_{ij}x_j = 0, \quad i = 1,\ldots,m. \tag{1.2.19}$$

This system of equations is exactly the system (1.2.13) that allows a nontrivial solution in view of Theorem 1.4. Hence the proof follows. □

We are now prepared to study in Section 1.3 several fundamental properties of vector spaces.

Exercises

1.2.1 Let U_1 and U_2 be subspaces of a vector space U. Show that $U_1 \cup U_2$ is a subspace of U if and only if $U_1 \subset U_2$ or $U_2 \subset U_1$.

1.2.2 Let $\mathcal{P}_n$ denote the vector space of the polynomials of degrees up to n over a field $\mathbb{F}$ expressed in terms of a variable t. Show that the vectors $1, t, \ldots, t^n$ in $\mathcal{P}_n$ are linearly independent.

1.2.3 Show that the vectors in $\mathbb{F}^n$ defined in (1.2.5) are linearly independent.

1.2.4 Show that S_0 defined in (1.2.2) may also be expressed as

$$S_0 = \mathrm{Span}\{e_1 - e_n, \ldots, e_{n-1} - e_n\} \tag{1.2.20}$$

and deduce that, in $\mathbb{R}^n$, the vectors

$$\left(1, \frac{1}{2}, \ldots, \frac{1}{2^{n-2}}, 2\left[\frac{1}{2^{n-1}} - 1\right]\right), e_1 - e_n, \ldots, e_{n-1} - e_n \tag{1.2.21}$$

are linearly dependent ($n \geq 4$).

1.2.5 Show that $\mathbb{F}_S(n,n)$, $\mathbb{F}_A(n,n)$, $\mathbb{F}_D(n,n)$, $\mathbb{F}_L(n,n)$, and $\mathbb{F}_U(n,n)$ are all subspaces of $\mathbb{F}(n,n)$.

1.2.6 Let $u_1, \ldots, u_n$ ($n \geq 2$) be linearly independent vectors in a vector space U and set

$$v_{i-1} = u_{i-1} + u_i, \quad i = 2, \ldots, n; \quad v_n = u_n + u_1. \tag{1.2.22}$$

Investigate whether $v_1, \ldots, v_n$ are linearly independent as well.

1.2.7 Let $\mathbb{F}$ be a field. For any two vectors $u = (a_1, \ldots, a_n)$ and $v = (b_1, \ldots, b_n)$ in $\mathbb{F}^n$ ($n \geq 2$), viewed as matrices, we see that the matrix product $A = u^t v$ lies in $\mathbb{F}(n,n)$. Prove that any two row vectors of A are linearly dependent. What happens to the column vectors of A?

1.2.8 Consider a slightly strengthened version of the second part of Problem 1.2.1: Let U_1 and U_2 be subspaces of a vector space U, $U_1 \neq U$, and $U_2 \neq U$. Show without using Problem 1.2.1 that there exists a vector in U which lies outside $U_1 \cup U_2$. Explain how you may apply the conclusion of this exercise to prove that of Problem 1.2.1.

1.2.9   (A challenging extension of Problem 1.2.8) Let $U_1, \ldots, U_k$ be k subspaces of a vector space U over a field of characteristic 0. If $U_i \neq U$ for $i = 1, \ldots, k$, show that there is a vector in U which lies outside $\cup_{i=1}^k U_i$.

1.2.10   For $A \in \mathbb{F}(m, n)$ and $B \in \mathbb{F}(n, m)$ with $m > n$ show that AB as an element in $\mathbb{F}(m, m)$ can never be invertible.

1.2.11   Consider the system of equations over $\mathbb{R}$:

$$x_1 - 2x_2 + 3x_3 - x_4 = 1, \tag{1.2.23}$$
$$-x_1 + x_2 - 2x_3 + x_4 = -5. \tag{1.2.24}$$

(a) Find all solutions of the system.

(b) Represent the set of solutions, $S \subset \mathbb{R}^4$, in the "coset" form

$$S = u + V, \tag{1.2.25}$$

where $u \in \mathbb{R}^4$ and V is a subspace of $\mathbb{R}^4$ given in the form

$$V = \text{Span}\{v, w\}.$$

1.2.12   Determine in $\mathbb{R}^3$ whether u is a linear combination (span) of v and w where

(a) $u = (5, 1, -5), v = (1, -2, -3), w = (-2, 3, -4)$.

(b) $u = (-2, 2, 2), v = (1, 2, -1), w = (-3, -3, 3)$.

1.2.13   Consider the following set of three vectors in $\mathbb{F}$:

$$S = \{(1, 1, 0), (1, 0, 1), (0, 1, 1)\}.$$

(a) Show that if $\mathbb{F} = \mathbb{Q}$ then S is linearly independent.

(b) Show that if $\mathbb{F} = \mathbb{Z}_2$ then S is linearly dependent.

(c) What happens when $\mathbb{F} = \mathbb{Z}_p$ where $p > 2$?

1.2.14   Let $\mathbb{F}$ be a field of characteristic *not* equal to 2 and U a vector space over $\mathbb{F}$.

(a) Show that if u and v are distinct vectors in U then u and v are linearly independent if and only if $u + v$ and $u - v$ are linearly independent.

(b) Show that if u, v, and w are distinct vectors in U then u, v, and w are linearly independent if and only if $u + v, u + w$, and $v + w$ are linearly independent.

1.2.15   Consider the vector space $C(-\infty, \infty)$ (the set of real-valued functions over the interval $(-\infty, \infty)$). Show that the functions

$$f_1(t) = e^{r_1 t}, \quad f_2(t) = e^{r_2 t}, \quad \ldots, \quad f_n(t) = e^{r_n t}, \quad t \in \mathbb{R},$$

where $r_1, r_2, \ldots, r_n$ are distinct constants, are linearly independent.

1.3 Bases, dimensionality, and coordinates

Let U be a vector space over a field $\mathbb{F}$, take $u_1, \ldots, u_n \in U$, and set $V = \mathrm{Span}\{u_1, \ldots, u_n\}$. Eliminating linearly dependent vectors from the set $\{u_1, \ldots, u_n\}$ if necessary, we can certainly assume that the vectors $u_1, \ldots, u_n$ are already made linearly independent. Thus, any vector $u \in V$ may take the form

$$u = a_1 u_1 + \cdots + a_n u_n, \quad a_1, \ldots, a_n \in \mathbb{F}. \tag{1.3.1}$$

It is not hard to see that the coefficients $a_1, \ldots, a_n$ in the above representation must be unique. In fact, if we also have

$$u = b_1 u_1 + \cdots + b_n u_n, \quad b_1, \ldots, b_n \in \mathbb{F}, \tag{1.3.2}$$

then, combining the above two relations, we have $(a_1 - b_1)u_1 + \cdots + (a_n - b_n)u_n = 0$. Since $u_1, \ldots, u_n$ are linearly independent, we obtain $a_1 = b_1, \ldots, a_n = b_n$ and the uniqueness follows.

Furthermore, if there is another set of vectors $v_1, \ldots, v_m$ in U such that

$$\mathrm{Span}\{v_1, \ldots, v_m\} = \mathrm{Span}\{u_1, \ldots, u_n\}, \tag{1.3.3}$$

then $m \geq n$ in view of Theorem 1.5. As a consequence, if $v_1, \ldots, v_m$ are also linearly independent, then $m = n$. This observation leads to the following.

Definition 1.6 If there are vectors $u_1, \ldots, u_n \in U$ such that $U = \mathrm{Span}\{u_1, \ldots, u_n\}$, then U is said to be *finitely generated* and the set of vectors $\{u_1, \ldots, u_n\}$ is called a *basis* of U if these vectors are linearly independent. The number of vectors in any basis of a finitely generated vector space, n, is independent of the choice of the basis and is referred to as the *dimension* of the finitely generated vector space, written as $\dim(U) = n$. A finitely generated vector space is also said to be of *finite-dimensionality* or *finite dimensional*. If a vector space U is not finite dimensional, it is said to be *infinite dimensional*, also written as $\dim(U) = \infty$. Thus, a zero vector space has no basis and its dimension is zero.

As an example of an infinite-dimensional vector space, we show that when $\mathbb{R}$ is regarded as a vector space over $\mathbb{Q}$, then $\dim(\mathbb{R}) = \infty$. In fact, recall that a real number is called an *algebraic number* if it is the zero of a polynomial with coefficients in $\mathbb{Q}$. We also know that there are many non-algebraic numbers in $\mathbb{R}$, called *transcendental numbers*. Let τ be such a transcendental number. Then for any $n = 1, 2, \ldots$ the numbers $1, \tau, \tau^2, \ldots, \tau^n$ are linearly independent in the vector space $\mathbb{R}$ over the field $\mathbb{Q}$. Indeed, if there are $r_0, r_1, r_2, \ldots, r_n \in \mathbb{Q}$ so that

$$r_0 + r_1 \tau + r_2 \tau^2 + \cdots + r_n \tau^n = 0, \tag{1.3.4}$$

and at least one number among $r_0, r_1, r_2, \ldots, r_n$ is nonzero, then τ is the zero of the nontrivial polynomial

$$p(t) = r_0 + r_1 t + r_2 t^2 + \cdots + r_n t^n, \tag{1.3.5}$$

which violates the assumption that τ is transcendental. Thus $\mathbb{R}$ is infinite dimensional over $\mathbb{Q}$.

The following theorem indicates that it is fairly easy to construct a basis for a finite-dimensional vector space.

Theorem 1.7 *Let U be an n-dimensional vector space over a field $\mathbb{F}$. Any n linearly independent vectors in U form a basis of U.*

Proof Let $u_1, \ldots, u_n \in U$ be linearly independent vectors. We only need to show that they span U. In fact, take any $u \in U$. We know that $u_1, \ldots, u_n, u$ are linearly dependent. So there is a nonzero vector $(a_1, \ldots, a_n, a) \in \mathbb{F}^{n+1}$ such that

$$a_1 u_1 + \cdots + a_n u_n + a u = 0. \tag{1.3.6}$$

Of course, $a \neq 0$, otherwise it contradicts the assumption that $u_1, \ldots, u_n$ are linearly independent. So $u = (-a^{-1})(a_1 u_1 + \cdots + a_n u_n)$. Thus $u \in \text{Span}\{u_1, \ldots, u_n\}$. $\square$

Definition 1.8 Let $\{u_1, \ldots, u_n\}$ be a basis of the vector space U. Given $u \in U$ there are unique scalars $a_1, \ldots, a_n \in \mathbb{F}$ such that

$$u = a_1 u_1 + \cdots + a_n u_n. \tag{1.3.7}$$

These scalars $a_1, \ldots, a_n$ are called the *coordinates*, and $(a_1, \ldots, a_n) \in \mathbb{F}^n$ the *coordinate vector*, of the vector u with respect to the basis $\{u_1, \ldots, u_n\}$.

It will be interesting to investigate the relation between the coordinate vectors of a vector under different bases.

Let $\mathcal{U} = \{u_1, \ldots, u_n\}$ and $\mathcal{V} = \{v_1, \ldots, v_n\}$ be two bases of the vector space U. For $u \in U$, let $(a_1, \ldots, a_n) \in \mathbb{F}^n$ and $(b_1, \ldots, b_n) \in \mathbb{F}^n$ be the coordinate vectors of u with respect to $\mathcal{U}$ and $\mathcal{V}$, respectively. Thus

$$u = a_1 u_1 + \cdots + a_n u_n = b_1 v_1 + \cdots + b_n v_n. \tag{1.3.8}$$

On the other hand, we have

$$v_j = \sum_{i=1}^{n} a_{ij} u_i, \quad j = 1, \ldots, n. \tag{1.3.9}$$

The $n \times n$ matrix $A = (a_{ij})$ is called a *basis transition matrix* or *basis change matrix*. Inserting (1.3.9) into (1.3.8), we have

$$\sum_{i=1}^{n} a_i u_i = \sum_{i=1}^{n} \left(\sum_{j=1}^{n} a_{ij} b_j \right) u_i. \tag{1.3.10}$$

Hence, by the linear independence of the basis vectors, we have

$$a_i = \sum_{j=1}^{n} a_{ij} b_j, \quad i = 1, \ldots, n. \tag{1.3.11}$$

Note that the relation (1.3.9) between bases may be formally and conveniently expressed in a "matrix form" as

$$(v_1, \ldots, v_n) = (u_1, \ldots, u_n)A, \tag{1.3.12}$$

or concisely $\mathcal{V} = \mathcal{U}A$, or

$$\begin{pmatrix} v_1 \\ \vdots \\ v_n \end{pmatrix} = A^t \begin{pmatrix} u_1 \\ \vdots \\ u_n \end{pmatrix}, \tag{1.3.13}$$

where multiplications between scalars and vectors are made in a well-defined manner. On the other hand, the relation (1.3.11) between coordinate vectors may be rewritten as

$$\begin{pmatrix} a_1 \\ \vdots \\ a_n \end{pmatrix} = A \begin{pmatrix} b_1 \\ \vdots \\ b_n \end{pmatrix}, \tag{1.3.14}$$

or

$$(a_1, \ldots, a_n) = (b_1, \ldots, b_n)A^t. \tag{1.3.15}$$

Exercises

1.3.1   Let U be a vector space with $\dim(U) = n \geq 2$ and V a subspace of U with a basis $\{v_1, \ldots, v_{n-1}\}$. Prove that for any $u \in U \setminus V$ the vectors $u, v_1, \ldots, v_{n-1}$ form a basis for U.

1.3.2   Show that $\dim(\mathbb{F}(m, n)) = mn$.

1.3.3   Determine $\dim(\mathbb{F}_S(n, n))$, $\dim(\mathbb{F}_A(n, n))$, and $\dim(\mathbb{F}_D(n, n))$.

1.3.4 (Lagrange polynomials) Let $\mathbb{F}$ be an infinite field and $c_0, c_1, \ldots, c_n$ be distinct elements in $\mathbb{F}$. Define the degree n polynomials, called the Lagrange polynomials associated with $c_0, c_1, \ldots, c_n$, as follows:

$$L_i(t) = \frac{(t-c_0)\cdots(t-c_{i-1})(t-c_{i+1})\cdots(t-c_n)}{(c_i-c_0)\cdots(c_i-c_{i-1})(c_i-c_{i+1})\cdots(c_i-c_n)}$$

$$= \prod_{j=0,\ldots,n,\, j\neq i} \frac{(t-c_j)}{(c_i-c_j)}, \quad i=0,1,\ldots,n. \qquad (1.3.16)$$

(a) Show that

$$L_i(c_j) = \delta_{ij}, \quad \text{where } \delta_{ij} = \begin{cases} 0, & i \neq j, \\ 1, & i = j. \end{cases} \qquad (1.3.17)$$

(b) Show that $\{L_0(t), L_1(t), \ldots, L_n(t)\}$ forms a basis of $\mathcal{P}_n$ $(n \geq 1)$.

(c) Use the Lagrange polynomials to prove that if the values of two polynomials in $\mathcal{P}_n$ agree at $n+1$ points then the two polynomials coincide.

(d) Let $p(t) \in \mathcal{P}_n$ and $a_i = p(c_i)$ for $i = 0, 1, \ldots, n$. Find the coordinates of $p(t)$ with respect to the Lagrange basis $\{L_0(t), L_1(t), \ldots, L_n(t)\}$.

(e) Let $f(t)$ be an arbitrary function over $\mathbb{F}$. Find a unique $p(t) \in \mathcal{P}_n$ such that

$$f(c_i) = p(c_i), \quad i = 0, 1, \ldots, n. \qquad (1.3.18)$$

Such a $p(t)$ is called the interpolation polynomial of $f(t)$.

(Hint: Use the Lagrange basis of $\mathcal{P}_n$ with respect to $c_0, c_1, \ldots, c_n$.)

1.3.5 Let $\mathcal{P}$ be the vector space of all polynomials with coefficients in a field $\mathbb{F}$. Show that $\dim(\mathcal{P}) = \infty$.

1.3.6 Consider the vector space $\mathbb{R}^3$ and the bases $\mathcal{U} = \{e_1, e_2, e_3\}$ and $\mathcal{V} = \{e_1, e_1+e_2, e_1+e_2+e_3\}$. Find the basis transition matrix A from $\mathcal{U}$ into $\mathcal{V}$ satisfying $\mathcal{V} = \mathcal{U}A$. Find the coordinate vectors of the given vector $(1, 2, 3) \in \mathbb{R}^3$ with respect to the bases $\mathcal{U}$ and $\mathcal{V}$, respectively, and relate these vectors with the matrix A.

1.3.7 Prove that a basis transition matrix must be invertible.

1.3.8 Let U be an n-dimensional vector space over a field $\mathbb{F}$ where $n \geq 2$ (say). Consider the following construction:

(a) Take $u_1 \in U \setminus \{0\}$.

(b) Take $u_2 \in U \setminus \mathrm{Span}\{u_1\}$.

(c) Take (if any) $u_3 \in U \setminus \mathrm{Span}\{u_1, u_2\}$.

(d) In general, take $u_i \in U \setminus \mathrm{Span}\{u_1, \ldots, u_{i-1}\}$ $(i \geq 2)$.

Show that this construction will terminate itself in exactly n steps, that is, it will not be possible anymore to get u_{n+1}, and that the vectors $u_1, u_2, \ldots, u_n$ so obtained form a basis of U.

1.3.9   Determine which of the following sets of vectors form a basis for $\mathbb{R}^3$:

$$S = \{(1,0,-1),(2,5,1),(0,-4,3)\}, \qquad (1.3.19)$$
$$T = \{(-1,3,2),(3,-1,-3),(1,5,1)\}. \qquad (1.3.20)$$

1.3.10 Let $\mathcal{P}_n$ be the vector space of polynomials of degrees up to $n \geq 1$ over the field $\mathbb{R}$.

(a) Determine $\dim(V_0)$ where

$$V_0 = \{p(t) \in \mathcal{P}_n \mid p(0) = 0\}. \qquad (1.3.21)$$

(b) Determine $\dim(V_a)$ for any $a \in \mathbb{R}$ where

$$V_a = \{p(t) \in \mathcal{P}_n \mid p(a) = 0\}. \qquad (1.3.22)$$

1.3.11 Let U be a vector space and V a subspace of U.

(a) Assume $\dim(U) < \infty$. Show that if $\dim(V) = \dim(U)$ then $V = U$.
(b) Assume $\dim(U) = \infty$ and $\dim(V) = \infty$. Give an example to show that it may happen that $V \neq U$.

1.3.12 It is clear that $\mathbb{C}$ may be regarded as a vector space over $\mathbb{R}$. Hence, in this situation, $\mathbb{C} = \mathrm{Span}\{1, i\}$ and $\dim(\mathbb{C}) = 2$. Let U be an n-dimensional vector space over $\mathbb{C}$. Then U may also be viewed as a vector space over $\mathbb{R}$. Prove that when U is viewed as a vector space over $\mathbb{R}$, then $\dim(U) = 2n$.

1.4 Dual spaces

Let U be an n-dimensional vector space over a field $\mathbb{F}$. A *functional* (also called a *form* or a *1-form*) f over U is a *linear function* $f : U \to \mathbb{F}$ satisfying

$$f(u+v) = f(u) + f(v), \quad u,v \in U; \quad f(au) = af(u), \quad a \in \mathbb{F}, u \in U. \quad (1.4.1)$$

Let f and g be two functionals. Then we can define another functional called the *sum* of f and g, denoted by $f + g$, by

$$(f+g)(u) = f(u) + g(u), \quad u \in U. \qquad (1.4.2)$$

Similarly, let f be a functional and $a \in \mathbb{F}$. We can define another functional called the *scalar multiple* of a with f, denoted by af, by

$$(af)(u) = af(u), \quad u \in U. \qquad (1.4.3)$$

It is a simple exercise to check that these two operations make the set of all functionals over U a vector space over $\mathbb{F}$. This vector space is called the *dual space* of U, denoted by U'.

Let $\{u_1, \ldots, u_n\}$ be a basis of U. For any $f \in U'$ and any $u = a_1 u_1 + \cdots + a_n u_n \in U$, we have

$$f(u) = f\left(\sum_{i=1}^n a_i u_i\right) = \sum_{i=1}^n a_i f(u_i). \tag{1.4.4}$$

Hence, f is uniquely determined by its values on the basis vectors,

$$f_1 = f(u_1), \ldots, f_n = f(u_n). \tag{1.4.5}$$

Conversely, for arbitrarily assigned values $f_1, \ldots, f_n$ in (1.4.5), we define

$$f(u) = \sum_{i=1}^n a_i f_i \quad \text{for any } u = \sum_{i=1}^n a_i u_i \in U. \tag{1.4.6}$$

It is clear that f is a functional. That is, $f \in U'$. Of course, such an f satisfies (1.4.5).

Thus, we have seen a well-defined 1-1 correspondence

$$U' \leftrightarrow \mathbb{F}^n, \quad f \leftrightarrow (f_1, \ldots, f_n). \tag{1.4.7}$$

Especially we may use $u'_1, \ldots, u'_n$ to denote the elements in U' corresponding to the vectors $e_1, \ldots, e_n$ in $\mathbb{F}^n$ given by (1.2.5). Then we have

$$u'_i(u_j) = \delta_{ij} = \begin{cases} 0, & i \neq j \\ 1, & i = j, \end{cases} \quad i, j = 1, \ldots, n. \tag{1.4.8}$$

It is clear that $u'_1, \ldots, u'_n$ are linearly independent and span U' because an element f of U' satisfying (1.4.5) is simply given by

$$f = f_1 u'_1 + \cdots + f_n u'_n. \tag{1.4.9}$$

In other words, $\{u'_1, \ldots, u'_n\}$ is a basis of U', commonly called the *dual basis* of U' with respect to the basis $\{u_1, \ldots, u_n\}$ of U. In particular, we have seen that U and U' are of the same dimensionality.

Let $\mathcal{U} = \{u_1, \ldots, u_n\}$ and $\mathcal{V} = \{v_1, \ldots, v_n\}$ be two bases of the vector space U. Let their dual bases be denoted by $\mathcal{U}' = \{u'_1, \ldots, u'_n\}$ and $\mathcal{V}' = \{v'_1, \ldots, v'_n\}$, respectively. Suppose that the bases $\mathcal{U}'$ and $\mathcal{V}'$ are related through

$$u'_j = \sum_{i=1}^n a'_{ij} v'_i, \quad j = 1, \ldots, n. \tag{1.4.10}$$

Using (1.3.9) and (1.4.10) to evaluate $u'_i(v_j)$, we obtain

$$u'_i(v_j) = u'_i \left(\sum_{k=1}^{n} a_{kj} u_k \right) = \sum_{k=1}^{n} a_{kj} u'_i(u_k) = \sum_{k=1}^{n} a_{kj} \delta_{ik} = a_{ij}, \qquad (1.4.11)$$

$$u'_i(v_j) = \sum_{k=1}^{n} a'_{ki} v'_k(v_j) = \sum_{k=1}^{n} a'_{ki} \delta_{kj} = a'_{ji}, \qquad (1.4.12)$$

which leads to $a'_{ij} = a_{ji}$ $(i, j = 1, \dots, n)$. In other words, we have arrived at the correspondence relation

$$u'_j = \sum_{i=1}^{n} a_{ji} v'_i, \quad j = 1, \dots, n. \qquad (1.4.13)$$

With matrix notation as before, we have

$$(u'_1, \dots, u'_n) = (v'_1, \dots, v'_n) A^t, \qquad (1.4.14)$$

or

$$\begin{pmatrix} u'_1 \\ \vdots \\ u'_n \end{pmatrix} = A \begin{pmatrix} v'_1 \\ \vdots \\ v'_n \end{pmatrix}. \qquad (1.4.15)$$

Besides, for any $u' \in U'$ written as

$$u' = a'_1 u'_1 + \cdots + a'_n u'_n = b'_1 v'_1 + \cdots + b'_n v'_n, \qquad (1.4.16)$$

the discussion in Section 1.3 and this section allow us to get

$$b'_i = \sum_{j=1}^{n} a_{ji} a'_j, \quad i = 1, \dots, n. \qquad (1.4.17)$$

Thus, in matrix form, we obtain the relation

$$(b'_1 \dots, b'_n) = (a'_1 \dots, a'_n) A, \qquad (1.4.18)$$

or

$$\begin{pmatrix} b'_1 \\ \vdots \\ b'_n \end{pmatrix} = A^t \begin{pmatrix} a'_1 \\ \vdots \\ a'_n \end{pmatrix}. \qquad (1.4.19)$$

Comparing the above results with those established in Section 1.3, we see that, with respect to bases and dual bases, the coordinates vectors in U and U' follow "opposite" rules of correspondence. For this reason, coordinate vectors in U are often called *contravariant vectors* and those in U' *covariant vectors*.

Using the relation stated in (1.4.8), we see that we may naturally view $u_1, \ldots, u_n$ as elements in $(U')' = U''$ so that they form a basis of U'' dual to $\{u'_1, \ldots, u'_n\}$.

$$u_i(u'_j) \equiv u'_j(u_i) = \delta_{ij} = \begin{cases} 0, & i \neq j, \\ 1, & i = j, \end{cases} \quad i, j = 1, \ldots, n. \tag{1.4.20}$$

Thus, for any $u \in U''$ satisfying $u(u'_i) = a_i$ $(i = 1, \ldots, n)$, we have

$$u = a_1 u_1 + \cdots + a_n u_n. \tag{1.4.21}$$

In this way, we see that U'' may be identified with U. In other words, we have seen that the identification just made spells out the relationship

$$U'' = U, \tag{1.4.22}$$

which is also referred to as *reflexivity* of U or U is said to be *reflexive*.

Notationally, for $u \in U$ and $u' \in U'$, it is often convenient to rewrite $u'(u)$, which is linear in both the u and u' arguments, as

$$u'(u) = \langle u', u \rangle. \tag{1.4.23}$$

Then our identification (1.4.22), made through setting

$$u'(u) = u(u'), \quad u \in U, \quad u' \in U', \tag{1.4.24}$$

simply says that the "pairing" $\langle \cdot, \cdot \rangle$ as given in (1.4.23) is symmetric:

$$\langle u', u \rangle = \langle u, u' \rangle, \quad u \in U, \quad u' \in U'. \tag{1.4.25}$$

For any non-empty subset $S \subset U$, the *annihilator* of S, denoted by S^0, is the subset of U' given by

$$S^0 = \{u' \in U' \mid \langle u', u \rangle = 0, \forall u \in S\}. \tag{1.4.26}$$

It is clear that S^0 is always a subspace of U' regardless of whether S is a subspace of U. Likewise, for any non-empty subset $S' \subset U'$, we can define the annihilator of S', S'^0, as the subset

$$S'^0 = \{u \in U \mid \langle u', u \rangle = 0, \forall u' \in S'\} \tag{1.4.27}$$

of U. Of course, S'^0 is always a subspace of U.

Exercises

1.4.1 Let $\mathbb{F}$ be a field. Describe the dual spaces $\mathbb{F}'$ and $(\mathbb{F}^2)'$.

1.4.2 Let U be a finite-dimensional vector space. Prove that for any vectors $u, v \in U$ $(u \neq v)$ there exists an element $f \in U'$ such that $f(u) \neq f(v)$.

1.4.3   Let U be a finite-dimensional vector space and $f, g \in U'$. For any $v \in U$, $f(v) = 0$ if and only if $g(v) = 0$. Show that f and g are linearly dependent.

1.4.4   Let $\mathbb{F}$ be a field and

$$V = \{(x_1, \ldots, x_n) \in \mathbb{F}^n \mid x_1 + \cdots + x_n = 0\}. \qquad (1.4.28)$$

Show that any $f \in V^0$ may be expressed as

$$f(x_1, \ldots, x_n) = c \sum_{i=1}^{n} x_i, \quad (x_1, \ldots, x_n) \in \mathbb{F}^n, \qquad (1.4.29)$$

for some $c \in \mathbb{F}$.

1.4.5   Let $U = \mathcal{P}_2$, where $\mathrm{ch}(\mathbb{F}) \neq 2$, and f, g, and $h \in U'$ be defined by

$$f(p) = p(-1), \quad g(p) = p(0), \quad h(p) = p(1), \quad p(t) \in \mathcal{P}_2. \quad (1.4.30)$$

(a) Show that $\mathcal{B}' = \{f, g, h\}$ is a basis for U'.
(b) Find a basis $\mathcal{B}$ of U which is dual to $\mathcal{B}'$.

1.4.6   Let U be an n-dimensional vector space and V be an m-dimensional subspace of U. Show that the annihilator V^0 is an $(n-m)$-dimensional subspace of U'. In other words, there holds the dimensionality equation

$$\dim(V) + \dim(V^0) = \dim(U). \qquad (1.4.31)$$

1.4.7   Let U be an n-dimensional vector space and V be an m-dimensional subspace of U. Show that $V^{00} = (V^0)^0 = V$.

1.5 Constructions of vector spaces

Let U be a vector space and V and W its subspaces. It is clear that $V \cap W$ is also a subspace of U but $V \cup W$ in general may fail to be a subspace of U. The smallest subspace of U that contains $V \cup W$ should contain all vectors in U of the form $v + w$, where $v \in V$ and $w \in W$. Such an observation motivates the following definition.

Definition 1.9 If U is a vector space and V and W are its subspaces, the *sum* of V and W, denoted by $V + W$, is the subspace of U given by

$$V + W \equiv \{u \in U \mid u = v + w, \, v \in V, w \in W\}. \qquad (1.5.1)$$

Checking that $V + W$ is a subspace of U that is also the smallest subspace of U containing $V \cup W$ will be assigned as an exercise.

Now let $\mathcal{B}_0 = \{u_1, \ldots, u_k\}$ be a basis of $V \cap W$. Expand it to obtain bases for V and W, respectively, of the forms

$$\mathcal{B}_V = \{u_1, \ldots, u_k, v_1, \ldots, v_l\}, \quad \mathcal{B}_W = \{u_1, \ldots, u_k, w_1, \ldots, w_m\}. \qquad (1.5.2)$$

From the definition of $V + W$, we get

$$V + W = \text{Span}\{u_1, \ldots, u_k, v_1, \ldots, v_l, w_1, \ldots, w_m\}. \qquad (1.5.3)$$

We can see that $\{u_1, \ldots, u_k, v_1, \ldots, v_l, w_1, \ldots, w_m\}$ is a basis of $V + W$. In fact, we only need to show that the vectors $u_1, \ldots, u_k, v_1, \ldots, v_l, w_1, \ldots, w_m$ are linearly independent. For this purpose, consider the linear relation

$$a_1 u_1 + \cdots + a_k u_k + b_1 v_1 + \cdots + b_l v_l + c_1 w_1 + \cdots + c_m w_m = 0, \qquad (1.5.4)$$

where $a_1, \ldots, a_k, b_1, \ldots, b_l, c_1, \ldots, c_m$ are scalars. We claim that

$$w = c_1 w_1 + \cdots + c_m w_m = 0. \qquad (1.5.5)$$

Otherwise, using (1.5.4) and (1.5.5), we see that $w \in V$. However, we already have $w \in W$. So $w \in V \cap W$ which is false since $u_1, \ldots, u_k, w_1, \ldots, w_m$ are linearly independent. Thus (1.5.5) follows and $c_1 = \cdots = c_m = 0$. Applying (1.5.5) in (1.5.4), we immediately have $a_1 = \cdots = a_k = b_1 = \cdots = b_l = 0$.

Therefore we can summarize the above discussion to conclude with the following theorem.

Theorem 1.10 *The following general dimensionality formula*

$$\dim(V + W) = \dim(V) + \dim(W) - \dim(V \cap W) \qquad (1.5.6)$$

is valid for the sum of any two subspaces in a vector space.

Of great importance is the situation when $\dim(V \cap W) = 0$ or $V \cap W = \{0\}$. In this situation, the sum is called *direct sum* and rewritten as $V \oplus W$. Thus, we have

$$\dim(V \oplus W) = \dim(V) + \dim(W). \qquad (1.5.7)$$

Direct sum has the following characteristic.

Theorem 1.11 *The sum of two subspaces V and W of U is a direct sum if and only if each vector u in $V + W$ may be expressed as the sum of a unique vector $v \in V$ and a unique vector $w \in W$.*

Proof   Suppose first $V \cap W = \{0\}$. For any $u \in V + W$, assume that it may be expressed as

$$u = v_1 + w_1 = v_2 + w_2, \quad v_1, v_2 \in V, \quad w_1, w_2 \in W. \qquad (1.5.8)$$

From (1.5.8), we have $v_1 - v_2 = w_2 - w_1$ which lies in $V \cap W$. So $v_1 - v_2 = w_2 - w_1 = 0$ and the stated uniqueness follows.

Suppose that any $u \in V + W$ can be expressed as $u = v + w$ for some unique $v \in V$ and $w \in W$. If $V \cap W \neq \{0\}$, take $x \in V \cap W$ with $x \neq 0$. Then zero vector 0 may be expressed as $0 = x + (-x)$ with $x \in V$ and $(-x) \in W$, which violates the stated uniqueness since $0 = 0 + 0$ with $0 \in V$ and $0 \in W$, as well. □

Let V be a subspace of an n-dimensional vector space U and $\mathcal{B}_V = \{v_1, \ldots, v_k\}$ be a basis of V. Extend $\mathcal{B}_V$ to get a basis of U, say $\{v_1, \ldots, v_k, w_1, \ldots, w_l\}$, where $k + l = n$. Define

$$W = \mathrm{Span}\{w_1, \ldots, w_l\}. \tag{1.5.9}$$

Then we obtain $U = V \oplus W$. The subspace W is called a *linear complement*, or simply *complement*, of V in U. Besides, the subspaces V and W are said to be *mutually complementary* in U.

We may also build up a vector space from any two vector spaces, say V and W, over the same field $\mathbb{F}$, as a direct sum of V and W. To see this, we construct vectors of the form

$$u = (v, w), \quad v \in V, \quad w \in W, \tag{1.5.10}$$

and define vector addition and scalar multiplication component-wise by

$$u_1 + u_2 = (v_1, w_1) + (v_2, w_2) = (v_1 + v_2, w_1 + w_2),$$
$$v_1, v_2 \in V, \quad w_1, w_2 \in W, \tag{1.5.11}$$

$$au = a(v, w) = (av, aw), \quad v \in V, \quad w \in W, \quad a \in \mathbb{F}. \tag{1.5.12}$$

It is clear that the set U of all vectors of the form (1.5.10) equipped with the vector addition (1.5.11) and scalar multiplication (1.5.12) is a vector space over $\mathbb{F}$. Naturally we may *identify* V and W with the subspaces of U given by

$$\tilde{V} = \{(v, 0) \mid v \in V\}, \quad \tilde{W} = \{(0, w) \mid w \in W\}. \tag{1.5.13}$$

Of course, $U = \tilde{V} \oplus \tilde{W}$. Thus, in a well-understood sense, we may also rewrite this relation as $U = V \oplus W$ as anticipated. Sometimes the vector space U so constructed is also referred to as the *direct product* of V and W and rewritten as $U = V \times W$. In this way, $\mathbb{R}^2$ may naturally be viewed as $\mathbb{R} \times \mathbb{R}$, for example.

More generally, let $V_1, \ldots, V_k$ be any k subspaces of a vector space U. In a similar manner we can define the sum

$$V = V_1 + \cdots + V_k, \tag{1.5.14}$$

which is of course a subspace of U. Suggested by the above discussion, if each $v \in V$ may be written as $v = v_1 + \cdots + v_k$ for uniquely determined

$v_1 \in V_1, \ldots, v_k \in V_k$, then we say that V is the direct sum of $V_1, \ldots, V_k$ and rewrite such a relation as

$$V = V_1 \oplus \cdots \oplus V_k. \tag{1.5.15}$$

It should be noted that, when $k \geq 3$, extra caution has to be exerted when checking whether the sum (1.5.14) is a direct sum. For example, the naive condition

$$V_i \cap V_j = \{0\}, \quad i \neq j, \quad i, j = 1, \ldots, k, \tag{1.5.16}$$

among $V_1, \ldots, V_k$, is not sufficient anymore to ensure (1.5.15).

To illustrate this subtlety, let us consider $V = \mathbb{F}^2$ and take

$$V_1 = \text{Span}\left\{ \begin{pmatrix} 1 \\ 0 \end{pmatrix} \right\}, \quad V_2 = \text{Span}\left\{ \begin{pmatrix} 1 \\ 1 \end{pmatrix} \right\}, \quad V_3 = \text{Span}\left\{ \begin{pmatrix} 1 \\ -1 \end{pmatrix} \right\}. \tag{1.5.17}$$

It is clear that V_1, V_2, and V_3 satisfy (1.5.16) and $V = V_1 + V_2 + V_3$ but V cannot be a direct sum of V_1, V_2, and V_3.

In fact, in such a general situation, a correct condition which replaces (1.5.16) is:

$$V_i \cap \left(\sum_{1 \leq j \leq k, j \neq i} V_j \right) = \{0\}, \quad i = 1, \ldots, k. \tag{1.5.18}$$

In other words, when the condition (1.5.18) is fulfilled, then (1.5.15) is valid. The proof of this fact is left as an exercise.

Exercises

1.5.1   For

$$U = \{(x_1, \ldots, x_n) \in \mathbb{F}^n \,|\, x_1 + \cdots + x_n = 0\}, \tag{1.5.19}$$
$$V = \{(x_1, \ldots, x_n) \in \mathbb{F}^n \,|\, x_1 = \cdots = x_n\},$$

where $\text{ch}(\mathbb{F}) = 0$, prove that $\mathbb{F}^n = U \oplus V$.

1.5.2   Consider the vector space of all $n \times n$ matrices over the field $\mathbb{F}$, denoted by $\mathbb{F}(n, n)$. As before, use $\mathbb{F}_S(n, n)$ and $\mathbb{F}_A(n, n)$ to denote the subspaces of symmetric and anti-symmetric matrices. Assume that the characteristic of $\mathbb{F}$ is not equal to 2. For any $M \in \mathbb{F}(n, n)$, rewrite M as

$$M = \frac{1}{2}(M + M^t) + \frac{1}{2}(M - M^t). \tag{1.5.20}$$

Check that $\frac{1}{2}(M + M^t) \in \mathbb{F}_S(n,n)$ and $\frac{1}{2}(M - M^t) \in \mathbb{F}_A(n,n)$. Use this fact to prove the decomposition

$$\mathbb{F}(n,n) = \mathbb{F}_S(n,n) \oplus \mathbb{F}_A(n,n). \tag{1.5.21}$$

What happens when the characteristic of $\mathbb{F}$ is 2 such as when $\mathbb{F} = \mathbb{Z}_2$?

1.5.3   Show that $\mathbb{F}_L(n,n) \cap \mathbb{F}_U(n,n) = \mathbb{F}_D(n,n)$.

1.5.4   Use $\mathbb{F}_L(n,n)$ and $\mathbb{F}_U(n,n)$ in $\mathbb{F}(n,n)$ to give an example for the dimensionality relation (1.5.6).

1.5.5   Let $X = C[a,b]$ $(a,b \in \mathbb{R}$ and $a < b)$ be the vector space of all real-valued continuous functions over the interval $[a,b]$ and

$$Y = \left\{ f \in X \,\middle|\, \int_a^b f(t)\, dt = 0 \right\}. \tag{1.5.22}$$

(a) Prove that if $\mathbb{R}$ is identified with the set of all constant functions over $[a,b]$ then $X = \mathbb{R} \oplus Y$.

(b) For $a = 0, b = 1$, and $f(t) = t^2 + t - 1$, find the unique $c \in \mathbb{R}$ and $g \in Y$ such that $f(t) = c + g(t)$ for all $t \in [0,1]$.

1.5.6   Let U be a vector space and V and W its subspace such that $U = V + W$. If X is subspace of U, is it true that $X = (X \cap V) + (X \cap W)$?

1.5.7   Let U be a vector space and V, W, X some subspaces of U such that

$$U = V \oplus W, \quad U = V \oplus X. \tag{1.5.23}$$

Can one infer $W = X$ from the condition (1.5.23)? Explain why or why not.

1.5.8   Let U be a vector space and V and W subspaces of U. Prove that the sum $V + W$ is the smallest subspace of U containing $V \cup W$. That is, if X is a subspace of U containing $V \cup W$, then $V + W \subset X$.

1.5.9   Let $V_1, \ldots, V_k$ be some subspaces of U and set $V = V_1 + \cdots + V_k$. Show that this sum is a direct sum if and only if one of the following statements is true:

(a) $V_1, \ldots, V_k$ satisfy (1.5.18).

(b) If non-overlapping sets of vectors $\{v_1^1, \ldots, v_{l_1}^1\}, \ldots, \{v_1^k, \ldots, v_{l_k}^k\}$ are bases of $V_1, \ldots, V_k$, respectively, then the union

$$\{v_1^1, \ldots, v_{l_1}^1\} \cup \cdots \cup \{v_1^k, \ldots, v_{l_k}^k\} \tag{1.5.24}$$

is a basis of V.

(c) There holds the dimensionality relation

$$\dim(V) = \dim(V_1) + \cdots + \dim(V_k). \tag{1.5.25}$$

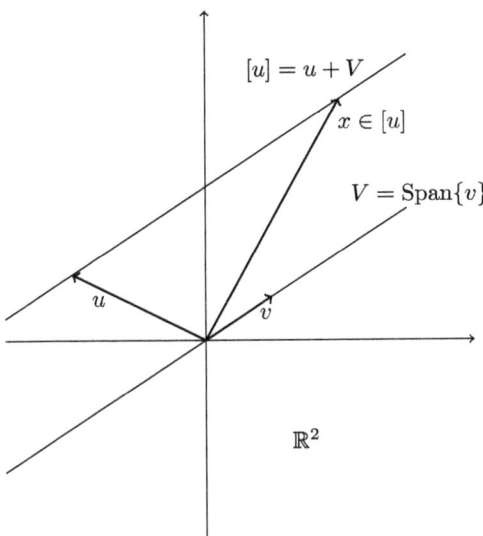

Figure 1.1 Plot of a quotient space construction in $\mathbb{R}^2$ from a subspace V spanned by a nonzero vector v. The coset $[u]$ represented by a vector u modulo V is depicted as a line passing through v and parallel to V such that any $x \in [u]$ is seen as the sum of u and a vector in V.

1.6 Quotient spaces

In order to motivate the introduction of the concept of quotient spaces, we first consider a concrete example in $\mathbb{R}^2$.

Let $v \in \mathbb{R}^2$ be any nonzero vector. Then

$$V = \text{Span}\{v\} \tag{1.6.1}$$

represents the line passing through the origin and along (or opposite to) the direction of v. More generally, for any $u \in \mathbb{R}^2$, the *coset*

$$[u] = \{u + w \mid w \in V\} = \{x \in \mathbb{R}^2 \mid x - u \in V\} \tag{1.6.2}$$

represents the line passing through the vector u and parallel to the vector v. See Figure 1.1 for an illustration.

Naturally, we define $[u_1] + [u_2] = \{x + y \mid x \in [u_1], y \in [u_2]\}$ and claim

$$[u_1] + [u_2] = [u_1 + u_2]. \tag{1.6.3}$$

In fact, let $z \in [u_1] + [u_2]$. Then there exist $x \in [u_1]$ and $y \in [u_2]$ such that $z = x + y$. Rewrite x and y as $x = u_1 + w_1$ and $y = u_2 + w_2$ for some w_1 and $w_2 \in V$.

Hence $z = (u_1 + u_2) + (w_1 + w_2)$ which implies $z \in [u_1 + u_2]$. Conversely, if $z \in [u_1 + u_2]$, then there is some $w \in V$ such that $z = (u_1 + u_2) + w = (u_1 + w) + u_2$. Since $u_1 + w \in [u_1]$ and $u_2 \in [u_2]$, we see that $z \in [u_1] + [u_2]$. Hence the claim follows.

From the property (1.6.3), we see clearly that the coset $[0] = V$ serves as an additive zero element among the set of all cosets.

Similarly, we may also naturally define $a[u] = \{ax \,|\, x \in [u]\}$ for $a \in \mathbb{R}$ where $a \neq 0$. Note that this last restriction is necessary because otherwise $0[u]$ would be a single-point set consisting of zero vector only. We claim

$$a[u] = [au], \quad a \in \mathbb{R} \setminus \{0\}. \tag{1.6.4}$$

In fact, if $z \in a[u]$, there is some $x \in [u]$ such that $z = ax$. Since $x \in [u]$, there is some $w \in V$ such that $x = u + w$. So $z = au + aw$ which implies $z \in [au]$. Conversely, if $z \in [au]$, then there is some $w \in V$ such that $z = au + w$. Since $z = a(u + a^{-1}w)$, we get $z \in a[u]$. So (1.6.4) is established.

Since the coset $[0]$ is already seen to the additive zero when adding cosets, we are prompted to define

$$0[u] = [0]. \tag{1.6.5}$$

Note that (1.6.4) and (1.6.5) may be collectively rewritten as

$$a[u] = [au], \quad a \in \mathbb{R}, \quad u \in \mathbb{R}^2. \tag{1.6.6}$$

We may examine that the above introduced addition between cosets and scalar multiplication with cosets make the set of all cosets into a vector space over $\mathbb{R}$, denoted by $\mathbb{R}^2/V$, and called the quotient space of $\mathbb{R}^2$ modulo V. As investigated, the geometric meaning of $\mathbb{R}^2/V$ is that it is the set of all the lines in $\mathbb{R}^2$ parallel to the vector v and that these lines can be added and multiplied by real scalars so that the set of lines enjoys the structure of a real vector space.

There is no difficulty extending the discussion to the case of $\mathbb{R}^3$ with V a line or plane through the origin.

More generally, the above quotient-space construction may be formulated as follows.

Definition 1.12 Let U be a vector space over a field $\mathbb{F}$ and V a subspace of U. The set of cosets represented by $u \in U$ given by

$$[u] = \{u + w \,|\, w \in V\} = \{u\} + V \equiv u + V, \tag{1.6.7}$$

equipped with addition and scalar multiplication defined by

$$[u] + [v] = [u + v], \quad u, v \in U, \quad a[u] = [au], \quad a \in \mathbb{F}, \quad u \in U, \tag{1.6.8}$$

forms a vector space over $\mathbb{F}$, called the *quotient space* of U modulo V, and is denoted by U/V.

Let $\mathcal{B}_V = \{v_1, \ldots, v_k\}$ be a basis of V. Extend $\mathcal{B}_V$ to get a basis of U, say

$$\mathcal{B}_U = \{v_1, \ldots, v_k, u_1, \ldots, u_l\}. \tag{1.6.9}$$

We claim that $\{[u_1], \ldots, [u_l]\}$ forms a basis for U/V. In fact, it is evident that these vectors span U/V. So we only need to show their linear independence.

Consider the relation

$$a_1 [u_1] + \cdots + a_l [u_l] = 0, \quad a_1, \ldots, a_l \in \mathbb{F}. \tag{1.6.10}$$

Note that $0 = [0]$ in U/V. So $a_1 u_1 + \cdots + a_l u_l \in V$. Thus there are scalars $b_1, \ldots, b_k \in \mathbb{F}$ such that

$$a_1 u_1 + \cdots + a_l u_l = b_1 w_1 + \cdots + b_k w_k. \tag{1.6.11}$$

Hence $a_1 = \cdots = a_l = b_1 = \cdots = b_k = 0$ and the claimed linear independence follows.

As a consequence of the afore-going discussion, we arrive at the following basic dimensionality relation:

$$\dim(U/V) + \dim(V) = \dim(U). \tag{1.6.12}$$

Note that the construction made to arrive at (1.6.12) demonstrates a practical way to find a basis for U/V. The quantity $\dim(U/V) = \dim(U) - \dim(V)$ is sometimes called the *codimension* of the subspace V in U. The quotient space U/V may be viewed as constructed from the space U after "collapsing" or "suppressing" its subspace V.

Exercises

1.6.1 Let U be a vector space and V a subspace of U. Let $[u]$ and $[v]$ denote the cosets mod V defined by

$$[u] = u + V, \quad [v] = v + V, \quad u, v \in U. \tag{1.6.13}$$

(a) Show that either $[u] = [v]$ or $[u] \cap [v] = \emptyset$.
(b) Show that U has the "partition by V" given by

$$U = \bigcup_{u \in U} [u]. \tag{1.6.14}$$

1.6.2   Let V be the subspace of $\mathbb{R}^2$ given by $V = \text{Span}\{(1,-1)\}$.

    (a) Draw V in $\mathbb{R}^2$.

    (b) Draw the cosets

$$S_1 = (1,1)+V, \quad S_2 = (2,1)+V. \tag{1.6.15}$$

    (c) Describe the quotient space $\mathbb{R}^2/V$.

    (d) Draw the coset $S_3 = (-2)S_1 + S_2$.

    (e) Determine whether S_3 is equal to $(-1,0)+V$ (explain why).

1.6.3   Let V be the subspace of $\mathcal{P}_2$ (set of polynomials of degrees up to 2 and with coefficients in $\mathbb{R}$) satisfying

$$\int_{-1}^{1} p(t)\,dt = 0, \quad p(t) \in \mathcal{P}_2. \tag{1.6.16}$$

    (a) Find a basis to describe V.

    (b) Find a basis for the quotient space $\mathcal{P}_2/V$ and verify (1.6.12).

1.6.4   Describe the plane given in terms of the coordinates $(x,y,z) \in \mathbb{R}^3$ by the equation

$$ax + by + cz = d, \quad (a,b,c) \neq (0,0,0), \tag{1.6.17}$$

where $a,b,c,d \in \mathbb{R}$ are constants, as a coset in $\mathbb{R}^3$ and as a point in a quotient space.

1.6.5   Let U be a vector spaces and V and W are some subspaces of U. For $u \in U$, use $[u]_V$ and $[u]_W$ to denote the cosets $u+V$ and $u+W$, respectively. Show that, if $V \subset W$, and $u_1, \ldots, u_k \in U$, then that $[u_1]_V, \ldots, [u_k]_V$ are linearly dependent implies that $[u_1]_W, \ldots, [u_k]_W$ are linearly dependent. In particular, if U is finite dimensional, then $\dim(U/V) \geq \dim(U/W)$, if $V \subset W$, as may also be seen from using (1.6.12).

1.6.6   Let U be a vector space over a field $\mathbb{F}$ and $\dim(U) = n$. Let V be a subspace of U with $\dim(V) = k$. Compute

$$\dim\left(U'/V^0\right) \tag{1.6.18}$$

and come up with a way to construct a basis for the quotient space U'/V^0.

1.6.7   Consider the subspace V of $\mathbb{R}^3$ defined by the equation

$$ax + by + cz = 0, \tag{1.6.19}$$

where $(a,b,c) \in \mathbb{R}^3$ is a fixed nonzero vector.

    (a) Show that

$$\mathbb{R}^3/V = \text{Span}\{[(a,b,c)]\}. \tag{1.6.20}$$

(b) Let (x_0, y_0, z_0) be a point in $\mathbb{R}^3$. Show that

$$(x_0, y_0, z_0) \in [(a, b, c)] \tag{1.6.21}$$

if and only if

$$ax_0 + by_0 + cz_0 = a^2 + b^2 + c^2. \tag{1.6.22}$$

(c) For $(a, b, c) = (1, 1, 1)$, plot V and $L = \text{Span}\{(1, 1, 1)\}$ in $\mathbb{R}^3$.
(d) (Continued from (c)) Explain how you may identify L with $\mathbb{R}^3/V$.

1.6.8 Consider $\mathbb{R}(3, 3)$ and its subspace $\mathbb{R}_S(3, 3)$. Obtain a basis of the quotient space

$$\mathbb{R}(3, 3)/\mathbb{R}_S(3, 3). \tag{1.6.23}$$

1.7 Normed spaces

It will be desirable to be able to evaluate the "length" or "magnitude" or "amplitude" of any vector in a vector space. In other words, it will be useful to associate to each vector a quantity that resembles the notion of length of a vector in (say) $\mathbb{R}^3$. Such a quantity is generically called *norm*.

In this section, we take the field $\mathbb{F}$ to be either $\mathbb{R}$ or $\mathbb{C}$.

Definition 1.13 Let U be a vector space over the field $\mathbb{F}$. A norm over U is a correspondence $\| \cdot \| : U \to \mathbb{R}$ such that we have the following:

(1) (Positivity) $\|u\| \geq 0$ for $u \in U$ and $\|u\| = 0$ only for $u = 0$.
(2) (Homogeneity) $\|au\| = |a| \|u\|$ for $a \in \mathbb{F}$ and $u \in U$.
(3) (Triangle inequality) $\|u + v\| \leq \|u\| + \|v\|$ for $u, v \in U$.

A vector space equipped with a norm is called a *normed space*. If $\| \cdot \|$ is the specific norm of the normed space U, we sometimes spell this fact out by stating "normed space $(U, \| \cdot \|)$."

Definition 1.14 Let $(U, \| \cdot \|)$ be a normed space and $\{u_k\}$ a sequence in U. For some $u_0 \in U$, we say that $u_k \to u_0$ or $\{u_k\}$ *converges* to u_0 as $k \to \infty$ if

$$\lim_{k \to \infty} \|u_0 - u_k\| = 0, \tag{1.7.1}$$

and the sequence $\{u_k\}$ is said to be a *convergent sequence*. u_0 is also said to be the *limit* of the sequence $\{u_k\}$.

The notions of convergence and limit are essential for carrying out calculus in a normed space.

Let U be a finite-dimensional space. We can easily equip U with a norm.

For example, assume that $\mathcal{B} = \{u_1, \ldots, u_n\}$ is a basis of U. For any $u \in U$, define

$$\|u\|_1 = \sum_{i=1}^{n} |a_i|, \quad \text{where} \quad u = \sum_{i=1}^{n} a_i u_i. \tag{1.7.2}$$

It is direct to verify that $\|\cdot\|_1$ indeed defines a norm.

More generally, for $p \geq 1$, we may set

$$\|u\|_p = \left(\sum_{i=1}^{n} |a_i|^p \right)^{\frac{1}{p}}, \quad \text{where} \quad u = \sum_{i=1}^{n} a_i u_i. \tag{1.7.3}$$

It can be shown that $\|\cdot\|_p$ also defines a norm over U. We will not check this fact. What interests us here, however, is the limit

$$\lim_{p \to \infty} \|u\|_p = \max\{|a_i| \,|\, i = 1, \ldots, n\}. \tag{1.7.4}$$

To prove (1.7.4), we note that the right-hand side of (1.7.4) is simply $|a_{i_0}|$ for some $i_0 \in \{1, \ldots, n\}$. Thus, in view of (1.7.3), we have

$$\|u\|_p \leq \left(\sum_{i=1}^{n} |a_{i_0}|^p \right)^{\frac{1}{p}} = |a_{i_0}| n^{\frac{1}{p}}. \tag{1.7.5}$$

From (1.7.5), we obtain

$$\limsup_{p \to \infty} \|u\|_p \leq |a_{i_0}|. \tag{1.7.6}$$

On the other hand, using (1.7.3) again, we have

$$\|u\|_p \geq (|a_{i_0}|^p)^{\frac{1}{p}} = |a_{i_0}|. \tag{1.7.7}$$

Thus,

$$\liminf_{p \to \infty} \|u\|_p \geq |a_{i_0}|. \tag{1.7.8}$$

Therefore (1.7.4) is established. As a consequence, we are motivated to adopt the notation

$$\|u\|_\infty = \max\{|a_i| \,|\, i = 1, \ldots, n\}, \quad \text{where} \quad u = \sum_{i=1}^{n} a_i u_i, \tag{1.7.9}$$

and restate our result (1.7.4) more elegantly as

$$\lim_{p \to \infty} \|u\|_p = \|u\|_\infty, \quad u \in U. \tag{1.7.10}$$

It is evident that $\|\cdot\|_\infty$ does define a norm over U.

Thus we have seen that there are many ways to introduce a norm over a vector space. So it will be important to be able to compare norms. For calculus, it is obvious that the most important thing with regard to norm is the notion of convergence. In particular, assume there are two norms, say $\|\cdot\|$ and $\|\cdot\|'$, equipped over the vector space U. We hope to know whether convergence with respect to norm $\|\cdot\|$ implies convergence with respect to norm $\|\cdot\|'$. This desire motivates the introduction of the following concept.

Definition 1.15 Let U be a vector space and $\|\cdot\|$ and $\|\cdot\|'$ are two norms over U. We say that $\|\cdot\|$ is *stronger* than $\|\cdot\|'$ if convergence with respect to norm $\|\cdot\|$ implies convergence with respect to norm $\|\cdot\|'$. More precisely, any convergent sequence $\{u_k\}$ with limit u_0 in $(U, \|\cdot\|)$ is also a convergent sequence with the same limit in $(U, \|\cdot\|')$.

Regarding the above definition, we have the following.

Theorem 1.16 *Let U be a vector space and $\|\cdot\|$ and $\|\cdot\|'$ are two norms over U. Then norm $\|\cdot\|$ is stronger than norm $\|\cdot\|'$ if and only if there is a constant $C > 0$ such that*

$$\|u\|' \leq C\|u\|, \quad \forall u \in U. \tag{1.7.11}$$

Proof If (1.7.11) holds, then it is clear that $\|\cdot\|$ is stronger than $\|\cdot\|'$.

Suppose that $\|\cdot\|$ is stronger than $\|\cdot\|'$ but (1.7.11) does not hold. Thus there is a sequence $\{u_k\}$ in U such that $u_k \neq 0$ $(k = 1, 2, \dots)$ and

$$\frac{\|u_k\|'}{\|u_k\|} \geq k, \quad k = 1, 2, \dots. \tag{1.7.12}$$

Define

$$v_k = \frac{1}{\|u_k\|'} u_k, \quad k = 1, 2, \dots. \tag{1.7.13}$$

Then (1.7.12) yields the bounds

$$\|v_k\| \leq \frac{1}{k}, \quad k = 1, 2, \dots. \tag{1.7.14}$$

Consequently, $v_k \to 0$ as $k \to \infty$ with respect to norm $\|\cdot\|$. However, according to (1.7.13), we have $\|v_k\|' = 1$, $k = 1, 2, \dots$, and $v_k \not\to 0$ as $k \to \infty$ with respect to norm $\|\cdot\|'$. This reaches a contradiction. □

Definition 1.17 Let U be a vector space and $\|\cdot\|$ and $\|\cdot\|'$ are two norms over U. We say that norms $\|\cdot\|$ and $\|\cdot\|'$ are *equivalent* if convergence in norm $\|\cdot\|$ is equivalent to convergence in norm $\|\cdot\|'$.

In view of Theorem 1.16, we see that norms $\|\cdot\|$ and $\|\cdot\|'$ are equivalent if and only if the inequality

$$C_1\|u\| \le \|u\|' \le C_2\|u\|, \quad \forall u \in U, \tag{1.7.15}$$

holds true for some suitable constants $C_1, C_2 > 0$.

The following theorem regarding norms over finite-dimensional spaces is of fundamental importance.

Theorem 1.18 *Any two norms over a finite-dimensional space are equivalent.*

Proof Let U be an n-dimensional vector space and $\mathcal{B} = \{u_1, \ldots, u_n\}$ a basis for U. Define the norm $\|\cdot\|_1$ by (1.7.2). Let $\|\cdot\|$ be any given norm over U. Then by the properties of norm we have

$$\|u\| \le \sum_{i=1}^n |a_i|\|u_i\| \le \alpha_2 \sum_{i=1}^n |a_i| = \alpha_2\|u\|_1, \tag{1.7.16}$$

where we have set

$$\alpha_2 = \max\{\|u_i\| \mid i = 1, \ldots, n\}. \tag{1.7.17}$$

In other words, we have shown that $\|\cdot\|_1$ is stronger than $\|\cdot\|$.

In order to show that $\|\cdot\|$ is also stronger than $\|\cdot\|_1$, we need to prove the existence of a constant $\alpha_1 > 0$ such that

$$\alpha_1\|u\|_1 \le \|u\|, \quad \forall u \in U. \tag{1.7.18}$$

Suppose otherwise that (1.7.18) fails to be valid. In other words, the set of ratios

$$\left\{ \frac{\|u\|}{\|u\|_1} \,\middle|\, u \in U, u \ne 0 \right\} \tag{1.7.19}$$

does not have a positive infimum. Then there is a sequence $\{v_k\}$ in U such that

$$\frac{1}{k}\|v_k\|_1 \ge \|v_k\|, \quad v_k \ne 0, \quad k = 1, 2, \ldots. \tag{1.7.20}$$

Now set

$$w_k = \frac{1}{\|v_k\|_1} v_k, \quad k = 1, 2, \ldots. \tag{1.7.21}$$

Then (1.7.20) implies that $w_k \to 0$ as $k \to \infty$ with respect to $\|\cdot\|$.

On the other hand, we have $\|w_k\|_1 = 1$ $(k = 1, 2, \ldots)$. Express w_k with respect to the basis $\mathcal{B}$ as

$$w_k = a_{1,k}u_1 + \cdots + a_{n,k}u_n, \quad a_{1,k}, \ldots, a_{n,k} \in \mathbb{F}, \quad k = 1, 2, \ldots. \tag{1.7.22}$$

The definition of norm $\|\cdot\|_1$ then implies $|a_{i,k}| \le 1$ $(i = 1, \ldots, n, k = 1, 2, \ldots)$. Hence, by the Bolzano–Weierstrass theorem, there is a subsequence of $\{k\}$, denoted by $\{k_s\}$, such that $k_s \to \infty$ as $s \to \infty$ and

$$a_{i,k_s} \to \text{ some } a_{i,0} \in \mathbb{F} \quad \text{as} \quad s \to \infty, \quad i = 1, \ldots, n. \tag{1.7.23}$$

Now set $w_0 = a_{1,0}u_1 + \cdots + a_{n,0}u_n$. Then $\|w_0 - w_{k_s}\|_1 \to 0$ as $s \to \infty$. Moreover, the triangle inequality gives us

$$\|w_0\|_1 \geq \|w_{k_s}\|_1 - \|w_{k_s} - w_0\|_1 = 1 - \|w_{k_s} - w_0\|_1, \quad s = 1, 2, \ldots. \quad (1.7.24)$$

Thus, letting $s \to \infty$ in (1.7.24), we obtain $\|w_0\|_1 \geq 1$.

However, substituting $u = w_0 - w_{k_s}$ in (1.7.16) and letting $s \to \infty$, we see that $w_{k_s} \to w_0$ as $s \to \infty$ with respect to norm $\|\cdot\|$ as well, which is false because we already know that $w_k \to 0$ as $k \to \infty$ with respect to norm $\|\cdot\|$.

Summarizing the above study, we see that there are some constants $\alpha_1, \alpha_2 > 0$ such that

$$\alpha_1 \|u\|_1 \leq \|u\| \leq \alpha_2 \|u\|_1, \quad \forall u \in U. \quad (1.7.25)$$

Finally, let $\|\cdot\|'$ be another norm over U. Then we have some constants $\beta_1, \beta_2 > 0$ such that

$$\beta_1 \|u\|_1 \leq \|u\|' \leq \beta_2 \|u\|_1, \quad \forall u \in U. \quad (1.7.26)$$

Combining (1.7.25) and (1.7.26), we arrive at the desired conclusion

$$\frac{\beta_1}{\alpha_2} \|u\| \leq \|u\|' \leq \frac{\beta_2}{\alpha_1} \|u\|, \quad \forall u \in U, \quad (1.7.27)$$

which establishes the equivalence of norms $\|\cdot\|$ and $\|\cdot\|'$ as stated. $\square$

As an immediate application of Theorem 1.18, we have the following.

Theorem 1.19 *A finite-dimensional normed space $(U, \|\cdot\|)$ is locally compact. That is, any bounded sequence in U contains a convergent subsequence.*

Proof Let $\{u_1, \ldots, u_n\}$ be any basis of U and introduce norm $\|\cdot\|_1$ by the expression (1.7.2). Let $\{v_k\}$ be a bounded sequence in $(U, \|\cdot\|)$. Then Theorem 1.18 says that $\{v_k\}$ is also bounded in $(U, \|\cdot\|_1)$. If we rewrite v_k as

$$v_k = a_{1,k}u_1 + \cdots + a_{n,k}u_n, \quad a_{1,k}, \ldots, a_{n,k} \in \mathbb{F}, \quad k = 1, 2, \ldots, \quad (1.7.28)$$

the definition of $\|\cdot\|_1$ then implies that the sequences $\{a_{i,k}\}$ ($i = 1, 2, \ldots$) are all bounded in $\mathbb{F}$. Thus the Bolzano–Weierstrass theorem indicates that there are subsequences $\{a_{i,k_s}\}$ ($i = 1, \ldots, n$), which converge to some $a_{i,0} \in \mathbb{F}$ ($i = 1, \ldots, n$) as $s \to \infty$. Consequently, setting $v_0 = a_{1,0}u_1 + \cdots + a_{n,0}u_n$, we have $\|v_0 - v_{k_s}\|_1 \to 0$ as $s \to \infty$. Thus $\|v_0 - v_{k_s}\| \leq C\|v_0 - v_{k_s}\|_1 \to 0$ as $s \to \infty$ as well. $\square$

Let $(U, \|\cdot\|)$ be a finite-dimensional normed vector space and V a subspace of U. Consider the quotient space U/V. As an exercise, it may be shown that

$$\|[u]\| = \inf\{\|v\| \,|\, v \in [u]\}, \quad [u] \in U/V, \quad u \in U, \quad (1.7.29)$$

defines a norm over U/V.

Exercises

1.7.1   Consider the vector space $C[a,b]$ of the set of all real-valued continuous functions over the interval $[a,b]$, where $a,b \in \mathbb{R}$ and $a < b$. Define two norms $\| \cdot \|_1$ and $\| \cdot \|_\infty$ by setting

$$\|u\|_1 = \int_a^b |u(t)|\,dt, \quad \|u\|_\infty = \max_{t \in [a,b]} |u(t)|, \quad \forall u \in C[a,b]. \quad (1.7.30)$$

Show that $\| \cdot \|_\infty$ is stronger than $\| \cdot \|_1$ but not vice versa.

1.7.2   Over the vector space $C[a,b]$ again, define norm $\| \cdot \|_p$ ($p \geq 1$) by

$$\|u\|_p = \left(\int_a^b |u(t)|^p\,dt \right)^{\frac{1}{p}}, \quad \forall u \in C[a,b]. \quad (1.7.31)$$

Show that $\| \cdot \|_\infty$ is stronger than $\| \cdot \|_p$ for any $p \geq 1$ and prove that

$$\lim_{p \to \infty} \|u\|_p = \|u\|_\infty, \quad \forall u \in C[a,b]. \quad (1.7.32)$$

1.7.3   Let $(U, \| \cdot \|)$ be a normed space and use U' to denote the dual space of U. For each $u' \in U'$, define

$$\|u'\|' = \sup\{|u'(u)|\,|\,u \in U, \|u\| = 1\}. \quad (1.7.33)$$

Show that $\| \cdot \|'$ defines a norm over U'.

1.7.4   Let U be a vector space and U' its dual space which is equipped with a norm, say $\| \cdot \|'$. Define

$$\|u\| = \sup\{|u'(u)|\,|\,u' \in U', \|u'\|' = 1\}. \quad (1.7.34)$$

Show that $\| \cdot \|$ defines a norm over U as well.

1.7.5   Prove that for any finite-dimensional normed space $(U, \| \cdot \|)$ with a given subspace V we may indeed use (1.7.29) to define a norm for the quotient space U/V.

1.7.6   Let $\| \cdot \|$ denote the Euclidean norm of $\mathbb{R}^2$ which is given by

$$\|u\| = \sqrt{a_1^2 + a_2^2}, \quad u = (a_1, a_2) \in \mathbb{R}^2. \quad (1.7.35)$$

Consider the subspace $V = \{(x,y) \in \mathbb{R}^2\,|\,2x - y = 0\}$ and the coset $[(-1,1)]$ in $\mathbb{R}^2$ modulo V.

(a)   Find the unique vector $v \in [(-1,1)]$ such that $\|v\| = \|[(-1,1)]\|$.

(b)   Draw the coset $[(-1,1)]$ in $\mathbb{R}^2$ and the vector v found in part (a) and explain the geometric content of the results.

1.7.7 Consider the space $U = C[0,1]$ equipped with the norm

$$\|f\| = \max\{|f(t)| \, | \, t \in [0,1]\}, \quad f \in U. \tag{1.7.36}$$

Show that the sequence $f_k(t) = t^k$ $(k = 1, 2, \ldots)$ is bounded in U but has no convergent subsequence. Explain why the Bolzano–Weierstrass theorem fails in U.

1.7.8 Show directly without using Theorem 1.18 that the norms $\|\ \|_1$ and $\|\ \|_\infty$ are equivalent over $\mathbb{R}^n$ or $\mathbb{C}^n$.

1.7.9 Let $\mathcal{P}_n$, where $n \geq 1$ is an integer, be the vector space of real polynomials of degrees up to n. Let

$$t_1 < \cdots < t_m \tag{1.7.37}$$

be m real numbers. For $p \in \mathcal{P}_n$, define

$$\|p\| = |p(t_1)| + \cdots + |p(t_m)|. \tag{1.7.38}$$

(a) Find the minimal possible integer m such that (1.7.38) equips $\mathcal{P}_n$ into a normed space.

(b) For m given, find the maximal possible n such that (1.7.38) makes $\mathcal{P}_n$ into a normed space.

(c) Assume $\mathcal{P}_n$ is a normed vector space with its norm being given by (1.7.38). Let $\{p_k\}$ be a sequence in $\mathcal{P}_n$ such that

$$\lim_{k \to \infty} p_k(t_1) = p(t_1), \quad \ldots, \quad \lim_{k \to \infty} p_k(t_m) = p(t_m), \tag{1.7.39}$$

for some $p \in \mathcal{P}_n$. Prove

$$\lim_{k \to \infty} (\sup\{|p_k(t) - p(t)| \, | \, a \leq t \leq b\}) = 0, \tag{1.7.40}$$

where $-\infty < a < b < \infty$.

(d) (Continued from (c)) Prove

$$\int_a^b |p_k(t) - p(t)| \, dt \to 0 \text{ as } k \to \infty. \tag{1.7.41}$$

2

Linear Mappings

In this chapter, we consider linear mappings over vector spaces. We begin by stating the definition and discussing the structural properties of linear mappings. Then, we introduce the notion of adjoint mappings and illustrate some of their applications. We next focus on linear mappings from a vector space into itself and study a series of important concepts such as invariance and reducibility, eigenvalues and eigenvectors, projections, nilpotent mappings, and polynomials of linear mappings. Finally, we discuss the use of norms of linear mappings and present a few analytic applications.

2.1 Linear mappings

A linear mapping may be regarded as a simplest correspondence between two vector spaces. In this section we start our study with the definition of linear mappings. We then discuss the matrix representation of a linear mapping, composition of linear mappings, and the rank and nullity of a linear mapping.

2.1.1 Definition, examples, and notion of associated matrices

Let U and V be two vector spaces over the same field $\mathbb{F}$. A *linear mapping* or *linear map* or *linear operator* is a correspondence T from U into V, written as $T : U \to V$, satisfying the following:

(1) (Additivity) $T(u_1 + u_2) = T(u_1) + T(u_2)$, $u_1, u_2 \in U$.
(2) (Homogeneity) $T(au) = aT(u)$, $a \in \mathbb{F}$, $u \in U$.

A special implication of the homogeneity condition is that $T(0) = 0$. One may also say that a linear mapping "respects" or preserves vector addition and scalar multiplication. Sometimes we also simply use Tu to denote $T(u)$ when there is no risk of confusion.

The set of all linear mappings from U into V will be denoted by $L(U,V)$. For S and $T \in L(U,V)$, we define $S+T$ to be a mapping from U into V satisfying

$$(S+T)(u) = S(u) + T(u), \quad \forall u \in U. \tag{2.1.1}$$

For any $a \in \mathbb{F}$ and $T \in L(U,V)$, we define aT to be the mapping from U into V satisfying

$$(aT)(u) = aT(u), \quad \forall u \in U. \tag{2.1.2}$$

We can directly check that the *mapping addition* (2.1.1) and *scalar-mapping multiplication* (2.1.2) make $L(U,V)$ a vector space over $\mathbb{F}$. We adopt the notation $L(U) = L(U,U)$.

As an example, consider the space of matrices, $\mathbb{F}(m,n)$. For $A = (a_{ij}) \in \mathbb{F}(m,n)$, define

$$T_A(x) = Ax = \begin{pmatrix} a_{11} & \cdots & a_{1n} \\ \cdots & \cdots & \cdots \\ a_{m1} & \cdots & a_{mn} \end{pmatrix} \begin{pmatrix} x_1 \\ \vdots \\ x_n \end{pmatrix}, \quad x = \begin{pmatrix} x_1 \\ \vdots \\ x_n \end{pmatrix} \in \mathbb{F}^n. \tag{2.1.3}$$

Then $T_A \in L(\mathbb{F}^n, \mathbb{F}^m)$. Besides, for the standard basis $\{e_1, \dots, e_n\}$ of $\mathbb{F}$, we have

$$T_A(e_1) = \begin{pmatrix} a_{11} \\ \vdots \\ a_{m1} \end{pmatrix}, \dots, T_A(e_n) = \begin{pmatrix} a_{1n} \\ \vdots \\ a_{mn} \end{pmatrix}. \tag{2.1.4}$$

In other words, the images of $e_1, \dots, e_n$ under the linear mapping T_A are simply the column vectors of the matrix A, respectively.

Conversely, take any element $T \in L(\mathbb{F}^n, \mathbb{F}^m)$. Let $v_1, \dots, v_n \in \mathbb{F}^m$ be images of $e_1, \dots, e_n$ under T such that

$$T(e_1) = v_1 = \begin{pmatrix} a_{11} \\ \vdots \\ a_{m1} \end{pmatrix} = \sum_{i=1}^{m} a_{i1} e_i, \dots, T(e_n) = v_n = \begin{pmatrix} a_{1n} \\ \vdots \\ a_{mn} \end{pmatrix} = \sum_{i=1}^{m} a_{in} e_i, \tag{2.1.5}$$

where $\{e_1, \dots, e_m\}$ is the standard basis of $\mathbb{F}^m$. Then any $x = x_1 e_1 + \cdots + x_n e_n$ has the image

$$T(x) = T\left(\sum_{j=1}^{n} x_j e_n\right) = \sum_{j=1}^{n} x_j v_j = (v_1, \ldots, v_n) \begin{pmatrix} x_1 \\ \vdots \\ x_n \end{pmatrix}$$

$$= \begin{pmatrix} a_{11} & \cdots & a_{1n} \\ \cdots & \cdots & \cdots \\ a_{m1} & \cdots & a_{mn} \end{pmatrix} \begin{pmatrix} x_1 \\ \vdots \\ x_n \end{pmatrix}. \tag{2.1.6}$$

In other words, T can be identified with T_A through the matrix A consisting of column vectors as images of $e_1, \ldots, e_n$ under T given in (2.1.5).

It may also be examined that mapping addition and scalar-mapping multiplication correspond to matrix addition and scalar-matrix multiplication.

In this way, as vector spaces, $\mathbb{F}(m,n)$ and $L(\mathbb{F}^n, \mathbb{F}^m)$ may be identified with each other.

In general, let $\mathcal{B}_U = \{u_1, \ldots, u_n\}$ and $\mathcal{B}_V = \{v_1, \ldots, v_m\}$ be bases of the finite-dimensional vector spaces U and V over the field $\mathbb{F}$, respectively. For any $T \in L(U,V)$, we can write

$$T(u_j) = \sum_{i=1}^{m} a_{ij} v_i, \quad a_{ij} \in \mathbb{F}, \quad i = 1, \ldots, m, \quad j = 1, \ldots, n. \tag{2.1.7}$$

Since the images $T(u_1), \ldots, T(u_n)$ completely determine the mapping T, we see that the matrix $A = (a_{ij})$ completely determines the mapping T. In other words, to each mapping in $L(U,V)$, there corresponds a unique matrix in $\mathbb{F}(m,n)$.

Conversely, for any $A = (a_{ij}) \in \mathbb{F}(m,n)$, we define $T(u_1), \ldots, T(u_n)$ by (2.1.7). Moreover, for any $x \in U$ given by $x = \sum_{j=1}^{n} x_j u_j$ for $x_1, \ldots, x_n \in \mathbb{F}$, we set

$$T(x) = T\left(\sum_{j=1}^{n} x_j u_j\right) = \sum_{j=1}^{n} x_j T(u_j). \tag{2.1.8}$$

It is easily checked that this makes T a well-defined element in $L(U,V)$.

Thus we again see, after specifying bases for U and V, we may identify $L(U,V)$ with $\mathbb{F}(m,n)$ in a natural way.

After the above general description of linear mappings, especially their identification with matrices, we turn our attention to some basic properties of linear mappings.

2.1.2 Composition of linear mappings

Let U, V, and W be vector spaces over a field $\mathbb{F}$ of respective dimensions n, l, and m. For $T \in L(U,V)$ and $S \in L(V,W)$, we can define the *composition* of T and S with the understanding:

$$(S \circ T)(x) = S(T(x)), \quad x \in U. \tag{2.1.9}$$

It is obvious that $S \circ T \in L(U,W)$. We now investigate the matrix of $S \circ T$ in terms of the matrices of S and T.

To this end, let $\mathcal{B}_U = \{u_1,\dots,u_n\}, \mathcal{B}_V = \{v_1,\dots,v_l\}$, and $\mathcal{B}_W = \{w_1,\dots,w_m\}$ be the bases of U,V, and W, respectively, and $A = (a_{ij}) \in \mathbb{F}(l,n)$ and $B = (b_{ij}) \in \mathbb{F}(m,l)$ be the correspondingly associated matrices of T and S, respectively. Then we have

$$(S \circ T)(u_j) = S(T(u_j)) = S\left(\sum_{i=1}^{l} a_{ij}v_i\right) = \sum_{i=1}^{l} a_{ij}S(v_i)$$

$$= \sum_{i=1}^{l}\sum_{k=1}^{m} a_{ij}b_{ki}w_k = \sum_{i=1}^{m}\sum_{k=1}^{l} b_{ik}a_{kj}w_i. \tag{2.1.10}$$

In other words, we see that if we take $C = (c_{ij}) \in \mathbb{F}(m,n)$ to be the matrix associated to the linear mapping $S \circ T$ with respect to the bases $\mathcal{B}_U$ and $\mathcal{B}_W$, then $C = BA$. Hence the composition of linear mappings corresponds to the multiplication of their associated matrices, in the same order. For this reason, it is also customary to use ST to denote $S \circ T$, when there is no risk of confusion.

The composition of linear mappings obviously enjoys the associativity property

$$R \circ (S \circ T) = (R \circ S) \circ T, \quad R \in L(W,X), \quad S \in L(V,W), \quad T \in L(U,V), \tag{2.1.11}$$

as can be seen from

$$(R \circ (S \circ T))(u) = R((S \circ T)(u)) = R(S(T(u))) \tag{2.1.12}$$

and

$$((R \circ S) \circ T)(u) = (R \circ S)(T(u)) = R(S(T(u))) \tag{2.1.13}$$

for any $u \in U$.

Thus, when applying (2.1.11) to the situation of linear mappings between finite-dimensional vector spaces and using their associated matrix representations, we obtain another proof of the associativity of matrix multiplication.

2.1.3 Null-space, range, nullity, and rank

For $T \in L(U,V)$, the subset

$$N(T) = \{x \in U \mid T(x) = 0\} \tag{2.1.14}$$

in U is a subspace of U called the *null-space* of T, which is sometimes called the *kernel* of T, and denoted as Kernel(T) or $T^{-1}(0)$. Here and in the sequel, note that, in general, $T^{-1}(v)$ denotes the set of *preimages* of $v \in V$ under T. That is,

$$T^{-1}(v) = \{x \in U \mid T(x) = v\}. \tag{2.1.15}$$

Besides, the subset

$$R(T) = \{v \in V \mid v = T(x) \text{ for some } x \in U\} \tag{2.1.16}$$

is a subspace of V called the *range* of T, which is sometimes called the *image* of U under T and denoted as Image(T) or $T(U)$.

For $T \in L(U,V)$, we say that T is *one-to-one*, or 1-1, or *injective*, if $T(x) \neq T(y)$ whenever $x,y \in U$ with $x \neq y$. It is not hard to show that T is 1-1 if and only if $N(T) = \{0\}$. We say that T is *onto* or *surjective* if $R(T) = V$.

A basic problem in linear algebra is to investigate whether the equation

$$T(x) = v \tag{2.1.17}$$

has a solution x in U for a given vector $v \in V$. From the meaning of $R(T)$, we see that (2.1.17) has a solution if and only if $v \in R(T)$, and the solution is unique if and only if $N(T) = \{0\}$. More generally, if $v \in R(T)$ and $u \in U$ is any particular solution of (2.1.17), then the set of all solutions of (2.1.17) is simply the coset

$$[u] = \{u\} + N(T) = u + N(T). \tag{2.1.18}$$

Thus, in terms of $N(T)$ and $R(T)$, we may understand the structure of the set of solutions of (2.1.17) completely.

First, as the set of cosets, we have

$$U/N(T) = \{T^{-1}(v) \mid v \in R(T)\}. \tag{2.1.19}$$

Furthermore, for v_1 and $v_2 \in R(T)$, take u_1 and $u_2 \in U$ such that $T(u_1) = v_1$ and $T(u_2) = v_2$. Then $[u_1] = T^{-1}(v_1), [u_2] = T^{-1}(v_2)$, and $[u_1] + [u_2] = [u_1 + u_2] = T^{-1}(v_1 + v_2)$; for $a \in \mathbb{F}, v \in R(T)$, and $u \in T^{-1}(v)$, we have $au \in T^{-1}(av)$ and $a[u] = [au] = T^{-1}(av)$, which naturally give us the construction of $U/N(T)$ as a vector space.

Next, we consider the quotient space $V/R(T)$, referred to as *cokernel* of T, written as Cokernel(T). Recall that $V/R(T)$ is the set of non-overlapping cosets in V modulo $R(T)$ so that exactly $[0] = R(T)$. There we have

$$V \setminus R(T) = \overset{\cdot}{\bigcup}[v], \tag{2.1.20}$$

where the union of cosets on the right-hand side of (2.1.20) is made over all $[v] \in V/R(T)$ except the zero element $[0]$. In other words, the union of all cosets in $V/R(T) \setminus \{[0]\}$ gives us the set of all such vectors v in V that Eq. (2.1.17) fails to have a solution in U. With set notation, this last statement is

$$\{v \in V \mid T^{-1}(v) = \emptyset\} = \bigcup[v]. \tag{2.1.21}$$

Thus, loosely speaking, the quotient space or cokernel $V/R(T)$ measures the "size" of the set of vectors in V that are "missed" by the image of U under T.

Definition 2.1 Let U and V be finite-dimensional vector spaces over the field $\mathbb{F}$. The *nullity* and *rank* of a linear mapping $T : U \to V$, denoted by $n(T)$ and $r(T)$, respectively, are the dimensions of $N(T)$ and $R(T)$. That is,

$$n(T) = \dim(N(T)), \quad r(T) = \dim(R(T)). \tag{2.1.22}$$

As a consequence of this definition, we see that T is 1-1 if and only if $n(T) = 0$ and T is onto if and only if $r(T) = \dim(V)$ or $\dim(\mathrm{Cokernel}(T)) = 0$. The following simple theorem will be of wide usefulness.

Theorem 2.2 *Let U and V be vector spaces over the field $\mathbb{F}$ and $T \in L(U,V)$.*

(1) *If $v_1, \ldots, v_k \in V$ are linearly independent and $u_1, \ldots, u_k \in U$ are such that $T(u_1) = v_1, \ldots, T(u_k) = v_k$, then $u_1, \ldots, u_k$ are linearly independent as well. In other words, the preimages of linearly independent vectors are also linearly independent.*

(2) *If $N(T) = \{0\}$ and $u_1, \ldots, u_k \in U$ are linearly independent, then $v_1 = T(u_1), \ldots, v_k = T(u_k) \in V$ are linearly independent as well. In other words, the images of linearly independent vectors under a 1-1 linear mapping are also linearly independent.*

Proof (1) Consider the relation $a_1 u_1 + \cdots + a_k u_k = 0$, $a_1, \ldots, a_k \in \mathbb{F}$. Applying T to the above relation, we get $a_1 v_1 + \cdots + a_k v_k = 0$. Hence $a_1 = \cdots = a_k = 0$.

(2) We similarly consider $a_1 v_1 + \cdots + a_k v_k = 0$. This relation immediately gives us $T(a_1 u_1 + \cdots + a_k u_k) = 0$. Using $N(T) = \{0\}$, we deduce $a_1 u_1 + \cdots + a_k u_k = 0$. Thus the linear independence of $u_1, \ldots, u_k$ implies $a_1 = \cdots = a_k = 0$. □

The fact stated in the following theorem is known as the *nullity-rank equation*, *rank-nullity equation*, or simply *rank equation*.

Theorem 2.3 *Let U and V be finite-dimensional vector spaces over a field $\mathbb{F}$ and $T \in L(U,V)$. Then*

$$n(T) + r(T) = \dim(U). \qquad (2.1.23)$$

Proof   Let $\{u_1,\ldots,u_k\}$ be a basis of $N(T)$. We then expand it to get a basis for the full space U written as $\{u_1,\ldots,u_k,w_1,\ldots,w_l\}$. Thus

$$R(T) = \text{Span}\{T(w_1),\ldots,T(w_l)\}. \qquad (2.1.24)$$

We now show that $T(w_1),\ldots,T(w_l)$ form a basis for $R(T)$ by establishing their linear independence. To this end, consider $b_1 T(w_1) + \cdots + b_l T(w_l) = 0$ for some $b_1,\ldots,b_l \in \mathbb{F}$. Hence $T(b_1 w_1 + \cdots + b_l w_l) = 0$ or $b_1 w_1 + \cdots + b_l w_l \in N(T)$. So there are $a_1,\ldots,a_k \in \mathbb{F}$ such that $b_1 w_1 + \cdots + b_l w_l = a_1 u_1 + \cdots + a_k u_k$. Since $u_1,\ldots,u_k,w_1,\ldots,w_l$ are linearly independent, we arrive at $a_1 = \cdots = a_k = b_1 = \cdots = b_l = 0$. In particular, $r(T) = l$ and the rank equation is valid. $\quad\square$

As an immediate application of the above theorem, we have the following.

Theorem 2.4 *Let U and V be finite-dimensional vector spaces over the field $\mathbb{F}$. If there is some $T \in L(U,V)$ such that T is both 1-1 and onto, then there must hold $\dim(U) = \dim(V)$. Conversely, if $\dim(U) = \dim(V)$, then there is some $T \in L(U,V)$ which is both 1-1 and onto.*

Proof   If T is 1-1 and onto, then $n(T) = 0$ and $r(T) = \dim(V)$. Thus $\dim(U) = \dim(V)$ follows from the rank equation.

Conversely, assume $\dim(U) = \dim(V) = n$ and let $\{u_1,\ldots,u_n\}$ and $\{v_1,\ldots,v_n\}$ be any bases of U and V, respectively. We can define a unique linear mapping $T : U \to V$ by specifying the images of $u_1,\ldots,u_n$ under T to be $v_1,\ldots,v_n$. That is,

$$T(u_1) = v_1,\ldots,T(u_n) = v_n, \qquad (2.1.25)$$

so that

$$T(u) = \sum_{i=1}^{n} a_i v_i, \quad \forall u = \sum_{i=1}^{n} a_i u_i \in U. \qquad (2.1.26)$$

Since $r(T) = n$, the rank equation implies $n(T) = 0$. So T is 1-1 as well. $\quad\square$

We can slightly modify the proof of Theorem 2.4 to prove the following.

Theorem 2.5 *Let U and V be finite-dimensional vector spaces over the field $\mathbb{F}$ so that $\dim(U) = \dim(V)$. If $T \in L(U,V)$ is either 1-1 or onto, then T must be both 1-1 and onto. In this situation, there is a unique $S \in L(V,U)$ such that $S \circ T : U \to U$ and $T \circ S : V \to V$ are identity mappings, denoted by I_U and I_V, respectively, satisfying $I_U(u) = u, \forall u \in U$ and $I_V(v) = v, \forall v \in V$.*

Proof Suppose $\dim(U) = \dim(V) = n$. If $T \in L(U,V)$ is 1-1, then $n(T) = 0$. So the rank equation gives us $r(T) = n = \dim(V)$. Hence T is onto. If $T \in L(U,V)$ is onto, then $r(T) = n = \dim(U)$. So the rank equation gives us $n(T) = 0$. Hence T is 1-1. Now let $\{u_1, \ldots, u_n\}$ be any basis of U. Since T is 1-1 and onto, we know that $v_1, \ldots, v_n \in V$ defined by (2.1.25) form a basis for V. Define now $S \in L(V,U)$ by setting

$$S(v_1) = u_1, \ldots, S(v_n) = u_n, \tag{2.1.27}$$

so that

$$S(v) = \sum_{i=1}^{n} b_i u_i, \quad \forall v = \sum_{i=1}^{n} b_i v_i \in V. \tag{2.1.28}$$

Then it is clear that $S \circ T = I_U$ and $T \circ S = I_V$. Thus the stated existence of $S \in L(V,U)$ follows.

If $R \in L(V,U)$ is another mapping such that $R \circ T = I_U$ and $T \circ R = I_V$. Then the associativity of the composition of linear mappings (2.1.11) gives us the result $R = R \circ I_V = R \circ (T \circ S) = (R \circ T) \circ S = I_U \circ S = S$. Thus the stated uniqueness of $S \in L(V,U)$ follows as well. $\qquad\square$

Let U and V be vector spaces over the field $\mathbb{F}$ and $T \in L(U,V)$. We say that T is *invertible* if there is some $S \in L(V,U)$ such that $S \circ T = I_U$ and $T \circ S = I_V$. Such a mapping S is necessarily unique and is called the *inverse* of T, denoted as T^{-1}.

The notion of inverse can be slightly relaxed to something referred to as a left or right inverse: We call an element $S \in L(V,U)$ a *left inverse* of $T \in L(U,V)$ if

$$S \circ T = I_U; \tag{2.1.29}$$

an element $R \in L(V,U)$ a *right inverse* of $T \in L(U,V)$ if

$$T \circ R = I_V. \tag{2.1.30}$$

It is interesting that if T is known to be invertible, then the left and right inverses coincide and are simply the inverse of T. To see this, we let $T^{-1} \in L(V,U)$ denote the inverse of T and use the associativity of composition of mappings to obtain from (2.1.29) and (2.1.30) the results

$$S = S \circ I_V = S \circ (T \circ T^{-1}) = (S \circ T) \circ T^{-1} = I_U \circ T^{-1} = T^{-1} \tag{2.1.31}$$

and

$$R = I_U \circ R = (T^{-1} \circ T) \circ R = T^{-1} \circ (T \circ R) = T^{-1} \circ I_V = T^{-1}, \tag{2.1.32}$$

respectively.

It is clear that, regarding $T \in L(U,V)$, the condition (2.1.29) implies that T is 1-1, and the condition (2.1.30) implies that T is onto. In view of Theorem 2.5, we see that when U and V are finite-dimensional and $\dim(U) = \dim(V)$, then T is invertible. Thus we have $S = R = T^{-1}$. On the other hand, if $\dim(U) \neq \dim(V)$, T can never be invertible and the notion of left and right inverses is of separate interest.

As an example, we consider the linear mappings $S : \mathbb{F}^3 \to \mathbb{F}^2$ and $T : \mathbb{F}^2 \to \mathbb{F}^3$ associated with the matrices

$$A = \begin{pmatrix} 1 & 0 & 0 \\ 0 & 1 & 0 \end{pmatrix}, \quad B = \begin{pmatrix} 1 & 0 \\ 0 & 1 \\ 0 & 0 \end{pmatrix}, \tag{2.1.33}$$

respectively according to (2.1.3). Then we may examine that S is a left inverse of T (thus T is a right inverse of S). However, it is absurd to talk about the invertibility of these mappings.

Let U and V be two vector spaces over the same field. An invertible element $T \in L(U,V)$ (if it exists) is also called an *isomorphism*. If there is an isomorphism between U and V, they are said to be *isomorphic* to each other, written as $U \approx V$.

As a consequence of Theorem 2.4, we see that two finite-dimensional vector spaces over the same field are isomorphic if and only if they have the same dimension.

Let U and V be finite-dimensional vector spaces over the field $\mathbb{F}$ and $\{u_1, \ldots, u_n\}$ and $\{v_1, \ldots, v_m\}$ are any bases of U and V, respectively. For fixed $i = 1, \ldots, m$, $j = 1, \ldots, n$, define $T_{ij} \in L(U,V)$ by setting

$$T_{ij}(u_j) = v_i; \quad T_{ij}(u_k) = 0, \quad 1 \leq k \leq n, \quad k \neq j. \tag{2.1.34}$$

It is clear that $\{T_{ij} \,|\, i = 1, \ldots, m, j = 1, \ldots, n\}$ is a basis of $L(U,V)$. In particular, we have

$$\dim(L(U,V)) = mn. \tag{2.1.35}$$

Thus naturally $L(U,V) \approx \mathbb{F}(m,n)$. In the special situation when $V = \mathbb{F}$ we have $L(U,\mathbb{F}) = U'$. Thus $\dim(L(U,\mathbb{F})) = \dim(U') = \dim(U) = n$ as obtained earlier.

Exercises

2.1.1   Consider $\mathcal{P}_n$ over $\mathbb{R}$ and the linear maps

$$D(p(t)) = p'(t), \quad D_1(p(t)) = tp'(t) + 4p(t), \quad p(t) \in \mathcal{P}_n. \tag{2.1.36}$$

Obtain the matrices that represent $D, D_1 \in L(\mathcal{P}_n, \mathcal{P}_n)$ with respect to the standard basis of $\mathcal{P}_n$, that is, $\{1, t, \ldots, t^n\}$.

2.1.2 Let $U = \mathbb{R}(2,2)$ and consider its bases

$$u_1 = \begin{pmatrix} 1 & 0 \\ 0 & 0 \end{pmatrix}, \quad u_2 = \begin{pmatrix} 0 & 1 \\ 0 & 0 \end{pmatrix},$$

$$u_3 = \begin{pmatrix} 0 & 0 \\ 1 & 0 \end{pmatrix}, \quad u_4 = \begin{pmatrix} 0 & 0 \\ 0 & 1 \end{pmatrix}, \tag{2.1.37}$$

and

$$v_1 = \begin{pmatrix} 1 & 0 \\ 0 & 1 \end{pmatrix}, \quad v_2 = \begin{pmatrix} 1 & 0 \\ 0 & -1 \end{pmatrix},$$

$$v_3 = \begin{pmatrix} 0 & 1 \\ 1 & 0 \end{pmatrix}, \quad v_4 = \begin{pmatrix} 0 & 1 \\ -1 & 0 \end{pmatrix}. \tag{2.1.38}$$

Consider $T \in L(U, U)$ defined by

$$T(u) = u^t, \tag{2.1.39}$$

where u^t is the transpose of $u \in U$.

(a) Obtain the matrix of T with respect to the bases $\{u_i\}$ and $\{v_i\}$.

(b) Obtain the matrix of T with respect to the bases $\{v_i\}$ and $\{u_i\}$.

2.1.3 For $A = (a_{ij}) \in \mathbb{F}(m,n)$, define a mapping $M_A : \mathbb{F}^m \to \mathbb{F}^n$ by setting

$$M_A(x) = xA = (x_1, \ldots, x_m) \begin{pmatrix} a_{11} & \cdots & a_{1n} \\ \cdots & \cdots & \cdots \\ a_{m1} & \cdots & a_{mn} \end{pmatrix}, \quad x \in \mathbb{F}^m, \tag{2.1.40}$$

where $\mathbb{F}^l$ is taken to be the vector space of $\mathbb{F}$-valued l-component row vectors. Show that M_A is linear and the row vectors of the matrix A are the images of the standard vectors $e_1, \ldots, e_m$ of $\mathbb{F}^m$ again taken to be row vectors. Describe, as vector spaces, how $\mathbb{F}(m,n)$ may be identified with $L(\mathbb{F}^m, \mathbb{F}^n)$.

2.1.4 Let U and V be vector spaces over the same field $\mathbb{F}$ and $T \in L(U, V)$. Prove that T is 1-1 if and only if $N(T) = \{0\}$.

2.1.5 Let U and V be vector spaces over the same field $\mathbb{F}$ and $T \in L(U, V)$. If Y is a subspace of V, is the subset of U given by

$$X = \{x \in U \mid T(x) \in Y\} \equiv T^{-1}(Y), \tag{2.1.41}$$

necessarily a subspace of U?

2.1.6 Let U and V be finite-dimensional vector spaces and $T \in L(U, V)$. Prove that $r(T) \leq \dim(U)$. In particular, if T is onto, then $\dim(U) \geq \dim(V)$.

2.1.7 Prove that if $T \in L(U,V)$ satisfies (2.1.29) or (2.1.30) then T is 1-1 or onto.

2.1.8 Consider $A \in \mathbb{F}(n,n)$. Prove that, if there are B or $C \in \mathbb{F}(n,n)$ such that either $AB = I_n$ or $CA = I_n$, then A is invertible and $B = A^{-1}$ or $C = A^{-1}$.

2.1.9 Let U be an n-dimensional vector space ($n \geq 2$) and U' its dual space. For $f \in U'$ recall that $f^0 = N(f) = \{u \in U \,|\, f(u) = 0\}$. Let $g \in U'$ be such that f and g are linearly independent. Hence $f^0 \neq g^0$.

(a) Show that $U = f^0 + g^0$ must hold.
(b) Establish $\dim(f^0 \cap g^0) = n - 2$.

2.1.10 For $T \in L(U)$, where U is a finite-dimensional vector space, show that $N(T^2) = N(T)$ if and only if $R(T^2) = R(T)$.

2.1.11 Let U and V be finite-dimensional vector spaces and $S, T \in L(U,V)$. Show that

$$r(S+T) \leq r(S) + r(T). \qquad (2.1.42)$$

2.1.12 Let U, V, and W be finite-dimensional vector spaces over the same field and $T \in L(U,V)$ and $S \in L(V,W)$. Establish the rank and nullity inequalities

$$r(S \circ T) \leq \min\{r(S), r(T)\}, \quad n(S \circ T) \leq n(S) + n(T). \qquad (2.1.43)$$

2.1.13 Let S and $T \in L(U)$ and S or T is non-singular or invertible. Show that $r(S \circ T) = \min\{r(S), r(T)\}$. In particular, if $S = T \circ R$ or $S = R \circ T$, where $R \in L(U)$ is non-singular, then $r(S) = r(T)$.

2.1.14 Let $A \in \mathbb{F}(m,n)$ and $B \in \mathbb{F}(n,m)$. Prove by using the nullity-rank equation that $AB \in \mathbb{F}(m,m)$ must be singular when $m > n$.

2.1.15 Let $T \in L(U,V)$ and $S \in L(V,W)$ where $\dim(U) = n$ and $\dim(V) = m$. Prove that

$$r(S \circ T) \geq r(S) + r(T) - m. \qquad (2.1.44)$$

(This result is also known as the *Sylvester inequality*.)

2.1.16 Let U, V, W, and X be finite-dimensional vector spaces over the same field and $T \in L(U,V), S \in L(V,W)$, and $R \in L(W,X)$. Establish the rank inequality

$$r(R \circ S) + r(S \circ T) \leq r(S) + r(R \circ S \circ T). \qquad (2.1.45)$$

(This result is also known as the *Frobenius inequality*.) Note that (2.1.44) may be deduced as a special case of (2.1.45).

2.1.17 Consider the vector space

$$\mathcal{P}_n = \{a_0 + a_1 t + \cdots + a_n t^n \,|\, a_0, a_1, \ldots, a_n \in \mathbb{R}\} \qquad (2.1.46)$$

with the standard basis $\{1,t,\ldots,t^n\}$ and maps $T : \mathcal{P}_n \to \mathcal{P}_{n+1}$ and $S :$ $\mathcal{P}_{n+1} \to \mathcal{P}_n$ defined by

$$T(p(t)) = p'(t) + 6 \int_0^t p(t)dt, \quad S(p(t)) = p''(t) - 2p'(t), \quad (2.1.47)$$

where $p'(t)$ is the derivative of $p(t)$.

For $n = 2$ and with respect to the standard bases, solve the following:

(a) Find $N(T), R(T), n(T)$, and $r(T)$.
(b) Find $N(S), R(S), n(S)$, and $r(S)$.

2.1.18 Let U and V be finite-dimensional vector spaces over $\mathbb{F}$. Investigate the problem whether for any pairs u_1 and $u_2 \in U$ $(u_1 \neq u_2)$ and v_1 and $v_2 \in V$ $(v_1 \neq v_2)$ there is some $T \in L(U,V)$ such that

$$T(u_1) = v_1, \quad T(u_2) = v_2. \quad (2.1.48)$$

You need to find out situations when this is possible and when this is not possible.

2.1.19 Let $T \in L(\mathbb{R}^2, \mathbb{R}^3)$ be such that

$$T(1,1) = (-1,1,1), \quad T(-1,1) = (0,2,1). \quad (2.1.49)$$

(a) Find $T(3,2)$
(b) Show that T is 1-1.
(c) Find $r(T)$.
(d) Find the matrix associated with T with respect to the standard bases of $\mathbb{R}^2$ and $\mathbb{R}^3$.
(e) Explain why you are able to determine T completely based on your knowledge about the values of T on two vectors $(1,1)$ and $(-1,1)$ only.

2.1.20 Let $\mathcal{P}$ be the vector space of all real polynomials.

(a) Define $T : \mathcal{P} \to \mathcal{P}$ by

$$T(p(t)) = p'(t) \quad \text{(derivative of } p(t)\text{).} \quad (2.1.50)$$

Show that T is onto but not 1-1.

(b) Define $S : \mathcal{P} \to \mathcal{P}$ by

$$S(p(t)) = \int_0^t p(t)\,dt. \quad (2.1.51)$$

Show that S is 1-1 but not onto.

(c) Explain why these findings do not conflict with Theorem 2.5.

2.1.21 Let U be a vector space over a field $\mathbb{F}$ and $\{v_1, \ldots, v_k\}$ and $\{w_1, \ldots, w_l\}$ be two sets of linearly independent vectors. That is, the dimensions of the subspaces $V = \text{Span}\{v_1, \ldots, v_k\}$ and $W = \text{Span}\{w_1, \ldots, w_l\}$ of U are k and l, respectively. Consider the subspace of $\mathbb{F}^{k+l}$ defined by

$$S = \left\{ (y_1, \ldots, y_k, z_1, \ldots, z_l) \in \mathbb{F}^{k+l} \,\middle|\, \sum_{i=1}^{k} y_i v_i + \sum_{j=1}^{l} z_j w_j = 0 \right\}. \quad (2.1.52)$$

Show that $\dim(S) = \dim(V \cap W)$.

2.1.22 Let U and V be finite-dimensional vector spaces, $T \in L(U,V)$ invertible, and $U = U_1 \oplus \cdots \oplus U_k$. Show that there holds $V = V_1 \oplus \cdots \oplus V_k$ where $V_i = T(U_i)$, $i = 1, \ldots, k$.

2.1.23 Let U be finite-dimensional and $T \in L(U)$. Show that $U = N(T) \oplus R(T)$ if and only if $N(T) \cap R(T) = \{0\}$.

2.2 Change of basis

It will be important to investigate how the associated matrix changes with respect to a change of basis for a linear mapping.

Let $\{u_1, \ldots, u_n\}$ and $\{\tilde{u}_1, \ldots, \tilde{u}_n\}$ be two bases of the n-dimensional vector space U over the field $\mathbb{F}$. A change of bases from $\{u_1, \ldots, u_n\}$ to $\{\tilde{u}_1, \ldots, \tilde{u}_n\}$ is simply a linear mapping $R \in L(U,U)$ such that

$$R(u_j) = \tilde{u}_j, \quad j = 1, \ldots, n. \quad (2.2.1)$$

Following the study of the previous section, we know that R is invertible. Moreover, if we rewrite (2.2.1) in a matrix form, we have

$$\tilde{u}_k = \sum_{j=1}^{n} b_{jk} u_j, \quad k = 1, \ldots, n, \quad (2.2.2)$$

where the matrix $B = (b_{jk}) \in \mathbb{F}(n,n)$ is called the *basis transition matrix* from the basis $\{u_1, \ldots, u_n\}$ to the basis $\{\tilde{u}_1, \ldots, \tilde{u}_n\}$, which is necessarily invertible.

Let V be an m-dimensional vector space over $\mathbb{F}$ and take $T \in L(U,V)$. Let $A = (a_{ij})$ and $\tilde{A} = (\tilde{a}_{ij})$ be the $m \times n$ matrices of the linear mapping T associated with the pairs of the bases

$$\{u_1, \ldots, u_n\} \quad \text{and} \quad \{v_1, \ldots, v_m\}, \quad \{\tilde{u}_1, \ldots, \tilde{u}_n\} \quad \text{and} \quad \{v_1, \ldots, v_m\}, \quad (2.2.3)$$

of the spaces U and V, respectively. Thus, we have

$$T(u_j) = \sum_{i=1}^{m} a_{ij} v_i, \quad T(\tilde{u}_j) = \sum_{i=1}^{m} \tilde{a}_{ij} v_i, \quad j = 1, \ldots, n. \quad (2.2.4)$$

Combining (2.2.2) and (2.2.4), we obtain

$$\sum_{i=1}^{m} \tilde{a}_{ij} v_i = T(\tilde{u}_j) = T\left(\sum_{k=1}^{n} b_{kj} u_k\right)$$

$$= \sum_{k=1}^{n} b_{kj} \sum_{i=1}^{m} a_{ik} v_i = \sum_{i=1}^{m} \left(\sum_{k=1}^{n} a_{ik} b_{kj}\right) v_i, \quad j = 1,\dots,n. \quad (2.2.5)$$

Therefore, we can read off the relation

$$\tilde{a}_{ij} = \sum_{k=1}^{n} a_{ik} b_{kj}, \quad i = 1,\dots,m, \quad j = 1,\dots,n, \quad (2.2.6)$$

or

$$\tilde{A} = AB. \quad (2.2.7)$$

Another way to arrive at (2.2.7) is to define $\tilde{T} \in L(U,V)$ by

$$\tilde{T}(u_j) = \sum_{i=1}^{m} \tilde{a}_{ij} v_i, \quad j = 1,\dots,n. \quad (2.2.8)$$

Then, with the mapping $R \in L(U,U)$ given in (2.2.1), the second relation in (2.2.4) simply says $\tilde{T} = T \circ R$, which leads to (2.2.7) immediately.

Similarly, we may consider a change of basis in V. Let $\{\hat{v}_1,\dots,\hat{v}_m\}$ be another basis of V which is related to the basis $\{v_1,\dots,v_m\}$ through an invertible mapping $S \in L(V,V)$ with

$$S(v_i) = \hat{v}_i, \quad i = 1,\dots,m, \quad (2.2.9)$$

given by the basis transition matrix $C = (c_{il}) \in \mathbb{F}(m,m)$ so that

$$\hat{v}_i = \sum_{l=1}^{m} c_{li} v_l, \quad i = 1,\dots,m. \quad (2.2.10)$$

For $T \in L(U,V)$, let $A = (a_{ij}), \hat{A} = (\hat{a}_{ij}) \in \mathbb{F}(m,n)$ be the matrices of T associated with the pairs of the bases

$$\{u_1,\dots,u_n\} \quad \text{and} \quad \{v_1,\dots,v_m\}, \quad \{u_1,\dots,u_n\} \quad \text{and} \quad \{\hat{v}_1,\dots,\hat{v}_m\}, \quad (2.2.11)$$

of the spaces U and V, respectively. Thus, we have

$$T(u_j) = \sum_{i=1}^{m} a_{ij} v_i = \sum_{i=1}^{m} \hat{a}_{ij} \hat{v}_i, \quad j = 1,\dots,n. \quad (2.2.12)$$

Inserting (2.2.10) into (2.2.12), we obtain the relation

$$a_{ij} = \sum_{l=1}^{m} c_{il} \hat{a}_{lj}, \quad i = 1,\dots,m, \quad j = 1,\dots,n. \quad (2.2.13)$$

Rewriting (2.2.13) in its matrix form, we arrive at

$$A = C\hat{A}. \tag{2.2.14}$$

As before, we may also define $\hat{T} \in L(U,V)$ by

$$\hat{T}(u_j) = \sum_{i=1}^{m} \hat{a}_{ij} v_i, \quad j = 1,\ldots,n. \tag{2.2.15}$$

Then

$$(S \circ \hat{T})(u_j) = \sum_{i=1}^{m} \hat{a}_{ij} S(v_i) = \sum_{i=1}^{m} \hat{a}_{ij} \hat{v}_i, \quad j = 1,\ldots,n. \tag{2.2.16}$$

In view of (2.2.12) and (2.2.16), we have $S \circ \hat{T} = T$, which leads again to (2.2.14).

An important special case is when $U = V$ with $\dim(U) = n$. We investigate how the associated matrices of an element $T \in L(U,U)$ with respect to two bases are related through the transition matrix between the two bases. For this purpose, let $\{u_1,\ldots,u_n\}$ and $\{\tilde{u}_1,\ldots,\tilde{u}_n\}$ be two bases of U related through the mapping $R \in L(U,U)$ so that

$$R(u_j) = \tilde{u}_j, \quad j = 1,\ldots,n. \tag{2.2.17}$$

Now let $A = (a_{ij}), \tilde{A} = (\tilde{a}_{ij}) \in \mathbb{F}(n,n)$ be the associated matrices of T with respect to the bases $\{u_1,\ldots,u_n\}$ and $\{\tilde{u}_1,\ldots,\tilde{u}_n\}$, respectively. Thus

$$T(u_j) = \sum_{i=1}^{n} a_{ij} u_i, \quad T(\tilde{u}_j) = \sum_{i=1}^{n} \tilde{a}_{ij} \tilde{u}_i, \quad j = 1,\ldots,n. \tag{2.2.18}$$

Define $\tilde{T} \in L(U,U)$ so that

$$\tilde{T}(u_j) = \sum_{i=1}^{n} \tilde{a}_{ij} u_i, \quad j = 1,\ldots,n. \tag{2.2.19}$$

Then (2.2.17)–(2.2.19) give us

$$R \circ \tilde{T} \circ R^{-1} = T. \tag{2.2.20}$$

Thus, if we choose $B = (b_{ij}) \in \mathbb{F}(n,n)$ to represent the invertible mapping (2.2.17), that is,

$$R(u_j) = \sum_{i=1}^{n} b_{ij} u_i, \quad j = 1,\ldots,n, \tag{2.2.21}$$

then (2.2.20) gives us the manner how the two matrices A and $\tilde{A}$ are related:

$$A = B\tilde{A}B^{-1}. \tag{2.2.22}$$

Such a relation spells out the following definition.

Definition 2.6 Two matrices $A, B \in \mathbb{F}(n,n)$ are said to be *similar* if there is an invertible matrix $C \in \mathbb{F}(n,n)$ such that $A = CBC^{-1}$, written as $A \sim B$.

It is clear that similarity of matrices is an *equivalence relation*. In other words, we have $A \sim A$, $B \sim A$ if $A \sim B$, and $A \sim C$ if $A \sim B$ and $B \sim C$, for $A, B, C \in \mathbb{F}(n,n)$.

Consequently, we see that the matrices associated to a linear mapping of a finite-dimensional vector space into itself with respect to different bases are similar.

Exercises

2.2.1 Consider the differentiation operation $D = \frac{\mathrm{d}}{\mathrm{d}t}$ as a linear mapping over $\mathcal{P}$ (the vector space of all polynomials of variable t) given by

$$D(p)(t) = \frac{\mathrm{d}p(t)}{\mathrm{d}t}, \quad p \in \mathcal{P}. \tag{2.2.23}$$

(a) Find the matrix that represents $D \in L(\mathcal{P}_2, \mathcal{P}_1)$ with respect to the standard bases

$$\mathcal{B}^1_{\mathcal{P}_2} = \{1, t, t^2\}, \quad \mathcal{B}^1_{\mathcal{P}_1} = \{1, t\} \tag{2.2.24}$$

of $\mathcal{P}_2$ and $\mathcal{P}_1$, respectively.

(b) If the basis of $\mathcal{P}_2$ is changed into

$$\mathcal{B}^2_{\mathcal{P}_2} = \{t - 1, t + 1, (t - 1)(t + 1)\}, \tag{2.2.25}$$

find the basis transition matrix from the basis $\mathcal{B}^1_{\mathcal{P}_2}$ into the basis $\mathcal{B}^2_{\mathcal{P}_2}$.

(c) Obtain the matrix that represents $D \in L(\mathcal{P}_2, \mathcal{P}_1)$ with respect to $\mathcal{B}^2_{\mathcal{P}_2}$ and $\mathcal{B}^1_{\mathcal{P}_1}$ directly first and then obtain it again by using the results of (a) and (b) in view of the relation stated as in (2.2.7).

(d) If the basis of $\mathcal{P}_1$ is changed into

$$\mathcal{B}^2_{\mathcal{P}_1} = \{t - 1, t + 1\}, \tag{2.2.26}$$

obtain the matrix that represents $D \in L(\mathcal{P}_2, \mathcal{P}_1)$ with respect to the bases $\mathcal{B}^1_{\mathcal{P}_2}$ and $\mathcal{B}^2_{\mathcal{P}_1}$, for $\mathcal{P}_2$ and $\mathcal{P}_1$, respectively.

(e) Find the basis transition matrix from $\mathcal{B}^1_{\mathcal{P}_1}$ into $\mathcal{B}^2_{\mathcal{P}_1}$.

(f) Obtain the matrix that represents $D \in L(\mathcal{P}_2, \mathcal{P}_1)$ with respect to $\mathcal{B}^1_{\mathcal{P}_2}$ and $\mathcal{B}^2_{\mathcal{P}_1}$ by using the results here in view of the relation stated as in (2.2.14).

2.2.2   (Continued from Problem 2.2.1) Now consider D as an element in $L(\mathcal{P}_2, \mathcal{P}_2)$.

    (a) Find the matrices that represent D with respect to $\mathcal{B}^1_{\mathcal{P}_2}$ and $\mathcal{B}^2_{\mathcal{P}_2}$, respectively.

    (b) Use the basis transition matrix from the basis $\mathcal{B}^1_{\mathcal{P}_2}$ into the basis $\mathcal{B}^2_{\mathcal{P}_2}$ of $\mathcal{P}_2$ to verify your results in (a) in view of the similarity relation stated as in (2.2.22).

2.2.3   Let $D: \mathcal{P}_n \to \mathcal{P}_n$ be defined by (2.2.23) and consider the linear mapping $T: \mathcal{P}_n \to \mathcal{P}_n$ defined by

$$T(p) = tD(p) - p, \quad p = p(t) \in \mathcal{P}_n. \tag{2.2.27}$$

Find $N(T)$ and $R(T)$ and show that $\mathcal{P}_n = N(T) \oplus R(T)$.

2.2.4   Show that the real matrices

$$\begin{pmatrix} a & b \\ c & d \end{pmatrix} \quad \text{and} \quad \frac{1}{2}\begin{pmatrix} a+b+c+d & a-b+c-d \\ a+b-c-d & a-b-c+d \end{pmatrix} \tag{2.2.28}$$

are similar through investigating the matrix representations of a certain linear mapping over $\mathbb{R}^2$ under the standard basis $\{e_1, e_2\}$ and the transformed basis $\{u_1 = e_1 + e_2, u_2 = e_1 - e_2\}$, respectively.

2.2.5   Prove that the matrices

$$\begin{pmatrix} a_{11} & a_{12} & \cdots & a_{1n} \\ a_{21} & a_{22} & \cdots & a_{2n} \\ \cdots & \cdots & \cdots & \cdots \\ a_{n1} & a_{n2} & \cdots & a_{nn} \end{pmatrix} \quad \text{and} \quad \begin{pmatrix} a_{nn} & \cdots & a_{n2} & a_{n1} \\ \cdots & \cdots & \cdots & \cdots \\ a_{2n} & \cdots & a_{22} & a_{21} \\ a_{1n} & \cdots & a_{12} & a_{11} \end{pmatrix}$$

$$\tag{2.2.29}$$

in $\mathbb{F}(n,n)$ are similar by realizing them as the matrix representatives of a certain linear mapping over $\mathbb{F}^n$ with respect to two appropriate bases of $\mathbb{F}^n$.

2.3 Adjoint mappings

Let U and V be finite-dimensional vector spaces over the field $\mathbb{F}$ and U' and V' their dual spaces. For $T \in L(U, V)$ and $v' \in V'$, we see that

$$\langle T(u), v' \rangle, \quad \forall u \in U, \tag{2.3.1}$$

defines a linear functional over U. Hence there is a unique vector $u' \in U'$ such that

$$\langle u, u' \rangle = \langle T(u), v' \rangle, \quad \forall u \in U. \tag{2.3.2}$$

Of course, u' depends on T and v'. So we may write this relation as

$$u' = T'(v').$$ (2.3.3)

Under such a notion, we can rewrite (2.3.2) as

$$\langle u, T'(v') \rangle = \langle T(u), v' \rangle, \quad \forall u \in U, \quad \forall v' \in V'.$$ (2.3.4)

Thus, in this way we have constructed a mapping $T' : V' \to U'$. We now show that T' is linear.

In fact, let v'_1 and $v'_2 \in V'$. Then (2.3.4) gives us

$$\langle u, T'(v'_i) \rangle = \langle T(u), v'_i \rangle, \quad \forall u \in U, \quad i = 1, 2.$$ (2.3.5)

Thus

$$\langle u, T'(v'_1) + T'(v'_2) \rangle = \langle T(u), v'_1 + v'_2 \rangle, \quad \forall u \in U.$$ (2.3.6)

In view of (2.3.4) and (2.3.6), we arrive at $T'(v'_1 + v'_2) = T'(v'_1) + T'(v'_2)$. Besides, for any $a \in \mathbb{F}$, we have from (2.3.4) that

$$\langle u, T'(av') \rangle = \langle T(u), av' \rangle = a \langle T(u), v' \rangle = a \langle u, T'(v') \rangle$$
$$= \langle u, aT'(v') \rangle, \quad \forall u \in U, \quad \forall v' \in V',$$ (2.3.7)

which yields $T'(av') = aT'(v')$. Thus the linearity of T' is established.

Definition 2.7 For any given $T \in L(U,V)$, the linear mapping $T' : V' \to U'$ defined by (2.3.4) is called the *adjoint* or *dual mapping*, or simply *adjoint* or *dual*, of T.

Using the fact $U'' = U, V'' = V$, and the relation (2.3.4), we have

$$T'' \equiv (T')' = T.$$ (2.3.8)

It will be interesting to study the matrices associated with T and T'. For this purpose, let $\{u_1, \ldots, u_n\}$ and $\{v_1, \ldots, v_m\}$ be bases of U and V, and $\{u'_1, \ldots, u'_n\}$ and $\{v'_1, \ldots, v'_m\}$ the corresponding dual bases, respectively. Take $A = (a_{ij}) \in \mathbb{F}(m,n)$ and $A' = (a'_{kl}) \in \mathbb{F}(n,m)$ such that

$$T(u_j) = \sum_{i=1}^{m} a_{ij} v_i, \quad j = 1, \ldots, n,$$ (2.3.9)

$$T'(v'_l) = \sum_{k=1}^{n} a'_{kl} u'_k, \quad l = 1, \ldots, m.$$ (2.3.10)

Inserting $u = u_j$ and $v' = v'_i$ in (2.3.4), we have $a'_{ji} = a_{ij}$ ($i = 1, \ldots, m$, $j = 1, \ldots, n$). In other words, we get $A' = A^t$. Thus the adjoint of a linear mapping corresponds to the transpose of a matrix.

We now investigate the relation between $N(T), R(T)$ and $N(T'), R(T')$.

Theorem 2.8 *Let U and V be finite-dimensional vector spaces and $T \in L(U,V)$. Then $N(T), N(T'), R(T), R(T')$, and their annihilators enjoy the following relations:*

(1) $N(T)^0 = R(T')$.
(2) $N(T) = R(T')^0$.
(3) $N(T')^0 = R(T)$.
(4) $N(T') = R(T)^0$.

Proof   We first prove (2). Let $u \in N(T)$. Then $\langle T'(v'), u \rangle = \langle v', T(u) \rangle = 0$ for any $v' \in V'$. So $u \in R(T')^0$. If $u \in R(T')^0$, then $\langle u', u \rangle = 0$ for any $u' \in R(T')$. So $\langle T'(v'), u \rangle = 0$ for any $v' \in V'$. That is, $\langle v', T(u) \rangle = 0$ for any $v' \in V'$. Hence $u \in N(T)$. So (2) is established.

We now prove (1). If $u' \in R(T')$, there is some $v' \in V$ such that $u' = T'(v')$. So $\langle u', u \rangle = \langle T'(v'), u \rangle = \langle v', T(u) \rangle = 0$ for any $u \in N(T)$. This shows $u' \in N(T)^0$. Hence $R(T') \subset N(T)^0$. However, using (2) and (1.4.31), we have

$$\dim(N(T)) = \dim(R(T')^0) = \dim(U') - \dim(R(T'))$$
$$= \dim(U) - \dim(R(T')), \qquad (2.3.11)$$

which then implies

$$\dim(R(T')) = \dim(U) - \dim(N(T)) = \dim(N(T)^0). \qquad (2.3.12)$$

This proves $N(T)^0 = R(T')$.

Finally, (3) and (4) follow from replacing T by T' and using $T'' = T$ in (1) and (2). □

We note that (1) in Theorem 2.8 may also follow from taking the annihilators on both sides of (2) and applying Problem 1.4.7. Thus, part (2) of Theorem 2.8 is the core of all the statements there.

We now use Theorem 2.8 to establish the following result.

Theorem 2.9 *Let U and V be finite-dimensional vector space. For any $T \in L(U,V)$, the rank of T and that of its dual $T' \in L(V', U')$ are equal. That is,*

$$r(T) = r(T'). \qquad (2.3.13)$$

Proof   Using (2) in Theorem 2.8, we have $n(T) = \dim(U) - r(T')$. However, applying the rank equation (2.1.23) or Theorem 2.3, we have $n(T) + r(T) = \dim(U)$. Combining these results, we have $r(T) = r(T')$. □

As an important application of Theorem 2.8, we consider the notion of rank of a matrix.

Let $T \in L(\mathbb{F}^n, \mathbb{F}^m)$ be defined by (2.1.6) by the matrix $A = (a_{ij}) \in \mathbb{F}(m,n)$. Then the images of the standard basis vectors $e_1, \ldots, e_n$ are the column vectors of A. Thus $R(T)$ is the vector space spanned by the column vectors of A whose dimension is commonly called the *column rank* of A, denoted as corank(A). Thus, we have

$$r(T) = \text{corank}(A). \qquad (2.3.14)$$

On the other hand, since the associated matrix of T' is A^t whose column vectors are the row vectors of A. Hence $R(T')$ is the vector space spanned by the row vectors of A (since $(\mathbb{F}^m)' = \mathbb{F}^m$) whose dimension is likewise called the *row rank* of A, denoted as rorank(A). Thus, we have

$$r(T') = \text{rorank}(A). \qquad (2.3.15)$$

Consequently, in view of (2.3.14), (2.3.15), and Theorem 2.9, we see that the column and row ranks of a matrix coincide, although they have different meanings.

Since the column and row ranks of a matrix are identical, they are jointly called the *rank* of the matrix.

Exercises

2.3.1 Let U be a finite-dimensional vector space over a field $\mathbb{F}$ and regard a given element $f \in U'$ as an element in $L(U, \mathbb{F})$. Describe the adjoint of f, namely, f', as an element in $L(\mathbb{F}', U')$, and verify $r(f) = r(f')$.

2.3.2 Let U, V, and W be finite-dimensional vector spaces over a field $\mathbb{F}$. Show that for $T \in L(U, V)$ and $S \in L(V, W)$ there holds $(S \circ T)' = T' \circ S'$.

2.3.3 Let U and V be finite-dimensional vector spaces over a field $\mathbb{F}$ and $T \in L(U, V)$. Prove that for given $v \in V$ the non-homogeneous equation

$$T(u) = v \qquad (2.3.16)$$

has a solution for some $u \in U$ if and only if for any $v' \in V'$ such that $T'(v') = 0$ one has $\langle v, v' \rangle = 0$. In particular, Eq. (2.3.16) has a solution for any $v \in V$ if and only if T' is injective. (This statement is also commonly known as the *Fredholm alternative*.)

2.3.4 Let U be a finite-dimensional vector space over a field $\mathbb{F}$ and $a \in \mathbb{F}$. Define $T \in L(U, U)$ by $T(u) = au$, where $u \in U$. Show that T' is then given by $T'(u') = au'$ for any $u' \in U'$.

2.3.5   Let $U = \mathbb{F}(n,n)$ and $B \in U$. Define $T \in L(U,U)$ by $T(A) = AB - BA$, where $A \in U$. For $f \in U'$ given by

$$f(A) = \langle f, A \rangle = \text{Tr}(A), \quad A \in U, \tag{2.3.17}$$

determine $T'(f)$.

2.3.6   Let $U = \mathbb{F}(n,n)$. Then $f(A) = \text{Tr}(AB^t)$ defines an element $f \in U'$.

(a) Use M_{ij} to denote the element in $\mathbb{F}(n,n)$ whose entry in the ith row and jth column is 1 but all other entries are zero. Verify the formula

$$\text{Tr}(M_{ij}M_{kl}) = \delta_{il}\delta_{jk}, \quad i,j,k,l = 1,\ldots,n. \tag{2.3.18}$$

(b) Apply (a) to show that for any $f \in U'$ there is a unique element $B \in U$ such that $f(A) = \text{Tr}(AB^t)$.

(c) For $T \in L(U,U)$, defined by $T(A) = A^t$, describe T'.

2.3.7   Let $U = C_c^\infty(-\infty,\infty)$ be the vector space of infinitely differentiable real-valued functions of compact supports (meaning that each element of U vanishes outside a certain bounded subinterval of $(-\infty,\infty)$). Assume we know that the dual space of U is identified with U such that each functional over U is of the form

$$\langle u,v \rangle = \int_{-\infty}^{\infty} u(t)v(t)\mathrm{d}t, \quad u \in U, \quad v \in U' = U. \tag{2.3.19}$$

Let D be the derivative map, $D : U \to U$, given by

$$D(u) = \frac{\mathrm{d}u}{\mathrm{d}t}, \quad u \in U. \tag{2.3.20}$$

Find the adjoint of D. (Hint: Explore what an integration by parts process will give you.)

2.3.8   (Continued from Problem 2.3.7) Find the adjoint of D^2.

2.3.9   (Continued from Problem 2.3.7) Find the adjoint of

$$T = a(t)D^2 + b(t)D + c(t)I, \tag{2.3.21}$$

where $a(t), b(t)$, and $c(t)$ are infinitely differentiable functions over $(-\infty,\infty)$ and I is the identity map.

2.4 Quotient mappings

Let U and V be two vector spaces over a field $\mathbb{F}$ and $T \in L(U,V)$. Suppose that X and Y are subspaces of U and V, respectively, which satisfy the property

$T(X) \subset Y$ or $T \in L(X,Y)$. We now show that such a property allows us to generate a linear mapping

$$\tilde{T} : U/X \to V/Y, \tag{2.4.1}$$

from T naturally.

As before, we use $[\cdot]$ to denote a coset in U/X or V/Y. Define $\tilde{T} : U/X \to V/Y$ by setting

$$\tilde{T}([u]) = [T(u)], \quad \forall [u] \in U/X. \tag{2.4.2}$$

We begin by showing that this definition does not suffer any ambiguity by verifying

$$\tilde{T}([u_1]) = \tilde{T}([u_2]) \quad \text{whenever } [u_1] = [u_2]. \tag{2.4.3}$$

In fact, if $[u_1] = [u_2]$, then $u_1 - u_2 \in X$. Thus $T(u_1) - T(u_2) = T(u_1 - u_2) \in Y$, which implies $[T(u_1)] = [T(u_2)]$. So (2.4.3) follows.

The linearity of $\tilde{T}$ can now be checked directly.

First, let u_1 and $u_2 \in U$. Then, by (2.4.2), we have

$$\tilde{T}([u_1] + [u_2]) = \tilde{T}([u_1 + u_2]) = [T(u_1 + u_2)] = [T(u_1) + T(u_2)]$$
$$= [T(u_1)] + [T(u_2)] = \tilde{T}([u_1]) + \tilde{T}([u_2]). \tag{2.4.4}$$

Next, let $a \in \mathbb{F}$ and $u \in U$. Then, again by (2.4.2), we have

$$\tilde{T}(a[u]) = \tilde{T}([au]) = [T(au)] = a[T(u)] = a\tilde{T}([u]). \tag{2.4.5}$$

It is not hard to show that the property $T(X) \subset Y$ is also necessary to ensure that (2.4.2) gives us a well-defined mapping $\tilde{T}$ from U/X into V/Y for $T \in L(U,V)$. Indeed, let $u \in X$. Then $[u] = [0]$. Thus $[T(u)] = \tilde{T}([u]) = \tilde{T}([0]) = [0]$ which implies $T(u) \in Y$.

In summary, we can state the following basic theorem regarding constructing quotient mappings between quotient spaces.

Theorem 2.10 *Let X and Y be subspaces of U and V, respectively, and $T \in L(U,V)$. Then (2.4.2) defines a linear mapping $\tilde{T}$ from U/X into V/Y if and only if the condition*

$$T(X) \subset Y \tag{2.4.6}$$

holds.

In the special situation when $X = \{0\} \subset U$ and Y is an arbitrary subspace of V, we see that $U/X = U$ any $T \in L(U,V)$ induces the quotient mapping

$$\tilde{T} : U \to V/Y, \quad \tilde{T}(u) = [T(u)]. \tag{2.4.7}$$

Exercises

2.4.1 Let V and W be subspaces of a vector space U. Then the quotient mapping $\tilde{I}: U/V \to U/W$ induced from the identity mapping $I: U \to U$ is well-defined and given by

$$\tilde{I}([u]_V) = [u]_W, \quad u \in U, \tag{2.4.8}$$

if $V \subset W$, where $[\cdot]_V$ and $[\cdot]_W$ denote the cosets in U/V and U/W, respectively. Show that the fact $\tilde{I}([0]_V) = [0]_W$ implies that if $[u_1]_V, \ldots, [u_k]_V$ are linearly dependent in U/V then $[u_1]_W, \ldots, [u_k]_W$ are linearly dependent in U/W.

2.4.2 Let U be a finite-dimensional vector space and V a subspace of U. Use $\tilde{I}$ to denote the quotient mapping $\tilde{I}: U \to U/V$ induced from the identity mapping $I: U \to U$ given as in (2.4.7). Apply Theorem 2.3 or the rank equation (2.1.23) to $\tilde{I}$ to reestablish the relation (1.6.12).

2.4.3 Let U and V be finite-dimensional vector spaces over a field and X and Y the subspaces of U and V, respectively. For $T \in L(U, V)$, assume $T(X) \subset Y$ and use $\tilde{T}$ to denote the quotient mapping from U/X into V/Y induced from T. Show that $r(\tilde{T}) \leq r(T)$.

2.5 Linear mappings from a vector space into itself

Let U be a vector space over a field $\mathbb{F}$. In this section, we study the important special situation when linear mappings are from U into itself. We shall denote the space $L(U, U)$ simply by $L(U)$. Such mappings are also often called *linear transformations*. First, we consider some general properties such as invariance and reducibility. Then we present some examples.

2.5.1 Invariance and reducibility

In this subsection, we consider some situations in which the complexity of a linear mapping may be "reduced" somewhat.

Definition 2.11 Let $T \in L(U)$ and V be a subspace of U. We say that V is an *invariant subspace* of T if $T(V) \subset V$.

Given $T \in L(U)$, it is clear that the null-space $N(T)$ and range $R(T)$ of T are both invariant subspaces of T.

To see how the knowledge about an invariant subspace reduces the complexity of a linear mapping, we assume that V is a nontrivial invariant subspace of $T \in L(U)$ where U is n-dimensional. Let $\{u_1, \ldots, u_k\}$ be any basis of V. We extend it to get a basis of U, say $\{u_1, \ldots, u_k, u_{k+1}, \ldots, u_n\}$. With respect to such a basis, we have

$$T(u_i) = \sum_{i'=1}^{k} b_{i'i} u_{i'}, \quad i = 1, \ldots, k, \quad T(u_j) = \sum_{j'=1}^{n} c_{j'j} u_{j'}, \quad j = k+1, \ldots, n,$$

$$(2.5.1)$$

where $B = (b_{i'i}) \in \mathbb{F}(k,k)$ and $C = (c_{j'j}) \in \mathbb{F}(n, n-k)$. With respect to this basis, the associated matrix $A \in \mathbb{F}(n,n)$ becomes

$$A = \begin{pmatrix} B & C_1 \\ 0 & C_2 \end{pmatrix}, \quad \begin{pmatrix} C_1 \\ C_2 \end{pmatrix} = C. \quad (2.5.2)$$

Such a matrix is sometimes referred to as *block* or *boxed upper triangular*.

Thus, we see that a linear mapping T over a finite-dimensional vector space U has a nontrivial invariant subspace if and only if there is a basis of U so that the associated matrix of T with respect to this basis is block or boxed upper triangular.

For the matrix A given in (2.5.2), the vanishing of the entries in the left-lower portion of the matrix indeed reduces the complexity of the matrix. We have seen clearly that such a "reduction" happens because of the invariance property:

$$T(\text{Span}\{u_1, \ldots, u_k\}) \subset \text{Span}\{u_1, \ldots, u_k\}. \quad (2.5.3)$$

Consequently, if we also have the following *additionally imposed* invariance property:

$$T(\text{Span}\{u_{k+1}, \ldots, u_n\}) \subset \text{Span}\{u_{k+1}, \ldots, u_n\}, \quad (2.5.4)$$

then $c_{j'j} = 0$ for $j' = 1, \ldots, k$ in (2.5.1) or $C_1 = 0$ in (2.5.2), which further reduces the complexity of the matrix A.

The above investigation motivates the introduction of the concept of *reducibility* of a linear mapping as follows.

Definition 2.12 We say that $T \in L(U)$ is *reducible* if there are nontrivial invariant subspaces V and W of T so that $U = V \oplus W$. In such a situation, we also say that T may be reduced over the subspaces V and W. Otherwise, if no such pair of invariant subspaces exist, we say that T is *irreducible*.

Thus, if U is n-dimensional and $T \in L(U)$ may be reduced over the nontrivial subspaces V and W, then we take $\{v_1, \ldots, v_k\}$ and $\{w_1, \ldots, w_l\}$ to be any bases of V and W, so that $\{v_1,\ldots,v_k,w_1,\ldots,w_l\}$ is a basis of U. Over such a basis, T has the representation

$$T(v_i) = \sum_{i'=1}^{k} b_{i'i}v_{i'}, \quad i=1,\ldots,k, \quad T(w_j) = \sum_{j'=1}^{l} c_{j'j}w_{j'}, \quad j=1,\ldots,l,$$

(2.5.5)

where the matrices $B = (b_{i'i})$ and $C = (c_{j'j})$ are in $\mathbb{F}(k,k)$ and $\mathbb{F}(l,l)$, respectively. Thus, we see that, with respect to such a basis of U, the associated matrix $A \in \mathbb{F}(n,n)$ assumes the form

$$A = \begin{pmatrix} B & 0 \\ 0 & C \end{pmatrix}, \quad (2.5.6)$$

which takes the form of a special kind of matrices called *block* or *boxed diagonal matrices*.

Thus, we see that a linear mapping T over a finite-dimensional vector space U is reducible if and only if there is a basis of U so that the associated matrix of T with respect to this basis is block or boxed diagonal.

An important and useful family of invariant subspaces are called eigenspaces which may be viewed as an extension of the notion of null-spaces.

Definition 2.13 For $T \in L(U)$, a scalar $\lambda \in \mathbb{F}$ is called an *eigenvalue* of T if the null-space $N(T - \lambda I)$ is not the zero space $\{0\}$. Any nonzero vector in $N(T - \lambda I)$ is called an *eigenvector* associated with the eigenvalue λ and $N(T - \lambda I)$ is called the *eigenspace* of T associated with the eigenvalue λ, and often denoted as E_λ. The integer $\dim(E_\lambda)$ is called the *geometric multiplicity* of the eigenvalue λ.

In particular, for $A \in \mathbb{F}(n,n)$, an eigenvalue, eigenspace, and the eigenvectors associated with and geometric multiplicity of an eigenvalue of the matrix A are those of the A-induced mapping $T_A : \mathbb{F}^n \to \mathbb{F}^n$, defined by $T_A(x) = Ax, x \in \mathbb{F}^n$.

Let λ be an eigenvalue of $T \in L(U)$. It is clear that E_λ is invariant under T and $T = \lambda I$ (I is the identity mapping) over E_λ. Thus, let $\{u_1,\ldots,u_k\}$ be a basis of E_λ and extend it to obtain a basis for the full space U. Then the associated matrix A of T with respect to this basis takes a block or boxed upper triangular form

$$A = \begin{pmatrix} \lambda I_k & C_1 \\ 0 & C_2 \end{pmatrix}, \quad (2.5.7)$$

where I_k denotes the identity matrix in $\mathbb{F}(k,k)$.

We may explore the concept of eigenspaces to pursue a further reduction of the associated matrix of a linear mapping.

Theorem 2.14 *Let $u_1, \ldots, u_k$ be eigenvectors associated with distinct eigenvalues $\lambda_1, \ldots, \lambda_k$ of some $T \in L(U)$. Then these vectors are linearly independent.*

Proof We use induction on k. If $k = 1$, the statement of the theorem is already valid since $u_1 \neq 0$. Assume the statement of the theorem is valid for $k = m \geq 1$. For $k = m + 1$, consider the relation

$$c_1 u_1 + \cdots + c_m u_m + c_{m+1} u_{m+1} = 0, \quad c_1, \ldots, c_m, c_{m+1} \in \mathbb{F}. \tag{2.5.8}$$

Applying T to (2.5.8), we have

$$c_1 \lambda_1 u_1 + \cdots + c_m \lambda_m u_m + c_{m+1} \lambda_{m+1} u_{m+1} = 0. \tag{2.5.9}$$

Multiplying (2.5.8) by λ_{m+1} and subtracting the result from (2.5.9), we obtain

$$c_1 (\lambda_1 - \lambda_{m+1}) u_1 + \cdots + c_m (\lambda_m - \lambda_{m+1}) u_m = 0. \tag{2.5.10}$$

Thus, by the assumption that $u_1, \ldots, u_m$ are linearly independent, we get $c_1 = \cdots = c_m = 0$ since $\lambda_1, \ldots, \lambda_m, \lambda_{m+1}$ are distinct. Inserting this result into (2.5.8), we find $c_{m+1} = 0$. So the proof follows. □

As a corollary of the theorem, we immediately conclude that a linear mapping over an n-dimensional vector space may have at most n distinct eigenvalues.

If $\lambda_1, \ldots, \lambda_k$ are the distinct eigenvalues of T, then Theorem 2.14 indicates that the sum of $E_{\lambda_1}, \ldots, E_{\lambda_k}$ is a direct sum,

$$E_{\lambda_1} + \cdots + E_{\lambda_k} = E_{\lambda_1} \oplus \cdots \oplus E_{\lambda_k}. \tag{2.5.11}$$

Thus, the equality

$$\dim(E_{\lambda_1}) + \cdots + \dim(E_{\lambda_k}) = \dim(U) = n \tag{2.5.12}$$

holds if and only if

$$U = E_{\lambda_1} \oplus \cdots \oplus E_{\lambda_k}. \tag{2.5.13}$$

In this situation T is reducible over $E_{\lambda_1}, \ldots, E_{\lambda_k}$ and may naturally be expressed as a *direct sum of mappings*

$$T = \lambda_1 I \oplus \cdots \oplus \lambda_k I, \tag{2.5.14}$$

where $\lambda_i I$ is understood to operate on E_{λ_i}, $i = 1, \ldots, k$. Furthermore, using the bases of $E_{\lambda_1}, \ldots, E_{\lambda_k}$ to form a basis of U, we see that the associated matrix A of T becomes diagonal

$$A = \mathrm{diag}\{\lambda_1 I_{n_1}, \ldots, \lambda_k I_{n_k}\}, \tag{2.5.15}$$

where $n_i = \dim(E_{\lambda_i})$ $(i = 1, \ldots, k)$. In particular, when T has n distinct eigenvalues, $\lambda_1, \ldots, \lambda_n$, then any n eigenvectors correspondingly associated with these eigenvalues form a basis of U. With respect to this basis, the associated matrix A of T is simply

$$A = \operatorname{diag}\{\lambda_1, \ldots, \lambda_n\}. \tag{2.5.16}$$

We now present some examples as illustrations.

Consider $T \in L(\mathbb{R}^2)$ defined by

$$T(x) = \begin{pmatrix} 2 & 1 \\ 1 & 2 \end{pmatrix} x, \quad x = \begin{pmatrix} x_1 \\ x_2 \end{pmatrix} \in \mathbb{R}^2. \tag{2.5.17}$$

Then it may be checked directly that T has 1 and 3 as eigenvalues and

$$E_1 = \operatorname{Span}\{(1, -1)^t\}, \quad E_3 = \operatorname{Span}\{(1, 1)^t\}. \tag{2.5.18}$$

Thus, with respect to the basis $\{(1, -1)^t, (1, 1)^t\}$, the associated matrix of T is

$$\begin{pmatrix} 1 & 0 \\ 0 & 3 \end{pmatrix}. \tag{2.5.19}$$

So T is reducible and reduced over the pair of eigenspaces E_1 and E_3.

Next, consider $S \in L(\mathbb{R}^2)$ defined by

$$S(x) = \begin{pmatrix} 0 & -1 \\ 1 & 0 \end{pmatrix} x, \quad x = \begin{pmatrix} x_1 \\ x_2 \end{pmatrix} \in \mathbb{R}^2. \tag{2.5.20}$$

We show that S has no nontrivial invariant spaces. In fact, if V is one, then $\dim(V) = 1$. Let $x \in V$ such that $x \neq 0$. Then the invariance of V requires $S(x) = \lambda x$ for some $\lambda \in \mathbb{R}$. Inserting this relation into (2.5.20), we obtain

$$-x_2 = \lambda x_1, \quad x_1 = \lambda x_2. \tag{2.5.21}$$

Hence $x_1 \neq 0$ and $x_2 \neq 0$. Iterating the two equations in (2.5.21), we obtain $\lambda^2 + 1 = 0$ which is impossible. So S is also irreducible.

2.5.2 Projections

In this subsection, we study an important family of reducible linear mappings called projections.

Definition 2.15 Let V and W be two complementary subspaces of U. That is, $U = V \oplus W$. For any $u \in U$, express u uniquely as $u = v + w$, where $v \in V$, $w \in W$, and define the mapping $P : U \to U$ by

$$P(u) = v. \tag{2.5.22}$$

Then $P \in L(U)$ and is called the *projection* of U onto V along W.

We need to check that the mapping P defined in Definition 2.15 is indeed linear. To see this, we take $u_1, u_2 \in U$ and express them as $u_1 = v_1 + w_1$, $u_2 = v_2 + w_2$, for unique $v_1, v_2 \in V$, $w_1, w_2 \in W$. Hence $P(u_1) = v_1$, $P(u_2) = v_2$. On the other hand, from $u_1 + u_2 = (v_1 + v_2) + (w_1 + w_2)$, we get $P(u_1 + u_2) = v_1 + v_2$. Thus $P(u_1 + u_2) = P(u_1) + P(u_2)$. Moreover, for any $a \in \mathbb{F}$ and $u \in U$, write $u = v + w$ for unique $v \in V$, $w \in W$. Thus $P(u) = v$ and $au = av + aw$ give us $P(au) = av = aP(u)$. So $P \in L(U)$ as claimed.

From Definition 2.15, we see that for $v \in V$ we have $P(v) = v$. Thus $P(P(u)) = P(u)$ for any $u \in U$. In other words, the projection P satisfies the special property $P \circ P = P$. For notational convenience, we shall use T^k to denote the k-fold composition $T \circ \cdots \circ T$ for any $T \in L(U)$. With this notation, we see that a projection P satisfies the condition $P^2 = P$. Any linear mapping satisfying such a condition is called *idempotent*.

We now show that being idempotent characterizes a linear mapping being a projection.

Theorem 2.16 *A linear mapping T over a vector space U is idempotent if and only if it is a projection. More precisely, if $T^2 = T$ and*

$$V = N(I - T), \quad W = N(T), \tag{2.5.23}$$

then $U = V \oplus W$ and T is simply the projection of U onto V along W.

Proof Let T be idempotent and define the subspaces V, W by (2.5.23). We claim

$$R(T) = N(I - T), \quad R(I - T) = N(T). \tag{2.5.24}$$

In fact, if $u \in R(T)$, then there is some $x \in U$ such that $u = T(x)$. Hence $(I - T)u = (I - T)(T(x)) = T(x) - T^2(x) = 0$. So $u \in N(I - T)$. If $u \in N(I - T)$, then $u = T(u)$ which implies $u \in R(T)$ already. If $u \in R(I - T)$, then there is some $x \in U$ such that $u = (I - T)(x)$. So $T(u) = T \circ (I - T)(x) = (T - T^2)(x) = 0$ and $u \in N(T)$. If $u \in N(T)$, then $T(u) = 0$ which allows us to rewrite u as $u = (I - T)(u)$. Thus $u \in R(I - T)$.

Now consider the identity

$$I = T + (I - T). \tag{2.5.25}$$

Applying (2.5.25) to U and using (2.5.24), we obtain $U = V + W$. Let $u \in V \cap W$. Using the definition of V, W in (2.5.23), we get $T(u) - u = 0$ and $T(u) = 0$. Thus $u = 0$. So $U = V \oplus W$.

Finally, for any $u \in U$, express u as $u = v + w$ for some unique $v \in V$ and $w \in W$ and let $P \in L(U)$ be the projection of U onto V along W. Then (2.5.23) indicates that $T(v) = v$ and $T(w) = 0$. So $T(u) = T(v+w) = T(v) + T(w) = v = P(u)$. That is, $T = P$ and the proof follows. $\quad\square$

An easy but interesting consequence of Theorem 2.16 is the following.

Theorem 2.17 *Let V and W be complementary subspaces of U. Then $P \in L(U)$ is the projection of U onto V along W if and only if $I - P$ is the projection of U onto W along V.*

Proof Since

$$(I - P)^2 = I - 2P + P^2, \tag{2.5.26}$$

we see that $(I - P)^2 = I - P$ if and only if $P^2 = P$. The rest of the conclusion follows from (2.5.23). $\quad\square$

From (2.5.23), we see that when $T \in L(U)$ is idempotent, it is reducible over the null-spaces $N(T)$ and $N(I - T)$. It may be shown that the converse is true, which is assigned as an exercise.

Recall that the null-spaces $N(T)$ and $N(I - T)$, if nontrivial, of an idempotent mapping $T \in L(U)$, are simply the eigenspaces E_0 and E_1 of T associated with the eigenvalues 0 and 1, respectively. So, with respect to a basis of U consisting of the vectors in any bases of E_0 and E_1, the associated matrix A of T is of the form

$$A = \begin{pmatrix} 0_k & 0 \\ 0 & I_l \end{pmatrix}, \tag{2.5.27}$$

where 0_k is the zero matrix in $\mathbb{F}(k,k)$, $k = \dim(N(T)) = n(T)$, $l = \dim(N(I - T)) = r(T)$.

Let $T \in L(U)$ and $\lambda_1, \ldots, \lambda_k$ be some distinct eigenvalues of T such that T reduces over $E_{\lambda_1}, \ldots, E_{\lambda_k}$. Use P_i to denote the projection of U onto E_{λ_i} along

$$E_{\lambda_1} \oplus \cdots \oplus \widehat{E_{\lambda_i}} \oplus \cdots \oplus E_{\lambda_k}, \tag{2.5.28}$$

where $\widehat{}$ indicates the item that is missing in the expression. Then we can represent T as

$$T = \lambda_1 P_1 + \cdots + \lambda_k P_k. \tag{2.5.29}$$

2.5.3 Nilpotent mappings

Consider the vector space $\mathcal{P}_n$ of the set of all polynomials of degrees up to n with coefficients in the field $\mathbb{F}$ and the differentiation operator

$$D = \frac{\mathrm{d}}{\mathrm{d}t} \quad \text{so that } D(a_0 + a_1 t + \cdots + a_n t^n) = a_1 + 2a_2 t + \cdots + n a_n t^{n-1}. \tag{2.5.30}$$

Then $D^{n+1} = 0$ (zero mapping). Such a linear mapping $D : \mathcal{P}_n \to \mathcal{P}_n$ is an example of *nilpotent mappings* we now study.

Definition 2.18 Let U be a finite-dimensional vector space and $T \in L(U)$. We say that T is *nilpotent* if there is an integer $k \geq 1$ such that $T^k = 0$. For a nilpotent mapping $T \in L(U)$, the smallest integer $k \geq 1$ such that $T^k = 0$ is called the *degree* or *index of nilpotence* of T.

The same definition may be stated for square matrices.

Of course, the degree of a nonzero nilpotent mapping is always at least 2.

Definition 2.19 Let U be a vector space and $T \in L(U)$. For any nonzero vector $u \in U$, we say that u is *T-cyclic* if there is an integer $m \geq 1$ such that $T^m(u) = 0$. The smallest such integer m is called the *period* of u under or relative to T. If each nonzero vector in U is T-cyclic, T is said to be *locally nilpotent*.

It is clear that a nilpotent mapping must be locally nilpotent. In fact, these two notions are equivalent in finite dimensions.

Theorem 2.20 *If U is finite-dimensional, then a mapping $T \in L(U)$ is nilpotent if and only if it is locally nilpotent.*

Proof Suppose that $\dim(U) = n \geq 1$ and $T \in L(U)$ is locally nilpotent. Let $\{u_1, \dots, u_n\}$ be any basis of U and $m_1, \dots, m_n$ be the periods of $u_1, \dots, u_n$, respectively. Set

$$k = \max\{m_1, \dots, m_n\}. \tag{2.5.31}$$

Then it is seen that T is nilpotent of degree k. $\qquad\square$

We now show how to use a cyclic vector to generate an invariant subspace.

Theorem 2.21 *For $T \in L(U)$, let $u \in U$ be a nonzero cyclic vector under T of period m. Then the vectors*

$$u, T(u), \dots, T^{m-1}(u) \tag{2.5.32}$$

are linearly independent so that they span an m-dimensional T-invariant subspace of U. In particular, $m \leq \dim(U)$.

Proof We only need to show that the set of vectors in (2.5.32) are linearly independent. If $m = 1$, the statement is self-evident. So we now assume $m \geq 2$.
Let $c_0, \dots, c_{m-1} \in \mathbb{F}$ so that

$$c_0 u + c_1 T(u) + \dots + c_{m-1} T^{m-1}(u) = 0. \tag{2.5.33}$$

Assume that there is some $l = 0, \ldots, m-1$ such that $c_l \neq 0$. Let $l \geq 0$ be the smallest such integer. Then $l \leq m-2$ so that (2.5.33) gives us

$$T^l(u) = -\frac{1}{c_l}(c_{l+1}T^{l+1}(u) + \cdots + c_{m-1}T^{m-1}(u)). \qquad (2.5.34)$$

This relation implies that

$$T^{m-1}(u) = T^{l+([m-1]-l)}(u)$$
$$= -\frac{1}{c_l}(c_{l+1}T^m(u) + \cdots + c_{m-1}T^{2(m-1)+l}(u)) = 0, \qquad (2.5.35)$$

which contradicts the fact that $T^{m-1}(u) \neq 0$. $\qquad \square$

We can apply the above results to study the reducibility of a nilpotent mapping.

Theorem 2.22 *Let U be a finite-dimensional vector space and $T \in L(U)$ a nilpotent mapping of degree k. If T is reducible over V and W, that is, $U = V \oplus W$, $T(V) \subset V$, and $T(W) \subset W$, then*

$$k \leq \max\{\dim(V), \dim(W)\}. \qquad (2.5.36)$$

In particular, if $k = \dim(U)$, then T is irreducible.

Proof   Let $\{v_1, \ldots, v_{n_1}\}$ and $\{w_1, \ldots, w_{n_2}\}$ be bases of V and W, respectively. Then, by Theorem 2.21, the periods of these vectors cannot exceed the right-hand side of (2.5.36). Thus (2.5.36) follows from (2.5.31).

If $k = \dim(U)$, then $\dim(V) = k$ and $\dim(W) = 0$ or $\dim(V) = 0$ and $\dim(W) = k$, which shows that T is irreducible. $\qquad \square$

Let $\dim(U) = n$ and $T \in L(U)$ be nilpotent of degree n. Then there is a vector $u \in U$ of period n. Set

$$U_1 = \text{Span}\{T^{n-1}(u)\}, \ldots, U_{n-1} = \text{Span}\{T(u), \ldots, T^{n-1}(u)\}. \qquad (2.5.37)$$

Then $U_1, \ldots, U_{n-1}$ are nontrivial invariant subspaces of T satisfying $U_1 \subset \cdots \subset U_{n-1}$ and $\dim(U_1) = 1, \ldots, \dim(U_{n-1}) = n-1$. Thus, T has invariant subspaces of all possible dimensions but yet is irreducible.

Again, let $\dim(U) = n$ and $T \in L(U)$ be nilpotent of degree n. Choose a cyclic vector $u \in U$ of period n and set

$$u_1 = u, \ldots, u_n = T^{n-1}(u). \qquad (2.5.38)$$

Then, with respect to the basis $\mathcal{B} = \{u_1, \ldots, u_n\}$, we have

$$T(u_i) = u_{i+1}, \quad i = 1, \ldots, n-1, \quad T(u_n) = 0. \qquad (2.5.39)$$

Hence, if we use $S = (s_{ij})$ to denote the matrix of T with respect to the basis $\mathcal{B}$, then

$$s_{ij} = \begin{cases} 1, & j = i-1, i = 2, \dots, n, \\ 0, & \text{otherwise.} \end{cases} \tag{2.5.40}$$

That is,

$$S = \begin{pmatrix} 0 & 0 & \cdots & & \cdots & 0 \\ 1 & 0 & \cdots & & \cdots & 0 \\ \vdots & \ddots & \ddots & & \ddots & \vdots \\ 0 & \cdots & 1 & & 0 & 0 \\ 0 & \cdots & & \cdots & 1 & 0 \end{pmatrix}. \tag{2.5.41}$$

Alternatively, we may also set

$$u_1 = T^{n-1}(u), \dots, u_n = u. \tag{2.5.42}$$

Then with respect to this reordered basis the matrix of T becomes

$$S = \begin{pmatrix} 0 & 1 & 0 & \cdots & 0 \\ 0 & 0 & 1 & \cdots & 0 \\ \vdots & \ddots & \ddots & \ddots & \vdots \\ \vdots & \cdots & \ddots & 0 & 1 \\ 0 & \cdots & & 0 & 0 \end{pmatrix}. \tag{2.5.43}$$

The $n \times n$ matrix S expressed in either (2.5.41) or (2.5.43) is also called a *shift matrix* which transparently indicates that T is nilpotent of degree n.

In general, we shall see later that, if $T \in L(U)$ is nilpotent and U is finite-dimensional, there are T-invariant subspaces $U_1, \dots, U_l$ such that

$$U = U_1 \oplus \cdots \oplus U_l, \quad \dim(U_1) = n_1, \dots, \dim(U_l) = n_l, \tag{2.5.44}$$

and the degree of T restricted to U_i is exactly n_i $(i = 1, \dots, l)$. Thus, we can choose T-cyclic vectors of respective periods $n_1, \dots, n_l$, say $u_1, \dots, u_l$, and use the vectors

$$u_1, \cdots, T^{n_1-1}(u_1), \dots, u_l, \dots, T^{n_l-1}(u_l), \tag{2.5.45}$$

say, as a basis of U. With respect to such a basis, the matrix S of T is

$$S = \begin{pmatrix} S_1 & 0 & \cdots & 0 \\ 0 & \ddots & \vdots & 0 \\ \vdots & \vdots & \ddots & \vdots \\ 0 & 0 & \cdots & S_l \end{pmatrix}, \tag{2.5.46}$$

where each S_i is an $n_i \times n_i$ shift matrix $(i = 1, \dots, l)$.

Let k be the degree of T or the matrix S. From (2.5.46), we see clearly that the relation

$$k = \max\{n_1, \ldots, n_l\} \tag{2.5.47}$$

holds.

2.5.4 Polynomials of linear mappings

In this section, we have seen that it is often useful to consider various powers of a linear mapping $T \in L(U)$ as well as some linear combinations of appropriate powers of T. These manipulations motivate the introduction of the notion of polynomials of linear mappings. Specifically, for any $p(t) \in \mathcal{P}$ with the form

$$p(t) = a_n t^n + \cdots + a_1 t + a_0, \quad a_0, a_1, \ldots, a_n \in \mathbb{F}, \tag{2.5.48}$$

we define $p(T) \in L(U)$ to be the linear mapping over U given by

$$p(T) = a_n T^n + \cdots + a_1 T + a_0 I. \tag{2.5.49}$$

It is straightforward to check that all usual operations over polynomials in variable t can be carried over correspondingly to those over polynomials in the powers of a linear mapping T over the vector space U. For example, if $f, g, h \in \mathcal{P}$ satisfy the relation $f(t) = g(t)h(t)$, then

$$f(T) = g(T)h(T), \tag{2.5.50}$$

because the powers of T follow the same rule as the powers of t. That is, $T^k T^l = T^{k+l}$, $k, l \in \mathbb{N}$.

For $T \in L(U)$, let $\lambda \in \mathbb{F}$ be an eigenvalue of T. Then, for any $p(t) \in \mathcal{P}$ given as in (2.5.48), $p(\lambda)$ is an eigenvalue of $p(T)$. To see this, we assume that $u \in U$ is an eigenvector of T associated with λ. We have

$$
\begin{aligned}
p(T)(u) &= (a_n T^n + \cdots + a_1 T + a_0 I)(u) \\
&= (a_n \lambda^n + \cdots + a_1 \lambda + a_0)(u) = p(\lambda)u,
\end{aligned}
\tag{2.5.51}
$$

as anticipated.

If $p \in \mathcal{P}$ is such that $p(T) = 0$, then we say that T is a root of p. Hence, if T is a root of p, any eigenvalue λ of T must also be a root of p, $p(\lambda) = 0$, by virtue of (2.5.51).

For example, an idempotent mapping is a root of the polynomial $p(t) = t^2 - t$, and a nilpotent mapping is a root of a polynomial of the form $p(t) = t^m$ ($m \in \mathbb{N}$). Consequently, the eigenvalues of an idempotent mapping can only be 0 and 1, and that of a nilpotent mapping, 0.

For $T \in L(U)$, let $\lambda_1, \ldots, \lambda_k$ be some distinct eigenvalues of T such that T reduces over $E_{\lambda_1}, \ldots, E_{\lambda_k}$. Then T must be a root of the polynomial

$$p(t) = (t - \lambda_1) \cdots (t - \lambda_k). \tag{2.5.52}$$

To see this, we rewrite any $u \in U$ as $u = u_1 + \cdots + u_k$, where $u_i \in E_{\lambda_i}$, $i = 1, \ldots, k$. Then

$$p(T)u = \sum_{i=1}^{k} p(T)u_i$$

$$= \sum_{i=1}^{k} (T - \lambda_1 I) \cdots \widehat{(T - \lambda_i I)} \cdots (T - \lambda_k I)(T - \lambda_i I)u_i$$

$$= 0, \tag{2.5.53}$$

which establishes $p(T) = 0$, as claimed.

It is clear that the polynomial $p(t)$, given in (2.5.52), is the lowest-degree nontrivial polynomial among all polynomials for which T is a root, which is often referred to as the *minimal polynomial* of T, a basic notion to be detailed later.

Exercises

2.5.1   Let S and $T \in L(U)$ and $S \circ T = T \circ S$. Show that the null-space $N(S)$ and range $R(S)$ of S are invariant subspaces under T. In particular, an eigenspace of T associated with eigenvalue λ is seen to be invariant under T when taking $S = T - \lambda I$.

2.5.2   Let S and $T \in L(U)$. Prove that the invertibility of the mappings $I + S \circ T$ and $I + T \circ S$ are equivalent by showing that if $I + S \circ T$ is invertible then so is $I + T \circ S$ with

$$(I + T \circ S)^{-1} = I - T \circ (I + S \circ T)^{-1} \circ S. \tag{2.5.54}$$

2.5.3   The rotation transformation over $\mathbb{R}^2$, denoted by $R_\theta \in L(\mathbb{R}^2)$, is given by

$$R_\theta(x) = \begin{pmatrix} \cos \theta & -\sin \theta \\ \sin \theta & \cos \theta \end{pmatrix} x, \quad x = \begin{pmatrix} x_1 \\ x_2 \end{pmatrix} \in \mathbb{R}^2. \tag{2.5.55}$$

Show that R_θ has no nontrivial invariant subspace in $\mathbb{R}^2$ unless $\theta = k\pi$ ($k \in \mathbb{Z}$).

2.5.4   Consider the vector space $\mathcal{P}$ of the set of all polynomials over a field $\mathbb{F}$. Define $T \in L(\mathcal{P})$ by setting

$$T(p) = tp(t), \quad \forall p(t) = a_0 + a_1 t + \cdots + a_n t^n \in \mathcal{P}. \tag{2.5.56}$$

Show that T cannot have an eigenvalue.

2.5.5   Let $A = (a_{ij}) \in \mathbb{F}(n,n)$ be invertible and satisfy

$$\sum_{j=1}^{n} a_{ij} = a \in \mathbb{F}, \quad i = 1,\dots,n. \qquad (2.5.57)$$

(a) Show that a must be an eigenvalue of A and that $a \neq 0$.
(b) Show that if $A^{-1} = (b_{ij})$ then

$$\sum_{j=1}^{n} b_{ij} = \frac{1}{a}, \quad i = 1,\dots,n. \qquad (2.5.58)$$

2.5.6   Let S and $T \in L(U)$ and assume that S and T are similar. That is, there is an invertible element $R \in L(U)$ such that $S = R \circ T \circ R^{-1}$. Show that $\lambda \in \mathbb{F}$ is an eigenvalue of S if and only if it is an eigenvalue of T.

2.5.7   Let $U = \mathbb{F}(n,n)$, where $\mathrm{ch}(\mathbb{F}) \neq 2$ and $n \geq 2$, and $T \in L(U)$ be defined by taking matrix transpose,

$$T(A) = A^t, \quad A \in U. \qquad (2.5.59)$$

Show that both ± 1 are the eigenvalues of T and identify E_1 and E_{-1}. Prove that T is reducible over the pair E_1 and E_{-1}. Can T have an eigenvalue different from ± 1? What is the minimal polynomial of T?

2.5.8   Let U be a vector space over a field $\mathbb{F}$ and $T \in L(U)$. If λ_1 and $\lambda_2 \in \mathbb{F}$ are distinct eigenvalues and u_1 and u_2 their respectively associated eigenvectors of T, show that, for any nonzero a_1 and $a_2 \in \mathbb{F}$, the vector $u = a_1 u_1 + a_2 u_2$ cannot be an eigenvector of T .

2.5.9   Let U be a finite-dimensional vector space over a field $\mathbb{F}$ and $T \in L(U)$. Assume that T is of rank 1.

(a) Prove that T must have an eigenvalue, say λ, in $\mathbb{F}$.
(b) If $\lambda \neq 0$, show that $R(T) = E_\lambda$.
(c) Is the statement in (b), that is, $R(T) = E_\lambda$, valid when $\lambda = 0$?

2.5.10   Let $A \in \mathbb{C}(n,n)$ be unitary. That is, $AA^\dagger = A^\dagger A = I_n$. Show that any eigenvalue of A is of unit modulus.

2.5.11   Show that, if $T \in L(U)$ is reducible over the pair of null-spaces $N(T)$ and $N(I - T)$, then T is idempotent.

2.5.12   Let S and $T \in L(U)$ be idempotent. Show that

(a) $R(S) = R(T)$ if and only if $S \circ T = T, T \circ S = S$.
(b) $N(S) = N(T)$ if and only if $S \circ T = S, T \circ S = T$.

2.5.13 Let $T \in L(\mathbb{R}^3)$ be defined by

$$T(x) = \begin{pmatrix} 1 & 1 & -1 \\ -1 & 1 & 1 \\ 1 & 3 & -1 \end{pmatrix} x, \quad x = \begin{pmatrix} x_1 \\ x_2 \\ x_3 \end{pmatrix} \in \mathbb{R}^3. \qquad (2.5.60)$$

Let $S \in L(\mathbb{R}^3)$ project $\mathbb{R}^3$ onto $R(T)$ along $N(T)$. Determine S by obtaining a matrix $A \in \mathbb{R}(3,3)$ that represents S. That is, $S(x) = Ax$ for $x \in \mathbb{R}^3$.

2.5.14 Let U be an n-dimensional vector space $(n \geq 2)$ and U' its dual space. Let u_1, u_2 and u_1', u_2' be independent vectors in U and U', respectively. Define a linear mapping $T \in L(U)$ by setting

$$T(u) = \langle u, u_1' \rangle u_1 + \langle u, u_2' \rangle u_2, \quad u \in U. \qquad (2.5.61)$$

(a) Find the condition(s) regarding u_1, u_2, u_1', u_2' under which T is a projection.

(b) When T is a projection, determine subspaces V and W of U in terms of u_1, u_2, u_1', u_2' so that T projects U onto V along W.

2.5.15 We may slightly generalize the notion of idempotent mappings to a mapping $T \in L(U)$ satisfying

$$T^2 = aT, \quad T \neq 0, \quad a \in \mathbb{F}, \quad a \neq 0, 1. \qquad (2.5.62)$$

Show that such a mapping T is reduced over the pair of subspaces $N(T)$ and $N(aI - T)$.

2.5.16 Let $T \in L(U)$ and $\lambda_1, \ldots, \lambda_k$ be some distinct eigenvalues of T such that T is reducible over the eigenspaces $E_{\lambda_1}, \ldots, E_{\lambda_k}$. Show that $\lambda_1, \ldots, \lambda_k$ are all the eigenvalues of T. In other words, if λ is any eigenvalue of T, then λ is among $\lambda_1, \ldots, \lambda_k$.

2.5.17 Consider the space $\mathcal{P}_n$ over a field $\mathbb{F}$ satisfying $\mathrm{ch}(\mathbb{F}) = 0$. Is the differential operator $D : \mathcal{P}_n \to \mathcal{P}_n$ reducible? Find an element in $\mathcal{P}_n$ that is of period $n + 1$ under D and use it to obtain a basis of $\mathcal{P}_n$.

2.5.18 Let α and β be nonzero elements in $\mathbb{F}(n, 1)$. Then $A = \alpha \beta^t \in \mathbb{F}(n, n)$.

(a) Prove that the necessary and sufficient condition for A to be nilpotent is $\alpha^t \beta = 0$. If A is nilpotent, what is the degree of A?

(b) Prove that the necessary and sufficient condition for A to be idempotent is $\alpha^t \beta = 1$.

2.5.19 Show that the linear mapping $T \in L(\mathcal{P})$ defined in (2.5.56) and the differential operator $D \in L(\mathcal{P})$ satisfy the identity $D \circ T - T \circ D = I$.

2.5.20 Let S and $T \in L(U)$ satisfy $S \circ T - T \circ S = I$. Establish the identity

$$S^k \circ T - T \circ S^k = k S^{k-1}, \quad k \geq 2. \qquad (2.5.63)$$

2.5.21 Let U be a finite-dimensional vector space over a field $\mathbb{F}$ and $T \in L(U)$ satisfying

$$T^3 = T \tag{2.5.64}$$

so that ± 1 cannot be the eigenvalues of T. Show that $T = 0$.

2.5.22 Let U be a finite-dimensional vector space over a field $\mathbb{F}$ and $S, T \in L(U)$ satisfying $S \sim T$. For any $p \in \mathcal{P}$, show that $p(S) \sim p(T)$.

2.5.23 Let U be a finite-dimensional vector space over a field $\mathbb{F}$ and $T \in L(U)$. For any $p \in \mathcal{P}$, show that $p(T)' = p(T')$.

2.5.24 Let U be an n-dimensional vector space over a field $\mathbb{F}$, $T \in L(U)$ nilpotent of degree $k \geq 2$, and $n(T) = l \geq 2$. Assume $k + l = n + 1$.

(a) Show that there are subspaces V and W of U, where

$$V = \text{Span}\{u, T(u), \ldots, T^{k-1}(u)\}, \tag{2.5.65}$$

with $u \in U$ a T-cyclic vector of period k, and W is an $(l-1)$-dimensional subspace of $N(T)$, such that T is reducible over the pair V and W.

(b) Describe $R(T)$ and determine $r(T)$.

2.5.25 Let $\mathcal{P}$ be the vector space of polynomials over $\mathbb{R}$ and consider the map $T \in L(\mathcal{P})$ defined by

$$T(p(t)) = \int_0^t p(\tau) d\tau, \quad p(t) \in \mathcal{P}. \tag{2.5.66}$$

Show that T has no cyclic vector.

2.5.26 Show that $T \in L(U)$ is nilpotent if and only if $I - T$ is invertible and given by

$$(I - T)^{-1} = \sum_{i=0}^{k-1} T^i \tag{2.5.67}$$

for some positive integer k, where we understand $T^0 = I$.

2.5.27 Let U be a finite-dimensional vector space over a field $\mathbb{F}$. If $\lambda \in \mathbb{F}$ is not an eigenvalue of $T \in L(U)$, then $\lambda I - T$ is invertible and

$$R(\lambda) = (\lambda I - T)^{-1} \tag{2.5.68}$$

is referred to as the resolvent of T at λ. If T is nilpotent of degree k, obtain $R(\lambda)$ as a polynomial of T.

2.5.28 Let U be a finite-dimensional vector space over a field $\mathbb{F}$ and $S, T \in L(U)$.

(a) Show that if S is invertible and T nilpotent, $S - T$ must also be invertible provided that S and T commute, $S \circ T = T \circ S$.

(b) Find an example to show that the condition in (a) that S and T commute cannot be removed.

2.5.29 Let U be finite-dimensional vector space, $T \in L(U)$, and V a subspace of U which is invariant under T. Show that if $U = R(T) \oplus V$ then $V = N(T)$.

2.6 Norms of linear mappings

In this section, we begin by considering general linear mappings between arbitrary finite-dimensional normed vector spaces. We then concentrate on mappings from a normed space into itself.

2.6.1 Definition and elementary properties of norms of linear mappings

Let U and V be finite-dimensional vector spaces over $\mathbb{R}$ or $\mathbb{C}$ and $\| \cdot \|_U$ and $\| \cdot \|_V$ the norms on U and V, respectively. Assume that $\mathcal{B} = \{u_1, \ldots, u_n\}$ is any basis of U. For any $u \in U$, write u as $u = \sum_{i=1}^{n} a_i u_i$, where $a_1, \ldots, a_n$ are the coordinates of u with respect to $\mathcal{B}$. Thus

$$
\begin{aligned}
\|T(u)\|_V = \left\| T\left(\sum_{i=1}^{n} a_i u_i \right) \right\|_V &\le \sum_{i=1}^{n} |a_i| \|T(u_i)\|_V \\
&\le \left(\max_{1 \le i \le n} \{\|T(u_i)\|\} \right) \sum_{i=1}^{n} |a_i| \\
&\equiv \left(\max_{1 \le i \le n} \{\|T(u_i)\|\} \right) \|u\|_1 \le C \|u\|_U,
\end{aligned}
\tag{2.6.1}
$$

where we have used the fact that norms over a finite-dimensional space are all equivalent. This estimate may also be restated as

$$
\frac{\|T(u)\|_V}{\|u\|_U} \le C, \quad u \in U, \quad u \ne 0.
\tag{2.6.2}
$$

This boundedness result enables us to formulate the definition of the norm of a linear mapping $T \in L(U,V)$ as follows.

Definition 2.23 For $T \in L(U,V)$, we define the norm of T, induced from the respective norms of U and V, by

$$
\|T\| = \sup \left\{ \frac{\|T(u)\|_V}{\|u\|_U} \,\middle|\, u \in U, u \ne 0 \right\}.
\tag{2.6.3}
$$

To show that (2.6.3) indeed defines a norm for the space $L(U,V)$, we need to examine that it fulfills all the properties required of a norm. To this end, we note from (2.6.3) that

$$\|T(u)\|_V \le \|T\|\|u\|_U, \quad u \in U. \tag{2.6.4}$$

For S and $T \in L(U,V)$, since the triangle inequality and (2.6.4) give us

$$\|(S+T)(u)\|_V \le \|S(u)\|_V + \|T(u)\|_V \le (\|S\| + \|T\|)\|u\|_U, \tag{2.6.5}$$

we obtain $\|S+T\| \le \|S\| + \|T\|$ which says the triangle inequality holds over $L(U,V)$.

Let a be any scalar. Then $\|aT(u)\|_V = |a|\|T(u)\|_V$ for any $u \in U$. Hence

$$
\begin{aligned}
\|aT\| &= \sup\left\{ \frac{\|aT(u)\|_V}{\|u\|_U} \,\middle|\, u \in U, u \ne 0 \right\} \\
&= |a|\left\{ \frac{\|T(u)\|_V}{\|u\|_U} \,\middle|\, u \in U, u \ne 0 \right\} = |a|\|T\|,
\end{aligned}
\tag{2.6.6}
$$

which indicates that homogeneity follows.

Finally, it is clear that $\|T\| = 0$ implies $T = 0$.

Let $T \in L(U,V)$ and $S \in L(V,W)$, where W is another normed vector space with the norm $\|\cdot\|_W$. Then $S \circ T \in L(U,W)$, and it follows from (2.6.4) that

$$\|(S\circ T)(u)\|_W \le \|S\|\|T(u)\|_V \le \|S\|\|T\|\|u\|_U, \quad u \in U. \tag{2.6.7}$$

Consequently, we get

$$\|S \circ T\| \le \|S\|\|T\|. \tag{2.6.8}$$

This simple but general inequality is of basic usefulness.

In the rest of the section, we will focus on mappings from U into itself. We note that, for the identity mapping $I \in L(U)$, it is clear that $\|I\| = 1$.

2.6.2 Invertibility of linear mappings as a generic property

Let U be a finite-dimensional vector space with norm $\|\cdot\|$. It has been seen that the space $L(U)$ may be equipped with an induced norm which may also be denoted by $\|\cdot\|$ since there is no risk of confusion. The availability of a norm of $L(U)$ allows one to perform analysis on $L(U)$ so that a deeper understanding of $L(U)$ may be achieved.

As an illustration, in this subsection, we will characterize invertibility of linear mappings using norm.

Theorem 2.24 *Let U be a finite-dimensional normed space. An element $T \in L(U)$ is invertible if and only if there is a constant $c > 0$ such that*

$$\|T(u)\| \geq c\|u\|, \quad u \in U. \tag{2.6.9}$$

Proof   Assume (2.6.9) is valid. Then it is clear that $N(T) = \{0\}$. Hence T is invertible. Conversely, assume that T is invertible and $T^{-1} \in L(U)$ is its inverse. Then $1 = \|I\| = \|T^{-1} \circ T\| \leq \|T^{-1}\|\|T\|$ implies that the norm of an invertible mapping can never be zero. Thus, for any $u \in U$, we have $\|u\| = \|(T^{-1} \circ T)(u)\| \leq \|T^{-1}\|\|T(u)\|$ or $\|T(u)\| \geq (\|T^{-1}\|)^{-1}\|u\|$, $u \in U$, which establishes (2.6.9). $\qquad\square$

We now show that invertibility is a *generic property* for elements in $L(U)$.

Theorem 2.25 *Let U be a finite-dimensional normed space and $T \in L(U)$.*

(1) *For any $\varepsilon > 0$ there exists an invertible element $S \in L(U)$ such that $\|S - T\| < \varepsilon$. This property says that the subset of invertible mappings in $L(U)$ is* dense *in $L(U)$ with respect to the norm of $L(U)$.*

(2) *If $T \in L(U)$ is invertible, then there is some $\varepsilon > 0$ such that $S \in L(U)$ is invertible whenever S satisfies $\|S - T\| < \varepsilon$. This property says that the subset of invertible mappings in $L(U)$ is* open *in $L(U)$ with respect to the norm of $L(U)$.*

Proof   For any scalar λ, consider $S_\lambda = T - \lambda I$. If $\dim(U) = n$, there are at most n possible values of λ for which S_λ is not invertible. Now $\|T - S_\lambda\| = |\lambda| \|I\| = |\lambda|$. So for any $\varepsilon > 0$, there is a scalar λ, $|\lambda| < \varepsilon$, such that S_λ is invertible. This proves (1).

We next consider (2). Let $T \in L(U)$ be invertible. Then (2.6.9) holds for some $c > 0$. Let $S \in L(U)$ be such that $\|S - T\| < \varepsilon$ for some $\varepsilon > 0$. Then, for any $u \in U$, we have

$$c\|u\| \leq \|T(u)\| = \|(T - S)(u) + S(u)\|$$
$$\leq \|T - S\|\|u\| + \|S(u)\| < \varepsilon\|u\| + \|S(u)\|, \quad u \in U, \tag{2.6.10}$$

or $\|S(u)\| \geq (c - \varepsilon)\|u\|$. Therefore, if we choose $\varepsilon < c$, we see in view of Theorem 2.24 that S is invertible when $\|S - T\| < \varepsilon$ provided that $\varepsilon < c$. Hence (2) follows as well. $\qquad\square$

2.6.3 Exponential of a linear mapping

Let $T \in L(U)$. For a positive integer m, we consider $T_m \in L(U)$ given by

$$T_m = \sum_{k=0}^{m} \frac{1}{k!} T^k, \tag{2.6.11}$$

with the understanding $T^0 = I$. Therefore, for $l < m$, we have the estimate

$$\|T_l - T_m\| \leq \sum_{k=l+1}^{m} \frac{\|T\|^k}{k!}. \tag{2.6.12}$$

In particular, $\|T_l - T_m\| \to 0$ as $l, m \to \infty$. Hence we see that the limit

$$\lim_{m \to \infty} \sum_{k=0}^{m} \frac{1}{k!} T^k \tag{2.6.13}$$

is a well-defined element in $L(U)$ and is naturally denoted as

$$e^T = \sum_{k=0}^{\infty} \frac{1}{k!} T^k, \tag{2.6.14}$$

called the exponential of $T \in L(U)$. Thus $e^0 = I$. As in calculus, if $S, T \in L(U)$ are commutative, we can verify the formula

$$e^S e^T \equiv e^S \circ e^T = e^{S+T}. \tag{2.6.15}$$

A special consequence of this simple property is that the exponential of any mapping $T \in L(U)$ is invertible. In fact, the relation (2.6.15) indicates that

$$(e^T)^{-1} = e^{-T}. \tag{2.6.16}$$

More generally, with the notation $\Phi(t) = e^{tT}$ ($t \in \mathbb{R}$), we have

(1) $\Phi(s)\Phi(t) = \Phi(s+t)$, $s, t \in \mathbb{R}$,
(2) $\Phi(0) = I$,

and we say that $\Phi : \mathbb{R} \to L(U)$ defines a *one-parameter group*.

Furthermore, we also have

$$\frac{1}{h}(\Phi(t+h) - \Phi(t)) = \frac{1}{h}(e^{(t+h)T} - e^{tT})$$

$$= \frac{1}{h}(e^{hT} - I)e^{tT} = e^{tT}\frac{1}{h}(e^{hT} - I)$$

$$= T\left(\sum_{k=0}^{\infty} \frac{h^k}{(k+1)!} T^k\right) e^{tT}$$

$$= e^{tT}\left(\sum_{k=0}^{\infty} \frac{h^k}{(k+1)!} T^k\right) T, \quad h \in \mathbb{R}, \quad h \neq 0. \tag{2.6.17}$$

Therefore, we obtain the limit

$$\lim_{h \to 0} \frac{1}{h}(\Phi(t+h) - \Phi(t)) = Te^{tT} = e^{tT}T. \tag{2.6.18}$$

In other words, the above result gives us

$$\Phi'(t) = \frac{d}{dt} e^{tT} = T e^{tT} = e^{tT} T = T \Phi(t) = \Phi(t) T, \quad t \in \mathbb{R}. \qquad (2.6.19)$$

In particular, $\Phi'(0) = T$, which intuitively says that T is the initial rate of change of the one-parameter group Φ. One also refers to this relationship as "T generates Φ" or "T is the generator of Φ."

The relation (2.6.19) suggests that it is legitimate to differentiate the series $\Phi(t) = \sum_{k=0}^{\infty} \frac{1}{k!} t^k T^k$ term by term:

$$\frac{d}{dt} \sum_{k=0}^{\infty} \frac{1}{k!} t^k T^k = \sum_{k=0}^{\infty} \frac{1}{k!} \frac{d}{dt} (t^k) T^k = T \sum_{k=0}^{\infty} \frac{1}{k!} t^k T^k, \quad t \in \mathbb{R}. \qquad (2.6.20)$$

With the exponential of a linear mapping, $T \in L(U)$, various elementary functions of T involving the exponentials of T may also be introduced accordingly. For example,

$$\cosh T = \frac{1}{2} \left(e^T + e^{-T} \right), \quad \sinh T = \frac{1}{2} \left(e^T - e^{-T} \right), \qquad (2.6.21)$$

$$\cos T = \frac{1}{2} \left(e^{iT} + e^{-iT} \right), \quad \sin T = \frac{1}{2i} \left(e^{iT} - e^{-iT} \right), \qquad (2.6.22)$$

are all well defined and enjoy similar properties of the corresponding classical functions.

The matrix version of the discussion here can also be easily formulated analogously and is omitted.

Exercises

2.6.1 Let U be a finite-dimensional normed space and $\{T_n\} \subset L(U)$ a sequence of invertible mappings which converges to a noninvertible mapping $T \in L(U)$. Show that $\|T_n^{-1}\| \to \infty$ as $n \to \infty$.

2.6.2 Let U and V be finite-dimensional normed spaces with the norms $\| \cdot \|_U$ and $\| \cdot \|_V$, respectively. For $T \in L(U,V)$, show that the induced norm of T may also be evaluated by the expression

$$\|T\| = \sup\{\|T(u)\|_V \mid u \in U, \|u\|_U = 1\}. \qquad (2.6.23)$$

2.6.3 Consider $T \in L(\mathbb{R}^n, \mathbb{R}^m)$ defined by

$$T(x) = Ax, \quad A = (a_{ij}) \in \mathbb{R}(m,n), \quad x = \begin{pmatrix} x_1 \\ \vdots \\ x_n \end{pmatrix} \in \mathbb{R}^n. \qquad (2.6.24)$$

Use the norm $\|\cdot\|_\infty$ for both $\mathbb{R}^m$ and $\mathbb{R}^n$, correspondingly, and denote by $\|T\|_\infty$ the induced norm of the linear mapping T given in (2.6.24). Show that

$$\|T\|_\infty = \max_{1\le i\le m}\left\{\sum_{j=1}^{n}|a_{ij}|\right\}. \tag{2.6.25}$$

(This quantity is also sometimes denoted as $\|A\|_\infty$.)

2.6.4 Let T be the linear mapping defined in (2.6.24). Show that, if we use the norm $\|\cdot\|_1$ for both $\mathbb{R}^m$ and $\mathbb{R}^n$, correspondingly, and denote by $\|T\|_1$ the induce norm of T, then

$$\|T\|_1 = \max_{1\le j\le n}\left\{\sum_{i=1}^{m}|a_{ij}|\right\}. \tag{2.6.26}$$

(This quantity is also sometimes denoted as $\|A\|_1$.)

2.6.5 Let U be a finite-dimensional normed space over $\mathbb{R}$ or $\mathbb{C}$ and use $\mathcal{N}$ to denote the subset of $L(U)$ consisting of all nilpotent mappings. Show that $\mathcal{N}$ is not an open subset of $L(U)$.

2.6.6 Let U be a finite-dimensional normed space and $T \in L(U)$. Prove that $\|e^T\| \le e^{\|T\|}$ (under the induced norm).

2.6.7 Let $A \in \mathbb{R}(n,n)$. Show that $(e^A)^t = e^{A^t}$ and that, if A is skew-symmetric, $A^t = -A$, then e^A is orthogonal.

2.6.8 Let $A \in \mathbb{C}(n,n)$. Show that $(e^A)^\dagger = e^{A^\dagger}$ and that, if A is anti-Hermitian, $A^\dagger = -A$, then e^A is unitary.

2.6.9 Let $A \in \mathbb{R}(n,n)$ and consider the initial value problem of the following system of differential equations:

$$\frac{dx}{dt} = Ax, \quad x = x(t) = \begin{pmatrix} x_1(t) \\ \vdots \\ x_n(t) \end{pmatrix}; \quad x(0) = x_0 = \begin{pmatrix} x_{1,0} \\ \vdots \\ x_{n,0} \end{pmatrix}. \tag{2.6.27}$$

(a) Show that, with the one-parameter group $\Phi(t) = e^{tA}$, the solution of the problem (2.6.27) is simply given by $x = \Phi(t)x_0$.

(b) Moreover, use $x_{(i)}(t)$ to denote the solution of (2.6.27) when $x_0 = e_i$ where $\{e_1,\ldots,e_n\}$ is the standard basis of $\mathbb{R}^n$, $i = 1,\ldots,n$. Show that $\Phi(t) = e^{tA}$ is the $n \times n$ matrix with $x_{(1)}(t),\ldots,x_{(n)}(t)$ as the n corresponding column vectors,

$$\Phi(t) = (x_{(1)}(t),\ldots,x_{(n)}(t)). \tag{2.6.28}$$

(This result provides a practical method for computing the *matrix exponential*, e^{tA}, which may also be viewed as the solution of the *matrix-valued initial value problem*

$$\frac{dX}{dt} = AX, \quad X(0) = I_n, \quad X \in \mathbb{R}(n,n).)$$ (2.6.29)

2.6.10 Consider the mapping $\Phi : \mathbb{R} \to \mathbb{R}(2,2)$ defined by

$$\Phi(t) = \begin{pmatrix} \cos t & -\sin t \\ \sin t & \cos t \end{pmatrix}, \quad t \in \mathbb{R}.$$ (2.6.30)

(a) Show that $\Phi(t)$ is a one-parameter group.
(b) Find the generator, say $A \in \mathbb{R}(2,2)$, of $\Phi(t)$.
(c) Compute e^{tA} directly by the formula

$$e^{tA} = \sum_{k=0}^{\infty} \frac{t^k}{k!} A^k$$ (2.6.31)

and verify $\Phi(t) = e^{tA}$.
(d) Use the practical method illustrated in Problem 2.6.9 to obtain the matrix exponential e^{tA} through solving two appropriate initial value problems as given in (2.6.27).

2.6.11 For the functions of $T \in L(U)$ defined in (2.6.21) and (2.6.22), establish the identities

$$\cosh^2 T - \sinh^2 T = I, \quad \cos^2 T + \sin^2 T = I.$$ (2.6.32)

2.6.12 For $T \in L(U)$, establish the formulas

$$\frac{d}{dt}(\sinh tT) = T \cosh tT, \quad \frac{d}{dt}(\cosh tT) = T \sinh tT.$$ (2.6.33)

2.6.13 For $T \in L(U)$, establish the formulas

$$\frac{d}{dt}(\sin tT) = T \cos tT, \quad \frac{d}{dt}(\cos tT) = -T \sin tT.$$ (2.6.34)

2.6.14 Let U be a finite-dimensional normed space and $T \in L(U)$. Show that if T is reducible so is e^T.

2.6.15 Let U be a finite-dimensional normed space and $T \in L(U)$. Use $\|T\|$ to denote the induced norm of the map T and assume $\|T\| < 1$.

(a) Show that $I - T$ is invertible. What happens if we do not assume $\|T\| < 1$ for the induced norm but assume $\|T\|' < 1$ under a certain other norm $\| \cdot \|'$? Substantiate your conclusions with reasoning.

(b) Establish the resolvent formula

$$(I - T)^{-1} = \sum_{k=0}^{\infty} T^k. \qquad (2.6.35)$$

You must justify convergence and meaningfulness of this expression.

(c) Is $I + T$ also invertible? If not, explain why. If yes, obtain the resolvent $(I + T)^{-1}$ as a power series as in (b).

3

Determinants

In this chapter, we introduce one of the most important computational tools in linear algebra – the determinants. First, we discuss some motivational examples. Next, we present the definition and basic properties of determinants. Then, we study some applications of determinants.

3.1 Motivational examples

We now present some examples occurring in geometry, algebra, and topology that use determinants as a natural underlying computational tool.

3.1.1 Area and volume

Let $u = (a_1, a_2)$ and $v = (b_1, b_2)$ be nonzero vectors in $\mathbb{R}^2$. We consider the area of the parallelogram formed by using these two vectors. First, we may express u in polar coordinates as

$$u = (a_1, a_2) = \|u\|(\cos \theta, \sin \theta). \tag{3.1.1}$$

Thus, we may easily resolve the vector v along the direction of u and the direction perpendicular to u as follows:

$$v = (b_1, b_2) = c_1(\cos \theta, \sin \theta) + c_2 \left(\cos \left[\theta \pm \frac{\pi}{2} \right], \sin \left[\theta \pm \frac{\pi}{2} \right] \right)$$
$$= (c_1 \cos \theta \mp c_2 \sin \theta, c_1 \sin \theta \pm c_2 \cos \theta), \quad c_1, c_2 \in \mathbb{R}. \tag{3.1.2}$$

Here c_2 may be interpreted as the length of the vector in the resolution that is taken to be perpendicular to u. Hence, from (3.1.2), we can read off the result

$$c_2 = \pm(b_2 \cos \theta - b_1 \sin \theta) = |b_2 \cos \theta - b_1 \sin \theta|. \tag{3.1.3}$$

Therefore, using (3.1.3) and then (3.1.1), we see that the area σ of the parallelogram under consideration is given by

$$\sigma = c_2 \|u\| = |\|u\| \cos \theta\, b_2 - \|u\| \sin \theta\, b_1| = |a_1 b_2 - a_2 b_1|. \qquad (3.1.4)$$

Thus, we see that the quantity $a_1 b_2 - a_2 b_1$ formed by the vectors (a_1, a_2) and (b_1, b_2) stands out, which is called the *determinant* of the matrix

$$A = \begin{pmatrix} a_1 & a_2 \\ b_1 & b_2 \end{pmatrix}, \qquad (3.1.5)$$

written as $\det(A)$ or denoted by

$$\begin{vmatrix} a_1 & a_2 \\ b_1 & b_2 \end{vmatrix}. \qquad (3.1.6)$$

Since $\det(A) = \pm\sigma$, it is also referred to as the *signed area* of the parallelogram formed by the vectors (a_1, a_2) and (b_1, b_2).

We now consider volume. We shall apply some vector algebra over $\mathbb{R}^3$ to facilitate our discussion.

We use $\cdot$ and $\times$ to denote the usual dot and cross products between vectors in $\mathbb{R}^3$. We use $\mathbf{i}, \mathbf{j}, \mathbf{k}$ to denote the standard mutually orthogonal unit vectors in $\mathbb{R}^3$ which form a right-hand system. For any vectors

$$u = (a_1, a_2, a_3) = a_1 \mathbf{i} + a_2 \mathbf{j} + a_3 \mathbf{k}, \quad v = (b_1, b_2, b_3) = b_1 \mathbf{i} + b_2 \mathbf{j} + b_3 \mathbf{k}, \quad (3.1.7)$$

in $\mathbb{R}^3$, we know that

$$u \times v = (a_2 b_3 - a_3 b_2)\mathbf{i} - (a_1 b_3 - a_3 b_1)\mathbf{j} + (a_1 b_2 - a_2 b_1)\mathbf{k} \qquad (3.1.8)$$

is perpendicular to both u and v, and $\|u \times v\|$ gives us the area of the parallelogram formed by using u and v as two adjacent edges, which generalizes the preceding discussion in $\mathbb{R}^2$. To avoid the trivial situation, we assume u and v are linearly independent. So $u \times v \neq 0$ and

$$\mathbb{R}^3 = \text{Span}\{u, v, u \times v\}. \qquad (3.1.9)$$

Let $w = (c_1, c_2, c_3)$ be another vector. Then (3.1.9) allows us to express w as

$$w = au + bv + c(u \times v), \quad a, b, c \in \mathbb{R}. \qquad (3.1.10)$$

From the geometry of the problem, we see that the volume of the parallelepiped formed by using $u, v,$ and w as adjacent edges is given by

$$\delta = \|u \times v\| \|c(u \times v)\| = |c| \|u \times v\|^2 \qquad (3.1.11)$$

because $\|c(u \times v)\|$ is the height of the parallelepiped, with the bottom area $\|u \times v\|$.

From (3.1.10), we have

$$w \cdot (u \times v) = c \|u \times v\|^2. \tag{3.1.12}$$

Inserting (3.1.12) into (3.1.11), we obtain the simplified volume formula

$$\delta = |w \cdot (u \times v)|$$
$$= |c_1(a_2b_3 - a_3b_2) - c_2(a_1b_3 - a_3b_1) + c_3(a_1b_2 - a_2b_1)|. \tag{3.1.13}$$

In analogy of the earlier discussion on the area formula in $\mathbb{R}^2$, we may set up the matrix

$$A = \begin{pmatrix} c_1 & c_2 & c_3 \\ a_1 & a_2 & a_3 \\ b_1 & b_2 & b_3 \end{pmatrix} \tag{3.1.14}$$

and defined the *signed volume* or determinant of the 3×3 matrix A as

$$\det(A) = \begin{vmatrix} c_1 & c_2 & c_3 \\ a_1 & a_2 & a_3 \\ b_1 & b_2 & b_3 \end{vmatrix}$$
$$= c_1(a_2b_3 - a_3b_2) - c_2(a_1b_3 - a_3b_1) + c_3(a_1b_2 - a_2b_1). \tag{3.1.15}$$

In view of the 2×2 determinants already defined, we may rewrite (3.1.15) in the decomposed form

$$\begin{vmatrix} c_1 & c_2 & c_3 \\ a_1 & a_2 & a_3 \\ b_1 & b_2 & b_3 \end{vmatrix} = c_1 \begin{vmatrix} a_2 & a_3 \\ b_2 & b_3 \end{vmatrix} - c_2 \begin{vmatrix} a_1 & a_3 \\ b_1 & b_3 \end{vmatrix} + c_3 \begin{vmatrix} a_1 & a_2 \\ b_1 & b_2 \end{vmatrix}. \tag{3.1.16}$$

3.1.2 Solving systems of linear equations

Next, we remark that a more standard motivational example for determinants is *Cramer's rule* or *Cramer's formulas* for solving systems of linear equations.

For example, consider the 2×2 system

$$\begin{cases} a_1x_1 + a_2x_2 &= c_1, \\ b_1x_1 + b_2x_2 &= c_2, \end{cases} \tag{3.1.17}$$

where $a_1, a_2, b_1, b_2, c_1, c_2 \in \mathbb{F}$. Multiplying the first equation by b_1, the second equation by a_1, and subtracting, we have

$$(a_1b_2 - a_2b_1)x_2 = a_1c_2 - b_1c_1; \tag{3.1.18}$$

multiplying the first equation by b_2, the second equation by a_2, and subtracting, we have

$$(a_1b_2 - a_2b_1)x_1 = b_2c_1 - a_2c_2. \tag{3.1.19}$$

Thus, with the notation of determinants and in view of (3.1.18) and (3.1.19), we may express the solution to (3.1.17) elegantly as

$$x_1 = \frac{\begin{vmatrix} c_1 & a_2 \\ c_2 & b_2 \end{vmatrix}}{\begin{vmatrix} a_1 & a_2 \\ b_1 & b_2 \end{vmatrix}}, \quad x_2 = \frac{\begin{vmatrix} a_1 & c_1 \\ b_1 & c_2 \end{vmatrix}}{\begin{vmatrix} a_1 & a_2 \\ b_1 & b_2 \end{vmatrix}}, \quad \text{if} \begin{vmatrix} a_1 & a_2 \\ b_1 & b_2 \end{vmatrix} \neq 0. \tag{3.1.20}$$

The extension of these formulas to 3×3 systems will be assigned as an exercise.

3.1.3 Topological invariants

Let f be a real-valued continuously differentiable function that maps the closed interval $[\alpha, \beta]$ ($\alpha < \beta$) into $[a, b]$ ($a < b$) so that the boundary points are mapped into boundary points as well,

$$f : \{\alpha, \beta\} \to \{a, b\}. \tag{3.1.21}$$

The function f maps the interval $[\alpha, \beta]$ to cover the interval $[a, b]$. Locally, if $f(t)$ increases, we consider that f covers its images positively there, and if $f(t)$ decreases, we consider that f covers its images negatively there, since the intervals $[\alpha, \beta]$ and $[a, b]$ are both oriented in the increasing order. In this way, since positive and negative local coverages undo each other in the global count of the number of coverages of the map $f : [\alpha, \beta] \to [a, b]$, say $N(f)$, we obtain the following three possibilities:

$$N(f) = \begin{cases} 1, & f(\alpha) = a, f(\beta) = b, \\ -1, & f(\alpha) = b, f(\beta) = a, \\ 0, & f(\alpha) = f(\beta). \end{cases} \tag{3.1.22}$$

This quantity may summarily be recast into the form of an integral,

$$N(f) = \frac{1}{b-a}(f(\beta) - f(\alpha)) = \frac{1}{b-a} \int_\alpha^\beta f'(t)\, dt, \tag{3.1.23}$$

and interpreted to be the number count for the orientation-preserving times minus the orientation-reversing times the function f maps the interval $[\alpha, \beta]$ to cover the interval $[a, b]$. Indeed, the right-hand side of (3.1.23) is well defined for any differentiable function f. In fact, the integral on the right-hand side of (3.1.23) is the simplest *topological invariant* called the *degree* of the function f, which may now be denoted by

$$\deg(f) = \frac{1}{b-a} \int_\alpha^\beta f'(t)\, dt. \tag{3.1.24}$$

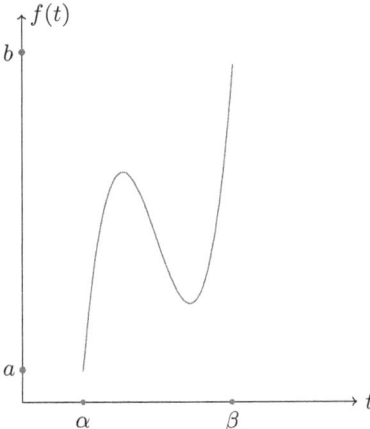

Figure 3.1 Plot of a single-variable function f that maps the interval $[\alpha, \beta]$ onto the interval $[a, b]$ resulting in the total coverage number $N(f) = 1$ as the topological degree of f, namely, $\deg(f)$.

The word "topological" is used to refer to the fact that a small alteration of f cannot perturb the value of $\deg(f)$ since $\deg(f)$ may only take integer values, and the right-hand side of (3.1.24), however, relies on the derivative of f continuously. Naturally, the quantity $f'(t)$ in (3.1.24) is also referred to as the topological density of the topological invariant $\deg(f)$.

In Figure 3.1, we present a plot of a function whose topological degree is 1.

As a simple application, we note that it is not hard to see that for any $c \in [a, b]$ the equation $f(t) = c$ has at least one solution when $\deg(f) \neq 0$.

We next extend our discussion of topological invariants to two-dimensional situations.

Let Γ and $\mathcal{C}$ be two closed differentiable curves in $\mathbb{R}^2$ oriented counterclockwise, and let

$$u : \Gamma \to \mathcal{C} \qquad (3.1.25)$$

be a differentiable map. In analogy with the case of a real-valued function over an interval discussed above, we may express the number count for the orientation-preserving times minus the orientation-reversing times u maps the curve Γ to cover the curve $\mathcal{C}$ in the form of a line integral,

$$\deg(u) = \frac{1}{|\mathcal{C}|} \int_\Gamma \tau \cdot du, \qquad (3.1.26)$$

where $|\mathcal{C}|$ denotes the length of the curve $\mathcal{C}$ and τ is the unit tangent vector along the positive direction of $\mathcal{C}$.

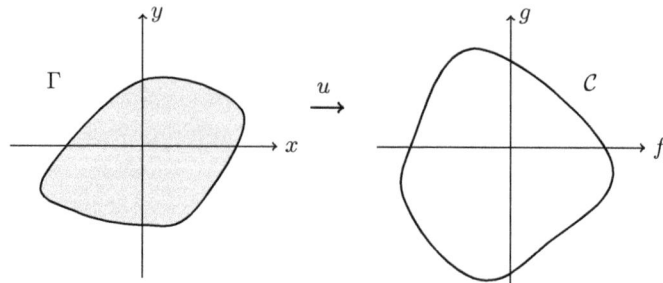

Figure 3.2 A simple closed curve Γ is mapped into another closed curve $\mathcal{C}$ through the differentiable map $u(x,y) = (f(x,y),g(x,y))$.

In Figure 3.2, we present a plot of such a map given as $u(x,y) = (f(x,y),g(x,y))$.

In the special situation when $\mathcal{C} = S^1$ (the unit circle in $\mathbb{R}^2$ centered at the origin), we write u as

$$u = (f,g), \quad f^2 + g^2 = 1, \tag{3.1.27}$$

where f,g are real-valued functions, so that

$$\tau = (-g,f). \tag{3.1.28}$$

Now assume further that the curve Γ is parametrized by a parameter t taken over the interval $[\alpha, \beta]$. Then, inserting (3.1.27) and (3.1.28) into (3.1.26) and using $|S^1| = 2\pi$, we have

$$\begin{aligned}
\deg(u) &= \frac{1}{2\pi} \int_\alpha^\beta (-g,f)\cdot(f',g')\,dt \\
&= \frac{1}{2\pi} \int_\alpha^\beta (fg' - gf')\,dt \\
&= \frac{1}{2\pi} \int_\alpha^\beta \begin{vmatrix} f & g \\ f' & g' \end{vmatrix} dt.
\end{aligned} \tag{3.1.29}$$

Thus we see that the concept of determinant arises naturally again.

Let v be a vector field over $\mathbb{R}^2$. Let Γ be a closed curve in $\mathbb{R}^2$ where $v \neq 0$. Then

$$v_\Gamma = \frac{1}{\|v\|}v \tag{3.1.30}$$

defines a map from Γ into S^1. The *index* of the vector field v along the curve Γ is then defined to be

$$\operatorname{ind}(v|_\Gamma) = \deg(v_\Gamma), \tag{3.1.31}$$

which is another useful topological invariant.

As an example, we consider a vector field v over $\mathbb{R}^2$ given by

$$v(x,y) = (x^2 - y^2, 2xy), \quad (x,y) \in \mathbb{R}^2. \tag{3.1.32}$$

Then $\|v(x,y)\|^2 = (x^2 + y^2)^2 > 0$ for $(x,y) \neq (0,0)$. So for any closed curve Γ not intersecting the origin, the quantity $\mathrm{ind}(v|_\Gamma)$ is well defined.

Let S_R^1 denote the circle of radius $R > 0$ in $\mathbb{R}^2$ centered at the origin. We may parametrize S_R^1 by the polar angle θ: $x = R\cos\theta, y = R\sin\theta$, $\theta \in [0, 2\pi]$. With (3.1.32), we have

$$
\begin{aligned}
v_{S_R^1} &= \frac{1}{R^2}(x^2 - y^2, 2xy) \\
&= (\cos^2\theta - \sin^2\theta, 2\cos\theta\sin\theta) = (\cos 2\theta, \sin 2\theta). \tag{3.1.33}
\end{aligned}
$$

Therefore, using (3.1.29), we get

$$
\begin{aligned}
\mathrm{ind}(v|_{S_R^1}) &= \deg(v_{S_R^1}) \\
&= \frac{1}{2\pi} \int_0^{2\pi} \begin{vmatrix} \cos 2\theta & \sin 2\theta \\ -2\sin 2\theta & 2\cos 2\theta \end{vmatrix} d\theta = 2. \tag{3.1.34}
\end{aligned}
$$

For any closed curve Γ enclosing but not intersecting the origin, we can continuously deform it into a circle S_R^1 ($R > 0$), while staying away from the origin in the process. By continuity or topological invariance, we obtain $\mathrm{ind}(v|_\Gamma) = \mathrm{ind}(v|_{S_R^1}) = 2$. The meaning of this result will be seen in the following theorem.

Theorem 3.1 *Let v be a vector field which is differentiable over a bounded and simply connected domain Ω in $\mathbb{R}^2$ and Γ a closed curve contained in Ω. If $v \neq 0$ on Γ and $\mathrm{ind}(v|_\Gamma) \neq 0$, then there must be at least one point enclosed inside Γ where $v = 0$.*

Proof Assume otherwise that $v \neq 0$ in the domain enclosed by Γ. Let γ be another closed curve enclosed inside Γ. Since γ may be obtained from Γ through a continuous deformation and $v = 0$ nowhere inside Γ, we have $\mathrm{ind}(v|_\Gamma) = \mathrm{ind}(v|_\gamma)$. On the other hand, if we parametrize the curve γ using its arclength s, then

$$\mathrm{ind}(v|_\gamma) = \deg(v_\gamma) = \frac{1}{2\pi} \int_\gamma \tau \cdot v_\gamma'(s)\, ds, \quad v_\gamma = \left(\frac{1}{\|v\|}v\right)\bigg|_\gamma, \tag{3.1.35}$$

where $\tau(s)$ is the unit tangent vector along the unit circle S^1 at the image point $v_\gamma(s)$ under the map $v_\gamma : \gamma \to S^1$. Rewrite v_γ as

$$v_\gamma(s) = (f(x(s), y(s)), g(x(s), y(s))). \tag{3.1.36}$$

Then

$$v_\gamma'(s) = (f_x x'(s) + f_y y'(s), g_x x'(s) + g_y y'(s)). \tag{3.1.37}$$

The assumption on v gives us the uniform boundedness $|f_x|, |f_y|, |g_x|, |g_y|$ inside Γ. Using this property and (3.1.37), we see that there is a γ-independent constant $C > 0$ such that

$$\|v'_\gamma(s)\| \le C\sqrt{x'(s)^2 + y'(s)^2} = C. \tag{3.1.38}$$

In view of (3.1.35) and (3.1.38), we have

$$1 \le |\mathrm{ind}(v|_\gamma)| \le \frac{1}{2\pi}\int_\gamma \|v'_\gamma(s)\|\,ds \le \frac{C}{2\pi}|\gamma|, \tag{3.1.39}$$

which leads to absurdness when the total arclength $|\gamma|$ of the curve γ is made small enough. $\qquad\square$

Thus, returning to the example (3.1.32), we conclude that the vector field v has a zero inside any circle S_R^1 $(R > 0)$ since we have shown that $\mathrm{ind}(v|_{S_R^1}) = 2 \ne 0$, which can only be the origin as seen trivially in (3.1.32) already.

We now use Theorem 3.1 to establish the celebrated *Fundamental Theorem of Algebra* as stated below.

Theorem 3.2 *Any polynomial of degree $n \ge 1$ with coefficients in $\mathbb{C}$ of the form*

$$f(z) = a_n z^n + a_{n-1} z^{n-1} + \cdots + a_0, \quad a_0, \dots, a_{n-1}, a_n \in \mathbb{C}, \quad a_n \ne 0, \tag{3.1.40}$$

must have a zero in $\mathbb{C}$. That is, there is some $z_0 \in \mathbb{C}$ such that $f(z_0) = 0$.

Proof Without loss of generality and for sake of simplicity, we may assume $a_n = 1$ otherwise we may divide $f(z)$ by a_n.

Let $z = x + iy$, $x, y \in \mathbb{R}$, and write $f(z)$ as

$$f(z) = P(x, y) + iQ(x, y), \tag{3.1.41}$$

where P and Q are real-valued functions of x and y. Consider the vector field

$$v(x, y) = (P(x, y), Q(x, y)). \tag{3.1.42}$$

Then it is clear that $\|v(x, y)\| = |f(z)|$, and it suffices to show that v vanishes at some $(x_0, y_0) \in \mathbb{R}^2$.

In order to simplify our calculation, we consider a one-parameter deformation of $f(z)$ given by

$$f^t(z) = z^n + t(a_{n-1} z^{n-1} + \cdots + a_0), \quad t \in [0, 1], \tag{3.1.43}$$

and denote the correspondingly constructed vector field by $v^t(x,y)$. So on the circle $S_R^1 = \{(x,y) \in \mathbb{R}^2 \mid \|(x,y)\| = |z| = R\}$ $(R > 0)$, we have the uniform lower estimate

$$\|v^t(x,y)\| = |f^t(z)|$$

$$\geq R^n \left(1 - |a_{n-1}|\frac{1}{R} - \cdots - |a_0|\frac{1}{R^n}\right)$$

$$\equiv C(R), \quad t \in [0,1]. \tag{3.1.44}$$

Thus, when R is sufficiently large, we have $C(R) \geq 1$ (say). For such a choice of R, by topological invariance, we have

$$\text{ind}(v|_{S_R^1}) = \text{ind}(v^1|_{S_R^1}) = \text{ind}(v^0|_{S_R^1}). \tag{3.1.45}$$

On the other hand, over S_R^1 we may again use the polar angle θ: $x = R\cos\theta$, $y = R\sin\theta$, or $z = Re^{i\theta}$, to represent f^0 as $f^0(z) = R^n e^{in\theta}$. Hence $v^0 = R^n(\cos n\theta, \sin n\theta)$. Consequently,

$$v_{S_R^1}^0 = \frac{1}{\|v^0\|}v^0 = (\cos n\theta, \sin n\theta). \tag{3.1.46}$$

Therefore, as before, we obtain

$$\text{ind}(v^0|_{S_R^1}) = \deg(v_{S_R^1}^0)$$

$$= \frac{1}{2\pi}\int_0^{2\pi}\begin{vmatrix}\cos n\theta & \sin n\theta \\ -n\sin n\theta & n\cos n\theta\end{vmatrix} d\theta = n. \tag{3.1.47}$$

In view of (3.1.45) and (3.1.47), we get $\text{ind}(v|_{S_R^1}) = n$. Thus, applying Theorem 3.1, we conclude that v must vanish somewhere inside the circle S_R^1. $\qquad\square$

Use Σ to denote a closed surface in $\mathbb{R}^3$ and S^2 the standard unit sphere in $\mathbb{R}^3$. We may also consider a map $u : \Sigma \to S^2$. Since the orientation of S^2 is given by its unit outnormal vector, say v, we may analogously express the number count, for the number of times that u covers S^2 in an orientation-preserving manner minus the number of times that u covers S^2 in an orientation-reversing manner, in the form of a surface integral, also called the *degree* of the map $u : \Sigma \to S^2$, by

$$\deg(u) = \frac{1}{|S^2|}\int_\Sigma v \cdot d\sigma, \tag{3.1.48}$$

where $d\sigma$ is the vector area element over S^2 induced from the map u.

To further facilitate computation, we may assume that Σ is parametrized by the parameters s, t over a two-dimensional domain Ω and $u = (f, g, h)$, where f, g, and h are real-valued functions of s and t so that $f^2 + g^2 + h^2 = 1$. At the image point u, the unit outnormal of S^2 at u, is simply u itself. Moreover, the vector area element at u under the mapping u can be represented as

$$d\sigma = \left(\frac{\partial u}{\partial s} \times \frac{\partial u}{\partial t} \right) dsdt. \tag{3.1.49}$$

Thus, inserting these and $|S^2| = 4\pi$ into (3.1.48), we arrive at

$$\deg(u) = \frac{1}{4\pi} \int_\Omega u \cdot \left(\frac{\partial u}{\partial s} \times \frac{\partial u}{\partial t} \right) dsdt$$

$$= \frac{1}{4\pi} \int_\Omega \begin{vmatrix} f & g & h \\ f_s & g_s & h_s \\ f_t & g_t & h_t \end{vmatrix} dsdt. \tag{3.1.50}$$

This gives another example of the use of determinants.

Exercises

3.1.1   For the 3×3 system of equations

$$\begin{cases} a_1 x_1 + a_2 x_2 + a_3 x_3 = d_1, \\ b_1 x_1 + b_2 x_2 + b_3 x_3 = d_2, \\ c_1 x_1 + c_2 x_2 + c_3 x_3 = d_3, \end{cases} \tag{3.1.51}$$

find similar solution formulas as those for (3.1.17) expressed as ratios of some 3×3 determinants.

3.1.2   Let $f : [\alpha, \beta] \to [a, b]$ be a real-valued continuously differentiable function satisfying $\{f(\alpha), f(\beta)\} \subset \{a, b\}$, where $\alpha, \beta, a, b \in \mathbb{R}$ and $\alpha < \beta$ and $a < b$. Show that if $\deg(f) \neq 0$ then $f(t) = c$ has a solution for any $c \in (a, b)$.

3.1.3   In (3.1.23), assume $f(\alpha) = a$ and regard $s = f(t)$ as giving the position and $f'(t)$ the velocity of an object at time t. Formulate a physical interpretation of the results stated in (3.1.22).

3.1.4   Consider the vector fields or maps $f, g : \mathbb{R}^2 \to \mathbb{R}^2$ defined by

$$f(x, y) = (ax, by), \quad g(x, y) = (a^2 x^2 - b^2 y^2, 2abxy), \quad (x, y) \in \mathbb{R}^2, \tag{3.1.52}$$

where $a, b > 0$ are constants.

(a) Compute the indices of f and g around a suitable closed curve around the origin of $\mathbb{R}^2$.

(b) What do you notice? Are the results depend on a, b? In particular, what are the numbers (counting multiplicities) of zeros of the maps f and g?

(c) If we make some small alterations of f and g by a positive parameter $\varepsilon > 0$ to get (say)

$$f_\varepsilon(x,y) = (ax - \varepsilon, by),$$
$$g_\varepsilon(x,y) = (a^2x^2 - b^2y^2 - \varepsilon, 2abxy), \quad (x,y) \in \mathbb{R}^2, \qquad (3.1.53)$$

what happens to the indices of f_ε and g_ε around the same closed curve? What happens to the zeros of f_ε and g_ε?

3.1.5 Consider the following simultaneous system of nonlinear equations:

$$\begin{cases} x^3 - 3xy^2 = 5\cos^2(x+y), \\ 3x^2y - y^3 = -2e^{-x^2y^2}. \end{cases} \qquad (3.1.54)$$

Use the topological method in this section to prove that the system has at least one solution.

3.1.6 Consider the complex plane $\mathbb{C}$ and complex variable $z \in \mathbb{C}$ with its complex conjugate $\bar{z}$:

$$z = x + iy, \quad \bar{z} = x - iy, \quad x, y \in \mathbb{R}, \quad i = \sqrt{-1}, \qquad (3.1.55)$$

and study the following generalized "polynomial" function of degree n:

$$f(z) = z^k\bar{z}^l + \sum_{m=0}^{n-1} \sum_{i+j=m} a_{ij}z^i\bar{z}^j, \quad a_{ij} \in \mathbb{C}, \quad i,j \geq 0, \qquad (3.1.56)$$

where k, l are non-negative integers and $k + l = n \geq 1$.

(a) Extend the proof of the Fundamental Theorem of Algebra to establish "your new theorem" that the function given in (3.1.56) has a root if $k \neq l$. Explain why Fundamental Theorem of Algebra is a special case of this theorem.

(b) Give an example to show that when $k = l$ (hence n is an even integer) the function (3.1.56) may not have a root.

(c) Explain why the Liouville theorem cannot be applied to study the function (3.1.56) such that the topological method we have learned is seen to be more powerful in its range of coverage of problems.

3.1.7 Consider the stereographic projection of S^2 sited in $\mathbb{R}^3$ with the Cartesian coordinates x, y, z onto the xy-plane through the south pole $(0, 0, -1)$, which induces a parameterization of S^2 by $\mathbb{R}^2$ given by

$$f = \frac{2x}{1+x^2+y^2}, \quad g = \frac{2y}{1+x^2+y^2}, \quad h = \frac{1-x^2-y^2}{1+x^2+y^2}, \quad (x,y) \in \mathbb{R}^2,$$
$$(3.1.57)$$

where $u = (f, g, h) \in S^2$. Regarded as the identity map $u : S^2 \to S^2$, we have $\deg(u) = 1$. Verify this result by computing the integral

$$\deg(u) = \frac{1}{4\pi} \int_{\mathbb{R}^2} \begin{vmatrix} f & g & h \\ f_x & g_x & h_x \\ f_y & g_y & h_y \end{vmatrix} \, dx dy. \qquad (3.1.58)$$

3.1.8 The *hedgehog map* is a map $S^2 \to S^2$ defined in terms of the parameterization of $\mathbb{R}^2$ by polar coordinates r, θ by the expression

$$u = (\cos(n\theta)\sin f(r), \sin(n\theta)\sin f(r), \cos f(r)), \qquad (3.1.59)$$

where $0 < r < \infty$, $\theta \in [0, 2\pi]$, $n \in \mathbb{Z}$, and f is a real-valued function satisfying the boundary condition $f(0) = \pi$ and $f(\infty) = 0$. Compute

$$\deg(u) = \frac{1}{4\pi} \int_0^\infty \int_0^{2\pi} u \cdot \left(\frac{\partial u}{\partial r} \times \frac{\partial u}{\partial \theta} \right) d\theta dr \qquad (3.1.60)$$

and explain your result.

3.2 Definition and properties of determinants

Motivated by the practical examples shown in Section 3.1, we now systematically develop the notion of determinants. There are many ways to define determinants. The inductive definition to be presented below is perhaps the simplest.

Definition 3.3 Consider $A \in \mathbb{F}(n, n)$ ($n \geq 1$). If $n = 1$, then $A = (a)$ ($a \in \mathbb{F}$) and the determinant of A, $\det(A)$, is defined to be $\det(A) = a$; if $A = (a_{ij})$ with $n \geq 2$, the *minor* M_{ij} of the entry a_{ij} is the determinant of the $(n-1) \times (n-1)$ submatrix of A obtained from deleting the ith row and jth column of A occupied by a_{ij} and the *cofactor* C_{ij} is given by

$$C_{ij} = (-1)^{i+j} M_{ij}, \quad i, j = 1, \ldots, n. \qquad (3.2.1)$$

The determinant of A is defined by the expansion formula

$$\det(A) = \sum_{i=1}^n a_{i1} C_{i1}. \qquad (3.2.2)$$

The formula (3.2.2) is also referred to as the *cofactor expansion* of the determinant according to the first column.

This definition indicates that if a column of an $n \times n$ matrix A is zero then $\det(A) = 0$. To show this, we use induction. When $n = 1$, it is trivial. Assume that the statement is true at $n-1$ $(n \geq 2)$. We now prove the statement at n $(n \geq 2)$. In fact, if the first column of A is zero, then $\det(A) = 0$ simply by the definition of determinant (see (3.2.2)); if another column rather than the first column of A is zero, then all the cofactors C_{i1} vanish by the inductive assumption, which still results in $\det(A) = 0$. The definition also implies that if $A = (a_{ij})$ is upper triangular then $\det(A)$ is the product of its diagonal entries, $\det(A) = a_{11} \cdots a_{nn}$, as may be shown by induction as well.

The above definition of determinant immediately leads to the following important properties.

Theorem 3.4 *Consider the $n \times n$ matrices A and B given as*

$$
A = \begin{pmatrix}
a_{11} & \cdots & a_{1n} \\
\cdots & \cdots & \cdots \\
a_{k1} & \cdots & a_{kn} \\
\cdots & \cdots & \cdots \\
a_{n1} & \cdots & a_{nn}
\end{pmatrix}, \quad
B = \begin{pmatrix}
a_{11} & \cdots & a_{1n} \\
\cdots & \cdots & \cdots \\
r a_{k1} & \cdots & r a_{kn} \\
\cdots & \cdots & \cdots \\
a_{n1} & \cdots & a_{nn}
\end{pmatrix}.
\tag{3.2.3}
$$

That is, B is obtained from A by multiplying the kth row of A by a scalar $r \in \mathbb{F}$, $k = 1, \ldots, n$. Then we have $\det(B) = r \det(A)$.

Proof We prove the theorem by induction.

When $n = 1$, the statement of the theorem is trivial.

Assume that the statement is true for $(n-1) \times (n-1)$ matrices when $n \geq 2$.

Now for the $n \times n$ matrices given in (3.2.3), use C_{ij}^A and C_{ij}^B to denote the cofactors of A and B, respectively, $i, j = 1, \ldots, n$. Then, (3.2.3) and the inductive assumption give us

$$
C_{k1}^B = C_{k1}^A; \quad C_{i1}^B = r C_{i1}^A, \quad i \neq k.
\tag{3.2.4}
$$

Therefore, we arrive at

$$
\det(B) = \sum_{i \neq k} a_{i1} C_{i1}^B + r a_{k1} C_{k1}^B = \sum_{i \neq k} a_{i1} r C_{i1}^A + r a_{k1} C_{k1}^A = r \det(A).
\tag{3.2.5}
$$

The proof is complete. □

This theorem implies that if a row of a matrix A is zero then $\det(A) = 0$.

As an application, we show that if an $n \times n$ matrix $A = (a_{ij})$ is lower triangular then $\det(A) = a_{11} \cdots a_{nn}$. In fact, when $n = 1$, there is nothing to show. Assume that the formula is true at $n-1$ $(n \geq 2)$. At $n \geq 2$, the first row of all the minor M_{i1} of A vanishes for each $i = 2, \ldots, n$. So $M_{i1} = 0$, $i = 2, \ldots, n$. However, the inductive assumption gives us $M_{11} = a_{22} \cdots a_{nn}$. Thus $\det(A) = a_{11}(-1)^{1+1} M_{11} = a_{11} \cdots a_{nn}$ as claimed.

Therefore, if an $n \times n$ matrix $A = (a_{ij})$ is either upper or lower triangular, we infer that there holds $\det(A^t) = \det(A)$, although, later, we will show that such a result is true for general matrices.

Theorem 3.5 *For the $n \times n$ matrices $A = (a_{ij}), B = (b_{ij}),$ and $C = (c_{ij})$, which have identical rows except the kth row in which $c_{kj} = a_{kj} + b_{kj}$, $j = 1, \ldots, n$, we have $\det(C) = \det(A) + \det(B)$.*

Proof   We again use induction on n.

The statement is clear when $n = 1$.

Assume that the statement is valid for the $n - 1$ case ($n \geq 2$).

For $A, B,$ and C given in the theorem with $n \geq 2$, with the notation in the proof of Theorem 3.4 and in view of the inductive assumption, we have

$$C_{k1}^C = C_{k1}^A; \quad C_{i1}^C = C_{i1}^A + C_{i1}^B, \quad i \neq k. \tag{3.2.6}$$

Consequently,

$$
\begin{aligned}
\det(C) &= \sum_{i \neq k} a_{i1} C_{i1}^C + c_{k1} C_{k1}^C \\
&= \sum_{i \neq k} a_{i1} (C_{i1}^A + C_{i1}^B) + (a_{k1} + b_{k1}) C_{k1}^A \\
&= \sum_{i \neq k} a_{i1} C_{i1}^A + a_{k1} C_{k1}^A + \sum_{i \neq k} a_{i1} C_{i1}^B + b_{k1} C_{k1}^A \\
&= \det(A) + \det(B), \tag{3.2.7}
\end{aligned}
$$

as asserted.   □

Theorem 3.6 *Let A and B be two $n \times n$ ($n \geq 2$) matrices so that B is obtained from interchanging any two rows of A. Then $\det(B) = -\det(A)$.*

Proof   We use induction on n.

At $n = 2$, we can directly check that the statement of the theorem is true.

Assume that the statement is true at $n - 1 \geq 2$.

Let A and B be $n \times n$ ($n \geq 3$) matrices given by

$$
A = \begin{pmatrix}
a_{11} & \cdots & a_{1n} \\
\cdots & \cdots & \cdots \\
a_{i1} & \cdots & a_{in} \\
\cdots & \cdots & \cdots \\
a_{j1} & \cdots & a_{jn} \\
\cdots & \cdots & \cdots \\
a_{n1} & \cdots & a_{nn}
\end{pmatrix}, \quad
B = \begin{pmatrix}
a_{11} & \cdots & a_{1n} \\
\cdots & \cdots & \cdots \\
a_{j1} & \cdots & a_{jn} \\
\cdots & \cdots & \cdots \\
a_{i1} & \cdots & a_{in} \\
\cdots & \cdots & \cdots \\
a_{n1} & \cdots & a_{nn}
\end{pmatrix}, \tag{3.2.8}
$$

where $j = i + k$ for some $k \geq 1$. We observe that it suffices to prove the adjacent case when $k = 1$ because when $k \geq 2$ we may obtain B from A simply by interchanging adjacent rows k times downwardly and then $k - 1$ times upwardly, which gives rise to an odd number of adjacent row interchanges.

For the adjacent row interchange, $j = i + 1$, the inductive assumption allows us to arrive at the following relations between the minors of the matrices A and B immediately:

$$M_{k1}^B = -M_{k1}^A, \quad k \neq i, i+1; \quad M_{i1}^B = M_{i+1,1}^A, \quad M_{i+1,1}^B = M_{i1}^A, \qquad (3.2.9)$$

which implies that the corresponding cofactors of A and B all differ by a sign,

$$C_{k1}^B = -C_{k1}^A, \quad k \neq i, i+1; \quad C_{i1}^B = -C_{i+1,1}^A, \quad C_{i+1,1}^B = -C_{i1}^A. \qquad (3.2.10)$$

Hence

$$
\begin{aligned}
\det(B) &= \sum_{k \neq i, i+1} a_{k1} C_{k1}^B + a_{i+1,1} C_{i1}^B + a_{i1} C_{i+1,1}^B \\
&= \sum_{k \neq i, i+1} a_{k1}(-C_{k1}^A) + a_{i+1,1}(-C_{i+1,1}^A) + a_{i1}(-C_{i1}^A) \\
&= -\det(A), \qquad (3.2.11)
\end{aligned}
$$

as expected. $\square$

This theorem indicates that if two rows of an $n \times n$ ($n \geq 2$) matrix A are identical then $\det(A) = 0$, provided that $\mathrm{ch}(\mathbb{F}) \neq 2$, although this property is true for any field $\mathbb{F}$ as may be proved by using a direct inductive argument. Thus adding a multiple of a row to another row of A does not alter the determinant of A:

$$
\begin{vmatrix}
a_{11} & \cdots & a_{1n} \\
\cdots & \cdots & \cdots \\
a_{i1} & \cdots & a_{in} \\
\cdots & \cdots & \cdots \\
ra_{i1}+a_{j1} & \cdots & ra_{in}+a_{jn} \\
\cdots & \cdots & \cdots \\
a_{n1} & \cdots & a_{nn}
\end{vmatrix}
=
\begin{vmatrix}
a_{11} & \cdots & a_{1n} \\
\cdots & \cdots & \cdots \\
a_{i1} & \cdots & a_{in} \\
\cdots & \cdots & \cdots \\
ra_{i1} & \cdots & ra_{in} \\
\cdots & \cdots & \cdots \\
a_{n1} & \cdots & a_{nn}
\end{vmatrix}
+
\begin{vmatrix}
a_{11} & \cdots & a_{1n} \\
\cdots & \cdots & \cdots \\
a_{i1} & \cdots & a_{in} \\
\cdots & \cdots & \cdots \\
a_{j1} & \cdots & a_{jn} \\
\cdots & \cdots & \cdots \\
a_{n1} & \cdots & a_{nn}
\end{vmatrix}
$$

$$= \det(A). \qquad (3.2.12)$$

The above results provide us with practical computational techniques when evaluating the determinant of an $n \times n$ matrix A. In fact, we may perform the following three types of *permissible row operations* on A:

(1) Multiply a row of A by a nonzero scalar. Such an operation may also be realized by multiplying A on the left by the matrix obtained from multiplying the corresponding row of the $n \times n$ identity matrix I by the same scalar.

(2) Interchange any two rows of A when $n \geq 2$. Such an operation may also be realized by multiplying A on the left by the matrix obtained from interchanging the corresponding two rows of the $n \times n$ identity matrix I.

(3) Add a multiple of a row to another row of A when $n \geq 2$. Such an operation may also be realized by multiplying A on the left by the matrix obtained from adding the same multiple of the row to another row, correspondingly, of the $n \times n$ identity matrix I.

The matrices constructed in the above three types of permissible row operations are called *elementary matrices* of types 1, 2, and 3. Let E be an elementary matrix of a given type. Then E is invertible and E^{-1} is of the same type. More precisely, if E is of type 1 and obtained from multiplying a row of I by the scalar r, then $\det(E) = r\det(I) = r$ and E^{-1} is simply obtained from multiplying the same row of I by r^{-1}, resulting in $\det(E^{-1}) = r^{-1}$; if E is of type 2, then $E^{-1} + E$ and $\det(E) = \det(E^{-1}) = -\det(I) = -1$; if E is of type 3 and obtained from adding an r multiple of the ith row to the jth row ($i \neq j$) of I, then E^{-1} is obtained from adding a $(-r)$ multiple of the ith row to the jth row of I and $\det(E) = \det(E^{-1}) = \det(I) = 1$. In all cases,

$$\det(E^{-1}) = \det(E)^{-1}. \tag{3.2.13}$$

In conclusion, the properties of determinant under permissible row operations may be summarized collectively as follows.

Theorem 3.7 *Let A be an $n \times n$ matrix and E be an elementary matrix of the same dimensions. Then*

$$\det(EA) = \det(E)\det(A). \tag{3.2.14}$$

For an $n \times n$ matrix A, we can perform a sequence of permissible row operations on A to reduce it into an upper triangular matrix, say U, whose determinant is simply the product of whose diagonal entries. Thus, if we express A as $E_k \cdots E_1 A = U$, where $E_1, \ldots, E_k$ are some elementary matrices, then Theorem 3.7 gives us the relation

$$\det(E_k) \cdots \det(E_1)\det(A) = \det(U). \tag{3.2.15}$$

We are now prepared to establish the similar properties of determinants with respect to column operations.

Theorem 3.8 *The conclusions of Theorems 3.4–3.6 hold when the statements about the row vectors are replaced correspondingly with the column vectors of matrices.*

Proof   Using induction, it is easy to see that the conclusions of Theorems 3.4 and 3.5 hold. We now prove that the conclusion of Theorem 3.6 holds when the row interchange there is replaced with column interchange. That is, we show that if two columns in an $n \times n$ matrix A are interchanged, its determinant will change sign. This property is not so obvious since our definition of determinant is based on the cofactor expansion by the first column vector and an interchange of the first column with another column alters the first column of the matrix. The effect of the value of determinant with respect to such an alteration needs to be examined closely, which will be our task below.

We still use induction.

At $n = 2$ the conclusion may be checked directly.

Assume the conclusion holds at $n - 1 \geq 1$.

We now prove the conclusion at $n \geq 3$. As before, it suffices to establish the conclusion for any adjacent column interchange.

If the column interchange does not involve the first column, we see that the conclusion about the sign change of the determinant clearly holds in view of the inductive assumption and the cofactor expansion formula (3.2.2) since all the cofactors C_{i1} ($i = 1, \ldots, n$) change their sign exactly once under any pair of column interchange.

Now consider the effect of an interchange of the first and second columns of A. It can be checked that such an operation may be carried out through multiplying A by the matrix F from the right, where F is obtained from the $n \times n$ identity matrix I by interchanging the first and second columns of I. Of course, $\det(F) = -1$.

Let $E_1, \ldots, E_k$ be a sequence of elementary matrices and $U = (u_{ij})$ an upper triangular matrix so that $E_k \cdots E_1 A = U$. Then we have

$$
UF = \begin{pmatrix}
u_{12} & u_{11} & u_{13} & \cdots & u_{1n} \\
u_{22} & 0 & u_{23} & \cdots & u_{2n} \\
0 & 0 & u_{33} & \cdots & u_{3n} \\
\cdots & \cdots & \cdots & \cdots & \cdots \\
0 & \cdots & \cdots & 0 & u_{nn}
\end{pmatrix}. \tag{3.2.16}
$$

Thus the cofactor expansion formula down the first column, as stated in Definition 3.3, leads us to the result

$$
\det(UF) = u_{12} \begin{vmatrix} 0 & u_{23} & \cdots & \cdots \\ 0 & u_{33} & \cdots & \cdots \\ \cdots & \cdots & \cdots & \cdots \\ 0 & \cdots & \cdots & u_{nn} \end{vmatrix} - u_{22} \begin{vmatrix} u_{11} & u_{13} & \cdots & \cdots \\ 0 & u_{33} & \cdots & \cdots \\ \cdots & \cdots & \cdots & \cdots \\ 0 & \cdots & \cdots & u_{nn} \end{vmatrix}
$$

$$
= -u_{11}u_{22}\cdots u_{nn} = -\det(U). \tag{3.2.17}
$$

Combining (3.2.15) and (3.2.17), we obtain

$$
\det(E_k \cdots E_1 AF) = \det(UF) = -\det(U)
$$
$$
= -\det(E_k)\cdots\det(E_1)\det(A). \tag{3.2.18}
$$

Thus, applying (3.2.14) on the left-hand side of (3.2.18), we arrive at $\det(AF) = -\det(A)$ as desired.

The interchange of the first and kth ($k \geq 2$) columns can be carried out by a sequence of $2(k-1) - 1 = 2k - 3$ adjacent column interchanges. Thus, by the inductive assumption, we arrive at the sign change property again, in general.
□

Theorem 3.8 gives us additional practical computational techniques when evaluating the determinant of an $n \times n$ matrix A because it allows us to perform the following three types of *permissible column operations* on A:

(1) Multiply a column of A by a nonzero scalar. Such an operation may also be realized by multiplying A on the right by the matrix obtained from multiplying the corresponding column of the $n \times n$ identity matrix I by the same scalar.
(2) Interchange any two columns of A when $n \geq 2$. Such an operation may also be realized by multiplying A on the right by the matrix obtained from interchanging the corresponding two columns of the $n \times n$ identity matrix I.
(3) Add a multiple of a column to another column of A when $n \geq 2$. Such an operation may also be realized by multiplying A on the right by the matrix obtained from adding the same multiple of the column to another column, correspondingly, of the $n \times n$ identity matrix I.

The matrices constructed in the above three types of permissible column operations are simply the elementary matrices defined and described earlier.

Like those about permissible row operations, the properties of determinant under permissible column operations may be summarized collectively as well as follows.

Theorem 3.9 *Let A be an n × n matrix and E be an elementary matrix of the same dimensions. Then*

$$\det(AE) = \det(A)\det(E). \tag{3.2.19}$$

With the above preparation, we are ready to harvest a series of important properties of determinants.

First we show that determinant is invariant under matrix transpose.

Theorem 3.10 *Let A be an n × n matrix. Then*

$$\det(A) = \det(A^t). \tag{3.2.20}$$

Proof Choose a sequence of elementary matrices $E_1, \ldots, E_k$ such that

$$E_k \cdots E_1 A = U \tag{3.2.21}$$

is an upper triangular matrix. Thus from $U^t = A^t E_1^t \cdots E_k^t$ and Theorem 3.9 we have

$$\begin{aligned}
\det(U) = \det(U^t) &= \det(A^t E_1^t \cdots E_k^t) \\
&= \det(A^t)\det(E_1^t)\cdots\det(E_k^t).
\end{aligned} \tag{3.2.22}$$

Comparing (3.2.22) with (3.2.15) and noting $\det(E_l) = \det(E_l^t)$, $l = 1, \ldots, k$, because an elementary matrix is either symmetric or lower or upper triangular, we arrive at $\det(A) = \det(A^t)$ as claimed. □

Next we show that determinant preserves matrix multiplication.

Theorem 3.11 *Let A and B be two n × n matrices. Then*

$$\det(AB) = \det(A)\det(B). \tag{3.2.23}$$

Proof Let $E_1, \ldots, E_k$ be a sequence of elementary matrices so that (3.2.21) holds for an upper triangular matrix U. There are two cases to be treated separately.

(1) U has a zero row. Then $\det(U) = 0$. Moreover, by the definition of matrix multiplication, we see that UB also has a zero row. Hence $\det(UB) = 0$. On the other hand (3.2.21) leads us to $E_k \cdots E_1 AB = UB$. So

$$\det(E_k)\cdots\det(E_1)\det(A) = \det(U) = 0, \quad \det(E_k)\cdots\det(E_1)\det(AB) = 0.$$

In particular, $\det(A) = 0$ and $\det(AB) = 0$. Thus (3.2.23) is valid.

(2) U has no zero row. Hence $u_{nn} \neq 0$. Using (type 3) row operations if necessary we may assume $u_{in} = 0$ for all $i \leq n-1$. Therefore $u_{n-1,n-1} \neq 0$. Thus, using more (type 3) row operations if necessary, we may eventually assume that U is made diagonal with $u_{11} \neq 0, \ldots, u_{nn} \neq 0$. Using (type 1)

row operations if necessary we may assume $u_{11} = \cdots = u_{nn} = 1$. That is, $U = I$. Thus we get $E_k \cdots E_1 A = I$ and $E_k \cdots E_1 AB = B$. Consequently,

$$\det(E_k) \cdots \det(E_1) \det(A) = 1, \quad \det(E_k) \cdots \det(E_1) \det(AB) = \det(B),$$

which immediately lead us to the anticipated conclusion (3.2.23).

The proof is complete. □

The formula (3.2.23) can be used immediately to derive a few simple but basic conclusions about various matrices.

For example, if $A \in \mathbb{F}(n,n)$ is invertible, then there is some $B \in \mathbb{F}(n,n)$ such that $AB = I_n$. Thus $\det(A) \det(B) = \det(AB) = \det(I_n) = 1$ which implies that $\det(A) \neq 0$. In other words, the condition $\det(A) \neq 0$ is necessary for any $A \in \mathbb{F}(n,n)$ to be invertible. In Section 3.3, we shall show that this condition is also sufficient. As another example, if $A \in \mathbb{R}(n,n)$ is orthogonal, then $AA^t = I_n$. Hence $(\det(A))^2 = \det(A) \det(A^t) = \det(AA^t) = \det(I_n) = 1$. In other words, the determinant of an orthogonal matrix can only take values ± 1. Similarly, if $A \in \mathbb{C}(n,n)$ is unitary, then the condition $AA^\dagger = I_n$ leads us to the conclusion

$$\begin{aligned} |\det(A)|^2 &= \det(A)\overline{\det(A)} \\ &= \det(A)\det(\overline{A}^t) = \det(AA^\dagger) = \det(I_n) = 1 \end{aligned} \quad (3.2.24)$$

(cf. (3.2.36)). That is, the determinant of a unitary matrix is of modulus 1.

Below we show that we can make a cofactor expansion along any column or row to evaluate a determinant.

Theorem 3.12 *Let $A = (a_{ij})$ be an $n \times n$ matrix and $C = (C_{ij})$ its cofactor matrix. Then*

$$\det(A) = \sum_{i=1}^{n} a_{ik} C_{ik}$$

$$= \sum_{j=1}^{n} a_{kj} C_{kj}, \quad k = 1, \ldots, n. \quad (3.2.25)$$

In other words, the determinant of A may be evaluated by a cofactor expansion along any column or any row of A.

Proof We first consider the column expansion case.

We will make induction on $k = 1, \ldots, n$.

If $k = 1$, there is nothing to show.

Assume the statement is true at $k \geq 1$ and $n \geq 2$. Interchanging the kth and $(k+1)$th columns and using the inductive assumption that a cofactor expansion can be made along the kth column of the matrix with the two columns already interchanged, we have

$$-\det(A) = a_{1,k+1}(-1)^{1+k}M_{1,k+1} + \cdots + a_{i,k+1}(-1)^{i+k}M_{i,k+1}$$
$$+ \cdots + a_{n,k+1}(-1)^{n+k}M_{n,k+1}, \tag{3.2.26}$$

which is exactly what was claimed at $k+1$:

$$\det(A) = \sum_{i=1}^{n} a_{i,k+1}(-1)^{i+(k+1)}M_{i,k+1}. \tag{3.2.27}$$

We next consider the row expansion case.

We use $M_{ij}^{A^t}$ to denote the minor of the (i,j)th entry of the matrix A^t ($i, j = 1, \ldots, n$). Applying Theorem 3.10 and Definition 3.3 we have

$$\det(A) = \det(A^t) = \sum_{j=1}^{n} a_{1j}(-1)^{j+1}M_{j1}^{A^t} = \sum_{j=1}^{n} a_{1j}(-1)^{1+j}M_{1j}, \tag{3.2.28}$$

which establishes the legitimacy of the cofactor expansion formula along the first row of A. The validity of the cofactor expansion along an arbitrary row may be proved by induction as done for the column case. □

Assume that $A \in \mathbb{F}(n,n)$ takes a *block* or *boxed upper triangular form*,

$$A = \begin{pmatrix} A_1 & A_3 \\ 0 & A_2 \end{pmatrix}, \tag{3.2.29}$$

where $A_1 \in \mathbb{F}(k,k)$, $A_2 \in \mathbb{F}(l,l)$, $A_3 \in \mathbb{F}(k,l)$, and $k+l = n$. Then we have the useful formula

$$\det(A) = \det(A_1)\det(A_2). \tag{3.2.30}$$

To prove (3.2.30), we use a few permissible row operations to reduce A_1 into an upper triangular form U_1 whose diagonal entries are $u_{11}, \ldots, u_{kk}$ (say). Thus $\det(U_1) = u_{11} \cdots u_{kk}$. Likewise, we may also use a few permissible row operations to reduce A_2 into an upper triangular form U_2 whose diagonal entries are $u_{k+1,k+1}, \ldots, u_{nn}$ (say). Thus $\det(U_2) = u_{k+1,k+1} \cdots u_{nn}$. Now apply the same sequences of permissible row operations on A. The boxed upper triangular form of A allows us to reduce A into the following upper triangular form:

$$U = \begin{pmatrix} U_1 & A_4 \\ 0 & U_2 \end{pmatrix}. \tag{3.2.31}$$

Since the diagonal entries of U are $u_{11}, \ldots, u_{kk}, u_{k+1,k+1}, \ldots, u_{nn}$, we have

$$\det(U) = u_{11} \cdots u_{kk} u_{k+1,k+1} \cdots u_{nn} = \det(U_1)\det(U_2). \tag{3.2.32}$$

Discounting the effects of the permissible row operations on A, A_1, and A_2, we see that (3.2.32) implies (3.2.30).

It is easy to see that if A takes a *block* or *boxed lower triangular form*,

$$A = \begin{pmatrix} A_1 & 0 \\ A_3 & A_2 \end{pmatrix}, \qquad (3.2.33)$$

where $A_1 \in \mathbb{F}(k,k)$, $A_2 \in \mathbb{F}(l,l)$, $A_3 \in \mathbb{F}(l,k)$, and $k+l=n$, then (3.2.30) still holds. Indeed, taking transpose in (3.2.33), we have

$$A^t = \begin{pmatrix} A_1^t & A_3^t \\ 0 & A_2^t \end{pmatrix}, \qquad (3.2.34)$$

which becomes a boxed upper triangular matrix studied earlier. Thus, using Theorem 3.10 and (3.2.30), we obtain

$$\det(A) = \det(A^t) = \det(A_1^t)\det(A_2^t) = \det(A_1)\det(A_2), \qquad (3.2.35)$$

as anticipated.

Exercises

3.2.1   For $A \in \mathbb{C}(n,n)$ use Definition 3.3 to establish the property

$$\overline{\det(A)} = \det(\overline{A}), \qquad (3.2.36)$$

where $\overline{A}$ is the matrix obtained from A by taking complex conjugate for all entries of A. For $A \in \mathbb{F}(n,n)$ $(n \geq 2)$ use Definition 3.3 to show that $\det(A) = 0$ if A contains two identical rows, without assuming $\mathrm{ch}(\mathbb{F}) \neq 2$.

3.2.2   Let $E \in \mathbb{F}(n,n)$ be such that each row and each column of E can only have exactly one nonzero entry which may either be 1 or -1. Show that $\det(E) = \pm 1$.

3.2.3   In $\mathbb{F}(n,n)$, *anti-upper triangular* and *anti-lower triangular* matrices are of the forms

$$\begin{pmatrix} a_{11} & a_{12} & \cdots & a_{1n} \\ \vdots & \vdots & \iddots & 0 \\ \vdots & \iddots & \iddots & \vdots \\ a_{n1} & 0 & \cdots & 0 \end{pmatrix}, \begin{pmatrix} 0 & \cdots & 0 & a_{1n} \\ \vdots & \iddots & \iddots & \vdots \\ 0 & \iddots & \iddots & \vdots \\ a_{n1} & a_{n2} & \cdots & a_{nn} \end{pmatrix}, \qquad (3.2.37)$$

respectively. Establish the formulas to express the determinants of these matrices in terms of the anti-diagonal entries $a_{n1}, \ldots, a_{1n}$.

3.2.4 Show that

$$\det \begin{pmatrix} x & 1 & 1 & 1 \\ 1 & y & 0 & 0 \\ 1 & 0 & z & 0 \\ 1 & 0 & 0 & t \end{pmatrix} = txyz - yz - tz - ty, \quad x,y,z,t \in \mathbb{R}. \quad (3.2.38)$$

3.2.5 Let A and $B \in \mathbb{F}(3,3)$ and assume that the first and second columns of A are same as the first and second columns of B. If $\det(A) = 5$ and $\det(B) = 2$, find $\det(3A - 2B)$ and $\det(3A + 2B)$.

3.2.6 (Extension of Problem 3.2.5) Let A and $B \in \mathbb{F}(n,n)$ such that only the jth columns of them are possibly different. Establish the formula

$$\det(aA + bB) = (a+b)^{n-1}(a\det(A) + b\det(B)), \quad a,b \in \mathbb{F}. \quad (3.2.39)$$

3.2.7 Let $A(t) = (a_{ij}(t)) \in \mathbb{R}(n,n)$ be such that each entry $a_{ij}(t)$ is a differentiable function of $t \in \mathbb{R}$. Establish the differentiation formula

$$\frac{d}{dt}\det(A(t)) = \sum_{i,j=1}^{n} \frac{da_{ij}(t)}{dt} C_{ij}(t), \quad (3.2.40)$$

where $C_{ij}(t)$ is the cofactor of the entry $a_{ij}(t)$, $i,j = 1,\dots,n$, in the matrix $A(t)$.

3.2.8 Let $A = (a_{ij}), B = (b_{ij})$, and $C = (c_{ij}) \in \mathbb{F}(n,n)$ whose differences only occur in their jth columns (j is fixed) such that

$$c_{ij} = a_{ij} + b_{ij}, \quad i = 1,\dots,n. \quad (3.2.41)$$

Establish the formula $\det(C) = \det(A) + \det(B)$ by induction.

3.2.9 Prove the formula

$$\det \begin{pmatrix} x & a_1 & a_2 & \cdots & a_n \\ a_1 & x & a_2 & \cdots & a_n \\ a_1 & a_2 & x & \cdots & a_n \\ \vdots & \vdots & \vdots & \ddots & \vdots \\ a_1 & a_2 & a_3 & \cdots & x \end{pmatrix} = \left(x + \sum_{i=1}^{n} a_i\right) \prod_{i=1}^{n}(x - a_i). \quad (3.2.42)$$

3.2.10 Let $p_1(t),\dots,p_{n+2}(t)$ be $n+2$ polynomials of degrees up to $n \in \mathbb{N}$ and with coefficients in $\mathbb{C}$. Show that for any $n+2$ numbers $c_1,\dots,c_{n+2}$ in $\mathbb{C}$, there holds

$$\det \begin{pmatrix} p_1(c_1) & p_1(c_2) & \cdots & p_1(c_{n+2}) \\ p_2(c_1) & p_2(c_2) & \cdots & p_2(c_{n+2}) \\ \vdots & \vdots & \ddots & \vdots \\ p_{n+2}(c_1) & p_{n+2}(c_2) & \cdots & p_{n+2}(c_{n+2}) \end{pmatrix} = 0. \quad (3.2.43)$$

3.2.11 (Determinant representation of a polynomial) Establish the formula

$$
\det\begin{pmatrix}
x & -1 & 0 & \cdots & 0 & 0 \\
0 & x & -1 & \cdots & 0 & 0 \\
0 & 0 & x & \ddots & 0 & 0 \\
\vdots & \vdots & \vdots & \ddots & \ddots & \vdots \\
0 & 0 & 0 & \cdots & x & -1 \\
a_0 & a_1 & a_2 & \cdots & a_{n-1} & a_n
\end{pmatrix}
$$
$$= a_n x^n + a_{n-1} x^{n-1} + \cdots + a_1 x + a_0. \tag{3.2.44}$$

3.2.12 For $n \geq 2$ establish the formula

$$
\det\begin{pmatrix}
a_1 - \lambda & a_2 & \cdots & a_n \\
a_1 & a_2 - \lambda & \cdots & a_n \\
\vdots & \vdots & \ddots & \vdots \\
a_1 & a_2 & \cdots & a_n - \lambda
\end{pmatrix} = (-1)^n \lambda^{n-1}\left(\lambda - \sum_{i=1}^n a_i\right).
$$
$$\tag{3.2.45}$$

3.2.13 Let $A \in \mathbb{R}(n,n)$ be such that $AA^t = I_n$ and $\det(A) < 0$. Show that $\det(A + I_n) = 0$.

3.2.14 Let $A \in \mathbb{F}(n,n)$ such that the entries of A are either 1 or -1. Show that $\det(A)$ is an even integer when $n \geq 2$.

3.2.15 Let $A = (a_{ij}) \in \mathbb{R}(n,n)$ satisfy the following *diagonally dominant* condition:

$$|a_{ii}| > \sum_{j \neq i} |a_{ij}|, \quad i = 1, \ldots, n. \tag{3.2.46}$$

Show that $\det(A) \neq 0$. (This result is also known as the *Levy–Desplanques theorem*.)

3.2.16 (A refined version of Problem 3.2.15) Let $A = (a_{ij}) \in \mathbb{R}(n,n)$ satisfy the following *positive diagonally dominant condition*:

$$a_{ii} > \sum_{j \neq i} |a_{ij}|, \quad i = 1, \ldots, n. \tag{3.2.47}$$

Show that $\det(A) > 0$. (This result is also known as the *Minkowski theorem*.)

3.2.17 If $A \in \mathbb{F}(n,n)$ is skew-symmetric and n is odd, where $\mathrm{ch}(\mathbb{F}) \neq 2$, then A must be singular. What happens when n is even or $\mathrm{ch}(\mathbb{F}) = 2$?

3.2.18 Let α and $\beta \in \mathbb{F}(1,n)$. Establish the formula

$$\det(I_n - \alpha^t \beta) = 1 - \alpha\beta^t. \tag{3.2.48}$$

3.2.19 Compute the determinant

$$f(x_1,\ldots,x_n) = \det \begin{pmatrix} 100 & x_1 & x_2 & \cdots & x_n \\ x_1 & 1 & 0 & \cdots & 0 \\ x_2 & 0 & 1 & \cdots & 0 \\ \vdots & \vdots & \vdots & \ddots & \vdots \\ x_n & 0 & 0 & \cdots & 1 \end{pmatrix} \qquad (3.2.49)$$

and describe what the equation $f(x_1,\ldots,x_n) = 0$ represents geometrically.

3.3 Adjugate matrices and Cramer's rule

Let $A = (a_{ij})$ be an $n \times n$ $(n \geq 2)$ matrix and $C = (C_{ij})$ be the associated cofactor matrix. For $k,l = 1,\ldots,n$, we may apply Theorem 3.12 to obtain the relations

$$\sum_{i=1}^{n} a_{ik}C_{il} = 0, \quad k \neq l, \qquad (3.3.1)$$

$$\sum_{j=1}^{n} a_{kj}C_{lj} = 0, \quad k \neq l. \qquad (3.3.2)$$

In fact, it is easy to see that the left-hand side of (3.3.1) is the cofactor expansion of the determinant along the lth column of such a matrix that is obtained from A through replacing the lth column by the kth column of A whose value must be zero and the left-hand side of (3.3.2) is the cofactor expansion of the determinant along the lth row of such a matrix that is obtained from A through replacing the lth row by the kth row of A whose value must also be zero.

We can summarize the properties stated in (3.2.25), (3.3.1), and (3.3.2) by the expressions

$$C^t A = \det(A)I_n, \quad A C^t = \det(A)I_n. \qquad (3.3.3)$$

These results motivate the following definition.

Definition 3.13 Let A be an $n \times n$ matrix and C its cofactor matrix. The *adjugate matrix* of A, denoted by $\text{adj}(A)$, is the transpose of the cofactor matrix of A:

$$\text{adj}(A) = C^t. \qquad (3.3.4)$$

Adjugate matrices are sometimes also called *adjoint* or *adjunct* matrices.
As a consequence of this definition and (3.3.3) we have

$$\text{adj}(A)A = A\,\text{adj}(A) = \det(A)I_n, \qquad (3.3.5)$$

which leads immediately to the following conclusion.

Theorem 3.14 *Let A be an $n \times n$ matrix. Then A is invertible if and only if* $\det(A) \neq 0$. *Furthermore, if A is invertible, then A^{-1} may be expressed as*

$$A^{-1} = \frac{1}{\det(A)} \operatorname{adj}(A). \tag{3.3.6}$$

Proof If $\det(A) \neq 0$, from (3.3.5) we arrive at (3.3.6). Conversely, if A^{-1} exists, then from $AA^{-1} = I_n$ and Theorem 3.11 we have $\det(A)\det(A^{-1}) = 1$. Thus $\det(A) \neq 0$. ☐

As an important application, we consider the unique solution of the system

$$Ax = b, \tag{3.3.7}$$

where $A = (a_{ij})$ is an invertible $n \times n$ matrix, $x = (x_1, \dots, x_n)^t$ the vector of unknowns, and $b = (b_1, \dots, b_n)^t$ a given non-homogeneous right-hand-side vector.

In such a situation we can use (3.3.6) to get

$$x = \frac{1}{\det(A)} \operatorname{adj}(A) b. \tag{3.3.8}$$

Therefore, with $\operatorname{adj}(A) = (A_{ij}) = C^t$ (where $C = (C_{ij})$ is the cofactor matrix of A), we may read off to obtain the result

$$x_i = \frac{1}{\det(A)} \sum_{j=1}^{n} A_{ij} b_j = \frac{1}{\det(A)} \sum_{j=1}^{n} b_j C_{ji}$$

$$= \frac{\det(A_i)}{\det(A)}, \quad i = 1, \dots, n, \tag{3.3.9}$$

where A_i is the matrix obtained from A after replacing the ith column of A by the vector b, $i = 1, \dots, n$.

The formulas stated in (3.3.9) are called *Cramer's formulas*. Such a solution method is also called *Cramer's rule*.

Let $A \in \mathbb{F}(m, n)$ and of rank k. Then there are k row vectors of A which are linearly independent. Use B to denote the submatrix of A consisting of those k row vectors. Since B is of rank k we know that there are k column vectors of B which are linearly independent. Use C to denote the submatrix of B consisting of those k column vectors. Then C is a submatrix of A which lies in $\mathbb{F}(k, k)$ and is of rank k. In particular $\det(C) \neq 0$. In other words, we have shown that if A is of rank k then A has a $k \times k$ submatrix of nonzero determinant.

To end this section, we consider a practical problem as an application of determinants: the unique determination of a polynomial by interpolation.

Let $p(t)$ be a polynomial of degree $(n-1) \geq 1$ over the field $\mathbb{F}$ given by

$$p(t) = a_{n-1}t^{n-1} + \cdots + a_1 t + a_0, \tag{3.3.10}$$

and $t_1, \ldots, t_n$ be n points in $\mathbb{F}$ so that $p(t_i) = p_i$ $(i = 1, \ldots, n)$. To ease the illustration to follow, we may assume that n is sufficiently large (say $n \geq 5$). Therefore we have the simultaneous system of equations

$$\begin{cases} a_0 + a_1 t_1 + \cdots + a_{n-2} t_1^{n-2} + a_{n-1} t_1^{n-1} & = & p_1, \\ a_0 + a_1 t_2 + \cdots + a_{n-2} t_2^{n-2} + a_{n-1} t_2^{n-1} & = & p_2, \\ \cdots\cdots\cdots\cdots\cdots\cdots\cdots\cdots\cdots\cdots\cdots\cdots & \cdots & \cdots, \\ a_0 + a_1 t_n + \cdots + a_{n-2} t_n^{n-2} + a_{n-1} t_n^{n-1} & = & p_n, \end{cases} \tag{3.3.11}$$

in the n unknowns $a_0, a_1, \ldots, a_{n-2}, a_{n-1}$, whose coefficient matrix, say A, has the determinant

$$\det(A) = \begin{vmatrix} 1 & t_1 & t_1^2 & \cdots & t_1^{n-2} & t_1^{n-1} \\ 1 & t_2 & t_2^2 & \cdots & t_2^{n-2} & t_2^{n-1} \\ \cdots & \cdots & \cdots & \cdots & \cdots & \cdots \\ 1 & t_n & t_n^2 & \cdots & t_n^{n-2} & t_n^{n-1} \end{vmatrix}. \tag{3.3.12}$$

Adding the $(-t_1)$ multiple of the second last column to the last column,..., the $(-t_1)$ multiple of the second column to the third column, and the $(-t_1)$ multiple of the first column to the second column, we get

$$\det(A) = \begin{vmatrix} 1 & 0 & 0 & \cdots & 0 & 0 \\ 1 & (t_2 - t_1) & t_2(t_2 - t_1) & \cdots & t_2^{n-3}(t_2 - t_1) & t_2^{n-2}(t_2 - t_1) \\ \cdots & \cdots & \cdots & \cdots & \cdots & \cdots \\ 1 & (t_n - t_1) & t_n(t_n - t_1) & \cdots & t_n^{n-3}(t_n - t_1) & t_n^{n-2}(t_n - t_1) \end{vmatrix}$$

$$= \prod_{i=2}^{n}(t_i - t_1) \begin{vmatrix} 1 & t_2 & t_2^2 & \cdots & t_2^{n-3} & t_2^{n-2} \\ 1 & t_3 & t_3^2 & \cdots & t_3^{n-3} & t_3^{n-2} \\ \cdots & \cdots & \cdots & \cdots & \cdots & \cdots \\ 1 & t_n & t_n^2 & \cdots & t_n^{n-3} & t_n^{n-2} \end{vmatrix}. \tag{3.3.13}$$

Continuing the same expansion, we eventually get

$$\det(A) = \left(\prod_{i=2}^{n}(t_i - t_1) \right) \left(\prod_{j=3}^{n}(t_j - t_2) \right) \cdots \left(\prod_{k=n-1}^{n}(t_k - t_{n-2}) \right) (t_n - t_{n-1})$$

$$= \prod_{1 \leq i < j \leq n}(t_j - t_i). \tag{3.3.14}$$

Hence the matrix A is invertible if and only if $t_1, t_2, \ldots, t_n$ are distinct. Under such a condition the coefficients $a_0, a_1, \ldots, a_{n-1}$ are uniquely determined.

The determinant (3.3.12) is called the *Vandermonde determinant*.

If A and $B \in \mathbb{F}(n,n)$ are similar, there is an invertible $C \in \mathbb{F}(n,n)$ such that $A = C^{-1}BC$, which gives us $\det(A) = \det(B)$. That is, similar matrices have the same determinant. In view of this fact and the fact that the matrix representations of a linear mapping from a finite-dimensional vector space into itself with respect to different bases are similar, we may define the *determinant of a linear mapping*, say T, denoted by $\det(T)$, to be the determinant of the matrix representation of T with respect to any basis. In this situation, we see that T is invertible if and only if $\det(T) \neq 0$.

Exercises

3.3.1   Show that $A \in \mathbb{F}(m,n)$ is of rank k if and only if there is a $k \times k$ submatrix of A whose determinant is nonzero and that the determinant of any square submatrix of a larger size (if any) is zero.

3.3.2   Let $A \in \mathbb{R}(n,n)$ be an orthogonal matrix. Prove that $\mathrm{adj}(A) = A^t$ or $-A^t$ depending on the sign of $\det(A)$.

3.3.3   For $A \in \mathbb{F}(n,n)$ $(n \geq 2)$ establish the formula

$$\det(\mathrm{adj}(A)) = (\det(A))^{n-1}. \tag{3.3.15}$$

In particular this implies that A is non-singular if and only if $\mathrm{adj}(A)$ is so.

3.3.4   For $A \in \mathbb{F}(n,n)$ $(n \geq 2)$ prove the rank relations

$$r(\mathrm{adj}(A)) = \begin{cases} n, & r(A) = n, \\ 1, & r(A) = n-1, \\ 0, & r(A) \leq n-2. \end{cases} \tag{3.3.16}$$

3.3.5   Let $A \in \mathbb{F}(n,n)$ be invertible. Show that $\mathrm{adj}(A^{-1}) = (\mathrm{adj}(A))^{-1}$.

3.3.6   For $A \in \mathbb{F}(n,n)$ where $n \geq 3$ show that $\mathrm{adj}(\mathrm{adj}(A)) = (\det(A))^{n-2}A$. What happens when $n = 2$?

3.3.7   For $A \in \mathbb{F}(n,n)$ where $n \geq 2$ show that $\mathrm{adj}(A^t) = (\mathrm{adj}(A))^t$.

3.3.8   Let $A \in \mathbb{R}(n,n)$ satisfy $\mathrm{adj}(A) = A^t$. Prove that $A \neq 0$ if and only if $\det(A) \neq 0$.

3.3.9   For A and $B \in \mathbb{R}(n,n)$ show that $\mathrm{adj}(AB) = \mathrm{adj}(B)\mathrm{adj}(A)$. Thus, if A is idempotent, so is $\mathrm{adj}(A)$, and if A is nilpotent, so is $\mathrm{adj}(A)$.

3.3.10 For $A = (a_{ij}) \in \mathbb{F}(n,n)$ consider the linear system:

$$\begin{cases} a_{11}x_1 + \cdots + a_{1n}x_n & = & 0, \\ \qquad \cdots\cdots\cdots\cdots\cdots & \cdots & \cdots \\ a_{n1}x_1 + \cdots + a_{nn}x_n & = & 0. \end{cases} \tag{3.3.17}$$

Show that if A is singular but $\mathrm{adj}(A) \neq 0$ then the space of solutions of (3.3.17) is spanned by any nonzero column vector of $\mathrm{adj}(A)$.

3.3.11 Use Cramer's rule and the Vandermonde determinant to find the quadratic polynomial $p(t)$ with coefficients in $\mathbb{C}$ whose values at $t = 1, 1+i, -3$ are $-2, 0, 5$, respectively.

3.3.12 Let $p(t)$ be a polynomial of degree $n-1$ given in (3.3.10). Use Cramer's rule and the Vandermonde determinant to prove that $p(t)$ cannot have n distinct zeros unless it is the zero polynomial.

3.3.13 Let $A \in \mathbb{F}(m,n)$, $B \in \mathbb{F}(n,m)$, and $m > n$. Show that $\det(AB) = 0$.

3.3.14 For $U = \mathbb{F}(n,n)$ define $T \in L(U)$ by $T(A) = AB - BA$ for $A \in U$, where $B \in U$ is fixed. Show that for such a linear mapping T we have $\det(T) = 0$ no matter how B is chosen.

3.4 Characteristic polynomials and Cayley–Hamilton theorem

We first consider the concrete case of matrices.

Let $A = (a_{ij})$ be an $n \times n$ matrix over the field $\mathbb{F}$. We first consider the linear mapping $T_A : \mathbb{F}^n \to \mathbb{F}^n$ induced from A defined by

$$T_A(x) = Ax, \quad x = \begin{pmatrix} x_1 \\ \vdots \\ x_n \end{pmatrix} \in \mathbb{F}^n. \tag{3.4.1}$$

Recall that an eigenvalue of T_A is a scalar $\lambda \in \mathbb{F}$ such that the null-space

$$E_\lambda = N(T_A - \lambda I) = \{x \in \mathbb{F}^n \mid T_A(x) = \lambda x\}, \tag{3.4.2}$$

where $I \in L(\mathbb{F}^n)$ is the identity mapping, is nontrivial, and E_λ is called the eigenspace of T_A, whose nonzero vectors are called eigenvectors. The eigenvalues, eigenspaces, and eigenvectors of T_A are also often simply referred to as those of the matrix A. The purpose of this section is to show how to use determinant as a tool to find the eigenvalues of A.

If λ is an eigenvalue of A and x an associated eigenvector, then $Ax = \lambda x$. In other words, x is a nonzero solution of the equation $(\lambda I_n - A)x = 0$. Hence the matrix $\lambda I_n - A$ is singular. Therefore λ satisfies

$$\det(\lambda I_n - A) = 0. \tag{3.4.3}$$

Of course the converse is true as well: If λ satisfies (3.4.3) then $(\lambda I_n - A)x = 0$ has a nontrivial solution, which indicates that λ is an eigenvalue of A. Consequently the eigenvalues of A are the roots of the function

$$p_A(\lambda) = \det(\lambda I_n - A), \tag{3.4.4}$$

which is seen to be a polynomial of degree n following the cofactor expansion formula (3.2.2) in Definition 3.3. The polynomial $p_A(\lambda)$ defined in (3.4.4) is called the *characteristic polynomial* associated with the matrix A, whose roots are called the *characteristic roots* of A. So the eigenvalues of A are the characteristic roots of A. In particular, A can have at most n distinct eigenvalues.

Theorem 3.15 *Let* $A = (a_{ij})$ *be an* $n \times n$ $(n \geq 2)$ *matrix and* $p_A(\lambda)$ *its characteristic polynomial. Then* $p_A(\lambda)$ *is of the form*

$$p_A(\lambda) = \lambda^n - \mathrm{Tr}(A)\lambda^{n-1} + \cdots + (-1)^n \det(A). \qquad (3.4.5)$$

Proof   Write $p_A(\lambda) = a_n\lambda^n + a_{n-1}\lambda^{n-1} + \cdots + a_0$. Then

$$a_0 = p_A(0) = \det(-A) = (-1)^n \det(A) \qquad (3.4.6)$$

as asserted. Besides, using Definition 3.3 and induction, we see that the two leading-degree terms in $p_A(\lambda)$ containing λ^n and λ^{n-1} can only appear in the product of the entry $\lambda - a_{11}$ and the cofactor $C_{11}^{\lambda I_n - A}$ of the entry $\lambda - a_{11}$ in the matrix $\lambda I_n - A$. Let A_{n-1} be the submatrix of A obtained by deleting the row and column vectors of A occupied by the entry a_{11}. Then $C_{11}^{\lambda I_n - A} = \det(\lambda I_{n-1} - A_{n-1})$ whose two leading-degree terms containing λ^{n-1} and λ^{n-2} can only appear in the product of $\lambda - a_{22}$ and its cofactor in the matrix $\lambda I_{n-1} - A_{n-1}$. Carrying out this process to the end we see that the two leading-degree terms in $p_A(\lambda)$ can only appear in the product

$$(\lambda - a_{11}) \cdots (\lambda - a_{nn}), \qquad (3.4.7)$$

which may be read off to give us the results

$$\lambda^n, \quad -(a_{11} + \cdots + a_{nn})\lambda^{n-1}. \qquad (3.4.8)$$

This completes the proof.   □

If $A \in \mathbb{C}(n,n)$ we may also establish Theorem 3.15 by means of calculus. In fact, we have

$$\frac{p_A(\lambda)}{\lambda^n} = \det\left(I_n - \frac{1}{\lambda}A\right). \qquad (3.4.9)$$

Thus, letting $\lambda \to \infty$ in (3.4.9), we obtain $a_n = \det(I_n) = 1$. Moreover, (3.4.9) also gives us the expression

$$a_n + a_{n-1}\frac{1}{\lambda} + \cdots + a_0\frac{1}{\lambda^n} = \det\left(I_n - \frac{1}{\lambda}A\right). \qquad (3.4.10)$$

Thus, replacing $\frac{1}{\lambda}$ by t, we get

$$
\begin{aligned}
q(t) &\equiv a_n + a_{n-1}t + \cdots + a_0 t^n \\
&= \begin{vmatrix}
1 - ta_{11} & -ta_{12} & \cdots & -ta_{1n} \\
-ta_{21} & 1 - ta_{22} & \cdots & -ta_{2n} \\
\cdots & \cdots & \cdots & \cdots \\
-ta_{n1} & -ta_{n2} & \cdots & 1 - ta_{nn}
\end{vmatrix}.
\end{aligned} \tag{3.4.11}
$$

Therefore

$$
a_{n-1} = q'(0) = -\sum_{i=1}^{n} a_{ii} \tag{3.4.12}
$$

as before.

We now consider the abstract case.

Let U be an n-dimensional vector space over a field $\mathbb{F}$ and $T \in L(U)$. Assume that $u \in U$ is an eigenvector of T associated to the eigenvalue $\lambda \in \mathbb{F}$. Given a basis $\mathcal{U} = \{u_1, \ldots, u_n\}$, we express u as

$$
u = x_1 u_1 + \cdots + x_n u_n, \quad x = (x_1, \ldots, x_n)^t \in \mathbb{F}^n. \tag{3.4.13}
$$

Let $A = (a_{ij}) \in \mathbb{F}(n,n)$ be the matrix representation of T with respect to the basis $\mathcal{U}$ so that

$$
T(u_j) = \sum_{i=1}^{n} a_{ij} u_i, \quad j = 1, \ldots, n. \tag{3.4.14}
$$

Then the relation $T(u) = \lambda u$ leads to

$$
\sum_{j=1}^{n} a_{ij} x_j = \lambda x_i, \quad i = 1, \ldots, n, \tag{3.4.15}
$$

which is exactly the matrix equation $Ax = \lambda x$ already discussed. Hence λ may be obtained from solving for the roots of the characteristic equation (3.4.3).

Let $\mathcal{V} = \{v_1, \ldots, v_n\}$ be another basis of U and $B = (b_{ij}) \in \mathbb{F}(n,n)$ be the matrix representation of T with respect to the basis $\mathcal{V}$ so that

$$
T(v_j) = \sum_{i=1}^{n} b_{ij} v_i, \quad i = 1, \ldots, n. \tag{3.4.16}
$$

Since $\mathcal{U}$ and $\mathcal{V}$ are two bases of U, there is an invertible matrix $C = (c_{ij}) \in \mathbb{F}(n,n)$ called the basis transition matrix such that

$$
v_j = \sum_{i=1}^{n} c_{ij} u_i, \quad j = 1, \ldots, n. \tag{3.4.17}
$$

Following the study of Chapter 2, we know that $A, B,$ and C are related through the similarity relation $A = C^{-1}BC$. Hence we have

$$
\begin{aligned}
p_A(\lambda) = \det(\lambda I_n - A) &= \det(\lambda I_n - C^{-1}BC) \\
&= \det(C^{-1}[\lambda I_n - B]C) \\
&= \det(\lambda I_n - B) = p_B(\lambda).
\end{aligned}
\tag{3.4.18}
$$

That is, two similar matrices have the same characteristic polynomial. Thus we may use $p_A(\lambda)$ to define the *characteristic polynomial of linear mapping* $T \in L(U)$, rewritten as $p_T(\lambda)$, where A is the matrix representation of T with respect to any given basis of U, since such a polynomial is independent of the choice of the basis.

The following theorem, known as the *Cayley–Hamilton Theorem*, is of fundamental importance in linear algebra.

Theorem 3.16 *Let $A \in \mathbb{C}(n,n)$ and $p_A(\lambda)$ be its characteristic polynomial. Then*

$$
p_A(A) = 0.
\tag{3.4.19}
$$

Proof   We split the proof into two situations.

First we assume that A has n distinct eigenvalues, say $\lambda_1, \dots, \lambda_n$. We can write $p_A(\lambda)$ as

$$
p_A(\lambda) = \prod_{i=1}^{n} (\lambda - \lambda_i).
\tag{3.4.20}
$$

Let $u_1, \dots, u_n \in \mathbb{C}^n$ be the eigenvectors associated to the eigenvalues $\lambda_1, \dots, \lambda_n$, respectively, which form a basis of $\mathbb{C}^n$. Then

$$
\begin{aligned}
p_A(A)u_i &= \left(\prod_{j=1}^{n} (A - \lambda_j I_n) \right) u_i \\
&= \left(\prod_{j \neq i} (A - \lambda_j I_n) \right) (A - \lambda_i I_n) u_i = 0,
\end{aligned}
\tag{3.4.21}
$$

for any $i = 1, \dots, n$. This proves $p_A(A) = 0$.

Next we consider the general situation when A may have multiple eigenvalues. We begin by showing that A may be approximated by matrices of n distinct eigenvalues. We show this by induction on n. When $n = 1$, there is nothing to do. Assume the statement is true at $n - 1 \geq 1$. We now proceed with $n \geq 2$.

In view of Theorem 3.2 there exists an eigenvalue of A, say λ_1. Let u_1 be an associated eigenvector. Extend $\{u_1\}$ to get a basis for $\mathbb{C}^n$, say $\{u_1, u_2, \ldots, u_n\}$. Then the linear mapping defined by (3.4.1) satisfies

$$T_A(u_j) = Au_j = \sum_{i=1}^{n} b_{ij} u_i, \ i = 1, \ldots, n;$$

$$b_{11} = \lambda_1, b_{i1} = 0, i = 2, \ldots, n. \tag{3.4.22}$$

That is, the matrix $B = (b_{ij})$ of T_A with respect to the basis $\{u_1, \ldots, u_n\}$ is of the form

$$B = \begin{pmatrix} \lambda_1 & b_0 \\ 0 & B_0 \end{pmatrix}, \tag{3.4.23}$$

where $B_0 \in \mathbb{C}(n-1, n-1)$ and $b_0 \in \mathbb{C}^{n-1}$ is a row vector. By the inductive assumption, for any $\varepsilon > 0$, there is some $C_0 \in \mathbb{C}(n-1, n-1)$ such that $\|B_0 - C_0\| < \varepsilon$ and C_0 has $n-1$ distinct eigenvalues, say $\lambda_2, \ldots, \lambda_n$, where and in the sequel we use $\| \cdot \|$ to denote any norm of the space of square matrices whenever there is no risk of confusion. Now set

$$C = \begin{pmatrix} \lambda_1 + \delta & b_0 \\ 0 & C_0 \end{pmatrix}. \tag{3.4.24}$$

It is clear that the eigenvalues of C are $\lambda_1 + \delta, \lambda_2, \ldots, \lambda_n$, which are distinct when $\delta > 0$ is small enough. Of course there is a constant $K > 0$ depending on n only such that $\|B - C\| < K\varepsilon$ when $\delta > 0$ is sufficiently small.

On the other hand, if we use the notation $u_j = (u_{1j}, \ldots, u_{nj})^t, j = 1, \ldots, n$, then we have

$$u_j = \sum_{i=1}^{n} u_{ij} e_i, \quad j = 1, \ldots, n. \tag{3.4.25}$$

Thus, with $U = (u_{ij})$, we obtain the relation $A = U^{-1} B U$. Therefore

$$\|A - U^{-1} C U\| = \|U^{-1}(B - C) U\| \leq \|U^{-1}\| \|U\| K \varepsilon. \tag{3.4.26}$$

Of course $U^{-1} C U$ has the same n distinct eigenvalues as the matrix C. Hence the asserted approximation property in the situation of $n \times n$ matrices is established.

Finally, let $\{A^{(l)}\}$ be a sequence in $\mathbb{C}(n, n)$ such that $A^{(l)} \to A$ as $l \to \infty$ and each $A^{(l)}$ has n distinct eigenvalues ($l = 1, 2, \ldots$). We have already shown that $p_{A^{(l)}}(A^{(l)}) = 0$ ($l = 1, 2, \ldots$). Consequently,

$$p_A(A) = \lim_{l \to \infty} p_{A^{(l)}}(A^{(l)}) = 0. \tag{3.4.27}$$

The proof of the theorem is now complete. $\qquad\qquad\qquad\qquad\qquad\square$

Of course Theorem 3.16 is valid for $A \in \mathbb{F}(n,n)$ whenever $\mathbb{F}$ is a subfield of $\mathbb{C}$, although A may not have an eigenvalue in $\mathbb{F}$. Important examples include $\mathbb{F} = \mathbb{Q}$ and $\mathbb{F} = \mathbb{R}$.

We have seen that the proof of Theorem 3.16 relies on the assumption of the existence of an eigenvalue which in general is only valid when the underlying field is $\mathbb{C}$. In fact, however, the theorem holds universally over any field. Below we give a proof of it without assuming $\mathbb{F} = \mathbb{C}$, which is of independent interest and importance.

To proceed, we need to recall and reexamine the definition of polynomials in the most general terms. Given a field $\mathbb{F}$, a polynomial p over $\mathbb{F}$ is an expression of the form

$$p(t) = a_0 + a_1 t + \cdots + a_n t^n, \quad a_0, a_1, \ldots, a_n \in \mathbb{F}, \tag{3.4.28}$$

where $a_0, a_1, \ldots, a_n$ are called the coefficients of p and the variable t is a formal symbol which "generates" the formal symbols $t^0 = 1, t^1 = t, \ldots, t^n = (t)^n$ (called the powers of t), which "participate" in all the algebraic operations following the usual associative, commutative, and distributive laws as if t were an $\mathbb{F}$-valued parameter or variable. For example, $(at^i)(bt^j) = abt^{i+j}$ for any $a, b \in \mathbb{F}$ and $i, j \in \mathbb{N}$. The set of all polynomials with coefficients in $\mathbb{F}$ and in the variable t is denoted by $\mathcal{P}$ which is a vector space over $\mathbb{F}$, with addition and scalar multiplication defined in the usual ways, whose zero element is any polynomial with zero coefficients. In other words, two polynomials in $\mathcal{P}$ are considered equal if and only if the coefficients of the same powers of t in the two polynomials all agree.

In the rest of this section, we shall assume that the variable of any polynomial we consider is such a formal symbol. It is clear that this assumption does not affect the computation we perform.

Let $A \in \mathbb{F}(n,n)$ and λ be a (formal) variable just introduced. Then we have

$$(\lambda I_n - A)\mathrm{adj}(\lambda I_n - A) = \det(\lambda I_n - A)I_n, \tag{3.4.29}$$

whose right-hand side is simply

$$\det(\lambda I_n - A)I_n = (\lambda^n + a_{n-1}\lambda^{n-1} + \cdots + a_1\lambda + a_0)I_n. \tag{3.4.30}$$

On the other hand, we may expand the left-hand side of (3.4.29) into the form

$$\begin{aligned}
&(\lambda I_n - A)\mathrm{adj}(\lambda I_n - A) \\
&= (\lambda I_n - A)(A_{n-1}\lambda^{n-1} + \cdots + A_1\lambda + A_0) \\
&= A_{n-1}\lambda^n + (A_{n-2} - AA_{n-1})\lambda^{n-1} + (A_0 - AA_1)\lambda - AA_0,
\end{aligned} \tag{3.4.31}$$

where $A_0, A_1, \ldots, A_{n-1} \in \mathbb{F}(n,n)$. Comparing the like powers of λ in (3.4.30) and (3.4.31) we get the relations

$$A_{n-1} = I_n, \; A_{n-2} - AA_{n-1} = a_{n-1}I_n, \; \ldots,$$
$$A_0 - AA_1 = a_1 I_n, \; -AA_0 = a_0 I_n. \tag{3.4.32}$$

Multiplying on the left the first relation in (3.4.32) by A^n, the second by $A^{n-1}, \ldots$, and the second last by A, and then summing up the results, we obtain

$$0 = A^n + a_{n-1}A^{n-1} + \cdots + a_1 A + a_0 I_n = p_A(A), \tag{3.4.33}$$

as anticipated.

Let U be an n-dimensional vector space over the field $\mathbb{F}$ and $T \in L(U)$. Given a basis $\mathcal{U} = \{u_1, \ldots, u_n\}$ let $A \in \mathbb{F}(n,n)$ be the matrix that represents T with respect to $\mathcal{U}$. Then we know that A^i represents T^i with respect to $\mathcal{U}$ for any $i = 1, 2 \ldots$. As a consequence, for any polynomial $p(t)$ with coefficients in $\mathbb{F}$, the matrix $p(A)$ represents $p(T)$ with respect to $\mathcal{U}$. Therefore the Cayley–Hamilton Theorem may be restated in terms of linear mappings as follows.

Theorem 3.17 *Let U be finite-dimensional vector space over an arbitrary field. Any $T \in L(U)$ is trivialized or annihilated by its characteristic polynomial $p_T(\lambda)$. That is,*

$$p_T(T) = 0. \tag{3.4.34}$$

For $A \in \mathbb{F}(n,n)$ $(n \geq 2)$ let $p_A(\lambda) = \lambda^n + a_{n-1}\lambda^{n-1} + \cdots + a_1 \lambda + a_0$ be the characteristic polynomial of A. Theorem 3.15 gives us $a_0 = (-1)^n \det(A)$. Inserting this result into the equation $p_A(A) = 0$ we have

$$A(A^{n-1} + a_{n-1}A^{n-2} + \cdots + a_1 I_n) = (-1)^{n+1} \det(A) I_n, \tag{3.4.35}$$

which leads us again to the conclusion that A is invertible whenever $\det(A) \neq 0$. Furthermore, in this situation, the relation (3.4.35) implies

$$A^{-1} = \frac{(-1)^{n+1}}{\det(A)} (A^{n-1} + a_{n-1}A^{n-2} + \cdots + a_1 I_n), \tag{3.4.36}$$

or alternatively,

$$\text{adj}(A) = (-1)^{n+1}(A^{n-1} + a_{n-1}A^{n-2} + \cdots + a_1 I_n), \quad \det(A) \neq 0. \tag{3.4.37}$$

We leave it as an exercise to show that the condition $\det(A) \neq 0$ above for (3.4.37) to hold is not necessary and can thus be dropped.

Exercises

3.4.1   Let $p_A(\lambda)$ be the characteristic polynomial of a matrix $A = (a_{ij}) \in \mathbb{F}(n,n)$, where $n \geq 2$ and $\mathbb{F} = \mathbb{R}$ or $\mathbb{C}$. Show that in $p_A(\lambda) = \lambda^n + \cdots + a_1\lambda + a_0$ we have

$$a_1 = (-1)^{n-1} \sum_{i=1}^{n} C_{ii}, \qquad (3.4.38)$$

where C_{ii} is the cofactor of the entry a_{ii} of the matrix A $(i = 1, \ldots, n)$.

3.4.2   Consider the subset $\mathcal{D}$ of $\mathbb{C}(n,n)$ defined by

$$\mathcal{D} = \{A \in \mathbb{C}(n,n) \,|\, A \text{ has } n \text{ distinct eigenvalues}\}. \qquad (3.4.39)$$

Prove that $\mathcal{D}$ is open in $\mathbb{C}(n,n)$.

3.4.3   Let $A \in \mathbb{F}(2,2)$ be given by

$$A = \begin{pmatrix} a & b \\ c & d \end{pmatrix}. \qquad (3.4.40)$$

Find the characteristic polynomial of A. Assume $\det(A) = ad - bc \neq 0$ and use (3.4.36) to derive the formula that gives A^{-1}.

3.4.4   Show that (3.4.37) is true for any $A \in \mathbb{F}(n,n)$. That is, the condition $\det(A) \neq 0$ in (3.4.37) may actually be removed.

3.4.5   Let U be an n-dimensional vector space over $\mathbb{F}$ and $T \in L(U)$ is nilpotent of degree n. What is the characteristic polynomial of T?

3.4.6   Let D be the derivative operator over $\mathcal{P}_n$ $(n \geq 1)$. For the map $T : \mathcal{P}_n \to \mathcal{P}_n$ given by

$$T(p(t)) = D^2 p(t) + 2t D p(t) - p(t), \quad p(t) \in \mathcal{P}_n, \qquad (3.4.41)$$

compute $\det(T)$.

3.4.7   Let U be an n-dimensional vector space over a field $\mathbb{F}$ and $T \in L(U)$ have n distinct eigenvalues, $\lambda_1, \ldots, \lambda_n \in \mathbb{F}$. Prove the formulas

$$\det(T) = \lambda_1 \cdots \lambda_n, \quad p_T(\lambda) = (\lambda - \lambda_1) \cdots (\lambda - \lambda_n). \qquad (3.4.42)$$

3.4.8   (Extension of Problem 3.4.7) Let U be a finite-dimensional vector space over a field $\mathbb{F}$ and $T \in L(U)$. Assume that T is reducible over its eigenspaces. That is, if $\lambda_1, \ldots, \lambda_k \in \mathbb{F}$ are all the distinct eigenvalues of T and $E_{\lambda_1}, \ldots, E_{\lambda_k}$ the correspondingly associated eigenspaces, then

$$U = E_{\lambda_1} \oplus \cdots \oplus E_{\lambda_k}. \qquad (3.4.43)$$

Establish the formulas

$$\det(T) = \lambda_1^{m_1} \cdots \lambda_k^{m_k}, \quad p_T(\lambda) = (\lambda - \lambda_1)^{m_1} \cdots (\lambda - \lambda_k)^{m_k}, \quad (3.4.44)$$

where $m_1, \ldots, m_k$ are the geometric multiplicities of the eigenvalues $\lambda_1, \ldots, \lambda_k$, respectively.

3.4.9 For A and $B \in \mathbb{F}(n,n)$, prove that the characteristic polynomials of AB and BA are identical. That is,

$$p_{AB}(\lambda) = p_{BA}(\lambda). \tag{3.4.45}$$

3.4.10 Let $\mathbb{F}$ be a field and α and $\beta \in \mathbb{F}(1,n)$. Find the characteristic polynomial of the matrix $\alpha^t \beta \in \mathbb{F}(n,n)$.

3.4.11 Consider the matrix

$$A = \begin{pmatrix} 1 & 2 & 0 \\ 0 & 2 & 0 \\ -2 & -1 & -1 \end{pmatrix}. \tag{3.4.46}$$

(a) Use the characteristic polynomial of A and the Cayley–Hamilton theorem to find A^{-1}.

(b) Use the characteristic polynomial of A and the Cayley–Hamilton theorem to find A^{10}.

4

Scalar Products

In this chapter, we consider vector spaces over a field which is either $\mathbb{R}$ or $\mathbb{C}$. We shall start from the most general situation of scalar products. We then consider the situations when scalar products are nondegenerate and positive definite, respectively.

4.1 Scalar products and basic properties

In this section, we use $\mathbb{F}$ to denote the field $\mathbb{R}$ or $\mathbb{C}$.

Definition 4.1 Let U be a vector space over $\mathbb{F}$. A *scalar product* over U is defined to be a bilinear symmetric function $f : U \times U \to \mathbb{F}$, written simply as $(u,v) \equiv f(u,v)$, $u,v \in U$. In other words the following properties hold:

(1) (Symmetry) $(u,v) = (v,u) \in \mathbb{F}$ for $u,v \in U$.
(2) (Additivity) $(u+v,w) = (u,w) + (v,w)$ for $u,v,w \in U$.
(3) (Homogeneity) $(au,v) = a(u,v)$ for $a \in \mathbb{F}$ and $u,v \in U$.

We say that u and $v \in U$ are mutually *perpendicular* or *orthogonal* to each other, written as $u \perp v$, if $(u,v) = 0$. More generally, for any non-empty subset S of U, we use the notation

$$S^{\perp} = \{u \in U \mid (u,v) = 0 \text{ for any } v \in S\}. \tag{4.1.1}$$

For $u \in U$, we say that u is a *null vector*, if $(u,u) = 0$.

It is obvious that $S^{\perp}$ is a subspace of U for any non-empty subset S of U. Moreover $\{0\}^{\perp} = U$. Furthermore, it is easy to show that, if the vectors $u_1,\dots,u_k$ are mutually perpendicular and not null, then they are linearly independent.

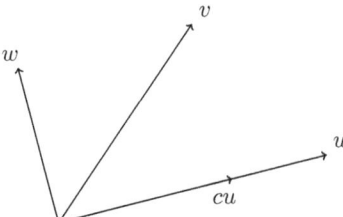

Figure 4.1 The orthogonal resolution of a vector v as the sum of its orthogonal projection along a non-null vector u and a vector w perpendicular to u.

Let u and $v \in U$ so that u is not null. Then, we can resolve v into the sum of two mutually perpendicular vectors, one in Span$\{u\}$, say cu for some scalar c, and one in Span$\{u\}^\perp$, say w. In fact, rewrite v as $v = w + cu$ and require $(u, w) = 0$. We obtain the unique solution $c = (u, v)/(u, u)$. In summary, we have obtained the orthogonal decomposition

$$v = w + \frac{(u, v)}{(u, u)} u, \quad w = v - \frac{(u, v)}{(u, u)} u \in \text{Span}\{u\}^\perp. \tag{4.1.2}$$

See Figure 4.1 for an illustration.

As the first application of the decomposition (4.1.2), we state the following:

Theorem 4.2 *Let U be a finite-dimensional vector space equipped with a scalar product $(\cdot, \cdot)$ and set $U_0 = U^\perp$. Any basis of U_0 can be extended to become an orthogonal basis of U. In other words any two vectors in such a basis of U are mutually perpendicular.*

Proof   If $U_0 = U$ there is nothing to show. Below we assume $U_0 \neq U$.

If $U_0 = \{0\}$ we may start from any basis of U. If $U_0 \neq \{0\}$ let $\{u_1, \ldots, u_k\}$ be a basis of U_0 and extend it to get a basis of U, say $\{u_1, \ldots, u_k, v_1, \ldots, v_l\}$. That is, $U = U_0 \oplus V$, where

$$V = \text{Span}\{v_1, \ldots, v_l\}. \tag{4.1.3}$$

If $(v_1, v_1) = 0$ then there is some v_i $(i \geq 2)$ such that $(v_1, v_i) \neq 0$ otherwise $v_1 \in U_0$ which is false. Thus, without loss of generality, we may assume $(v_1, v_2) \neq 0$ and consider $\{v_2, v_1, \ldots, v_l\}$ instead if $(v_2, v_2) \neq 0$ otherwise we may consider $\{v_1 + v_2, v_2, \ldots, v_l\}$ as a basis of V because now $(v_1 + v_2, v_1 + v_2) = 2(v_1, v_2) \neq 0$. So we have seen that we may assume $(v_1, v_1) \neq 0$ to start with after renaming the basis vectors $\{v_1, \ldots, v_l\}$ of V if necessary. Now let $w_1 = v_1$ and set

$$w_i = v_i - \frac{(w_1, v_i)}{(w_1, w_1)} w_1, \quad i = 2, \ldots, l. \tag{4.1.4}$$

Then $w_i \neq 0$ since v_1, v_i are linearly independent for all $i = 2, \ldots, l$. It is clear that $w_i \perp w_1$ $(i = 2, \ldots, l)$. If $(w_i, w_i) \neq 0$ for some $i = 2, \ldots, l$, we may assume $i = 2$ after renaming the basis vectors $\{v_1, \ldots, v_l\}$ of V if necessary. If $(w_i, w_i) = 0$ for all $i = 2, \ldots, l$, there must be some $j \neq i, i, j = 2, \ldots, l$, such that $(w_i, w_j) \neq 0$, otherwise $w_i \in U_0$ for $i = 2, \ldots, l$, which is false. Without loss of generality, we may assume $(w_2, w_3) \neq 0$ and consider

$$w_2 + w_3 = (v_2 + v_3) - \frac{(w_1, v_2 + v_3)}{(w_1, w_1)} w_1. \tag{4.1.5}$$

It is clear that $(w_2 + w_3, w_2 + w_3) = 2(w_2, w_3) \neq 0$ and $(w_2 + w_3) \perp w_1$. Since $\{v_1, v_2 + v_3, v_3, \ldots, v_l\}$ is also a basis of V, the above procedure indicates that we may rename the basis vectors $\{v_1, \ldots, v_l\}$ of V if necessary so that we obtain

$$w_2 = v_2 - \frac{(w_1, v_2)}{(w_1, w_1)} w_1, \tag{4.1.6}$$

which satisfies $(w_2, w_2) \neq 0$. Of course $\mathrm{Span}\{v_1, \ldots, v_l\} = \mathrm{Span}\{w_1, w_2, v_3, \ldots, v_l\}$. Now set

$$w_i = v_i - \frac{(w_2, v_i)}{(w_2, w_2)} w_2 - \frac{(w_1, v_i)}{(w_1, w_1)} w_1, \quad i = 3, \ldots, l. \tag{4.1.7}$$

Then $w_i \perp w_2$ and $w_i \perp w_1$ for $i = 3, \ldots, l$. If $(w_i, w_i) \neq 0$ for some $i = 3, \ldots, l$, by renaming $\{v_1, \ldots, v_l\}$ if necessary, we may assume $(w_3, w_3) \neq 0$. If $(w_i, w_i) = 0$ for all $i = 3, \ldots, l$, then there is some $i = 4, \ldots, l$ such that $(w_3, w_i) \neq 0$. We may assume $(w_3, w_4) \neq 0$. Thus $(w_3 + w_4, w_3 + w_4) = 2(w_3, w_4) \neq 0$ and $(w_3 + w_4) \perp w_1, (w_3 + w_4) \perp w_2$. Of course, $\{v_1, v_2, v_3 + v_4, v_4 \ldots, v_l\}$ is also a basis of V. Thus we see that we may rename the basis vectors $\{v_1, \ldots, v_l\}$ of V so that we obtain

$$w_3 = v_3 - \frac{(w_2, v_3)}{(w_2, w_2)} w_2 - \frac{(w_1, v_3)}{(w_1, w_1)} w_1, \tag{4.1.8}$$

which again satisfies $(w_3, w_3) \neq 0$ and

$$\mathrm{Span}\{v_1, \ldots, v_l\} = \mathrm{Span}\{w_1, w_2, w_3, v_4, \ldots, v_l\}. \tag{4.1.9}$$

Therefore, by renaming the basis vectors $\{v_1, \ldots, v_l\}$ if necessary, we will be able to carry the above procedure out and obtain a new set of vectors $\{w_1, \ldots, w_l\}$ given by

$$w_1 = v_1, \quad w_i = v_i - \sum_{j=1}^{i-1} \frac{(w_j, v_i)}{(w_j, w_j)} w_j, \quad i = 2, \ldots, l, \tag{4.1.10}$$

and having the properties

$$
\begin{aligned}
(w_i, w_i) &\neq 0, && i = 1, \ldots, l, \\
(w_i, w_j) &= 0, && i \neq j, i, j = 1, \ldots, l, && (4.1.11) \\
\mathrm{Span}\{w_1, \ldots, w_l\} &= \mathrm{Span}\{v_1, \ldots, v_l\}.
\end{aligned}
$$

In other words, $\{u_1, \ldots, u_k, w_1, \ldots, w_l\}$ is seen to be an orthogonal basis of U.

$\square$

The method described in the proof of Theorem 4.2, especially the scheme given by the formulas in (4.1.10)–(4.1.11), is known as the *Gram–Schmidt procedure* for *basis orthogonalization*.

In the rest of this section, we assume $\mathbb{F} = \mathbb{R}$.

If $U_0 = U^\perp$ is a proper subspace of U, that is, $U_0 \neq \{0\}$ and $U_0 \neq U$, we have seen from Theorem 4.2 that we may express an orthogonal basis of U by (say)

$$
\{u_1, \ldots, u_{n_0}, v_1, \ldots, v_{n_+}, w_1, \ldots, w_{n_-}\}, \qquad (4.1.12)
$$

so that $\{u_1, \ldots, u_{n_0}\}$ is a basis of U_0 and that (if any)

$$
(v_i, v_i) > 0, \quad i = 1, \ldots, n_+, \qquad (w_i, w_i) < 0, \quad i = 1, \ldots, n_-. \qquad (4.1.13)
$$

It is clear that, with

$$
U_+ = \mathrm{Span}\{v_1, \ldots, v_{n_+}\}, \quad U_- = \mathrm{Span}\{w_1, \ldots, w_{n_-}\}, \qquad (4.1.14)
$$

we have the following elegant orthogonal subspace decomposition:

$$
U = U_0 \oplus U_+ \oplus U_-, \quad \dim(U_0) = n_0, \dim(U_+) = n_+, \dim(U_-) = n_-. \quad (4.1.15)
$$

It is interesting that the integers n_0, n_+, and n_- are independent of the choice of an orthogonal basis. Such a statement is also known as the *Sylvester theorem*.

In fact, it is obvious that n_0 is independent of the choice of an orthogonal basis since n_0 is the dimension of $U_0 = U^\perp$. Assume that

$$
\{\tilde{u}_1, \ldots, \tilde{u}_{n_0}, \tilde{v}_1, \ldots, \tilde{v}_{m_+}, \tilde{w}_1, \ldots, \tilde{w}_{m_-}\} \qquad (4.1.16)
$$

is another orthogonal basis of U so that $\{\tilde{u}_1, \ldots, \tilde{u}_{n_0}\}$ is a basis of U_0 and that (if any)

$$
(\tilde{v}_i, \tilde{v}_i) > 0, \quad i = 1, \ldots, m_+, \qquad (\tilde{w}_i, \tilde{w}_i) < 0, \quad i = 1, \ldots, m_-. \qquad (4.1.17)
$$

To proceed, we assume $n_+ \geq m_+$ for definiteness and we need to establish $n_+ \leq m_+$. For this purpose, we show that $u_1, \ldots, u_{n_0}, v_1, \ldots, v_{n_+}, \tilde{w}_1, \ldots, \tilde{w}_{m_-}$ are linearly independent. Indeed, if there are scalars $a_1, \ldots, a_{n_0}, b_1, \ldots, b_{n_+}$, $c_1, \ldots, c_{m_-}$ in $\mathbb{R}$ such that

$$
a_1 u_1 + \cdots + a_{n_0} u_{n_0} + b_1 v_1 + \cdots + b_{n_+} v_{n_+} = c_1 \tilde{w}_1 + \cdots + c_{m_-} \tilde{w}_{m_-}, \qquad (4.1.18)
$$

then we may take the scalar products of both sides of (4.1.18) with themselves to get

$$b_1^2(v_1,v_1)+\cdots+b_{n_+}^2(v_{n_+},v_{n_+}) = c_1^2(\tilde{w}_1,\tilde{w}_1)+\cdots+c_{m_-}^2(\tilde{w}_{m_-},\tilde{w}_{m_-}). \quad (4.1.19)$$

Thus, applying the properties (4.1.13) and (4.1.17) in (4.1.19), we conclude that $b_1 = \cdots = b_{n_+} = c_1 = \cdots = c_{m_-} = 0$. Inserting this result into (4.1.18) and using the linear independence of $u_1,\ldots,u_{n_0}$, we arrive at $a_1 = \cdots = a_{n_0} = 0$. So the asserted linear independence follows. As a consequence, we have

$$n_0 + n_+ + m_- \le \dim(U). \quad (4.1.20)$$

In view of (4.1.20) and $n_0 + m_+ + m_- = \dim(U)$ we find $n_+ \le m_+$ as desired.

Thus the integers n_0, n_+, n_- are determined by the scalar product and independent of the choice of an orthogonal basis. These integers are sometimes referred to as the *indices of nullity, positivity,* and *negativity* of the scalar product, respectively.

It is clear that for the orthogonal basis (4.1.12) of U we can further rescale the vectors $v_1,\ldots,v_{n_+}$ and $w_1,\ldots,w_{n_-}$ to make them satisfy

$$(v_i,v_i) = 1, \quad i = 1,\ldots,n_+, \qquad (w_i,w_i) = -1, \quad i = 1,\ldots,n_-. \quad (4.1.21)$$

Such an orthogonal basis is called an *orthonormal basis*.

Exercises

4.1.1   Let S be a non-empty subset of a vector space U equipped with a scalar product $(\cdot,\cdot)$. Show that $S^\perp$ is a subspace of U and $S \subset (S^\perp)^\perp$.

4.1.2   Let $u_1,\ldots,u_k$ be mutually perpendicular vectors of a vector space U equipped with a scalar product $(\cdot,\cdot)$. Show that if these vectors are not null then they must be linearly independent.

4.1.3   Let S_1 and S_2 be two non-empty subsets of a vector space U equipped with a scalar product $(\cdot,\cdot)$. If $S_1 \subset S_2$, show that $S_1^\perp \supset S_2^\perp$.

4.1.4   Consider the vector space $\mathbb{R}^n$ and define

$$(u,v) = u^t A v, \quad u = \begin{pmatrix} a_1 \\ \vdots \\ a_n \end{pmatrix}, v = \begin{pmatrix} b_1 \\ \vdots \\ b_n \end{pmatrix} \in \mathbb{R}^n, \quad (4.1.22)$$

where $A \in \mathbb{R}(n,n)$.

(a) Show that (4.1.22) defines a scalar product over $\mathbb{R}^n$ if and only if A is symmetric.

(b) Show that the subspace $U_0 = (\mathbb{R}^n)^\perp$ is in fact the null-space of the matrix A given by

$$N(A) = \{x \in \mathbb{R}^n \,|\, Ax = 0\}. \qquad (4.1.23)$$

4.1.5   In *special relativity*, one equips the space $\mathbb{R}^4$ with the *Minkowski scalar product* or *Minkowski metric* given by

$$(u,v) = a_1 b_1 - a_2 b_2 - a_3 b_3 - a_4 b_4, \quad u = \begin{pmatrix} a_1 \\ a_2 \\ a_3 \\ a_4 \end{pmatrix}, v = \begin{pmatrix} b_1 \\ b_2 \\ b_3 \\ b_4 \end{pmatrix} \in \mathbb{R}^4.$$
$$(4.1.24)$$

Find an orthogonal basis and determine the indices of nullity, positivity, and negativity of $\mathbb{R}^4$ equipped with this scalar product.

4.1.6   With the notation of the previous problem, consider the following modified scalar product:

$$(u,v) = a_1 b_1 - a_2 b_3 - a_3 b_2 - a_4 b_4. \qquad (4.1.25)$$

(a) Use the Gram–Schmidt procedure to find an orthonormal basis of $\mathbb{R}^4$.

(b) Compute the indices of nullity, positivity, and negativity.

4.1.7   Let U be a vector space with a scalar product and V and W two subspaces of U. Show that

$$(V + W)^\perp = V^\perp \cap W^\perp. \qquad (4.1.26)$$

4.1.8   Let U be an n-dimensional vector space over $\mathbb{C}$ with a scalar product $(\cdot, \cdot)$. Show that if $n \geq 2$ then there must be a vector $u \in U$, $u \neq 0$, such that $(u, u) = 0$.

4.2 Nondegenerate scalar products

Let U be a vector space equipped with the scalar product $(\cdot, \cdot)$. In this section, we examine the special situation when $U_0 = U^\perp = \{0\}$.

Definition 4.3 A vector space equipped with a scalar product $(\cdot, \cdot)$ is said to be *nondegenerate* if $U_0 = U^\perp = \{0\}$. Or equivalently, if $(u, v) = 0$ for all $v \in U$ then $u = 0$.

The most important consequence of a nondegenerate scalar product is that it allows us to identity U with its dual space U' naturally through the pairing given by the scalar product. To see this, we note that for each $v \in U$,

$$f(u) = (u, v), \quad u \in U, \tag{4.2.1}$$

defines an element $f \in U'$. We now show that all elements of U' may be defined this way.

In fact, assume $\dim(U) = n$ and let $\{u_1, \ldots, u_n\}$ be an orthogonal basis of U. Since

$$(u_i, u_i) \equiv c_i \neq 0, \quad i = 1, \ldots, n, \tag{4.2.2}$$

we may take

$$v_i = \frac{1}{c_i} u_i, \quad i = 1, \ldots, n, \tag{4.2.3}$$

to achieve

$$(u_i, v_j) = \delta_{ij}, \quad i, j = 1, \ldots, n. \tag{4.2.4}$$

Thus, if we define $f_i \in U'$ by setting

$$f_i(u) = (u, v_i), \quad u \in U, \quad i = 1, \ldots, n, \tag{4.2.5}$$

then $\{f_1, \ldots, f_n\}$ is seen to be a basis of U' dual to $\{u_1, \ldots, u_n\}$. Therefore, for each $f \in U'$, there are scalars $a_1, \ldots, a_n$ such that

$$f = a_1 f_1 + \cdots + a_n f_n. \tag{4.2.6}$$

Consequently, for any $u \in U$, we have

$$\begin{aligned} f(u) = (a_1 f_1 + \cdots + a_n f_n)(u) &= a_1 f_1(u) + \cdots + a_n f_n(u) \\ &= a_1(u, v_1) + \cdots + a_n(u, v_n) \\ &= (u, a_1 v_1 + \cdots + a_n v_n) \equiv (u, v), \end{aligned} \tag{4.2.7}$$

which proves that any element f of U' is of the form (4.2.1). Such a statement, that is, any functional over U may be represented as a scalar product, in the context of infinite-dimensional spaces, is known as the *Riesz representation theorem*.

In order to make our discussion more precise, we denote the dependence of f on v in (4.2.1) explicitly by $f \equiv v' \in U'$ and use $\rho : U \to U'$ to express this correspondence,

$$\rho(v) = v'. \tag{4.2.8}$$

Therefore, we may summarize various relations discussed above as follows,

$$\langle u, \rho(v) \rangle = \langle u, v' \rangle = (u, v), \quad u, v \in U. \tag{4.2.9}$$

Then we can check to see that ρ is linear. In fact, for $v, w \in U$, we have

$$\langle u, \rho(v+w) \rangle = (u, v+w) = (u, v) + (u, w)$$
$$= \langle u, \rho(v) \rangle + \langle u, \rho(w) \rangle$$
$$= \langle u, \rho(v) + \rho(w) \rangle, \quad u \in U, \qquad (4.2.10)$$

which implies $\rho(v+w) = \rho(v) + \rho(w)$. Besides, for $a \in \mathbb{F}$ and $v \in U$, we have

$$\langle u, \rho(av) \rangle = (u, av) = a(u, v) = a\langle u, \rho(v) \rangle = \langle u, a\rho(v) \rangle, \qquad (4.2.11)$$

which establishes $\rho(av) = a\rho(v)$. Thus the linearity of $\rho : U \to U'$ follows.

Since we have seen that $\rho : U \to U'$ is onto, we conclude that $\rho : U \to U'$ is an isomorphism, which may rightfully be called the *Riesz isomorphism*. As a consequence, we can rewrite (4.2.9) as

$$\langle u, \rho(v) \rangle = \langle u, v' \rangle = (u, v) = (u, \rho^{-1}(v')), \quad u, v \in U, \quad v' \in U'. \qquad (4.2.12)$$

On the other hand, for $T \in L(U)$, recall that the dual of T, $T' \in L(U')$, satisfies

$$\langle u, T'(v') \rangle = \langle T(u), v' \rangle, \quad u \in U, \quad v' \in U'. \qquad (4.2.13)$$

So in view of (4.2.12) and (4.2.13), we arrive at

$$(T(u), v) = (u, (\rho^{-1} \circ T' \circ \rho)(v)), \quad u, v \in U. \qquad (4.2.14)$$

In other words, for any $T \in L(U)$, there is a unique element $T^* \in L(U)$, called the *dual of T with respect to the nondegenerate scalar product* $(\cdot, \cdot)$ and determined by the relation

$$T^* = \rho^{-1} \circ T' \circ \rho, \qquad (4.2.15)$$

via the Riesz isomorphism $\rho : U \to U'$, such that

$$(T(u), v) = (u, T^*(v)), \quad u, v \in U. \qquad (4.2.16)$$

Through the Riesz isomorphism, we may naturally identify U' with U. In this way, we may view U as its own dual space and describe U as a *self-dual space*. Thus, for $T \in L(U)$, we may naturally identify T^* with T' as well, without spelling out the Riesz isomorphism, which leads us to formulate the following definition.

Definition 4.4 Let U be a vector space with a nondegenerate scalar product $(\cdot, \cdot)$. For a mapping $T \in L(U)$, the unique mapping $T' \in L(U)$ satisfying

$$(u, T(v)) = (T'(u), v), \quad u, v \in U, \qquad (4.2.17)$$

is called the *dual* or *adjoint* mapping of T, with respect to the scalar product $(\cdot, \cdot)$.

If $T = T'$, T is said to be a *self-dual* or *self-adjoint mapping* with respect to the scalar product $(\cdot, \cdot)$.

Definition 4.5 Let $T \in L(U)$ where U is a vector space equipped with a scalar product $(\cdot, \cdot)$. We say that T is an *orthogonal mapping* if $(T(u), T(v)) = (u, v)$ for any $u, v \in U$.

As an immediate consequence of the above definition, we have the following basic results.

Theorem 4.6 *That $T \in L(U)$ is an orthogonal mapping is equivalent to one of the following statements:*

(1) $(T(u), T(u)) = (u, u)$ *for any $u \in U$.*
(2) *For any orthogonal basis $\{u_1, \ldots, u_n\}$ of U the vectors $T(u_1), \ldots, T(u_n)$ are mutually orthogonal and $(T(u_i), T(u_i)) = (u_i, u_i)$ for $i = 1, \ldots, n$.*
(3) $T' \circ T = T \circ T' = I$, *the identity mapping over U.*

Proof   If T is orthogonal, it is clear that (1) holds.

Now assume (1) is valid. Using the properties of the scalar product, we have the identity

$$2(u, v) = (u + v, u + v) - (u, u) - (v, v), \quad u, v \in U. \tag{4.2.18}$$

So $2(T(u), T(v)) = (T(u + v), T(u + v)) - (T(u), T(u)) - (T(v), T(v)) = 2(u, v)$ for any $u, v \in U$. Thus T is orthogonal.

That T being orthogonal implies (2) is trivial.

Assume (2) holds. We express any $u, v \in U$ as

$$u = \sum_{i=1}^{n} a_i u_i, \quad v = \sum_{i=1}^{n} b_i u_i, \quad a_i, b_i \in \mathbb{F}, \quad i = 1, \ldots, n. \tag{4.2.19}$$

Therefore we have

$$(T(u), T(v)) = \left(T\left(\sum_{i=1}^{n} a_i u_i \right), T\left(\sum_{j=1}^{n} b_j u_j \right) \right)$$

$$= \sum_{i,j=1}^{n} a_i b_j (T(u_i), T(u_j)) = \sum_{i=1}^{n} a_i b_i (T(u_i), T(u_i))$$

$$= \sum_{i=1}^{n} a_i b_i (u_i, u_i) = \left(\sum_{i=1}^{n} a_i u_i, \sum_{j=1}^{n} b_j u_j \right) = (u, v), \tag{4.2.20}$$

which establishes the orthogonality of T.

We now show that T being orthogonal and (3) are equivalent. In fact, if T is orthogonal, then $(u, (T' \circ T)(v)) = (u, v)$ or $(u, (T' \circ T - I)(v)) = 0$ for any

$u \in U$. By the non-degeneracy of the scalar product we get $(T' \circ T - I)(v) = 0$ for any $v \in U$ which proves $T' \circ T = I$. In other words, T' is a left inverse of T. In view of the discussion in Section 2.1, T' is also a right inverse of T. That is, $T \circ T' = I$. So (3) follows. That (3) implies the orthogonality of T is obvious. $\qquad\qquad\qquad\qquad\qquad\qquad\qquad\qquad\qquad\qquad\qquad\qquad\qquad$ $\square$

As an example, we consider $\mathbb{R}^2$ equipped with the standard *Euclidean scalar product*, that is, the dot product, given as

$$(u,v)_+ \equiv u \cdot v = a_1 b_1 + a_2 b_2 \quad \text{for } u = \begin{pmatrix} a_1 \\ a_2 \end{pmatrix}, v = \begin{pmatrix} b_1 \\ b_2 \end{pmatrix} \in \mathbb{R}^2. \quad (4.2.21)$$

It is straightforward to check that the rotation mapping $R_\theta : \mathbb{R}^2 \to \mathbb{R}^2$ defined by

$$R_\theta(u) = \begin{pmatrix} \cos\theta & -\sin\theta \\ \sin\theta & \cos\theta \end{pmatrix} u, \quad \theta \in \mathbb{R}, \quad u \in \mathbb{R}^2, \quad (4.2.22)$$

is an orthogonal mapping with respect to the scalar product (4.2.21). However, it fails to be orthogonal with respect to the Minkowski scalar product

$$(u,v)_- \equiv a_1 b_1 - a_2 b_2 \quad \text{for } u = \begin{pmatrix} a_1 \\ a_2 \end{pmatrix}, v = \begin{pmatrix} b_1 \\ b_2 \end{pmatrix} \in \mathbb{R}^2. \quad (4.2.23)$$

Nevertheless, if we modify R_θ into $\rho_\theta : \mathbb{R}^2 \to \mathbb{R}^2$ using hyperbolic cosine and sine functions and dropping the negative sign, by setting

$$\rho_\theta(u) = \begin{pmatrix} \cosh\theta & \sinh\theta \\ \sinh\theta & \cosh\theta \end{pmatrix} u, \quad \theta \in \mathbb{R}, \quad u \in \mathbb{R}^2, \quad (4.2.24)$$

we see that ρ_θ is orthogonal with respect to the scalar product (4.2.23), although it now fails to be orthogonal with respect to (4.2.21), of course. This example clearly illustrates the dependence of the form of an orthogonal mapping on the underlying scalar product.

As another example, consider the space $\mathbb{R}^2$ with the scalar product

$$(u,v)_* = a_1 b_1 + a_1 b_2 + a_2 b_1 - a_2 b_2, \quad u = \begin{pmatrix} a_1 \\ a_2 \end{pmatrix}, \quad v = \begin{pmatrix} b_1 \\ b_2 \end{pmatrix} \in \mathbb{R}^2. \quad (4.2.25)$$

It is clear that $(\cdot,\cdot)_*$ is nondegenerate and may be rewritten as

$$(u,v)_* = u^t \begin{pmatrix} 1 & 1 \\ 1 & -1 \end{pmatrix} v. \quad (4.2.26)$$

Thus, for a mapping $T \in L(\mathbb{R}^2)$ defined by

$$T(u) = \begin{pmatrix} a & b \\ c & d \end{pmatrix} u \equiv Au, \quad u \in \mathbb{R}^2, \quad a,b,c,d \in \mathbb{R}, \quad (4.2.27)$$

we have

$$(T'(u),v)_* = (u,T(v))_* = u^t \begin{pmatrix} 1 & 1 \\ 1 & -1 \end{pmatrix} Av$$

$$= u^t \begin{pmatrix} 1 & 1 \\ 1 & -1 \end{pmatrix} A \begin{pmatrix} 1 & 1 \\ 1 & -1 \end{pmatrix}^{-1} \begin{pmatrix} 1 & 1 \\ 1 & -1 \end{pmatrix} v, \qquad (4.2.28)$$

which implies

$$T'(u) = \begin{pmatrix} 1 & 1 \\ 1 & -1 \end{pmatrix}^{-1} A^t \begin{pmatrix} 1 & 1 \\ 1 & -1 \end{pmatrix} u$$

$$= \frac{1}{2} \begin{pmatrix} a+b+c+d & a+b-c-d \\ a-b+c-d & a-b-c+d \end{pmatrix} u, \quad u \in \mathbb{R}^2. \qquad (4.2.29)$$

Consequently, if T is self-adjoint with respect to the scalar product $(\cdot,\cdot)_*$, the condition $a = b+c+d$ holds for a but b,c,d are arbitrary.

Exercises

4.2.1 Let $A \in \mathbb{R}(n,n)$ and set

$$(u,v)_A = u^t Av, \quad u,v \in \mathbb{R}^n, \qquad (4.2.30)$$

where A is symmetric so that (4.2.30) defines a scalar product.

(a) Show that (4.2.30) defines a nondegenerate scalar product if and only if A is invertible.

(b) Let $T : \mathbb{R}^n \to \mathbb{R}^n$ be defined by $T(u) = Bu$ for $u \in \mathbb{R}^n$ where $B \in \mathbb{R}(n,n)$. Find the condition on B that makes T self-adjoint with respect to the scalar product (4.2.30).

4.2.2 Let U be a vector space equipped with a nondegenerate scalar product, $(\cdot,\cdot)$, and $T \in L(U)$.

(a) Show that the pairing $(u,v)_T = (u,T(v))$ $(u,v \in U)$ defines a scalar product on U if and only if T is self-adjoint.

(b) Show that for a self-adjoint mapping $T \in L(U)$ the pairing $(\cdot,\cdot)_T$ given above defines a nondegenerate scalar product over U if and only if T is invertible.

4.2.3 Consider the vector space $\mathbb{R}^2$ equipped with the nondegenerate scalar product

$$(u,v) = a_1 b_2 + a_2 b_1, \quad u = \begin{pmatrix} a_1 \\ a_2 \end{pmatrix}, v = \begin{pmatrix} b_1 \\ b_2 \end{pmatrix} \in \mathbb{R}^2. \qquad (4.2.31)$$

Let

$$V = \text{Span}\left\{ \begin{pmatrix} 1 \\ 0 \end{pmatrix} \right\}. \tag{4.2.32}$$

Show that $V^\perp = V$. This provides an example that in general $V + V^\perp$ may fail to make up the full space.

4.2.4 Let $A = (a_{ij}) \in \mathbb{R}(2,2)$ and define $T_A \in L(\mathbb{R}^2)$ by

$$T_A(u) = \begin{pmatrix} a_{11} & a_{12} \\ a_{21} & a_{22} \end{pmatrix} u, \quad u = \begin{pmatrix} a_1 \\ a_2 \end{pmatrix} \in \mathbb{R}^2. \tag{4.2.33}$$

Find conditions on A such that T_A is self-adjoint with respect to the scalar product $(\cdot,\cdot)_-$ defined in (4.2.23).

4.2.5 Define the scalar product

$$(u,v)_0 = a_1 b_2 + a_2 b_1, \quad u = \begin{pmatrix} a_1 \\ a_2 \end{pmatrix}, v = \begin{pmatrix} b_1 \\ b_2 \end{pmatrix} \in \mathbb{R}^2, \quad (4.2.34)$$

over $\mathbb{R}^2$.

(a) Show that the scalar product $(\cdot,\cdot)_0$ is nondegenerate.
(b) For $T_A \in L(\mathbb{R}^2)$ given in (4.2.33), obtain the adjoint mapping T_A' of T_A and find conditions on the matrix A so that T_A is self-adjoint with respect to the scalar product $(\cdot,\cdot)_0$.

4.2.6 Use U to denote the vector space of real-valued functions with all orders of derivatives over the real line $\mathbb{R}$ which vanish outside bounded intervals. Equip U with the scalar product

$$(u,v) = \int_{-\infty}^{\infty} u(t)v(t)\,dt, \quad u,v \in U. \tag{4.2.35}$$

Show that the linear mapping $D = \frac{d}{dt} : U \to U$ is *anti-self-dual* or *anti-self-adjoint*. That is, $D' = -D$.

4.2.7 Let U be a finite-dimensional vector space equipped with a scalar product, $(\cdot,\cdot)$. Decompose U into the direct sum as stated in (4.1.15). Show that we may use the scalar product $(\cdot,\cdot)$ of U to make the quotient space U/U_0 into a space with a nondegenerate scalar product when $U_0 \neq U$, still denoted by $(\cdot,\cdot)$, given by

$$([u],[v]) = (u,v), \quad [u],[v] \in U/U_0. \tag{4.2.36}$$

4.2.8 Let U be a finite-dimensional vector space equipped with a scalar product, $(\cdot,\cdot)$, and V a subspace of U. Show that $(\cdot,\cdot)$ is a nondegenerate scalar product over V if and only if $V \cap V^\perp = \{0\}$.

4.2.9 Let U be a finite-dimensional vector space with a nondegenerate scalar product $(\cdot,\cdot)$. Let $\rho : U \to U'$ be the Riesz isomorphism. Show that for any subspace V of U there holds

$$\rho(V^\perp) = V^0. \qquad (4.2.37)$$

In other words, the mapping ρ is an isomorphism from $V^\perp$ onto V^0. In particular, $\dim(V^\perp) = \dim(V^0)$. Thus, in view of (1.4.31) and (4.2.37), we have the dimensionality equation

$$\dim(V) + \dim(V^\perp) = \dim(U). \qquad (4.2.38)$$

4.2.10 Let U be a finite-dimensional space with a nondegenerate scalar product and V a subspace of U. Use (4.2.38) to establish $V = (V^\perp)^\perp$.

4.2.11 Let U be a finite-dimensional space with a nondegenerate scalar product and V, W two subspaces of U. Establish the relation

$$(V \cap W)^\perp = V^\perp + W^\perp. \qquad (4.2.39)$$

4.2.12 Let U be an n-dimensional vector space over $\mathbb{C}$ with a nondegenerate scalar product $(\cdot,\cdot)$. Show that if $n \geq 2$ then there must be linearly independent vectors $u, v \in U$ such that $(u,u) = 0$ and $(v,v) = 0$ but $(u,v) = 1$.

4.3 Positive definite scalar products

In this section we consider two types of positive definite scalar products: real ones and complex ones. Real ones are modeled over the standard *Euclidean scalar product* on $\mathbb{R}^n$:

$$(u,v) = u \cdot v = u^t v = a_1 b_1 + \cdots + a_n b_n,$$

$$u = \begin{pmatrix} a_1 \\ \vdots \\ a_n \end{pmatrix}, \quad v = \begin{pmatrix} b_1 \\ \vdots \\ b_n \end{pmatrix} \in \mathbb{R}^n, \qquad (4.3.1)$$

and complex ones over the standard *Hermitian scalar product* on $\mathbb{C}^n$:

$$(u,v) = \bar{u} \cdot v = u^\dagger v = \bar{a}_1 b_1 + \cdots + \bar{a}_n b_n,$$

$$u = \begin{pmatrix} a_1 \\ \vdots \\ a_n \end{pmatrix}, \quad v = \begin{pmatrix} b_1 \\ \vdots \\ b_n \end{pmatrix} \in \mathbb{C}^n. \qquad (4.3.2)$$

The common feature of these products is the positivity property $(u,u) \geq 0$ for any vector u and $(u,u) = 0$ only when $u = 0$. The major difference is that the former is symmetric but the latter fails to be so. Instead, there holds the adjusted property $(u,v) = \overline{(v,u)}$ for any $u,v \in \mathbb{C}^n$, which is seen to be naturally implemented to ensure the positivity property. As a consequence, in both the $\mathbb{R}^n$ and $\mathbb{C}^n$ cases, we are able to define the norm of a vector u to be $\|u\| = \sqrt{(u,u)}$.

Motivated from the above examples, we can bring forth the following definition.

Definition 4.7 A *positive definite scalar product over a real vector space U* is scalar product $(\cdot,\cdot)$ satisfying $(u,u) \geq 0$ for $u \in U$ and $(u,u) = 0$ only when $u = 0$.

A *positive definite scalar product over a complex vector space U* is a scalar function $(u,v) \in \mathbb{C}$, defined for each pair of vectors u and $v \in U$, which satisfies the following conditions:

(1) (Hermitian symmetry) $(u,v) = \overline{(v,u)}$ for $u,v \in U$.
(2) (Additivity) $(u+v,w) = (u,w) + (v,w)$ for $u,v,w \in U$.
(3) (Partial homogeneity) $(u,av) = a(u,v)$ for $a \in \mathbb{C}$ and $u,v \in U$.
(4) (Positivity) $(u,u) \geq 0$ for $u \in U$ and $(u,u) = 0$ only when $u = 0$.

Since the real case is contained as a special situation of the complex case, we shall focus our discussion on the complex case, unless otherwise stated.

Needless to say additivity regarding the second argument in $(\cdot,\cdot)$ still holds since

$$(u,v+w) = \overline{(v+w,u)}$$
$$= \overline{(v,u)} + \overline{(w,u)} = (u,v) + (u,w), \quad u,v,w \in U. \quad (4.3.3)$$

On the other hand, homogeneity regarding the first argument takes a modified form,

$$(au,v) = \overline{(v,au)} = \overline{a}\,\overline{(v,u)} = \overline{a}(u,v), \quad a \in \mathbb{C}, \quad u \in U. \quad (4.3.4)$$

We will extend our study carried out in Sections 4.1 and 4.2 for general scalar products to the current situation of a positive definite scalar product which is necessarily nondegenerate since $(u,u) > 0$ for any nonzero vector u in U.

First, we see that for u and $v \in U$ we can still use the condition $(u,v) = 0$ to define u and v to be mutually perpendicular vectors. Next, since for any $u \in U$ we have $(u,u) \geq 0$, we can formally define the norm of u as in $\mathbb{C}^n$ by

$$\|u\| = \sqrt{(u,u)}. \quad (4.3.5)$$

It is clearly seen that the norm so defined enjoys the positivity and homogeneity conditions required of a norm. That it also satisfies the triangle inequality will be established shortly. Thus (4.3.5) indeed gives rise to a norm of the space U which is said to be induced from the positive definite scalar product $(\cdot, \cdot)$.

Let u and $v \in U$ be perpendicular. Then we have

$$\|u+v\|^2 = \|u\|^2 + \|v\|^2. \tag{4.3.6}$$

This important expression is also known as the *Pythagoras theorem*, which may be proved by a simple expansion

$$\|u+v\|^2 = (u+v, u+v)$$
$$= (u,u) + (u,v) + (v,u) + (v,v) = \|u\|^2 + \|v\|^2, \tag{4.3.7}$$

since $(u,v) = 0$ and $(v,u) = \overline{(u,v)} = 0$.

For u and $v \in U$ with $u \neq 0$, we may use the orthogonal decomposition formula (4.1.2) to resolve v into the form

$$v = \left(v - \frac{(u,v)}{(u,u)} u \right) + \frac{(u,v)}{(u,u)} u \equiv w + \frac{(u,v)}{(u,u)} u, \tag{4.3.8}$$

so that $(u,w) = 0$. Hence, in view of the Pythagoras theorem, we get

$$\|v\|^2 = \|w\|^2 + \left| \frac{(u,v)}{(u,u)} \right|^2 \|u\|^2 \geq \frac{|(u,v)|^2}{\|u\|^2}. \tag{4.3.9}$$

In other words, we have

$$|(u,v)| \leq \|u\| \|v\|, \tag{4.3.10}$$

with equality if and only if $w = 0$ or equivalently, $v \in \text{Span}\{u\}$. Of course (4.3.10) is valid when $u = 0$. Hence in summary we may state that (4.3.10) holds for any $u, v \in U$ and that the equality is achieved if and only if u, v are linearly dependent.

The inequality (4.3.10) is the celebrated *Schwarz inequality*, or the *Cauchy–Schwarz inequality*, whose derivation is seen to be another direct application of the vector orthogonal decomposition formula (4.1.2).

We now apply the Schwarz inequality to establish the triangle inequality for the norm $\| \cdot \|$ induced from a positive definite scalar product $(\cdot, \cdot)$.

Let u and $v \in U$. Then in view of (4.3.10) we have

$$\|u+v\|^2 = \|u\|^2 + \|v\|^2 + (u,v) + (v,u)$$
$$\leq \|u\|^2 + \|v\|^2 + 2|(u,v)|$$
$$\leq \|u\|^2 + \|v\|^2 + 2\|u\| \|v\|$$
$$= (\|u\| + \|v\|)^2. \tag{4.3.11}$$

Hence the triangle inequality $\|u+v\| \leq \|u\| + \|v\|$ follows.

If $\{u_1,\ldots,u_n\}$ is a basis of U, we may invoke the Gram–Schmidt procedure

$$v_1 = u_1, \quad v_i = u_i - \sum_{j=1}^{i-1} \frac{(v_j, u_i)}{(v_j, v_j)} v_j, \quad i = 2,\ldots,n, \tag{4.3.12}$$

as before to obtain an orthogonal basis for U. In fact, we may examine the validity of this procedure by a simple induction.

When $n = 1$, there is nothing to show.

Assume the procedure is valid at $n = k \geq 1$.

At $n = k+1$, by the inductive assumption, we know that we may construct $\{v_1,\ldots,v_k\}$ to get an orthogonal basis for $\mathrm{Span}\{u_1,\ldots,u_k\}$. Define

$$v_{k+1} = u_{k+1} - \sum_{j=1}^{k} \frac{(v_j, u_{k+1})}{(v_j, v_j)} v_j. \tag{4.3.13}$$

Then we can check that $(v_{k+1}, v_i) = 0$ for $i = 1,\ldots,k$ and

$$v_{k+1} \in \mathrm{Span}\{u_{k+1}, v_1,\ldots,v_k\} \subset \mathrm{Span}\{u_1,\ldots,u_k,u_{k+1}\}. \tag{4.3.14}$$

Of course $v_{k+1} \neq 0$ otherwise $u_{k+1} \in \mathrm{Span}\{v_1,\ldots,v_k\} = \mathrm{Span}\{u_1,\ldots,u_k\}$. Thus we have obtained $k+1$ nonzero mutually orthogonal vectors $v_1,\ldots,v_k,v_{k+1}$ which make up a basis for $\mathrm{Span}\{u_1,\ldots,u_k,u_{k+1}\}$ as asserted.

Thus, we have seen that, in the positive definite scalar product situation, from any basis $\{u_1,\ldots,u_n\}$ of U, the Gram–Schmidt procedure (4.3.12) provides a scheme of getting an orthogonal basis $\{v_1,\ldots,v_n\}$ of U so that each of its subset $\{v_1,\ldots,v_k\}$ is an orthogonal basis of $\mathrm{Span}\{u_1,\ldots,u_k\}$ for $k = 1,\ldots,n$.

Let $\{v_1,\ldots,v_n\}$ be an orthogonal basis for U. The positivity property allows us to modify the basis further by setting

$$w_i = \frac{1}{\|v_i\|} v_i, \quad i = 1,\ldots,n, \tag{4.3.15}$$

so that $\{w_1,\ldots,w_n\}$ is an orthogonal basis of U consisting of unit vectors (that is, $\|w_i\| = 1$ for $i = 1,\ldots,n$). Such an orthogonal basis is called an *orthonormal basis*.

We next examine the dual space U' of U in view of the positive definite scalar product $(\cdot,\cdot)$ over U.

For any $u \in U$, it is clear that

$$f(v) = (u,v), \quad v \in U, \tag{4.3.16}$$

defines an element in U'. Now let $\{u_1,\ldots,u_n\}$ be an orthonormal basis of U and define $f_i \in U'$ by setting

$$f_i(v) = (u_i,v), \quad v \in U. \tag{4.3.17}$$

Then $\{f_1,\dots,f_i\}$ is a basis of U' which is dual to $\{u_1,\dots,u_n\}$. Hence, for any $f \in U'$, there are scalars $a_1,\dots,a_n$ such that

$$f = a_1 f_1 + \cdots + a_n f_n. \tag{4.3.18}$$

Consequently, we have

$$
\begin{aligned}
f(v) &= a_1 f_1(v) + \cdots + a_n f_n(v) \\
&= a_1(u_1,v) + \cdots + a_n(u_n,v) \\
&= (\bar{a}_1 u_1 + \cdots + \bar{a}_n u_n, v) \\
&\equiv (u,v), \quad u = \bar{a}_1 u_1 + \cdots + \bar{a}_n u_n.
\end{aligned}
\tag{4.3.19}
$$

Thus, each element in U' may be represented by an element in U in the form of a scalar product. In other words, the Riesz representation theorem still holds here, although homogeneity of such a representation takes an adjusted form,

$$
\begin{aligned}
& f_i \mapsto u_i, \quad i = 1,\dots,n; \\
& f = a_1 f_1 + \cdots + a_n f_n \mapsto \bar{a}_1 u_1 + \cdots + \bar{a}_n u_n.
\end{aligned}
\tag{4.3.20}
$$

Nevertheless, we now show that adjoint mappings are well defined.

Let U and V be finite-dimensional vector spaces over $\mathbb{C}$ with positive definite scalar products, both denoted by $(\cdot,\cdot)$. For $T \in L(U,V)$ and any $v \in V$, the expression

$$f(u) = (v,T(u)), \quad u \in U, \tag{4.3.21}$$

defines an element $f \in U'$. Hence there is a unique element $w \in U$ such that $f(u) = (w,u)$. Since w depends on v, we may denote this relation by $w = T'(v)$. Hence

$$(v,T(u)) = (T'(v),u). \tag{4.3.22}$$

We will prove $T' \in L(V,U)$.

In fact, for v_1 and $v_2 \in V$, we have

$$
\begin{aligned}
(T'(v_1+v_2),u) &= (v_1+v_2,T(u)) \\
&= (v_1,T(u)) + (v_2,T(u)) \\
&= (T'(v_1),u) + (T'(v_2),u) \\
&= (T'(v_1)+T'(v_2),u), \quad u \in U.
\end{aligned}
\tag{4.3.23}
$$

Thus $T'(v_1+v_2) = T'(v_1) + T'(v_2)$ and additivity follows.

For $a \in \mathbb{C}$ and $v \in V$, we have

$$(T'(av), u) = (av, T(u)) = \bar{a}(v, T(u))$$
$$= \bar{a}(T'(v), u) = (aT'(v), u), \quad u \in U. \tag{4.3.24}$$

This shows $T'(av) = aT'(v)$ and homogeneity also follows.

Of particular interest is a mapping from U into itself. In this case we can define a self-dual or self-adjoint mapping T with respect to the positive definite scalar product $(\cdot, \cdot)$ to be such that $T' = T$. Similar to Definition 4.5, we also have the following:

Definition 4.8 Let U be a real or complex vector space equipped with a positive definite scalar product $(\cdot, \cdot)$. Assume that $T \in L(U)$ satisfies $(T(u), T(v)) = (u, v)$ for any u and $v \in U$.

(1) T is called *orthogonal* when U is real.
(2) T is called *unitary* when U is complex.

In analogue to Theorem 4.6, we have the following:

Theorem 4.9 *That $T \in L(U)$ is orthogonal or unitary is equivalent to one of the following statements:*

(1) *T is norm-preserving. That is, $\|T(u)\| = \|u\|$ for any $u \in U$, where $\|\cdot\|$ is the norm of U induced from its positive definite scalar product $(\cdot, \cdot)$.*
(2) *T maps an orthonormal basis to another orthonormal basis of U.*
(3) *$T' \circ T = T \circ T' = I$.*

Proof We only need to carry out the proof in the complex case because now the scalar product $(\cdot, \cdot)$ fails to be symmetric and the relation (4.2.18) is invalid.

That T being unitary implies (1) is trivial since $\|T(u)\|^2 = (T(u), T(u)) = (u, u) = \|u\|^2$ for any $u \in U$.

Assume (1) holds. From the expansions

$$\|u + v\|^2 = \|u\|^2 + \|v\|^2 + 2\Re\{(u, v)\}, \tag{4.3.25}$$

$$\|iu + v\|^2 = \|u\|^2 + \|v\|^2 + 2\Im\{(u, v)\}, \tag{4.3.26}$$

we obtain the following *polarization identity* in the complex situation:

$$(u, v) = \frac{1}{2}(\|u + v\|^2 - \|u\|^2 - \|v\|^2) + \frac{1}{2}i(\|iu + v\|^2 - \|u\|^2 - \|v\|^2),$$
$$u, v \in U. \tag{4.3.27}$$

Applying (4.3.27), we obtain

$$
\begin{aligned}
(T(u), T(v)) &= \frac{1}{2}(\|T(u+v)\|^2 - \|T(u)\|^2 - \|T(v)\|^2) \\
&\quad + \frac{1}{2}i(\|T(iu+v)\|^2 - \|T(u)\|^2 - \|T(v)\|^2) \\
&= \frac{1}{2}(\|u+v\|^2 - \|u\|^2 - \|v\|^2) + \frac{1}{2}i(\|iu+v\|^2 - \|u\|^2 - \|v\|^2) \\
&= (u, v), \quad u, v \in U.
\end{aligned}
\tag{4.3.28}
$$

Hence T is unitary.

The rest of the proof is similar to that of Theorem 4.6 and thus skipped. □

If $T \in L(U)$ is orthogonal or unitary and $\lambda \in \mathbb{C}$ an eigenvalue of T, then it is clear that $|\lambda| = 1$ since T is norm-preserving.

Let $A \in \mathbb{F}(n,n)$ and define $T_A \in L(\mathbb{F}^n)$ in the usual way $T_A(u) = Au$ for any column vector $u \in \mathbb{F}^n$.

When $\mathbb{F} = \mathbb{R}$, let the positive define scalar product be the Euclidean one given in (4.3.1). That is,

$$
(u, v) = u^t v, \quad u, v \in \mathbb{R}^n.
\tag{4.3.29}
$$

Thus

$$
(u, T_A(v)) = u^t A v = (A^t u)^t v = (T_A'(u), v), \quad u, v \in \mathbb{R}^n.
\tag{4.3.30}
$$

Therefore $T_A'(u) = A^t u$ ($u \in \mathbb{R}^n$). If T_A is orthogonal, then $T_A' \circ T_A = T_A \circ T_A' = I$ which leads to $A^t A = AA^t = I_n$. Besides, a self-adjoint mapping $T_A = T_A'$ is defined by a symmetric matrix, $A = A^t$.

The above discussion may be carried over to the abstract setting as follows.

Let U be a real vector space with a positive definite scalar product $(\cdot, \cdot)$ and $\mathcal{B} = \{u_1, \dots, u_n\}$ an orthonormal basis. Assume $T \in L(U)$ is represented by the matrix $A \in \mathbb{R}(n,n)$ with respect to the basis $\mathcal{B}$ so that

$$
T(u_j) = \sum_{i=1}^n a_{ij} u_i, \quad j = 1, \dots, n.
\tag{4.3.31}
$$

Similarly $T' \in L(U)$ is represented by $A' = (a_{ij}') \in \mathbb{R}(n,n)$. Then we have

$$
a_{ij} = (u_i, T(u_j)) = (T'(u_i), u_j) = a_{ji}', \quad i, j = 1, \dots, n.
\tag{4.3.32}
$$

So we again have $A' = A^t$. If T is orthogonal, then $T \circ T' = T' \circ T = I$, which gives us $AA^t = A^t A = I_n$ as before. If T is self-adjoint mapping $T = T'$, then A is again a symmetric matrix.

When $\mathbb{F} = \mathbb{C}$, let the positive define scalar product be the Hermitian one given in (4.3.2). Then

$$(u,v) = \bar{u}^t v, \quad u,v \in \mathbb{C}^n. \tag{4.3.33}$$

Thus

$$(u, T_A(v)) = \bar{u}^t A v = \overline{\left(\overline{A}^t u\right)}^t v = (T'_A(u), v), \quad u,v \in \mathbb{C}^n. \tag{4.3.34}$$

Therefore $T'_A(u) = \overline{A}^t u$ $(u \in \mathbb{C}^n)$. If T_A is unitary, then $T'_A \circ T_A = T_A \circ T'_A = I$ which leads to $\overline{A}^t A = A\overline{A}^t = I_n$. Besides, a self-adjoint mapping $T = T'$ is defined by a matrix which is symmetric under complex conjugation and transpose of A. That is, $A = \overline{A}^t$.

Similar to the real vector space situation, we leave it as an exercise to examine that, in the abstract setting, the matrix representing the adjoint mapping with respect to an orthonormal basis is obtained by taking matrix transpose and complex conjugate of the matrix of the original mapping with respect to the same basis.

The above calculations lead us to formulate the following concepts, which were originally introduced in Section 1.1 without explanation.

Definition 4.10 A real matrix $A \in \mathbb{R}(n,n)$ is said to be *orthogonal* if its transpose A^t is its inverse. That is, $AA^t = A^tA = I_n$. It is easily checked that A is orthogonal if and only if its sets of column and row vectors both form orthonormal bases of $\mathbb{R}^n$ with the standard Euclidean scalar product.

A complex matrix $A \in \mathbb{C}(n,n)$ is said to be *unitary* if the complex conjugate of its transpose $\overline{A}^t$, also called its *Hermitian conjugate* denoted as $A^\dagger = \overline{A}^t$, is the inverse of A. That is $AA^\dagger = A^\dagger A = I_n$. It is easily checked that A is unitary if and only if its sets of column and row vectors both form orthonormal bases of $\mathbb{C}^n$ with the standard Hermitian scalar product.

A complex matrix $A \in \mathbb{C}(n,n)$ is called *Hermitian* if it satisfies the property $A = A^\dagger$, or equivalently, if it defines a self-adjoint mapping $T_A = T'_A$.

With the above terminology and the Gram–Schmidt procedure, we may establish a well-known matrix factorization result, commonly referred to as the *QR factorization*, for a non-singular matrix.

In fact, let $A \in \mathbb{C}(n,n)$ be non-singular and use $u_1, \ldots, u_n$ to denote the n corresponding column vectors of A which form a basis of $\mathbb{C}^n$. Use $(\cdot, \cdot)$ to denote the Hermitian scalar product on $\mathbb{C}^n$. That is, $(u,v) = u^\dagger v$, where $u, v \in \mathbb{C}^n$ are column vectors. Apply the Gram–Schmidt procedure to set

$$\begin{cases} v_1 &=& u_1, \\ v_2 &=& u_2 - \frac{(v_1, u_2)}{(v_1, v_1)} v_1, \\ \cdots && \cdots \cdots \\ v_n &=& u_n - \frac{(v_1, u_n)}{(v_1, v_1)} v_1 - \cdots - \frac{(v_{n-1}, u_n)}{(v_{n-1}, v_{n-1})} v_{n-1}. \end{cases} \tag{4.3.35}$$

Then $\{v_1,\ldots,v_n\}$ is an orthogonal basis of $\mathbb{C}^n$. Set $w_i = (1/\|v_i\|)v_i$ for $i = 1,\ldots,n$. We see that $\{w_1,\ldots,w_n\}$ is an orthonormal basis of $\mathbb{C}^n$. Therefore, inverting (4.3.35) and rewriting the resulting relations in terms of $\{w_1,\ldots,w_n\}$, we get

$$\left\{ \begin{array}{rcl} u_1 &=& \|v_1\|w_1, \\ u_2 &=& \frac{(v_1,u_2)}{(v_1,v_1)}\|v_1\|w_1 + \|v_2\|w_2, \\ \cdots & \cdots & \cdots\cdots \\ u_n &=& \frac{(v_1,u_n)}{(v_1,v_1)}\|v_1\|w_1 + \cdots + \frac{(v_{n-1},u_n)}{(v_{n-1},v_{n-1})}\|v_{n-1}\|w_{n-1} + \|v_n\|w_n. \end{array} \right.$$

$$(4.3.36)$$

For convenience, we may express (4.3.36) in the compressed form

$$\left\{ \begin{array}{rcl} u_1 &=& r_{11}w_1, \\ u_2 &=& r_{12}w_1 + r_{22}w_2, \\ \cdots & \cdots & \cdots\cdots \\ u_n &=& r_{1n}w_1 + \cdots + r_{n-1,n}w_{n-1} + r_{nn}w_n, \end{array} \right.$$

$$(4.3.37)$$

which implies the matrix relation

$$A = (u_1,\ldots,u_n) = (w_1,\ldots,w_n) \begin{pmatrix} r_{11} & r_{12} & \cdots & r_{1n} \\ 0 & r_{22} & \cdots & r_{2n} \\ \vdots & \vdots & \ddots & \vdots \\ 0 & 0 & \cdots & r_{nn} \end{pmatrix}$$

$$= QR,$$

$$(4.3.38)$$

where $Q = (w_1,\ldots,w_n) \in \mathbb{C}(n,n)$ is unitary since its column vectors are mutually perpendicular and of unit length and $R = (r_{ij}) \in \mathbb{C}(n,n)$ is upper triangular with positive diagonal entries since $r_{ii} = \|v_i\|$ for $i = 1,\ldots,n$. This explicit construction is the desired QR factorization for A. It is clear that if A is real then the matrices Q and R are also real and Q is orthogonal.

To end this section, we note that the norm of a vector u in a vector space U equipped with a positive definite scalar product $(\cdot,\cdot)$ may be evaluated through the following useful expression:

$$\|u\| = \sup\{|(u,v)| \,|\, v \in U, \|v\| = 1\}. \qquad (4.3.39)$$

Indeed, let η denote the right-hand side of (4.3.39). From the Schwarz inequality (4.3.10), we get $|(u,v)| \le \|u\|$ for any $v \in U$ satisfying $\|v\| = 1$. So $\eta \le \|u\|$. To show that $\eta \ge \|u\|$, it suffices to consider the nontrivial situation $u \ne 0$. In this case, we have

$$\eta \ge \left| \left(u, \frac{1}{\|u\|}u \right) \right| = \|u\|. \qquad (4.3.40)$$

Hence, in conclusion, $\eta = \|u\|$ and (4.3.39) follows.

Exercises

4.3.1 Let U be a vector space with a positive definite scalar product $(\cdot,\cdot)$. Show that $u_1,\ldots,u_k \in U$ are linearly independent if and only if their associated *metric matrix*, also called the *Gram matrix*,

$$M = \begin{pmatrix} (u_1,u_1) & \cdots & (u_1,u_k) \\ \cdots & \cdots & \cdots \\ (u_k,u_1) & \cdots & (u_k,u_k) \end{pmatrix} \tag{4.3.41}$$

is non-singular.

4.3.2 (Continued from Problem 4.3.1) Show that if $u \in U$ lies in Span$\{u_1,\ldots,u_k\}$ then the column vector $((u_1,u),\ldots,(u_k,u))^t$ lies in the column space of the metric matrix M. However, the converse is not true when $k < \dim(U)$.

4.3.3 Let U be a complex vector space with a positive definite scalar product and $\mathcal{B} = \{u_1,\ldots,u_n\}$ an orthonormal basis of U. For $T \in L(U)$, let A and $A' \in \mathbb{C}(n,n)$ be the matrices that represent T and T', respectively, with respect to the basis $\mathcal{B}$. Show that $A' = A^\dagger$.

4.3.4 Let U be a finite-dimensional complex vector space with a positive definite scalar product and $S \in L(U)$ be anti-self-adjoint. That is, $S' = -S$. Show that $I \pm S$ must be invertible.

4.3.5 Consider the complex vector space $\mathbb{C}(m,n)$. Show that

$$(A,B) = \mathrm{Tr}(A^\dagger B), \quad A,B \in \mathbb{C}(m,n) \tag{4.3.42}$$

defines a positive-definite scalar product over $\mathbb{C}(m,n)$ which extends the traditional Hermitian scalar product over $\mathbb{C}^m = \mathbb{C}(m,1)$. With such a scalar product, the quantity $\|A\| = \sqrt{(A,A)}$ is sometimes called the *Hilbert–Schmidt norm* of the matrix $A \in \mathbb{C}(m,n)$.

4.3.6 Let $(\cdot,\cdot)$ be the standard Hermitian scalar product on $\mathbb{C}^m$ and $A \in \mathbb{C}(m,n)$. Establish the following statement known as the *Fredholm alternative* for complex matrix equations: Given $b \in \mathbb{C}^m$ the non-homogeneous equation $Ax = b$ has a solution for some $x \in \mathbb{C}^n$ if and only if $(y,b) = 0$ for any solution $y \in \mathbb{C}^m$ of the homogeneous equation $A^\dagger y = 0$.

4.3.7 For a and $b \in \mathbb{R}$ $(a < b)$, consider the vector space $U = C^1[a,b]$ of all real-valued continuously differentiable functions over the interval $[a,b]$ and set

$$(u,v) = \int_a^b (u(t)v(t) + u'(t)v'(t))dt, \quad u,v \in U. \tag{4.3.43}$$

(a) Show that (4.3.43) defines a positive-definite scalar product over U.

(b) Define now $V = \{1\}^{\perp}$. Show that the average value of any function u in V must be zero.

4.3.8   For $A \in \mathbb{C}(n,n)$ show that if the column vectors of A form an orthonormal basis of $\mathbb{C}^n$ with respect to the standard Hermitian scalar product so do the row vectors of A.

4.3.9   Let $A \in \mathbb{R}(n,n)$ be symmetric and non-singular such that

$$(u,v) = u^t A v, \quad u,v \in \mathbb{R}^n \tag{4.3.44}$$

defines a positive-definite scalar product over $\mathbb{R}^n$.

Consider the linear function f over $\mathbb{R}^n$ defined by

$$f(v) = a_1 x_1 + \cdots + a_n x_n, \quad a_1, \ldots, a_n \in \mathbb{R}; \quad v = \begin{pmatrix} x_1 \\ \vdots \\ x_n \end{pmatrix} \in \mathbb{R}^n.$$
$$\tag{4.3.45}$$

Obtain the unique vector $u \in \mathbb{R}^n$ such that

$$f(v) = (u,v), \quad \forall v \in \mathbb{R}^n. \tag{4.3.46}$$

(Note that this problem provides a concrete appreciation of the Riesz representation theorem.)

4.3.10   Let U be the real vector space of all smooth and compactly supported real-valued functions over $\mathbb{R}^3$. Define

$$(u,v) = \int_{\mathbb{R}^3} u(x,y,z)v(x,y,z)\mathrm{d}x\mathrm{d}y\mathrm{d}z, \quad u,v \in U, \tag{4.3.47}$$

which is a positive-definite scalar product over U. Consider the Laplace map $\Delta : U \to U$ defined by

$$\Delta u = u_{xx} + u_{yy} + u_{zz}. \tag{4.3.48}$$

Find the dual map of Δ, namely a map $\Delta' : U \to U$ such that

$$(\Delta' u, v) = (u, \Delta v), \quad u,v \in U. \tag{4.3.49}$$

(Hint: Use the divergence theorem.)

4.3.11   Let U be a finite-dimensional complex vector space with a positive-definite scalar product. Show that $T \in L(U)$ is self-adjoint if and only if $S = iT$ is anti-self-adjoint.

4.3.12   For the real matrix

$$A = \begin{pmatrix} 1 & -1 & 1 \\ -1 & 1 & 2 \\ 2 & 1 & -2 \end{pmatrix}, \tag{4.3.50}$$

obtain a QR factorization.

4.4 Orthogonal resolutions of vectors

In this section, we continue to study a finite dimensional vector space U with a positive definite scalar product $(\cdot,\cdot)$. We focus our attention on the problem of resolving a vector into the span of a given set of orthogonal vectors. To avoid the trivial situation, we always assume that the set of the orthogonal vectors concerned never contains zero vector unless otherwise stated.

We begin with the following basic orthogonal decomposition theorem.

Theorem 4.11 *Let V be a subspace of U. Then there holds the orthogonal decomposition*

$$U = V \oplus V^{\perp}. \tag{4.4.1}$$

Proof   If $V = \{0\}$, there is nothing to show. Assume $V \neq \{0\}$. Let $\{v_1,\ldots,v_k\}$ be an orthogonal basis of V. We can then expand $\{v_1,\ldots,v_k\}$ into an orthogonal basis for U, which may be denoted as $\{v_1,\ldots,v_k,w_1,\ldots,w_l\}$, where $k+l = n = \dim(U)$. Of course, $w_1,\ldots,w_l \in V^{\perp}$. For any $u \in V^{\perp}$, we rewrite u as

$$u = a_1 v_1 + \cdots + a_k v_k + b_1 w_1 + \cdots + b_l w_l, \tag{4.4.2}$$

with some scalars $a_1,\ldots,a_k,b_1,\ldots,b_l$. From $(u,v_i) = a_i(v_i,v_i) = 0$ ($i = 1,\ldots,k$), we obtain $a_1 = \cdots = a_k = 0$, which establishes $u \in \mathrm{Span}\{w_1,\ldots,w_l\}$. So $V^{\perp} \subset \mathrm{Span}\{w_1,\ldots,w_l\}$. Therefore $V^{\perp} = \mathrm{Span}\{w_1,\ldots,w_l\}$ and $U = V + V^{\perp}$.

Finally, take $u \in V \cap V^{\perp}$. Then $(u,u) = 0$. By the positivity condition, we get $u = 0$. Thus (4.4.1) follows. □

We now follow (4.4.1) to concretely construct the orthogonal decomposition of a given vector.

Theorem 4.12 *Let V be a nonzero subspace of U with an orthogonal basis $\{v_1,\ldots,v_k\}$. Any $u \in U$ may be uniquely decomposed into the form*

$$u = v + w, \quad v \in V, \quad w \in V^{\perp}, \tag{4.4.3}$$

where v is given by the expression

$$v = \sum_{i=1}^{k} \frac{(v_i,u)}{(v_i,v_i)} v_i. \tag{4.4.4}$$

Moreover, the vector v given in (4.4.4) is the unique solution of the minimization problem

$$\eta \equiv \inf\{\|u - x\| \,|\, x \in V\}. \tag{4.4.5}$$

Proof   The validity of the expression (4.4.3) for some unique $v \in V$ and $w \in V^{\perp}$ is already ensured by Theorem 4.11.

We rewrite v as

$$v = \sum_{i=1}^{k} a_i v_i. \tag{4.4.6}$$

Then $(v_i, v) = a_i(v_i, v_i)$ $(i = 1, \ldots, k)$. That is,

$$a_i = \frac{(v_i, v)}{(v_i, v_i)}, \quad i = 1, \ldots, k, \tag{4.4.7}$$

which verifies (4.4.4).

For the scalars $a_1, \ldots, a_k$ given in (4.4.7) and v in (4.4.6), we have from (4.4.3) the relation

$$u - \sum_{i=1}^{k} b_i v_i = w + \sum_{i=1}^{k} (a_i - b_i) v_i, \quad x = \sum_{i=1}^{k} b_i v_i \in V. \tag{4.4.8}$$

Consequently,

$$
\begin{aligned}
\|u - x\|^2 &= \left\| u - \sum_{i=1}^{k} b_i v_i \right\|^2 \\
&= \left\| w + \sum_{i=1}^{k} (a_i - b_i) v_i \right\|^2 \\
&= \|w\|^2 + \sum_{i=1}^{k} |a_i - b_i|^2 \|v_i\|^2 \\
&\geq \|w\|^2,
\end{aligned}
\tag{4.4.9}
$$

and the lower bound $\|w\|^2$ is attained only when $b_i = a_i$ for all $i = 1, \ldots, k$, or $x = v$.

So the proof is complete.   □

Definition 4.13 Let $\{v_1, \ldots, v_k\}$ be a set of orthogonal vectors in U. For $u \in U$, the sum

$$\sum_{i=1}^{k} a_i v_i, \quad a_i = \frac{(v_i, u)}{(v_i, v_i)}, \quad i = 1, \ldots, k, \tag{4.4.10}$$

is called the *Fourier expansion* of u and $a_1, \ldots, a_k$ are the *Fourier coefficients* , with respect to the orthogonal set $\{v_1, \ldots, v_k\}$.

Of particular interest is when a set of orthogonal vectors becomes a basis.

Definition 4.14 Let $\{v_1, \ldots, v_n\}$ be a set of orthogonal vectors in U. The set is said to be *complete* if it is a basis of U.

The completeness of a set of orthogonal vectors is seen to be characterized by the norms of vectors in relation to their Fourier coefficients.

Theorem 4.15 *Let U be a vector space with a positive definite scalar product and $V = \{v_1, \ldots, v_n\}$ a set of orthogonal vectors in U. For any $u \in U$, let the scalars $a_1, \ldots, a_n$ be the Fourier coefficients of u with respect to V.*

(1) *There holds the inequality*

$$\sum_{i=1}^{n} |a_i|^2 \|v_i\|^2 \leq \|u\|^2 \tag{4.4.11}$$

(which is often referred to as the Bessel inequality).

(2) *That $u \in \mathrm{Span}\{v_1, \ldots, v_n\}$ if and only if the equality in (4.4.11) is attained. That is,*

$$\sum_{i=1}^{n} |a_i|^2 \|v_i\|^2 = \|u\|^2 \tag{4.4.12}$$

(which is often referred to as the Parseval identity). Therefore, the set V is complete if and only if the Parseval identity (4.4.12) holds for any $u \in U$.

Proof For given u, use v to denote the Fourier expansion of v as stated in (4.4.3) and (4.4.4) with $k = n$. Then $\|u\|^2 = \|w\|^2 + \|v\|^2$. In particular, $\|v\|^2 \leq \|u\|^2$, which is (4.4.11).

It is clear that $u \in \mathrm{Span}\{v_1, \ldots, v_n\}$ if and only if $w = 0$, which is equivalent to the fulfillment of the equality $\|u\|^2 = \|v\|^2$, which is (4.4.12). $\qquad\square$

If $\{u_1, \ldots, u_n\}$ is a set of orthogonal unit vectors in U, the Fourier expansion and Fourier coefficients of a vector $u \in U$ with respect to $\{u_1, \ldots, u_n\}$ take the elegant forms

$$\sum_{i=1}^{n} a_i u_i, \quad a_i = (u_i, u), \quad i = 1, \ldots, n, \tag{4.4.13}$$

such that the Bessel inequality becomes

$$\sum_{i=1}^{n} |(u_i, u)|^2 \leq \|u\|^2. \tag{4.4.14}$$

Therefore $\{u_1, \ldots, u_n\}$ is complete if and only if the Parseval identity

$$\sum_{i=1}^{n} |(u_i, u)|^2 = \|u\|^2 \tag{4.4.15}$$

holds for any $u \in U$.

It is interesting to note that, since $a_i = (v_i, u)/\|v_i\|^2$ $(i = 1, \ldots, n)$ in (4.4.11) and (4.4.12), the inequalities (4.4.11) and (4.4.14), and the identities (4.4.12) and (4.4.15), are actually of the same structures, respectively.

Exercises

4.4.1 Let U be a finite-dimensional vector space with a positive definite scalar product, $(\cdot, \cdot)$, and $S = \{u_1, \ldots, u_n\}$ an orthogonal set of vectors in U. Show that the set S is complete if and only if $S^\perp = \{0\}$.

4.4.2 Let V be a nontrivial subspace of a vector space U with a positive definite scalar product. Show that if $\{v_1, \ldots, v_k\}$ is an orthogonal basis of V, then the mapping $P : U \to U$ given by its Fourier expansion,

$$P(u) = \sum_{i=1}^{k} \frac{(v_i, u)}{(v_i, v_i)} v_i, \quad u \in U, \tag{4.4.16}$$

is the projection of U along $V^\perp$ onto V. That is, $P \in L(U)$, $P^2 = P$, $N(P) = V^\perp$, and $R(P) = V$.

4.4.3 Let U be a finite-dimensional vector space with a positive definite scalar product, $(\cdot, \cdot)$, and V a subspace of U. Use $P_V : U \to U$ to denote the projection of U onto V along $V^\perp$. Prove that if W is a subspace of U containing V then

$$\|u - P_V(u)\| \geq \|u - P_W(u)\|, \quad u \in U, \tag{4.4.17}$$

and $V = W$ if and only if equality in (4.4.17) holds for any $u \in U$, where the norm $\|\cdot\|$ of U is induced from the positive definite scalar product $(\cdot, \cdot)$. In other words, orthogonal projection of a vector into a larger subspace provides a better approximation of the vector.

4.4.4 Let V be a nontrivial subspace of a finite-dimensional vector space U with a positive definite scalar product. Recall that over U/V we may define the norm

$$\|[u]\| = \inf\{\|x\| \,|\, x \in [u]\}, \quad [u] \in U/V, \tag{4.4.18}$$

for the quotient space U/V.

(a) Prove that for each $[u] \in U/V$ there is a unique $w \in [u]$ such that $\|[u]\| = \|w\|$.

(b) Find a practical method to compute the vector w shown to exist in part (a) among the coset $[u]$.

4.4.5   Consider the Euclidean space $\mathbb{R}^3$ with the usual scalar product and vector length or norm defined. Let V be the subspace of $\mathbb{R}^3$ which is the plane defined by the equation

$$V = \{(x,y,z) \in \mathbb{R}^3 \,|\, ax + by + cz = 0\}, \qquad (4.4.19)$$

where $(a,b,c) \in \mathbb{R}^3$ is a nonzero vector. Use the idea of orthogonal resolutions of vectors to establish that the distance from the point $(x_0, y_0, z_0) \in \mathbb{R}^3$ to the plane V is given by the formula

$$d = \frac{|ax_0 + by_0 + cz_0|}{\sqrt{a^2 + b^2 + c^2}}. \qquad (4.4.20)$$

(Hint: Set $u = (x_0, y_0, z_0) = v + w$, where $v \in V$ and $w \in V^\perp$. Then

$$d = \inf\{\|u - (x,y,z)\| \,|\, (x,y,z) \in V\}, \qquad (4.4.21)$$

and, in fact, $d = \|w\|$.)

4.4.6   (An extension of Problem 4.4.5) Consider the plane in $\mathbb{R}^3$:

$$V_h = \{(x,y,z) \in \mathbb{R}^3 \,|\, ax + by + cz = h\}, \quad h \in \mathbb{R}, \qquad (4.4.22)$$

where $(a,b,c) \in \mathbb{R}^3$ is a nonzero vector. For any $(x_0, y_0, z_0) \in \mathbb{R}^3$, establish that the distance formula from the point $(x_0, y_0, z_0) \in \mathbb{R}^3$ to the plane V_h is given by

$$d = \frac{|ax_0 + by_0 + cz_0 - h|}{\sqrt{a^2 + b^2 + c^2}}. \qquad (4.4.23)$$

4.4.7   Formulate and establish an n-dimensional version for the result in Problem 4.4.5. That is, develop a generalization of the result of Problem 4.4.5 in $\mathbb{R}^3$ to the situation of $\mathbb{R}^n$.

4.4.8   Formulate and establish an n-dimensional version for the result in Problem 4.4.6. That is, develop a generalization of the result of Problem 4.4.6 in $\mathbb{R}^3$ to the situation of $\mathbb{R}^n$.

4.4.9   Consider the vector space $\mathcal{P}_n$ of real-coefficient polynomials in variable t of degrees up to $n \geq 1$ with the positive-definite scalar product

$$(u,v) = \int_{-1}^{1} u(t)v(t)\,dt, \quad u,v \in \mathcal{P}_n. \qquad (4.4.24)$$

Applying the Gram–Schmidt procedure to the standard basis $\{1, t, \ldots, t^n\}$ of $\mathcal{P}_n$, we may construct an orthonormal basis, say $\{L_0, L_1, \ldots, L_n\}$, of $\mathcal{P}_n$. The set of polynomials $\{L_i(t)\}$ are the well-known *Legendre polynomials*. Explain why the degree of each $L_i(t)$ must be i ($i = 0, 1, \ldots, n$) and find $L_i(t)$ for $i = 0, 1, 2, 3, 4$.

4.4.10 In $\mathcal{P}_2$, find the Fourier expansion of the polynomial $u(t) = -3t^2 + t - 5$ in terms of the Legendre polynomials.

4.4.11 Let $\mathcal{P}_n$ be the vector space with the scalar product defined in (4.4.24). For $f \in \mathcal{P}'_3$ satisfying

$$f(1) = -1, \quad f(t) = 2, \quad f(t^2) = 6, \quad f(t^3) = -5, \qquad (4.4.25)$$

find an element $v \in \mathcal{P}_3$ such that v is the preimage of the Riesz mapping $\rho : \mathcal{P}_3 \to \mathcal{P}'_3$ of $f \in \mathcal{P}'_3$, or $\rho(v) = f$. That is, there holds

$$f(u) = (u, v), \quad u \in \mathcal{P}_3. \qquad (4.4.26)$$

4.5 Orthogonal and unitary versus isometric mappings

Let U be a vector space with a positive definite scalar product, $(\cdot, \cdot)$, which induces a norm $\| \cdot \|$ on U. If $T \in L(U)$ is orthogonal or unitary, then it is clear that

$$\|T(u) - T(v)\| = \|u - v\|, \quad u, v \in U. \qquad (4.5.1)$$

In other words, the distance of the images of any two vectors in U under T is the same as that between the two vectors. A mapping from U into itself satisfying such a property is called an *isometry* or *isometric*. In this section, we show that any mapping from a real vector space U with a positive definite scalar product into itself satisfying the property (4.5.1) must be linear. Therefore, in view of Theorems 4.6, it is orthogonal. In other words, in the real setting, being isometric characterizes a mapping being orthogonal.

Theorem 4.16 *Let U be a real vector space with a positive definite scalar product. A mapping T from U into itself satisfying the isometric property (4.5.1) and $T(0) = 0$ if and only if it is orthogonal.*

Proof   Assume T satisfies $T(0) = 0$ and (4.5.1). We show that T must be linear. To this end, from (4.5.1) and replacing v by 0, we get $\|T(u)\| = \|u\|$ for any $u \in U$. On the other hand, the symmetry of the scalar product $(\cdot, \cdot)$ gives us the identity

$$(u, v) = \frac{1}{2}(\|u + v\|^2 - \|u\|^2 - \|v\|^2), \quad u, v \in U. \qquad (4.5.2)$$

Replacing u, v in (4.5.2) by $T(u), -T(v)$, respectively, we get

$$
\begin{aligned}
-(T(u), T(v)) &= \frac{1}{2}(\|T(u) - T(v)\|^2 - \|T(u)\|^2 - \| - T(v)\|^2) \\
&= \frac{1}{2}(\|u - v\|^2 - \|u\|^2 - \|v\|^2) \\
&= -(u, v).
\end{aligned}
\tag{4.5.3}
$$

Hence $(T(u), T(v)) = (u, v)$ for any $u, v \in U$. Using this result, we have

$$
\begin{aligned}
&\|T(u + v) - T(u) - T(v)\|^2 \\
&= \|T(u + v)\|^2 + \|T(u)\|^2 + \|T(v)\|^2 \\
&\quad -2(T(u + v), T(u)) - 2(T(u + v), T(v)) + 2(T(u), T(v)) \\
&= \|u + v\|^2 + \|u\|^2 + \|v\|^2 - 2(u + v, u) - 2(u + v, v) + 2(u, v) \\
&= \|(u + v) - u - v\|^2 = 0,
\end{aligned}
\tag{4.5.4}
$$

which proves the additivity condition

$$
T(u + v) = T(u) + T(v), \quad u, v \in U.
\tag{4.5.5}
$$

Besides, setting $v = -u$ in (4.5.5) and using $T(0) = 0$, we have $T(-u) = -T(u)$ for any $u \in U$.

Moreover, (4.5.5) also implies that, for any integer m, we have $T(mu) = mT(u)$. Replacing u by $\frac{1}{m}u$ where m is a nonzero integer, we also have $T\left(\frac{1}{m}u\right) = \frac{1}{m}T(u)$. Combining these results, we conclude that

$$
T(ru) = rT(u), \quad r \in \mathbb{Q}, \quad u \in U.
\tag{4.5.6}
$$

Finally, for any $a \in \mathbb{R}$, let $\{r_k\}$ be a sequence in $\mathbb{Q}$ such that $r_k \to a$ as $k \to \infty$. Then we find

$$
\begin{aligned}
\|T(au) - aT(u)\| &\leq \|T(au) - T(r_k u)\| + \|r_k T(u) - aT(u)\| \\
&= |a - r_k|\|u\| + |r_k - a|\|T(u)\| \to 0 \quad \text{as } k \to \infty.
\end{aligned}
\tag{4.5.7}
$$

That is, $T(au) = aT(u)$ for any $a \in \mathbb{R}$ and $u \in U$. So homogeneity is established and the proof follows. □

We next give an example showing that, in the complex situation, being isometric alone is not sufficient to ensure a mapping to be unitary.

The vector space we consider here is taken to be $\mathbb{C}^n$ with the standard Hermitian scalar product (4.3.2) and the mapping $T : \mathbb{C}^n \to \mathbb{C}^n$ is given by $T(u) = \bar{u}$ for $u \in \mathbb{C}^n$. Then T satisfies (4.5.1) and $T(0) = 0$. However, T is not homogeneous with respect to scalar multiplication. More precisely, $T(au) \neq aT(u)$ whenever $a \in \mathbb{C}$ with $\Im(a) \neq 0$ and $u \in \mathbb{C}^n \setminus \{0\}$. Hence $T \notin L(U)$.

It will be interesting to spell out some conditions in the complex situation in addition to (4.5.1) which would when put together with (4.5.1) ensure a mapping to be unitary. The following theorem is such a result.

Theorem 4.17 *Let U be a complex vector space with a positive definite scalar product* $(\cdot, \cdot)$. *A mapping T from U into itself satisfying the isometric property (4.5.1), $T(0) = 0$, and*

$$\|iT(u) - T(v)\| = \|iu - v\|, \quad u, v \in U, \tag{4.5.8}$$

if and only if it is unitary, where $\|\cdot\|$ *is induced from* $(\cdot, \cdot)$.

Proof It is obvious that if $T \in L(U)$ is unitary then both (4.5.1) and (4.5.8) are fulfilled. We now need to show that the converse is true.

First, setting $v = iu$ in (4.5.8), we obtain

$$T(iu) = iT(u), \quad u \in U. \tag{4.5.9}$$

Next, replacing u, v in (4.3.27) by $T(u), -T(v)$, respectively, and using (4.5.1) and (4.5.8), we have

$$
\begin{aligned}
-(T(u), T(v)) &= \frac{1}{2}(\|T(u) - T(v)\|^2 - \|T(u)\|^2 - \|T(v)\|^2) \\
&\quad + \frac{1}{2}i(\|iT(u) - T(v)\|^2 - \|T(u)\|^2 - \|T(v)\|^2) \\
&= \frac{1}{2}(\|u - v\|^2 - \|u\|^2 - \|v\|^2) \\
&\quad + \frac{1}{2}i(\|iu - v\|^2 - \|u\|^2 - \|u\|^2) \\
&= -(u, v), \quad u, v \in U.
\end{aligned} \tag{4.5.10}
$$

That is, $(T(u), T(v)) = (u, v)$ for $u, v \in U$. Thus a direct expansion gives us the result

$$
\begin{aligned}
&\|T(u + v) - T(u) - T(v)\|^2 \\
&= \|T(u + v)\|^2 + \|T(u)\|^2 + \|T(v)\|^2 \\
&\quad - 2\Re(T(u + v), T(u)) - 2\Re(T(u + v), T(v)) + 2\Re(T(u), T(v)) \\
&= \|u + v\|^2 + \|u\|^2 + \|v\|^2 - 2\Re(u + v, u) - 2\Re(u + v, v) + 2\Re(u, v) \\
&= \|(u + v) - u - v\|^2 = 0, \tag{4.5.11}
\end{aligned}
$$

which establishes the additivity property $T(u + v) = T(u) + T(v)$ for $u, v \in U$ as before. Therefore (4.5.6) holds in the current complex formalism.

In view of the additivity property of T, (4.5.6), and (4.5.9), we have

$$T((p+iq)u) = T(pu) + T(iqu) = pT(u) + iqT(u)$$
$$= (p+iq)T(u), \quad p,q \in \mathbb{Q}, \quad u \in U. \qquad (4.5.12)$$

Finally, for any $a = b + ic \in \mathbb{C}$ where $b, c \in \mathbb{R}$, we may choose a sequence $\{r_k\}$ ($r_k = p_k + iq_k$ with $p_k, q_k \in \mathbb{Q}$) such that $p_k \to b$ and $q_k \to c$ or $r_k \to a$ as $k \to \infty$. In view of these and (4.5.12), we see that (4.5.7) holds in the complex situation as well. In other words, $T(au) = aT(u)$ for $a \in \mathbb{C}$ and $u \in U$.

The proof is complete. $\qquad\qquad\qquad\qquad\qquad\qquad\qquad\qquad\qquad\qquad$ □

We note that a more direct but perhaps less intuitive proof of the homogeneity of T without resorting to a limiting argument in the spirit of (4.5.7) is as follows.

In fact, for any $a \in \mathbb{C}$, using $(T(u), T(v)) = (u, v)$ for any $u, v \in U$, we have

$$\|T(au) - aT(u)\|^2 = (T(au) - aT(u), T(au) - aT(u))$$
$$= (T(au), T(au)) + |a|^2(T(u), T(u)) - a(T(u), T(au)) - \bar{a}(T(au), T(u))$$
$$= (au, au) + |a|^2(u, u) - a(u, au) - \bar{a}(au, u) = 0, \quad u \in U, \qquad (4.5.13)$$

as anticipated.

It is worth noting that Theorems 4.16 and 4.17 are valid in general without restricting to finite-dimensional spaces.

Exercises

4.5.1 Let U be a real vector space with a positive definite scalar product $(\cdot, \cdot)$. Establish the following variant of Theorem 4.16: A mapping T from U into itself satisfying the property

$$\|T(u) + T(v)\| = \|u + v\|, \quad u, v \in U, \qquad (4.5.14)$$

if and only if it is orthogonal, where $\| \cdot \|$ is induced from $(\cdot, \cdot)$.

4.5.2 Show that the mapping $T : \mathbb{C}^n \to \mathbb{C}^n$ given by $T(u) = \bar{u}$ for $u \in \mathbb{C}^n$, which is equipped with the standard Hermitian scalar product, satisfies (4.5.14) but that it is not unitary.

4.5.3 Let U be a complex vector space with a positive definite scalar product $(\cdot, \cdot)$. Establish the following variant of Theorem 4.17: A mapping T from U into itself satisfying the property (4.5.14) and

$$\|iT(u) + T(v)\| = \|iu + v\|, \quad u, v \in U, \qquad (4.5.15)$$

if and only if it is unitary, where $\| \cdot \|$ is induced from $(\cdot, \cdot)$.

4.5.4   Check to see why the mapping T defined in Problem 4.5.2 fails to satisfy the property (4.5.15) with $U = \mathbb{C}^n$.

4.5.5   Let $T : \mathbb{R}^n \to \mathbb{R}^n$ $(n \geq 2)$ be defined by

$$T(x) = \begin{pmatrix} x_n \\ \vdots \\ x_1 \end{pmatrix}, \quad x = \begin{pmatrix} x_1 \\ \vdots \\ x_n \end{pmatrix} \in \mathbb{R}^n, \tag{4.5.16}$$

where $\mathbb{R}^n$ is equipped with the standard Euclidean scalar product.

(a) Show that T is an isometry.
(b) Determine all the eigenvalues of T.
(c) Show that eigenvectors associated to different eigenvalues are mutually perpendicular.
(d) Determine the minimal polynomial of T.

5

Real Quadratic Forms and Self-Adjoint Mappings

In this chapter, we exclusively consider vector spaces over the field of reals unless otherwise stated. First, we present a general discussion on bilinear and quadratic forms and their matrix representations. We also show how a symmetric bilinear form may be uniquely represented by a self-adjoint mapping. Then, we establish the main spectrum theorem for self-adjoint mappings based on a proof of the existence of an eigenvalue using calculus. Next, we focus on characterizing the positive definiteness of self-adjoint mappings. After these we study the commutativity of self-adjoint mappings. Finally, we show the effectiveness of using self-adjoint mappings in computing the norm of a mapping between different spaces and in the formalism of least-squares approximations.

5.1 Bilinear and quadratic forms

Let U be a finite-dimensional vector space over $\mathbb{R}$. The simplest real-valued functions over U are linear functions, which are also called functionals earlier and have been studied. The next simplest real-valued functions to be studied are *bilinear forms* whose definition is given as follows.

Definition 5.1 A function $f : U \times U \to \mathbb{R}$ is called a *bilinear form* if it satisfies for any $u, v, w \in U$ and $a \in \mathbb{R}$ the following conditions:

(1) $f(u+v,w) = f(u,w) + f(v,w)$, $f(au,v) = af(u,v)$.
(2) $f(u,v+w) = f(u,v) + f(u,w)$, $f(u,av) = af(u,v)$.

Let $\mathcal{B} = \{u_1, \ldots, u_n\}$ be a basis of U. For $u, v \in U$ with coordinate vectors $x = (x_1, \ldots, x_n)^t, y = (y_1, \ldots, y_n)^t \in \mathbb{R}^n$ with respect to $\mathcal{B}$, we have

$$f(u,v) = f\left(\sum_{i=1}^{n} x_i u_i, \sum_{j=1}^{n} y_j u_j\right) = \sum_{i,j=1}^{n} x_i f(u_i, u_j) y_j = x^t A y, \qquad (5.1.1)$$

153

where $A = (a_{ij}) = (f(u_i, u_j)) \in \mathbb{R}(n, n)$ is referred to as the matrix representation of the bilinear form f with respect to the basis $\mathcal{B}$.

Let $\tilde{\mathcal{B}} = \{\tilde{u}_1, \ldots, \tilde{u}_n\}$ be another basis of U so that $\tilde{A} = (\tilde{a}_{ij}) = (f(\tilde{u}_i, \tilde{u}_j)) \in \mathbb{R}(n, n)$ is the matrix representation of f with respect to $\tilde{\mathcal{B}}$. If $\tilde{x}$ and $\tilde{y} \in \mathbb{R}^n$ are the coordinate vectors of u and $v \in U$ with respect to $\tilde{\mathcal{B}}$ and the basis transition matrix between $\mathcal{B}$ and $\tilde{\mathcal{B}}$ is $B = (b_{ij})$ so that

$$\tilde{u}_j = \sum_{i=1}^{n} b_{ij} u_i, \quad j = 1, \ldots, n, \tag{5.1.2}$$

then $x = B\tilde{x}$ and $y = B\tilde{y}$ (cf. Section 1.3). Hence we arrive at

$$f(u, v) = \tilde{x}^t \tilde{A} \tilde{y} = x^t A y = \tilde{x}^t (B^t A B) \tilde{y}, \tag{5.1.3}$$

which leads to the relation $\tilde{A} = B^t A B$ and gives rise to the following concept:

Definition 5.2 For A and $B \in \mathbb{F}(n, n)$, we say that A and B are *congruent* if there is an invertible element $C \in \mathbb{F}(n, n)$ such that

$$A = C^t B C. \tag{5.1.4}$$

Therefore, our calculation above shows that the matrix representations of a bilinear form with respect to different bases are congruent.

For a bilinear form $f : U \times U \to \mathbb{R}$, we can set

$$q(u) = f(u, u), \quad u \in U, \tag{5.1.5}$$

which is called the *quadratic form* associated with the bilinear form f. The quadratic form q is *homogeneous* of degree 2 since $q(tu) = t^2 q(u)$ for any $t \in \mathbb{R}$ and $u \in U$.

Of course q is uniquely determined by f through (5.1.5). However, the converse is not true, which will become clear after the following discussion.

To proceed, let $\mathcal{B} = \{u_1, \ldots, u_n\}$ be a basis of U and $u \in U$ any given vector whose coordinate vector with respect to $\mathcal{B}$ is $x = (x_1, \ldots, x_n)^t$. Then, from (5.1.1), we have

$$q(u) = x^t A x = x^t \left(\frac{1}{2}(A + A^t) + \frac{1}{2}(A - A^t) \right) x$$
$$= \frac{1}{2} x^t (A + A^t) x, \tag{5.1.6}$$

since $x^t A x$ is a scalar which results in $x^t A x = (x^t A x)^t = x^t A^t x$. In other words, the quadratic form q constructed from (5.1.5) can only capture the information contained in the symmetric part, $\frac{1}{2}(A + A^t)$, but nothing in the skew-symmetric part, $\frac{1}{2}(A - A^t)$, of the matrix $A = (f(u_i, u_j))$. Consequently, the quadratic form q cannot determine the bilinear form f completely, in general situations, unless the skew-symmetric part of A is absent or $A - A^t = 0$. In other words, $A = (f(u_i, u_j))$ is symmetric. This observation motivates the following definition.

Definition 5.3 A bilinear form $f : U \times U \to \mathbb{R}$ is *symmetric* if it satisfies

$$f(u, v) = f(v, u), \quad u, v \in U. \tag{5.1.7}$$

If f is a symmetric bilinear form, then we have the expansion

$$f(u + v, u + v) = f(u, u) + f(v, v) + 2f(u, v), \quad u, v \in U. \tag{5.1.8}$$

Thus, if q is the quadratic form associated with f, we derive from (5.1.8) the relation

$$f(u, v) = \frac{1}{2}(q(u + v) - q(u) - q(v)), \quad u, v \in U, \tag{5.1.9}$$

which indicates how f is uniquely determined by q. In a similar manner, we also have

$$f(u, v) = \frac{1}{4}(q(u + v) - q(u - v)), \quad u, v \in U. \tag{5.1.10}$$

As in the situation of scalar products, the relations of the types (5.1.9) and (5.1.10) are often referred to as *polarization identities* for symmetric bilinear forms.

From now on we will concentrate on symmetric bilinear forms.

Let $f : U \times U \to \mathbb{R}$ be a symmetric bilinear form. If x and $y \in \mathbb{R}^n$ are the coordinate vectors of u and $v \in U$ with respect to a basis $\mathcal{B}$, then $f(u, v)$ is given by (5.1.1) so that matrix $A \in \mathbb{R}(n, n)$ is symmetric. Recall that $(x, y) = x^t y$ is the Euclidean scalar product over $\mathbb{R}^n$. Thus, if we view A as a linear mapping $\mathbb{R}^n \to \mathbb{R}^n$ given by $x \mapsto Ax$, then the right-hand side of (5.1.1) is simply (x, Ay). Since $A = A^t$, the right-hand side of (5.1.1) is also (Ax, y). In other words, A defines a self-adjoint mapping over $\mathbb{R}^n$ with respect to the standard Euclidean scalar product over $\mathbb{R}^n$.

Conversely, if U is a vector space equipped with a positive definite scalar product $(\cdot, \cdot)$ and $T \in L(U)$ is a self-adjoint or symmetric mapping, then

$$f(u, v) = (u, T(v)), \quad u, v \in U, \tag{5.1.11}$$

is a symmetric bilinear form. Thus, in this way, we see that symmetric bilinear forms are completely characterized.

In a more precise manner, we have the following theorem, which relates symmetric bilinear forms and self-adjoint mappings over a vector space with a positive definite scalar product.

Theorem 5.4 *Let U be a finite-dimensional vector space with a positive definite scalar product $(\cdot, \cdot)$. For any symmetric bilinear form $f : U \times U \to \mathbb{R}$, there* *is a unique self-adjoint or symmetric linear mapping, say $T \in L(U)$, such that the relation (5.1.11) holds.*

Proof   For each $v \in U$, the existence of a unique vector, say $T(v)$, so that (5.1.11) holds, is already shown in Section 4.2. Since f is bilinear, we have $T \in L(U)$. The self-adjointness or symmetry of T follows from the symmetry of f and the scalar product.   □

Note that, in view of Section 4.1, a symmetric bilinear form is exactly a kind of scalar product as well, not necessarily positive definite, though.

Exercises

5.1.1   Let $f : U \times U \to \mathbb{R}$ be a bilinear form such that $f(u,u) > 0$ and $f(v,v) < 0$ for some u and $v \in U$.

(a) Show that u and v are linearly independent.
(b) Show that there is some $w \in U, w \neq 0$ such that $f(w,w) = 0$.

5.1.2   Let A and $B \in \mathbb{F}(n,n)$. Show that if A, B are congruent then A, B must have the same rank.

5.1.3   Let A and $B \in \mathbb{F}(n,n)$. Show that if A, B are congruent and A is symmetric then so is B.

5.1.4   Are the matrices

$$A = \begin{pmatrix} 2 & 1 \\ 1 & 1 \end{pmatrix} \text{ and } B = \begin{pmatrix} 0 & 1 \\ 1 & 0 \end{pmatrix} \tag{5.1.12}$$

congruent in $\mathbb{R}(2,2)$?

5.1.5   Consider the identity matrix $I_n \in \mathbb{F}(n,n)$. Show that I_n and $-I_n$ are not congruent if $\mathbb{F} = \mathbb{R}$ but are congruent if $\mathbb{F} = \mathbb{C}$.

5.1.6   Consider the quadratic form

$$q(x) = x_1^2 + 2x_2^2 - x_3^2 + 2x_1 x_2 - 4x_1 x_3, \quad x = \begin{pmatrix} x_1 \\ x_2 \\ x_3 \end{pmatrix} \in \mathbb{R}^3, \tag{5.1.13}$$

where $\mathbb{R}^3$ is equipped with the standard Euclidean scalar product $(x,y) = x^t y$ for $x, y \in \mathbb{R}^3$.

(a) Find the unique symmetric bilinear form $f : \mathbb{R}^3 \times \mathbb{R}^3 \to \mathbb{R}$ such that $q(x) = f(x,x)$ for $x \in \mathbb{R}^3$.
(b) Find the unique self-adjoint mapping $T \in L(\mathbb{R}^3)$ such that $f(x,y) = (x, T(y))$ for any x and $y \in \mathbb{R}^3$.

5.1.7   Let $f : U \times U \to \mathbb{R}$ be a bilinear form.

   (a) Show that f is skew-symmetric, that is, $f(u,v) = -f(v,u)$ for any
       u and $v \in U$, if and only if $f(u,u) = 0$ for any $u \in U$.
   (b) Show that if f is skew-symmetric then the matrix representation of
       f with respect to an arbitrary basis of U must be skew-symmetric.

5.1.8   Prove that Theorem 5.4 is still true if the positive definite scalar product
        $(\cdot,\cdot)$ of U is only assumed to be nondegenerate instead to ensure the
        existence and uniqueness of a self-adjoint element $T \in L(U)$.

5.2 Self-adjoint mappings

Let U be a finite-dimensional vector space with a positive definite scalar
product, $(\cdot,\cdot)$. For self-adjoint mappings, we have the following foundational
theorem:

Theorem 5.5 *Assume that $T \in L(U)$ is self-adjoint with respect to the scalar
product of U. Then*

(1) *T must have a real eigenvalue.*
(2) *There is an orthonormal basis of U consisting of eigenvectors of T.*

Proof   We use induction on $\dim(U)$.
   There is nothing to show at $\dim(U) = 1$.
   Assume that the theorem is true at $\dim(U) = n - 1 \geq 1$.
   We investigate the situation when $\dim(U) = n \geq 2$.
   For convenience, let $\mathcal{B} = \{u_1, \dots, u_n\}$ be an orthonormal basis of U. For any
vector $u \in U$, let $x = (x_1, \dots, x_n)^t \in \mathbb{R}^n$ be its coordinate vector. Use $A = (a_{ij}) \in$
$\mathbb{R}(n,n)$ to denote the matrix representation of T with respect to $\mathcal{B}$ so that

$$T(u_j) = \sum_{i=1}^{n} a_{ij} u_i, \quad i = 1, \dots, n. \tag{5.2.1}$$

Then set

$$Q(x) = (u, T(u)) = \left(\sum_{i=1}^{n} x_i u_i, \sum_{j=1}^{n} x_j T(u_j) \right)$$

$$= \left(\sum_{i=1}^{n} x_i u_i, \sum_{j=1}^{n} x_j \sum_{k=1}^{n} a_{kj} u_k \right) = \sum_{i,j,k=1}^{n} x_i x_j a_{kj} \delta_{ik}$$

$$= \sum_{i,j=1}^{n} x_i x_j a_{ij} = x^t A x. \tag{5.2.2}$$

Consider the unit sphere in U given as

$$S = \left\{ u = \sum_{i=1}^{n} x_i u_i \in U \,\middle|\, \|u\|^2 = (u,u) = \sum_{i=1}^{n} x_i^2 = 1 \right\}, \qquad (5.2.3)$$

which may also be identified as the unit sphere in $\mathbb{R}^n$ centered at the origin and commonly denoted by S^{n-1}. Since S^{n-1} is compact, the function Q given in (5.2.2) attains its minimum over S^{n-1} at a certain point on S^{n-1}, say $x^0 = (x_1^0, \dots, x_n^0)^t$.

We may assume $x_n^0 \neq 0$. Without loss of generality, we may also assume $x_n^0 > 0$ (the case $x_n^0 < 0$ can be treated similarly). Hence, near x^0, we may represent the points on S^{n-1} by the formulas

$$x_n = \sqrt{1 - x_1^2 - \cdots - x_{n-1}^2}, \text{ where } (x_1, \dots, x_{n-1}) \text{ is near } (x_1^0, \dots, x_{n-1}^0).$$
$$(5.2.4)$$

Therefore, with

$$P(x_1, \dots, x_{n-1}) = Q\left(x_1, \dots, x_{n-1}, \sqrt{1 - x_1^2 - \cdots - x_{n-1}^2}\right), \qquad (5.2.5)$$

we see that $(x_1^0, \dots, x_{n-1}^0)$ is a critical point of P. Thus we have

$$(\nabla P)(x_1^0, \dots, x_{n-1}^0) = 0. \qquad (5.2.6)$$

In order to carry out the computation involved in (5.2.6), we rewrite the function P as

$$P(x_1, \dots, x_{n-1}) = \sum_{i,j=1}^{n-1} a_{ij} x_i x_j + 2 \sum_{i=1}^{n-1} a_{in} x_i \sqrt{1 - x_1^2 - \cdots - x_{n-1}^2}$$
$$+ a_{nn}(1 - x_1^2 - \cdots - x_{n-1}^2). \qquad (5.2.7)$$

Thus

$$\frac{\partial P}{\partial x_i} = 2 \sum_{j=1}^{n-1} a_{ij} x_j + 2 a_{in} \sqrt{1 - x_1^2 - \cdots - x_{n-1}^2}$$

$$- 2 \frac{x_i}{\sqrt{1 - x_1^2 - \cdots - x_{n-1}^2}} \sum_{j=1}^{n-1} a_{jn} x_j - 2 a_{nn} x_i, \quad i = 1, \dots, n-1. \quad (5.2.8)$$

Using (5.2.8) in (5.2.6), we arrive at

$$\sum_{j=1}^{n} a_{ij} x_j^0 = \frac{1}{x_n^0} \left(\sum_{j=1}^{n} a_{jn} x_j^0 \right) x_i^0, \quad i = 1, \dots, n-1. \qquad (5.2.9)$$

Note that we also have

$$\sum_{j=1}^{n} a_{nj} x_j^0 = \frac{1}{x_n^0} \left(\sum_{j=1}^{n} a_{jn} x_j^0 \right) x_n^0 \qquad (5.2.10)$$

automatically since A is symmetric. Combining (5.2.9) and (5.2.10), we get

$$Ax^0 = \lambda_0 x^0, \quad \lambda_0 = \frac{1}{x_n^0} \left(\sum_{j=1}^{n} a_{jn} x_j^0 \right), \qquad (5.2.11)$$

which establishes that λ_0 is an eigenvalue and x^0 an eigenvector associated with λ_0, of A.

Let $v_1 \in U$ be such that its coordinate vector with respect to the basis $\mathcal{B}$ is x^0. Then we have

$$
\begin{aligned}
T(v_1) = T \left(\sum_{j=1}^{n} x_j^0 u_j \right) &= \sum_{j=1}^{n} x_j^0 T(u_j) \\
&= \sum_{i,j=1}^{n} a_{ij} u_i x_j^0 = \sum_{i=1}^{n} \left(\sum_{j=1}^{n} a_{ij} x_j^0 \right) u_i \\
&= \lambda_0 \sum_{i=1}^{n} x_i^0 u_i = \lambda_0 v_1,
\end{aligned}
\qquad (5.2.12)
$$

which verifies that λ_0 is an eigenvalue of T and v_1 is an associated eigenvector.

Now set $U_1 = \text{Span}\{v_1\}$ and make the decomposition $U = U_1 \oplus U_1^\perp$. We claim that $U_1^\perp$ is invariant under T. In fact, for any $u \in U_1^\perp$, we have $(u, v_1) = 0$. Hence, $(T(u), v_1) = (u, T(v_1)) = (u, \lambda_0 v_1) = \lambda_0 (u, v_1) = 0$, which indicates $T(u) \in U_1^\perp$.

Of course, $\dim(U_1^\perp) = n - 1$. Using the inductive assumption, we know that $U_1^\perp$ has an orthonormal basis $\{v_2, \dots, v_n\}$, so that each v_i is an eigenvector associated with a real eigenvalue of T, $i = 2, \dots, n$.

Finally, we may rescale v_1 to make it a unit vector. Thus $\{v_1, v_2, \dots, v_n\}$ is an orthonormal basis of U so that each vector v_i is an eigenvector associated with a real eigenvalue of T, $i = 1, 2, \dots, n$, as desired.

With respect to the basis $\{v_1, \dots, v_n\}$, the matrix representation of T is diagonal, whose diagonal entries are the eigenvalues of T which are shown to be all real. In other words, all eigenvalues of T are real. $\qquad \square$

Let $T \in L(U)$ be self-adjoint, use $\lambda_1, \dots, \lambda_k$ to denote all the distinct eigenvalues of T, and denote by $E_{\lambda_1}, \dots, E_{\lambda_k}$ the corresponding eigenspaces, which are of course invariant subspaces of T. Using Theorem 5.5, we see that there holds the direct sum

$$U = E_{\lambda_1} \oplus \cdots \oplus E_{\lambda_k}. \qquad (5.2.13)$$

In particular, we may use E_0 to denote the eigenspace corresponding to the eigenvalue 0 (if any). Moreover, we may set

$$E_+ = \bigoplus_{\lambda_i > 0} E_{\lambda_i}, \quad E_- = \bigoplus_{\lambda_i < 0} E_{\lambda_i}. \qquad (5.2.14)$$

The associated numbers

$$n_0 = \dim(E_0), \quad n_+ = \dim(E_+), \quad n_- = \dim(E_-), \qquad (5.2.15)$$

are exactly what were previously called the indices of nullity, positivity, and negativity, respectively, in the context of a real scalar product in Section 4.2, which is simply a symmetric bilinear form here and may always be represented by a self-adjoint mapping over U. It is clear that n_0 is simply the nullity of T, or $n_0 = n(T)$, and $n_+ + n_-$ is the rank of T, $n_+ + n_- = r(T)$. Furthermore, for $u_i \in E_{\lambda_i}, u_j \in E_{\lambda_j}$, we have

$$\lambda_i(u_i, u_j) = (T(u_i), u_j) = (u_i, T(u_j)) = \lambda_j(u_i, u_j), \qquad (5.2.16)$$

which leads to $(\lambda_i - \lambda_j)(u_i, u_j) = 0$. Thus, for $i \neq j$, we have $(u_i, u_j) = 0$. In other words, the eigenspaces associated with distinct eigenvalues of T are mutually perpendicular. (This observation suggests a practical way to construct an orthogonal basis of U consisting of eigenvectors of T: First find all eigenspaces of T. Then obtain an orthogonal basis for each of these eigenspaces by using the Gram–Schmidt procedure. Finally put all these orthogonal bases together to get an orthogonal basis of the full space.)

A useful matrix version of Theorem 5.5 may be stated as follows.

Theorem 5.6 *A matrix $A \in \mathbb{R}(n,n)$ is symmetric if and only if there is an orthogonal matrix $P \in \mathbb{R}(n,n)$ such that*

$$A = P^t D P, \qquad (5.2.17)$$

where $D \in \mathbb{R}(n,n)$ is a diagonal matrix and the diagonal entries of D are the eigenvalues of A.

Proof   The right-hand side of (5.2.17) is of course symmetric.

Conversely, let A be symmetric. The linear mapping $T \in L(\mathbb{R}^n)$ defined by $T(x) = Ax$, with $x \in \mathbb{R}^n$ any column vector, is self-adjoint with respect to the standard scalar product over $\mathbb{R}^n$. Hence there are column vectors $u_1, \ldots, u_n$ in $\mathbb{R}^n$ consisting of eigenvectors of T or A associated with real eigenvalues, say $\lambda_1, \ldots, \lambda_n$, which form an orthonormal basis of $\mathbb{R}^n$. The relations $Au_1 = \lambda_1 u_1, \ldots, Au_n = \lambda_n u_n$, may collectively be rewritten as $AQ = QD$, where $D = \text{diag}\{\lambda_1, \ldots, \lambda_n\}$ and $Q \in \mathbb{R}(n,n)$ is made of taking $u_1, \ldots, u_n$ as the respective column vectors. Since $u_i^t u_j = \delta_{ij}, i, j = 1,\ldots,n$, we see that Q is an orthogonal matrix. Setting $P = Q^t$, we arrive at (5.2.17).   □

Note that the proof of Theorem 5.6 gives us a practical way to construct the matrices P and D in (5.2.17).

Exercises

5.2.1   The fact that a symmetric matrix $A \in \mathbb{R}(n,n)$ has and can only have real eigenvalues may also be proved algebraically more traditionally as follows: Consider A as an element in $\mathbb{C}(n,n)$ and let $\lambda \in \mathbb{C}$ be any of its eigenvalue whose existence is ensured by Fundamental Theorem of Algebra. Let $u \in \mathbb{C}^n$ be an eigenvector of A associated to λ. Then, using $(\cdot,\cdot)$ to denote the standard Hermitian scalar product over $\mathbb{C}^n$, we have

$$\lambda(u,u) = \lambda u^\dagger u = (u, Au). \qquad (5.2.18)$$

Use (5.2.18) to show that λ must be real, or equivalently, $\lambda = \overline{\lambda}$. Then show that A has an eigenvector in $\mathbb{R}^n$ associated to λ.

5.2.2   Show that for any $B \in \mathbb{R}(m,n)$ all eigenvalues of $A = B^t B \in \mathbb{R}(n,n)$ are real and non-negative.

5.2.3   Let $A \in \mathbb{R}(n,n)$ and $A^t = -A$.

(a) Show that if n is odd, then A must be singular.
(b) Show that if A is non-singular, then A has no real eigenvalue.
(c) Show that if A is viewed has an element in $\mathbb{C}(n,n)$, then any eigenvalue of A must be either 0 or imaginary.

5.2.4   Consider the quadratic form

$$q(x) = x_1^2 + 2x_2^2 - 2x_3^2 + 4x_1 x_3, \quad x = \begin{pmatrix} x_1 \\ x_2 \\ x_3 \end{pmatrix} \in \mathbb{R}^3. \qquad (5.2.19)$$

(a) Find a symmetric matrix $A \in \mathbb{R}(3,3)$ such that $q(x) = x^t A x$.
(b) Find an orthonormal basis of $\mathbb{R}^3$ (with respect to the standard Euclidean scalar product of $\mathbb{R}^3$) consisting of the eigenvectors of A.
(c) Find an orthogonal matrix $P \in \mathbb{R}(3,3)$ so that the substitution of the variable, $y = Px$, transforms the quadratic form (5.2.19) into the diagonal form

$$\lambda_1 y_1^2 + \lambda_2 y_2^2 + \lambda_3 y_3^2 = y^t \operatorname{diag}\{\lambda_1, \lambda_2, \lambda_3\} y, \quad y = \begin{pmatrix} y_1 \\ y_2 \\ y_3 \end{pmatrix} \in \mathbb{R}^3,$$

(5.2.20)

where λ_1, λ_2, and λ_3 are the eigenvalues of A, counting multiplicities.

5.2.5 Show that any symmetric matrix $A \in \mathbb{R}(n,n)$ must be congruent with a diagonal matrix of the form $D = \operatorname{diag}\{d_1, \ldots, d_n\}$ where $d_i = \pm 1$ or 0 for $i = 1, \ldots, n$.

5.2.6 Let $A \in \mathbb{R}(n,n)$ be symmetric and $\det(A) < 0$. Show that there is a column vector $x \in \mathbb{R}^n$ such that $x^t A x < 0$.

5.2.7 Show that if $A \in \mathbb{R}(n,n)$ is orthogonal then $\operatorname{adj}(A) = \pm A^t$.

5.2.8 Show that if $A_1, \ldots, A_k \in \mathbb{R}(n,n)$ are orthogonal, so is their product $A = A_1 \cdots A_k$.

5.2.9 Assume that $A \in \mathbb{R}(n,n)$ is symmetric and all of its eigenvalues are ± 1. Prove that A is orthogonal.

5.2.10 Let $A \in \mathbb{R}(n,n)$ be an upper or lower triangular matrix. If A is orthogonal, show that A must be diagonal and the diagonal entries of A can only be ± 1.

5.2.11 Show that if $T \in L(U)$ is self-adjoint and $T^m = 0$ for some integer $m \geq 1$ then $T = 0$.

5.2.12 Show that if $T \in L(U)$ is self-adjoint and m an odd positive integer then there is a self-adjoint element $S \in L(U)$ such that $T = S^m$.

5.2.13 Let $A \in \mathbb{R}(n,n)$ be symmetric and satisfy the equation $A^3 + A^2 + 4I_n = 0$. Prove that $A = -2I_n$.

5.2.14 Let $A \in \mathbb{R}(n,n)$ be orthogonal.

(a) Show that the real eigenvalues of A can only be ± 1.
(b) If $n = $ odd and $\det(A) = 1$, then 1 is an eigenvalue of A.
(c) If $\det(A) = -1$, then -1 is an eigenvalue of A.

5.2.15 Let $x \in \mathbb{R}^n$ be a nonzero column vector. Show that

$$P = I_n - \left(\frac{2}{x^t x}\right) x x^t \tag{5.2.21}$$

is an orthogonal matrix.

5.2.16 Let $A \in \mathbb{R}(n,n)$ be an idempotent symmetric matrix such that $r(A) = r$. Show that the characteristic polynomial of A is

$$p_A(\lambda) = \det(\lambda I_n - A) = (\lambda - 1)^r \lambda^{n-r}. \tag{5.2.22}$$

5.2.17 Let $A \in \mathbb{R}(n,n)$ be a symmetric matrix whose eigenvalues are all non-negative. Show that $\det(A + I_n) > 1$ if $A \neq 0$.

5.2.18 Let $u \in \mathbb{R}^n$ be a nonzero column vector. Show that there is an orthogonal matrix $Q \in \mathbb{R}(n,n)$ such that

$$Q^t(uu^t)Q = \text{diag}\{u^t u, 0, \dots, 0\}. \qquad (5.2.23)$$

5.3 Positive definite quadratic forms, mappings, and matrices

Let q be a quadratic form over a finite-dimensional vector space U with a positive definite scalar product $(\cdot, \cdot)$ and f a symmetric bilinear form that induces q: $q(u) = f(u,u), u \in U$. Then there is a self-adjoint mapping $T \in L(U)$ such that

$$q(u) = (u, T(u)), \quad u \in U. \qquad (5.3.1)$$

In this section, we apply the results of Section 5.2 to investigate the situation when q stays positive which is important in applications.

Definition 5.7 The positive definiteness of various subjects of concern is defined as follows:

(1) A quadratic form q over U is said to be *positive definite* if

$$q(u) > 0, \quad u \in U, \quad u \neq 0. \qquad (5.3.2)$$

(2) A self-adjoint mapping $T \in L(U)$ is said to be *positive definite* if

$$(u, T(u)) > 0, \quad u \in U, \quad u \neq 0. \qquad (5.3.3)$$

(3) A symmetric matrix $A \in \mathbb{R}(n,n)$ is said to be *positive definite* if for any nonzero column vector $x \in \mathbb{R}^n$ there holds

$$x^t A x > 0. \qquad (5.3.4)$$

Thus, when a quadratic form q and a self-adjoint mapping T are related through (5.3.1), then the positive definiteness of q and T are equivalent. Therefore it will be sufficient to study the positive definiteness of self-adjoint mappings, which will be shown to be equivalent to the positive definiteness of any matrix representations of the associated bilinear forms of the self-adjoint mappings. Recall that the matrix representation of the symmetric bilinear form induced from T with respect to an arbitrary basis $\{u_1, \dots, u_n\}$ of U is a matrix $A \in \mathbb{R}(n,n)$ defined by

$$A = (a_{ij}), \quad a_{ij} = (u_i, T(u_j)), \quad i, j = 1, \dots, n. \qquad (5.3.5)$$

The following theorem links various notions of positive definiteness.

Theorem 5.8 *That a self-adjoint mapping $T \in L(U)$ is positive definite is equivalent to any of the following statements:*

(1) *All the eigenvalues of T are positive.*
(2) *There is a positive constant, $\lambda_0 > 0$, such that*

$$(u, T(u)) \geq \lambda_0 \|u\|^2, \quad u \in U. \tag{5.3.6}$$

(3) *There is a positive definite mapping $S \in L(U)$ such that $T = S^2$.*
(4) *The matrix A defined in (5.3.5) with respect to an arbitrary basis is positive definite.*
(5) *The eigenvalues of the matrix A defined in (5.3.5) with respect to an arbitrary basis are all positive.*
(6) *The matrix A defined in (5.3.5) with respect to an arbitrary basis enjoys the factorization $A = B^2$ for some positive definite matrix $B \in \mathbb{R}(n,n)$.*

Proof   Assume T is positive definite. Let λ be any eigenvalue of T and $u \in U$ an associated eigenvector. Then $\lambda \|u\|^2 = (u, Tu) > 0$. Thus $\lambda > 0$ and (1) follows.

Assume (1) holds. Let $\{u_1, \ldots, u_n\}$ be an orthonormal basis of U so that u_i is an eigenvector associated with the eigenvalue λ_i $(i = 1, \ldots, n)$. Set $\lambda_0 = \min\{\lambda_1, \ldots, \lambda_n\}$. Then $\lambda_0 > 0$. Moreover, for any $u \in U$ with $u = \sum_{i=1}^{n} a_i u_i$, where $a_i \in \mathbb{R}$ $(i = 1, \ldots, n)$, we have

$$(u, T(u)) = \left(\sum_{i=1}^{n} a_i u_i, \sum_{j=1}^{n} \lambda_j a_j u_j \right) = \sum_{i=1}^{n} \lambda_i a_i^2 \geq \lambda_0 \sum_{i=1}^{n} a_i^2 = \lambda_0 \|u\|^2, \tag{5.3.7}$$

which establishes (2). It is obvious that (2) implies the positive definiteness of T.

We now show that the positive definiteness of T is equivalent to the statement (3). In fact, let $\{u_1, \ldots, u_n\}$ be an orthonormal basis of U consisting of eigenvectors with $\{\lambda_1, \ldots, \lambda_n\}$ the corresponding eigenvalues. In view of (1), $\lambda_i > 0$ for $i = 1, \ldots, n$. Now define $S \in L(U)$ to satisfy

$$S(u_i) = \sqrt{\lambda_i} u_i, \quad i = 1, \ldots, n. \tag{5.3.8}$$

It is clear that $T = S^2$.

Conversely, if $T = S^2$ for some positive definite mapping S, then 0 is not an eigenvalue of S. So S is invertible. Therefore

$$(u, T(u)) = (u, S^2(u)) = (S(u), S(u))$$
$$= \|S(u)\|^2 > 0, \quad u \in U, \quad u \neq 0, \tag{5.3.9}$$

and the positive definiteness of T follows.

Let $\{u_1,\dots,u_n\}$ be an arbitrary basis of U and $x = (x_1,\dots,x_n)^t \in \mathbb{R}^n$ a nonzero vector. For $u = \sum_{i=1}^n x_i u_i$, we have

$$
(u, T(u)) = \left(\sum_{i=1}^n x_i u_i, \sum_{j=1}^n x_j T(u_j) \right)
$$
$$
= \sum_{i,j=1}^n x_i (u_i, T(u_j)) x_j = x^t A x, \tag{5.3.10}
$$

which is positive if T is positive definite, and vice versa. So the equivalence of the positive definiteness of T and the statement (4) follows.

Suppose that A is positive definite. Let λ be any eigenvalue of A and x an associated eigenvector. Then $x^t A x = \lambda x^t x > 0$ implies $\lambda > 0$ since $x^t x > 0$. So (5) follows.

Now assume (5). Using Theorem 5.6, we see that $A = P^t D P$, where D is a diagonal matrix in $\mathbb{R}(n,n)$ whose diagonal entries are the positive eigenvalues of A, say $\lambda_1,\dots,\lambda_n$, and $P \in \mathbb{R}(n,n)$ is an orthogonal matrix. Then, with $y = Px = (y_1,\dots,y_n)^t \in \mathbb{R}^n$ where $x \in \mathbb{R}^n$, we have

$$
x^t A x = (Px)^t D Px = y^t D y = \sum_{i=1}^n \lambda_i y_i^2 > 0, \tag{5.3.11}
$$

whenever $x \neq 0$ since P is non-singular. Hence (5) implies (4) as well.

Finally, assume (5) holds and set

$$
D_1 = \mathrm{diag}\{ \sqrt{\lambda_1}, \dots, \sqrt{\lambda_n} \}. \tag{5.3.12}
$$

Then $D_1^2 = D$. Thus

$$
A = P^t D P = P^t D_1^2 P = (P^t D_1 P)(P^t D_1 P) = B^2, \quad B = P^t D_1 P. \tag{5.3.13}
$$

Using (5) we know that B is positive definite. So (6) follows. Conversely, if (6) holds, we may use the symmetry of B to get $x^t A x = x^t B^2 x = (Bx)^t (Bx) > 0$ whenever $x \in \mathbb{R}^n$ with $x \neq 0$ because B is non-singular which implies $Bx \neq 0$. Thus (4) holds.

The proof is complete. □

We note that, if in (5.3.5) the basis $\{u_1,\dots,u_n\}$ is orthonormal, then

$$
T(u_j) = \sum_{i=1}^n (u_i, T(u_j)) u_i = \sum_{i=1}^n a_{ij} u_i, \quad j = 1,\dots,n, \tag{5.3.14}
$$

which coincides with our notation adapted earlier.

For practical convenience, it is worth stating the following matrix version of Theorem 5.8.

Theorem 5.9 *That a symmetric matrix $A \in \mathbb{R}(n,n)$ is positive definite is equivalent to any of the following statements:*

(1) *All the eigenvalues of A are positive.*
(2) *There is a positive definite matrix $B \in \mathbb{R}(n,n)$ such that $A = B^2$.*
(3) *A is congruent to the identity matrix. That is, there is a non-singular matrix $B \in \mathbb{R}(n,n)$ such that $A = B^t I_n B = B^t B$.*

Proof   That A is positive definite is equivalent to either (1) or (2) has already been demonstrated in the proof of Theorem 5.8 and it remains to establish (3).

If there is a non-singular matrix $B \in \mathbb{R}(n,n)$ such that $A = B^t B$, then $x^t A x = (Bx)^t (Bx) > 0$ for $x \in \mathbb{R}^n$ with $x \neq 0$, which implies that A is positive definite.

Now assume A is positive definite. Then, combining Theorem 5.6 and (1), we can rewrite A as $A = P^t D P$ where $P \in \mathbb{R}(n,n)$ is orthogonal and $D \in \mathbb{R}(n,n)$ is diagonal whose diagonal entries, say $\lambda_1, \ldots, \lambda_n$, are all positive. Define the diagonal matrix D_1 by (5.3.12) and set $D_1 P = B$. Then B is non-singular and $A = B^t B$ as asserted. $\square$

Similarly, we say that the quadratic form q, self-adjoint mapping T, or symmetric matrix $A \in \mathbb{R}(n,n)$ is *positive semi-definite* or *non-negative* if in the condition (5.3.2), (5.3.3), or (5.3.4), respectively, the "greater than" sign ($>$) is replaced by a "greater than or equal to" sign ($\geq$). For positive semi-definite or non-negative mappings and matrices, the corresponding versions of Theorems 5.8 and 5.9 can similarly be stated, simply with the word "positive" there being replaced by "non-negative" and I_n in Theorem 5.9 (3) by $\mathrm{diag}\{1, \ldots, 1, 0, \ldots, 0\}$.

Besides, we can define a quadratic form q, self-adjoint mapping T, or symmetric matrix A to be *negative definite* or *negative semi-definite* (*non-positive*), if $-q$, $-T$, or $-A$ is positive definite or positive semi-definite (non-negative).

Thus, a quadratic form, self-adjoint mapping, or symmetric matrix is called *indefinite* or *non-definite* if it is neither positive semi-definite nor negative semi-definite.

As an illustration, we consider the quadratic form

$$q(x) = a(x_1^2 + x_2^2 + x_3^2 + x_4^2) + 2x_1 x_2 + 2x_1 x_3 + 2x_1 x_4, \quad x = \begin{pmatrix} x_1 \\ \vdots \\ x_4 \end{pmatrix} \in \mathbb{R}^4,$$

$$(5.3.15)$$

where a is a real number. It is clear that q may be represented by the matrix

$$A = \begin{pmatrix} a & 1 & 1 & 1 \\ 1 & a & 0 & 0 \\ 1 & 0 & a & 0 \\ 1 & 0 & 0 & a \end{pmatrix}, \tag{5.3.16}$$

with respect to the standard basis of $\mathbb{R}^n$, whose characteristic equation is

$$(\lambda - a)^2([\lambda - a]^2 - 3) = 0. \tag{5.3.17}$$

Consequently the eigenvalues of A are $\lambda_1 = \lambda_2 = a$, $\lambda_3 = a + \sqrt{3}$, and $\lambda_4 = a - \sqrt{3}$. Therefore, when $a > \sqrt{3}$, the matrix A is positive definite; when $a = \sqrt{3}$, A is positive semi-definite but not positive definite; when $a < -\sqrt{3}$, A is negative definite; when $a = -\sqrt{3}$, A is negative semi-definite but not negative definite; when $a \in (-\sqrt{3}, \sqrt{3})$, A is indefinite.

Exercises

5.3.1 Let $T \in L(U)$ be self-adjoint and $S \in L(U)$ be anti-self-adjoint. Show that if T is positive definite then so is $T - S^2$.

5.3.2 Prove that if A and B are congruent matrices then A being positive definite or positive semi-definite is equivalent to B being so.

5.3.3 Let $A \in \mathbb{R}(n,n)$ be symmetric.

(a) Show that e^A is positive definite.
(b) Establish the formula $\det(e^A) = e^{\text{Tr}(A)}$. (In fact, this formula is true for any complex square matrices known as the Jacobi formula.)

5.3.4 Let U be a vector space with a positive definite scalar product $(\cdot, \cdot)$. Show that, for any set of linearly independent vectors $\{u_1, \ldots, u_k\}$ in U, the metric matrix

$$M = \begin{pmatrix} (u_1, u_1) & \cdots & (u_1, u_k) \\ \cdots & \cdots & \cdots \\ (u_k, u_1) & \cdots & (u_k, u_k) \end{pmatrix}, \tag{5.3.18}$$

must be positive definite.

5.3.5 Show that a necessary and sufficient condition for $A \in \mathbb{R}(n,n)$ to be symmetric is that there exists an invertible matrix $B \in \mathbb{R}(n,n)$ such that $A = B^t B + a I_n$ for some $a \in \mathbb{R}$.

5.3.6 Let $A \in \mathbb{R}(n,n)$ be positive definite. Show that $\det(A) > 0$.

5.3.7 Show that the inverse of a positive definite matrix is also positive definite.

5.3.8 Show that if $A \in \mathbb{R}(n,n)$ is positive definite then so is $\text{adj}(A)$.

5.3.9 Let $A \in \mathbb{R}(m,n)$ where $m \neq n$. Then $AA^t \in \mathbb{R}(m,m)$ and $A^t A \in \mathbb{R}(n,n)$ are both symmetric. Prove that if A is of full rank, that is, $r(A) = \min\{m,n\}$, then one of the matrices, AA^t and $A^t A$, is positive definite and the other is positive semi-definite but can never be positive definite.

5.3.10 Let $A = (a_{ij}) \in \mathbb{R}(n,n)$ be positive definite.

(a) Prove that $a_{ii} > 0$ for $i = 1, \ldots, n$.

(b) Establish the inequalities

$$|a_{ij}| < \sqrt{a_{ii} a_{jj}}, \quad i \neq j, \quad i,j = 1, \ldots, n. \qquad (5.3.19)$$

5.3.11 Let $T \in L(U)$ be self-adjoint. Show that, if T is positive semi-definite and $k \geq 1$ any integer, then there is a positive semi-definite element $S \in L(U)$ such that $T = S^k$.

5.3.12 Let $A \in \mathbb{R}(n,n)$ be positive semi-definite and consider the null set

$$S = \{x \in \mathbb{R}^n \,|\, x^t A x = 0\}. \qquad (5.3.20)$$

Prove that $x \in S$ if and only if $Ax = 0$. Hence S is the null-space of the matrix A. What happens when A is indefinite such that there are nonzero vectors x and $y \in \mathbb{R}^n$ so that $x^t A x > 0$ and $y^t A y < 0$?

5.3.13 Let A and $B \in \mathbb{R}(n,n)$ be symmetric and A positive definite. Prove that there is a non-singular matrix $C \in \mathbb{R}(n,n)$ so that both $C^t A C$ and $C^t B C$ are diagonal matrices (that is, A and B are simultaneously congruent to diagonal matrices).

5.3.14 Assume that A and $B \in \mathbb{R}(n,n)$ are symmetric and the eigenvalues of A and B are greater than or equal to $a, b \in \mathbb{R}$, respectively. Show that the eigenvalues of $A + B$ are greater than or equal to $a + b$.

5.3.15 Let A and $B \in \mathbb{R}(n,n)$ be positive definite. Prove that the eigenvalues of AB are all positive.

5.3.16 Consider the quadratic form

$$q(x) = (n+1) \sum_{i=1}^{n} x_i^2 - \left(\sum_{i=1}^{n} x_i \right)^2, \quad x = \begin{pmatrix} x_1 \\ \vdots \\ x_n \end{pmatrix} \in \mathbb{R}^n. \qquad (5.3.21)$$

(a) Find a symmetric matrix $A \in \mathbb{R}(n,n)$ such that $q(x) = x^t A x$ for $x \in \mathbb{R}^n$.

(b) Compute the eigenvalues of A to determine whether q or A is positive definite.

5.3.17 Let A and $B \in \mathbb{R}(n,n)$ where A is positive definite and B is positive semi-definite. Establish the inequality

$$\det(A + B) \geq \det(A) + \det(B) \qquad (5.3.22)$$

and show that equality in (5.3.22) holds only when $B = 0$.

5.3.18 Let A and $B \in \mathbb{R}(n,n)$ be positive definite such that $B - A$ is positive semi-definite. Prove that any solution λ of the equation

$$\det(\lambda A - B) = 0, \qquad (5.3.23)$$

must be real and satisfy $\lambda \geq 1$.

5.3.19 Let U be an n-dimensional vector space with a positive definite scalar product $(\cdot, \cdot)$ and $T \in L(U)$ a positive definite self-adjoint mapping. Given $b \in U$, consider the function

$$f(u) = \frac{1}{2}(u, T(u)) - (u, b), \quad u \in U. \qquad (5.3.24)$$

(a) Show that the non-homogeneous equation

$$T(x) = b \qquad (5.3.25)$$

enjoys the following *variational principle*: $x \in U$ solves (5.3.25) if and only if it is the minimum point of f, that is, $f(u) \geq f(x)$ for any $u \in U$.

(b) Show without using the invertibility of T that the solution to (5.3.25) is unique.

5.3.20 Use the notation of the previous problem and consider the quadratic form $q(u) = (u, T(u))$ $(u \in U)$, where $T \in L(U)$ is positive semi-definite. Prove that q is *convex*. That is, for any $\alpha, \beta \geq 0$ satisfying $\alpha + \beta = 1$, there holds

$$q(\alpha u + \beta v) \leq \alpha q(u) + \beta q(v), \quad u, v \in U. \qquad (5.3.26)$$

5.4 Alternative characterizations of positive definite matrices

In this section, we present two alternative characterizations of positive definite matrices, which are useful in applications. The first one involves using determinants and the second one involves an elegant matrix decomposition.

Theorem 5.10 *Let $A = (a_{ij}) \in \mathbb{R}(n,n)$ be symmetric. The matrix A is positive definite if and only if all its leading principal minors are positive, that is, if and only if*

$$a_{11} > 0, \quad \begin{vmatrix} a_{11} & a_{12} \\ a_{21} & a_{22} \end{vmatrix} > 0, \quad \ldots, \quad \begin{vmatrix} a_{11} & \cdots & a_{1n} \\ \cdots & \cdots & \cdots \\ a_{n1} & \cdots & a_{nn} \end{vmatrix} > 0. \qquad (5.4.1)$$

Proof   First assume that A is positive definite. For any $k = 1, \ldots, n$, set

$$A_k = \begin{pmatrix} a_{11} & \cdots & a_{1k} \\ \cdots & \cdots & \cdots \\ a_{k1} & \cdots & a_{kk} \end{pmatrix}. \tag{5.4.2}$$

For any $y = (y_1, \ldots, y_k)^t \in \mathbb{R}^k$, $y \neq 0$, we take $x = (y_1, \ldots, y_k, 0, \ldots, 0)^t \in \mathbb{R}^n$. Thus, there holds

$$y^t A_k y = \sum_{i,j=1}^k a_{ij} y_i y_j = x^t A x > 0. \tag{5.4.3}$$

So $A_k \in \mathbb{R}(k,k)$ is positive definite. Using Theorem 5.9, there is a non-singular matrix $B \in \mathbb{R}(k,k)$ such that $A_k = B^t B$. Thus, $\det(A_k) = \det(B^t B) = \det(B^t)\det(B) = \det(B)^2 > 0$, and (5.4.1) follows.

We next assume that (5.4.1) holds and we show that A is positive definite by an inductive argument.

If $n = 1$, there is nothing to show.

Assume that the assertion is valid at $n-1$ ($n \geq 2$).

We prove that A is positive definite at $n \geq 2$.

To proceed, we rewrite A in a blocked form as

$$A = \begin{pmatrix} A_{n-1} & \alpha \\ \alpha^t & a_{nn} \end{pmatrix}, \tag{5.4.4}$$

where $\alpha = (a_{1n}, \ldots, a_{n-1,n})^t \in \mathbb{R}^{n-1}$. By the inductive assumption at $n-1$, we know that A_{n-1} is positive definite. Hence, applying Theorem 5.9, we have a non-singular matrix $B \in \mathbb{R}(n-1, n-1)$ such that $A_{n-1} = B^t B$. Thus

$$A = \begin{pmatrix} A_{n-1} & \alpha \\ \alpha^t & a_{nn} \end{pmatrix} = \begin{pmatrix} B^t & 0 \\ 0 & 1 \end{pmatrix} \begin{pmatrix} I_{n-1} & \beta \\ \beta^t & a_{nn} \end{pmatrix} \begin{pmatrix} B & 0 \\ 0 & 1 \end{pmatrix} \tag{5.4.5}$$

if we choose $\beta = (B^t)^{-1}\alpha \equiv (b_1, \ldots, b_{n-1})^t \in \mathbb{R}^{n-1}$. Since $\det(A) > 0$, we obtain, after making some suitable row operations, the result

$$\frac{\det(A)}{\det(A_{n-1})} = \det \begin{pmatrix} I_{n-1} & \beta \\ \beta^t & a_{nn} \end{pmatrix} = a_{nn} - b_1^2 - \cdots - b_{n-1}^2 > 0. \tag{5.4.6}$$

Now for $x \in \mathbb{R}^n$ so that

$$y = \begin{pmatrix} B & 0 \\ 0 & 1 \end{pmatrix} x \equiv (y_1, \ldots, y_{n-1}, y_n)^t, \tag{5.4.7}$$

we have

$$x^t A x = y^t \begin{pmatrix} I_{n-1} & \beta \\ \beta^t & a_{nn} \end{pmatrix} y$$

$$= y_1^2 + \cdots + y_{n-1}^2 + 2(b_1 y_1 + \cdots + b_{n-1} y_{n-1}) y_n + a_{nn} y_n^2$$

$$= (y_1 + b_1 y_n)^2 + \cdots + (y_{n-1} + b_{n-1} y_n)^2$$

$$+ (a_{nn} - b_1^2 - \cdots - b_{n-1}^2) y_n^2, \tag{5.4.8}$$

which is clearly positive whenever $y \neq 0$, or equivalently, $x \neq 0$, in view of the condition (5.4.6).

Therefore the proof is complete. ☐

Note that, if we take $x = (x_1, \ldots, x_n)^t \in \mathbb{R}^n$ so that all the nonvanishing components of x, if any, are given by

$$x_{i_1} = y_1, \ldots, x_{i_k} = y_k, \quad i_1, \ldots, i_k = 1, \ldots, n, \quad i_1 < \cdots < i_k, \tag{5.4.9}$$

(if $k = 1$ it is understood that x has only one nonvanishing component at $x_{i_1} = y_1$, if any) then for $y = (y_1, \ldots, y_k)^t \in \mathbb{R}^k$ there holds

$$y^t A_{i_1, \ldots, i_k} y = \sum_{l,m=1}^{k} a_{i_l i_m} y_l y_m = x^t A x, \tag{5.4.10}$$

where

$$A_{i_1, \ldots, i_k} = \begin{pmatrix} a_{i_1 i_1} & \cdots & a_{i_1 i_k} \\ \cdots & \cdots & \cdots \\ a_{i_k i_1} & \cdots & a_{i_k i_k} \end{pmatrix} \tag{5.4.11}$$

is a submatrix of A obtained from deleting all the ith rows and jth columns of A for $i, j = 1, \ldots, n$ and $i, j \neq i_1, \ldots, i_k$. The quantity $\det(A_{i_1, \ldots, i_k})$ is referred to as a *principal minor* of A of order k. Such a principal minor becomes a leading principal minor when $i_1 = 1, \ldots, i_k = k$ with $A_{1, \ldots, k} = A_k$. For $A = (a_{ij})$ the principal minors of order 1 are all its diagonal entries, $a_{11}, \ldots, a_{nn}$. It is clear that if A is positive definite then so is $A_{i_1, \ldots, i_k}$. Hence $\det(A_{i_1, \ldots, i_k}) > 0$. Therefore we arrive at the following slightly strengthened version of Theorem 5.10.

Theorem 5.11 *Let $A = (a_{ij}) \in \mathbb{R}(n,n)$ be symmetric. The matrix A is positive definite if and only if all its principal minors are positive, that is, if and only if*

$$a_{ii} > 0, \quad \begin{vmatrix} a_{i_1 i_1} & \cdots & a_{i_1 i_k} \\ \cdots & \cdots & \cdots \\ a_{i_k i_1} & \cdots & a_{i_k i_k} \end{vmatrix} > 0, \tag{5.4.12}$$

for $i, i_1, \ldots, i_k = 1, \ldots, n, i_1 < \cdots < i_k, k = 2, \ldots, n$.

We now pursue the decomposition of a positive definite matrix as another characterization of this kind of matrices.

Theorem 5.12 *Let $A \in \mathbb{R}(n,n)$ be symmetric. The matrix A is positive definite if and only if there is a non-singular lower triangular matrix $L \in \mathbb{R}(n,n)$ such that $A = LL^t$. Moreover, when A is positive definite, there is a unique such L with the property that all the diagonal entries of L are positive.*

Proof   If there is a non-singular $L \in \mathbb{R}(n,n)$ such that $A = LL^t$, we see in view of Theorem 5.9 that A is of course positive definite.

Now we assume $A = (a_{ij}) \in \mathbb{R}(n,n)$ is positive definite. We look for a unique lower triangular matrix $L = (l_{ij}) \in \mathbb{R}(n,n)$ such that $A = LL^t$ so that all the diagonal entries of L are positive, $l_{11} > 0, \ldots, l_{nn} > 0$.

We again use induction.

When $n = 1$, then $A = (a_{11})$ with $a_{11} > 0$. So the unique choice is $L = (\sqrt{a_{11}})$.

Assume that the assertion is valid at $n - 1$ $(n \geq 2)$.

We establish the decomposition at n $(n \geq 2)$.

To proceed, we rewrite A in the form (5.4.4). In view of Theorem 5.10, the matrix A_{n-1} is positive definite. Thus there is a unique lower triangular matrix $L_1 \in \mathbb{R}(n-1, n-1)$, with positive diagonal entries, so that $A_{n-1} = L_1 L_1^t$.

Now take $L \in \mathbb{R}(n,n)$ with

$$L = \begin{pmatrix} L_2 & 0 \\ \gamma^t & a \end{pmatrix}, \qquad (5.4.13)$$

where $L_2 \in \mathbb{R}(n-1, n-1)$ is a lower triangular matrix, $\gamma \in \mathbb{R}^{n-1}$ is a column vector, and $a \in \mathbb{R}$ is a suitable number. Then, if we set $A = LL^t$, we obtain

$$A = \begin{pmatrix} A_{n-1} & \alpha \\ \alpha^t & a_{nn} \end{pmatrix} = \begin{pmatrix} L_2 & 0 \\ \gamma^t & a \end{pmatrix} \begin{pmatrix} L_2^t & \gamma \\ 0 & a \end{pmatrix}$$

$$= \begin{pmatrix} L_2 L_2^t & L_2 \gamma \\ (L_2 \gamma)^t & \gamma^t \gamma + a^2 \end{pmatrix}. \qquad (5.4.14)$$

Therefore we arrive at the relations

$$A_{n-1} = L_2 L_2^t, \quad \alpha = L_2 \gamma, \quad \gamma^t \gamma + a^2 = a_{nn}. \qquad (5.4.15)$$

If we require that all the diagonal entries of L_2 be positive, then the inductive assumption leads to $L_2 = L_1$. Hence the vector γ is also uniquely determined, $\gamma = L_1^{-1} \alpha$. So it remains to show that the number a may be uniquely determined as well. For this purpose, we need to show in (5.4.15) that

$$a_{nn} - \gamma^t \gamma > 0. \qquad (5.4.16)$$

In fact, in (5.4.5), the matrix B is any non-singular element in $\mathbb{R}(n-1, n-1)$ that gives us $A_{n-1} = B^t B$. Thus, from the relation $A_{n-1} = L_1 L^t$ here we may specify $L_1 = B^t$ in (5.4.5) so that $\beta = \gamma$ there. That is,

$$A = \begin{pmatrix} A_{n-1} & \alpha \\ \alpha^t & a_{nn} \end{pmatrix} = \begin{pmatrix} L_1 & 0 \\ 0 & 1 \end{pmatrix} \begin{pmatrix} I_{n-1} & \gamma \\ \gamma^t & a_{nn} \end{pmatrix} \begin{pmatrix} L_1^t & 0 \\ 0 & 1 \end{pmatrix}. \quad (5.4.17)$$

Hence (5.4.6), namely, (5.4.16), follows immediately. Therefore the last relation in (5.4.15) leads to the unique determination of the number a:

$$a = \sqrt{a_{nn} - \gamma^t \gamma} > 0. \quad (5.4.18)$$

The proof follows. □

The above-described characterization that a positive definite $n \times n$ matrix can be decomposed as the product of a unique lower triangular matrix L, with positive diagonal entries, and its transpose, is known as the *Cholesky decomposition theorem*, which has wide applications in numerous areas. Through resolving the *Cholesky relation* $A = LL^t$, it may easily be seen that the lower triangular matrix $L = (l_{ij})$ can actually be constructed from $A = (a_{ij})$ explicitly following the formulas

$$\begin{cases} l_{11} = \sqrt{a_{11}}, \\ l_{i1} = \frac{1}{l_{11}} a_{i1}, \quad i = 2, \ldots, n, \\ l_{ii} = \sqrt{a_{ii} - \sum_{j=1}^{i-1} l_{ij}^2}, \quad i = 2, \ldots, n, \\ l_{ij} = \frac{1}{l_{jj}} \left(a_{ij} - \sum_{k=1}^{j-1} l_{ik} l_{jk} \right), \quad i = j+1, \ldots, n, \; j = 2, \ldots, n. \end{cases} \quad (5.4.19)$$

Thus, if $A \in \mathbb{R}(n,n)$ is non-singular, the product AA^t may always be reduced into the form LL^t for a unique lower triangular matrix L with positive diagonal entries.

Exercises

5.4.1   Let a_1, a_2, and a_3 be real numbers and set

$$A = \begin{pmatrix} a_1 & a_2 & a_3 \\ a_2 & a_3 & a_1 \\ a_3 & a_1 & a_2 \end{pmatrix}. \quad (5.4.20)$$

Prove that A can never be positive definite no matter how a_1, a_2, and a_3 are chosen.

5.4.2   Consider the quadratic form

$$q(x) = (x_1 + a_1 x_2)^2 + (x_2 + a_2 x_3)^2 + (x_3 + a_3 x_1)^2, \quad x = \begin{pmatrix} x_1 \\ x_2 \\ x_3 \end{pmatrix} \in \mathbb{R}^3,$$

(5.4.21)

where $a_1, a_2, a_3 \in \mathbb{R}$. Find a necessary and sufficient condition on a_1, a_2, and a_3 for q to be positive definite. Can you extend your finding to the case over $\mathbb{R}^n$?

5.4.3   Let $A = (a_{ij}) \in \mathbb{R}(n, n)$ be symmetric. Prove that if the matrix A is positive semi-definite then all its leading principal minors are non-negative,

$$a_{11} \geq 0, \quad \begin{vmatrix} a_{11} & a_{12} \\ a_{21} & a_{22} \end{vmatrix} \geq 0, \quad \dots, \quad \begin{vmatrix} a_{11} & \cdots & a_{1n} \\ \cdots & \cdots & \cdots \\ a_{n1} & \cdots & a_{nn} \end{vmatrix} \geq 0.$$

(5.4.22)

Is the converse true?

5.4.4   Investigate the possibility whether Theorem 5.12 can be extended so that the lower triangular matrix may be replaced by an upper triangular matrix so that the theorem may now read: A symmetric matrix A is positive definite if and only if there is a non-singular upper triangular matrix U such that $A = UU^t$. Furthermore, if A is positive definite, then there is a unique upper triangular matrix U with positive diagonal entries such that $A = UU^t$.

5.4.5   Assume that $A = (a_{ij}) \in \mathbb{R}(n, n)$ is positive definite and rewrite A as

$$A = \begin{pmatrix} A_{n-1} & \alpha \\ \alpha^t & a_{nn} \end{pmatrix},$$

(5.4.23)

where $A_{n-1} \in \mathbb{R}(n, n)$ is positive definite by Theorem 5.10 and α is a column vector in $\mathbb{R}^{n-1}$.

(a) Show that the matrix equation

$$\begin{pmatrix} I_{n-1} & \beta \\ 0 & 1 \end{pmatrix}^t \begin{pmatrix} A_{n-1} & \alpha \\ \alpha^t & a_{nn} \end{pmatrix} \begin{pmatrix} I_{n-1} & \beta \\ 0 & 1 \end{pmatrix} = \begin{pmatrix} A_{n-1} & 0 \\ 0 & a_{nn} + a \end{pmatrix}$$

(5.4.24)

has a unique solution for some $\beta \in \mathbb{R}^{n-1}$ and $a \leq 0$.

(b) Use (a) to establish the inequality

$$\det(A) \leq \det(A_{n-1}) a_{nn}.$$

(5.4.25)

(c) Use (b) and induction to establish the general conclusion

$$\det(A) \leq a_{11} \cdots a_{nn}.$$

(5.4.26)

(d) Show that (5.4.26) still holds when A is positive semi-definite.

5.4.6   Let $A \in \mathbb{R}(n,n)$ and view A as made of n column vectors $u_1, \ldots, u_n$ in $\mathbb{R}^n$ which is equipped with the standard Euclidean scalar product.

(a) Show that $A^t A$ is the metric or Gram matrix of $u_1, \ldots, u_n \in \mathbb{R}^n$.

(b) Use the fact that $A^t A$ is positive semi-definite and (5.4.26) to prove that

$$|\det(A)| \leq \|u_1\| \cdots \|u_n\|. \tag{5.4.27}$$

5.4.7   Let $A = (a_{ij}) \in \mathbb{R}(n,n)$ and $a \geq 0$ is a bound of the entries of A satisfying

$$|a_{ij}| \leq a, \quad i,j = 1, \ldots, n. \tag{5.4.28}$$

Apply the conclusion of the previous problem to establish the following *Hadamard inequality* for determinants:

$$|\det(A)| \leq a^n n^{\frac{n}{2}}. \tag{5.4.29}$$

5.4.8   Consider the matrix

$$A = \begin{pmatrix} 2 & 1 & -1 \\ 1 & 3 & 1 \\ -1 & 1 & 2 \end{pmatrix}. \tag{5.4.30}$$

(a) Apply Theorem 5.10 to show that A is positive definite.

(b) Find the unique lower triangular matrix $L \in \mathbb{R}(3,3)$ stated in Theorem 5.12 such that $A = LL^t$.

5.5 Commutativity of self-adjoint mappings

We continue our discussion about self-adjoint mappings over a real finite-dimensional vector space U with a positive definite scalar product $(\cdot, \cdot)$.

The main focus of this section is to characterize a situation when two mappings may be simultaneously diagonalized.

Theorem 5.13 *Let S and $T \in L(U)$ be self-adjoint. Then there is an orthonormal basis of U consisting of eigenvectors of both S and T if and only if S and T commute, or $S \circ T = T \circ S$.*

Proof   Let $\lambda_1, \ldots, \lambda_k$ be all the distinct eigenvalues of T and $E_{\lambda_1}, \ldots, E_{\lambda_k}$ the associated eigenspaces which are known to be mutually perpendicular. Assume that S and $T \in L(U)$ commute. Then, for any $u \in E_{\lambda_i}$ ($i = 1, \ldots, k$), we have

$$T(S(u)) = S(T(u)) = S(\lambda_i u) = \lambda_i S(u). \tag{5.5.1}$$

Thus $S(u) \in E_{\lambda_i}$. In other words, each E_{λ_i} is invariant under S. Since S is self-adjoint, each E_{λ_i} has an orthonormal basis, say $\{u_{i,1}, \ldots, u_{i,m_i}\}$, consisting of the eigenvectors of S. Therefore, the set of vectors

$$\{u_{1,1}, \ldots, u_{1,m_1}, \ldots, u_{k,1}, \ldots, u_{k,m_k}\} \qquad (5.5.2)$$

is an orthonormal basis of U consisting of the eigenvectors of both S and T.

Conversely, if $\{u_1, \ldots, u_n\}$ is an orthonormal basis of U consisting of the eigenvectors of both S and T, then

$$S(u_i) = \varepsilon_i u_i, \quad T(u_i) = \lambda_i u_i, \quad \varepsilon_i, \lambda_i \in \mathbb{R}, \quad i = 1, \ldots, n. \qquad (5.5.3)$$

Thus $T(S(u_i)) = \varepsilon_i \lambda_i u_i = S(T(u_i))$ $(i = 1, \ldots, n)$, which establishes $S \circ T = T \circ S$ as anticipated. $\qquad\square$

Note that although the theorem says that, when S and T commute and $n = \dim(U)$, there are n mutually orthogonal vectors which are the eigenvectors of S and T simultaneously, it does not say that an eigenvector of S or T must also be an eigenvector of T or S. For example, we may take $S = I$ (the identity mapping). Then S commute with any mapping and any nonzero vector u is an eigenvector of S but u cannot be an eigenvector of all self-adjoint mappings.

The matrix version of Theorem 5.13 is easily stated and proved: Two symmetric matrices A and $B \in \mathbb{R}(n,n)$ are commutative, $AB = BA$, if and only if there is an orthogonal matrix $P \in \mathbb{R}(n,n)$ such that

$$A = P'D_A P, \quad B = P'D_B P, \qquad (5.5.4)$$

where D_A and D_B are diagonal matrices in $\mathbb{R}(n,n)$ whose diagonal entries are the eigenvalues of A and B, respectively.

Exercises

5.5.1   Let U be a finite-dimensional vector space over a field $\mathbb{F}$. Show that for $T \in L(U)$, if T commutes with any $S \in L(U)$, then there is some $a \in \mathbb{F}$ such that $T = aI$.

5.5.2   If A and $B \in \mathbb{R}(n,n)$ are positive definite matrices and $AB = BA$, show that AB is also positive definite.

5.5.3   Let U be an n-dimensional vector space over a field $\mathbb{F}$ where $\mathbb{F} = \mathbb{R}$ or $\mathbb{C}$ and T and $S \in L(U)$. Show that, if T has n distinct eigenvalues in $\mathbb{F}$ and S commutes with T, then there is a polynomial $p(t)$, of degree at most $(n-1)$, of the form

$$p(t) = a_0 + a_1 t + \cdots + a_{n-1} t^{n-1}, \quad a_0, a_1, \ldots, a_{n-1} \in \mathbb{F}, \qquad (5.5.5)$$

such that $S = p(T)$.

5.5.4 (Continued from Problem 5.5.3) If $T \in L(U)$ has n distinct eigenvalues in $\mathbb{F}$, show that

$$\mathcal{C}_T = \{S \in L(U) \mid S \circ T = T \circ S\}, \tag{5.5.6}$$

that is, the subset of linear mappings over U and commutative with T, as a subspace of $L(U)$, is exactly n-dimensional.

5.5.5 Let U be a finite-dimensional vector space with a positive definite scalar product and $T \in L(U)$.

(a) Show that if T is *normal* satisfying $T \circ T' = T' \circ T$ then $\|T(u)\| = \|T'(u)\|$ for any $u \in U$. In particular, $N(T) = N(T')$.

(b) Show that if T is normal and idempotent, $T^2 = T$, then $T = T'$.

5.6 Mappings between two spaces

In this section, we briefly illustrate how to use self-adjoint mappings to study general mappings between two vector spaces with positive definite scalar products.

Use U and V to denote two real vector spaces of finite dimensions equipped with positive definite scalar products $(\cdot,\cdot)_U$ and $(\cdot,\cdot)_V$, respectively. For $T \in L(U,V)$, since

$$f(u) = (T(u), v)_V, \quad u \in U, \tag{5.6.1}$$

defines an element f in U', we know that there is a unique element in U depending on v, say $T'(v)$, such that $f(u) = (u, T'(v))_U$. That is,

$$(T(u), v)_V = (u, T'(v))_U, \quad u \in U, v \in V. \tag{5.6.2}$$

Hence we have obtained a well-defined mapping $T' : V \to U$.

It is straightforward to check that T' is linear. Thus $T' \in L(V,U)$. This construction allows us to consider the composed mappings $T' \circ T \in L(U)$ and $T \circ T' \in L(V)$, which are both seen to be self-adjoint.

Moreover, since

$$(u, (T' \circ T)(u))_U = (T(u), T(u))_V \geq 0, \quad u \in U, \tag{5.6.3}$$

$$(v, (T \circ T')(v))_V = (T'(v), T'(v))_U \geq 0, \quad v \in V, \tag{5.6.4}$$

we see that $T' \circ T \in L(U)$ and $T \circ T' \in L(V)$ are both positive semi-definite.

Let $\| \cdot \|_U$ and $\| \cdot \|_V$ be the norms induced from the positive definite scalar products $(\cdot,\cdot)_U$ and $(\cdot,\cdot)_V$, respectively. Recall that the norm of T with respect to $\| \cdot \|_U$ and $\| \cdot \|_V$ is given by

$$\|T\| = \sup\{\|T(u)\|_V \mid \|u\|_U = 1, u \in U\}. \tag{5.6.5}$$

On the other hand, since $T' \circ T \in L(U)$ is self-adjoint and positive semi-definite, there is an orthonormal basis $\{u_1, \ldots, u_n\}$ of U consisting of eigenvectors of $T' \circ T$, associated with the corresponding non-negative eigenvalues $\sigma_1, \ldots, \sigma_n$. Therefore, for $u = \sum_{i=1}^n a_i u_i \in U$ with $\|u\|_U^2 = \sum_{i=1}^n a_i^2 = 1$, we have

$$\|T(u)\|_V^2 = (T(u), T(u))_V = (u, (T' \circ T)(u))_U = \sum_{i=1}^n \sigma_i a_i^2 \le \sigma_0, \qquad (5.6.6)$$

where

$$\sigma_0 = \max_{1 \le i \le n} \{\sigma_i\}, \qquad (5.6.7)$$

which proves $\|T\| \le \sqrt{\sigma_0}$. Furthermore, let $i = 1, \ldots, n$ be such that $\sigma_0 = \sigma_i$. Then (5.6.5) leads to

$$\|T\|^2 \ge \|T(u_i)\|_V^2 = (u_i, (T' \circ T)(u_i))_U = \sigma_i = \sigma_0. \qquad (5.6.8)$$

Consequently, we may conclude with

$$\|T\| = \sqrt{\sigma_0}, \quad \sigma_0 \text{ is the largest eigenvalue of the mapping } T' \circ T. \quad (5.6.9)$$

In particular, the above also gives us a practical method to compute the norm of a linear mapping T from U into itself by using the generated self-adjoint mapping $T' \circ T$.

If $T \in L(U)$ is already self-adjoint, then, since $\|T\|$ is the radical root of the largest eigenvalue of T^2, we have the expression

$$\|T\| = \max_{1 \le i \le n} \{|\lambda_i|\}, \qquad (5.6.10)$$

where $\lambda_1, \ldots, \lambda_n$ are the eigenvalues of T.

An immediate consequence of (5.6.9) and (5.6.10) is the elegant formula

$$\|T\|^2 = \|T' \circ T\|, \quad T \in L(U, V). \qquad (5.6.11)$$

Assume that $T \in L(U)$ is self-adjoint. From (5.6.10), it is not hard to see that $\|T\|$ may also be assessed according to

$$\|T\| = \sup\{|(u, T(u))| \,|\, u \in U, \|u\| = 1\}. \qquad (5.6.12)$$

In fact, let η denote the right-hand side of (5.6.12). Then the Schwarz inequality (4.3.10) implies that $|(u, T(u))| \le \|T(u)\| \le \|T\|$ for $u \in U$ satisfying $\|u\| = 1$. So $\eta \le \|T\|$. On the other hand, let λ be the eigenvalue of T such that $\|T\| = |\lambda|$ and $u \in U$ an associated eigenvector with $\|u\| = 1$. Then $\|T\| = |\lambda| = |(u, T(u))| \le \eta$. Hence (5.6.12) is verified.

We now show how to extend (5.6.12) to evaluate the norm of a general linear mapping between vector spaces with scalar products.

Theorem 5.14 *Let U and V be finite-dimensional vector spaces equipped with positive definite scalar products* $(\cdot,\cdot)_U$ *and* $(\cdot,\cdot)_V$, *respectively. For* $T \in L(U,V)$, *we have*

$$\|T\| = \sup\{|(T(u),v)_V| \mid u \in U, v \in V, \|u\|_U = 1, \|v\|_V = 1\}. \qquad (5.6.13)$$

Proof   Recall that there holds

$$\|T\| = \sup\{\|T(u)\|_V \mid u \in U, \|u\|_U = 1\}. \qquad (5.6.14)$$

Thus, for any $\varepsilon > 0$, there is some $u_\varepsilon \in U$ with $\|u_\varepsilon\| = 1$ such that

$$\|T(u_\varepsilon)\|_V \geq \|T\| - \varepsilon. \qquad (5.6.15)$$

Furthermore, for $T(u_\varepsilon) \in V$, we have

$$\|T(u_\varepsilon)\|_V = \sup\{|(T(u_\varepsilon),v)_V| \mid v \in V, \|v\|_V = 1\}. \qquad (5.6.16)$$

Thus, there is some $v_\varepsilon \in V$ with $\|v_\varepsilon\|_V = 1$ such that

$$|(T(u_\varepsilon),v_\varepsilon)_V| \geq \|T(u_\varepsilon)\|_V - \varepsilon. \qquad (5.6.17)$$

As a consequence, if we use η to denote the right-hand side of (5.6.13), we may combine (5.6.15) and (5.6.17) to obtain $\eta \geq \|T\| - 2\varepsilon$. Since $\varepsilon > 0$ is arbitrary, we have $\eta \geq \|T\|$.

On the other hand, for $u \in U$ and $v \in V$ with $\|u\|_U = 1$ and $\|v\|_V = 1$, we may use the Schwarz inequality (4.3.10) to get $|(T(u),v)_V| \leq \|T(u)\|_V \leq \|T\|$. Hence $\eta \leq \|T\|$.

Therefore we arrive at $\|T\| = \eta$ and the proof follows. $\qquad\qquad\square$

An important consequence of Theorem 5.14 is that the norms of a linear mapping, over two vector spaces with positive definite scalar products, and its dual assume the same value.

Theorem 5.15 *Let U and V be finite-dimensional vector spaces equipped with positive definite scalar products* $(\cdot,\cdot)_U$ *and* $(\cdot,\cdot)_V$, *respectively. For* $T \in L(U,V)$ *and its dual* $T' \in L(V,U)$, *we have* $\|T\| = \|T'\|$. *Thus* $\|T' \circ T\| = \|T \circ T'\|$ *and the largest eigenvalues of the positive semi-definite mappings* $T' \circ T \in L(U)$ *and* $T \circ T' \in L(V)$ *are the same.*

Proof   The fact that $\|T\| = \|T'\|$ may be deduced from applying (5.6.13) to T' and the relation $(T(u),v)_V = (u,T'(v))_U$ ($u \in U$ and $v \in V$). The conclusion $\|T' \circ T\| = \|T \circ T'\|$ follows from (5.6.11) and that the largest eigenvalues of the positive semi-definite mappings $T' \circ T$ and $T \circ T'$ are the same is a consequence of the eigenvalue characterization of the norm of a self-adjoint mapping stated in (5.6.10) and $\|T' \circ T\| = \|T \circ T'\|$. $\qquad\square$

As an application, we see that, for any matrix $A \in \mathbb{R}(m,n)$, the largest eigenvalues of the symmetric matrices $A^t A \in \mathbb{R}(n,n)$ and $AA^t \in \mathbb{R}(m,m)$ must coincide. In fact, regarding eigenvalues, it is not hard to establish a more general result as stated as follows.

Theorem 5.16 *Let U and V be vector spaces over a field $\mathbb{F}$ and $T \in L(U,V)$ and $S \in L(V,U)$. Then the nonzero eigenvalues of $S \circ T$ and $T \circ S$ are the same.*

Proof   Let $\lambda \in \mathbb{F}$ be a nonzero eigenvalue of $S \circ T$ and $u \in U$ an associated eigenvector. We show that λ is also an eigenvalue of $T \circ S$. In fact, from

$$(S \circ T)(u) = \lambda u, \qquad (5.6.18)$$

we have $(T \circ S)(T(u)) = \lambda T(u)$. In order to show that λ is an eigenvalue of $T \circ S$, it suffices to show $T(u) \neq 0$. However, such a property is already seen in (5.6.18) since $\lambda \neq 0$ and $u \neq 0$.

Interchanging S and T, we see that, if λ is a nonzero eigenvalue of $T \circ S$, then it is also an eigenvalue of $S \circ T$.   □

It is interesting that we do not require any additional properties for the vector spaces U, V in order for Theorem 5.16 to hold.

We next study another important problem in applications known as the *least-squares approximation*.

Let $T \in L(U,V)$ and $v \in V$. Consider the optimization problem

$$\eta \equiv \inf\left\{\|T(u) - v\|_V^2 \mid u \in U\right\}. \qquad (5.6.19)$$

For simplicity, we first assume that (5.6.19) has a solution, that is, there is some $x \in U$ such that $\|T(x) - v\|_V^2 = \eta$, and we look for some appropriate condition the solution x will fulfill. For this purpose, we set

$$f(\varepsilon) = \|T(x + \varepsilon y) - v\|_V^2, \quad y \in U, \quad \varepsilon \in \mathbb{R}. \qquad (5.6.20)$$

Then $f(0) = \eta \leq f(\varepsilon)$. Hence we may expand the right-hand side of (5.6.20) to obtain

$$\begin{aligned} 0 = \left(\frac{\mathrm{d}f}{\mathrm{d}\varepsilon}\right)_{\varepsilon=0} &= (T(x) - v, T(y))_V + (T(y), T(x) - v)_V \\ &= 2((T' \circ T)(x) - T'(v), y)_U, \quad y \in U, \end{aligned} \qquad (5.6.21)$$

which implies that $x \in U$ is a solution to the equation

$$(T' \circ T)(x) = T'(v), \quad v \in V. \qquad (5.6.22)$$

This equation is commonly referred to as the *normal equation*. It is not hard to see that the equation (5.6.22) is always consistent for any $v \in V$. In fact, it is clear that $R(T' \circ T) \subset R(T')$. On the other hand, if $u \in N(T' \circ T)$, then $\|T(u)\|_V^2 = (u, (T' \circ T)(u)) = 0$. Thus $u \in N(T)$. This establishes $N(T) = N(T' \circ T)$. So by the rank equation we deduce $\dim(U) = n(T) + r(T) = n(T' \circ T) + r(T' \circ T)$. Therefore, in view of Theorem 2.9, we get $r(T' \circ T) = r(T) = r(T')$. That is, $R(T' \circ T) = R(T')$, as desired, which proves the solvability of (5.6.22) for any $v \in V$.

Next, let x be a solution to (5.6.22). We show that x solves (5.6.19). In fact, if y is another solution to (5.6.22), then $z = x - y \in N(T' \circ T) = N(T)$. Thus

$$\|T(y) - v\|_V^2 = \|T(x+z) - v\|_V^2 = \|T(x) - v\|_V^2, \tag{5.6.23}$$

which shows that the quantity $\|T(x) - v\|_V^2$ is independent of the solution x of (5.6.22). Besides, for any test element $u \in U$, we rewrite u as $u = x + w$, where x is a solution to (5.6.22) and $w \in U$. Then we have

$$\begin{aligned}
\|T(u) - v\|_V^2 &= \|T(x) - v\|_V^2 + 2(T(x) - v, T(w))_V + \|T(w)\|_V^2 \\
&= \|T(x) - v\|_V^2 + 2((T' \circ T)(x) - T'(v), w)_U + \|T(w)\|_V^2 \\
&= \|T(x) - v\|_V^2 + \|T(w)\|_V^2 \\
&\geq \|T(x) - v\|_V^2.
\end{aligned} \tag{5.6.24}$$

Consequently, x solves (5.6.19) as anticipated.

Using knowledge about self-adjoint mappings, we can express a solution to the normal equation (5.6.22) explicitly in terms of the eigenvectors and eigenvalues of $T' \circ T$. In fact, let $\{u_1, \ldots, u_k, \ldots, u_n\}$ be an orthonormal basis of U consisting of the eigenvectors of the positive semi-definite mapping $T' \circ T \in L(U)$ associated with the non-negative eigenvalues $\lambda_1, \ldots, \lambda_k, \ldots, \lambda_n$ among which $\lambda_1, \ldots, \lambda_k$ are positive (if any). Then a solution x of (5.6.22) may be written $x = \sum_{i=1}^{n} a_i u_i$ for some $a_1, \ldots, a_n \in \mathbb{R}$. Inserting this into (5.6.22), we obtain

$$\sum_{i=1}^{k} a_i \lambda_i u_i = T'(v). \tag{5.6.25}$$

Thus, taking scalar product on both sides of the above, we find

$$a_i = \frac{1}{\lambda_i} (u_i, T'(v))_U, \quad i = 1, \ldots, k, \tag{5.6.26}$$

which leads to the following general solution formula for the equation (5.6.22):

$$x = \sum_{i=1}^{k} \frac{1}{\lambda_i} (u_i, T'(v))_U u_i + x_0, \quad x_0 \in N(T), \tag{5.6.27}$$

since $N(T' \circ T) = N(T)$.

Exercises

5.6.1 Show that the mapping $T' : V \to U$ defined in (5.6.2) is linear.

5.6.2 Let U and V be finite-dimensional vector spaces with positive definite scalar products. For any $T \in L(U,V)$ show that the mapping $T' \circ T \in L(U)$ is positive definite if and only if $n(T) = 0$.

5.6.3 Let U and V be finite-dimensional vector spaces with positive definite scalar products, $(\cdot,\cdot)_U$ and $(\cdot,\cdot)_V$, respectively. For any $T \in L(U,V)$, define $T' \in L(V,U)$ by (5.6.2).

   (a) Establish the relations $R(T)^\perp = N(T')$ and $R(T) = N(T')^\perp$ (the latter is also known as the *Fredholm alternative*).
   (b) Prove directly, without resorting to Theorem 2.9, that $r(T) = r(T')$.
   (c) Prove that $r(T \circ T') = r(T' \circ T) = r(T)$.

5.6.4 Apply Problem 5.6.3 to verify the validity of the rank relation

$$r(A^t A) = r(AA^t) = r(A) = r(A^t), \quad A \in \mathbb{R}(m,n). \qquad (5.6.28)$$

5.6.5 Let $T \in L(\mathbb{R}^2, \mathbb{R}^3)$ be defined by

$$T\begin{pmatrix} x_1 \\ x_2 \end{pmatrix} = \begin{pmatrix} 1 & 2 \\ -1 & 1 \\ 2 & 3 \end{pmatrix} \begin{pmatrix} x_1 \\ x_2 \end{pmatrix}, \quad \begin{pmatrix} x_1 \\ x_2 \end{pmatrix} \in \mathbb{R}^2, \qquad (5.6.29)$$

where $\mathbb{R}^2$ and $\mathbb{R}^3$ are equipped with the standard Euclidean scalar products.

   (a) Find the eigenvalues of $T' \circ T$ and compute $\|T\|$.
   (b) Find the eigenvalues of $T \circ T'$ and verify that $T \circ T'$ is positive semi-definite but not positive definite (cf. Problem 5.6.2).
   (c) Check to see that the largest eigenvalues of $T' \circ T$ and $T \circ T'$ are the same.
   (d) Compute all eigenvalues of $T \circ T'$ and $T' \circ T$ and explain the results in view of Theorem 5.16.

5.6.6 Let $A \in \mathbb{F}(m,n)$ and $B \in \mathbb{F}(n,m)$ where $m < n$ and consider $AB \in \mathbb{F}(m,m)$ and $BA \in \mathbb{F}(n,n)$. Show that BA has at most $m+1$ distinct eigenvalues in $\mathbb{F}$.

5.6.7 (A specialization of Problem 5.6.6) Let $u, v \in \mathbb{F}^n$ be column vectors. Hence $u^t v \in \mathbb{F}$ and $vu^t \in \mathbb{F}(n,n)$.

   (a) Show that the matrix vu^t has a nonzero eigenvalue in $\mathbb{F}$ if and only if $u^t v \neq 0$.
   (b) Show that, when $u^t v \neq 0$, the only nonzero eigenvalue of vu^t in $\mathbb{F}$ is $u^t v$ so that v is an associated eigenvector.

(c) Show that, when $u^t v \neq 0$, the eigenspace of vu^t associated with the eigenvalue $u^t v$ is one-dimensional.

5.6.8   Consider the Euclidean space $\mathbb{R}^k$ equipped with the standard inner product and let $A \in \mathbb{R}(m,n)$. Formulate a solution of the following optimization problem:

$$\eta \equiv \inf\{\|Ax - b\|^2 \mid x \in \mathbb{R}^n\}, \quad b \in \mathbb{R}^m, \tag{5.6.30}$$

by deriving a matrix version of the normal equation.

5.6.9   Consider a parametrized plane in $\mathbb{R}^3$ given by

$$x_1 = y_1 + y_2, \quad x_2 = y_1 - y_2, \quad x_3 = y_1 + y_2, \quad y_1, y_2 \in \mathbb{R}. \tag{5.6.31}$$

Using the least-squares approximation to find a point in the plane which is the closest to the point in $\mathbb{R}^3$ with the coordinates $x_1 = 2, x_2 = 1$, and $x_3 = 3$.

5.6.10   Let $A \in \mathbb{R}(n,n)$ be symmetric and $\|A\|$ denote the induced norm of A. Show that

$$\|A^k\| = \|A\|^k, \quad k = 1,2,\ldots. \tag{5.6.32}$$

5.6.11   Let $A \in \mathbb{R}(n,n)$ be symmetric and $\|A\|_1$ denote the Hilbert–Schmidt norm of A associated with the positive definite scalar product $(A,B) = \text{Tr}(A^t B)$ for A and $B \in \mathbb{R}(n,n)$.

(a) Show that if $\lambda_1, \ldots, \lambda_n \in \mathbb{R}$ are all the eigenvalues of A (counting multiplicities) then

$$\|A\|_1 = \sqrt{\lambda_1^2 + \cdots + \lambda_n^2}. \tag{5.6.33}$$

(b) Obtain your formula for $\|A^k\|_1$ for $k = 1,2,\ldots.$
(c) Show that

$$\|A\|_1^k \geq \|A^k\|_1, \quad k = 1,2,\ldots, \tag{5.6.34}$$

with equality at any $k \geq 2$ if and only if $r(A) \leq 1$.

5.6.12   (Continued from Problems 5.6.10 and 5.6.11) Verify the Gelfand formula

$$\lim_{k \to \infty} \|A^k\|_1^{\frac{1}{k}} = \|A\| = \rho(A) = \max_{1 \leq i \leq n}\{|\lambda_i|\}. \tag{5.6.35}$$

5.6.13   Let U and V be finite-dimensional real vector spaces equipped with positive definite scalar products. For given $v \in V$ and $T \in L(U,V)$, consider the minimization problem

$$\eta = \inf\{\|Tu - v\|^2 \mid u \in U\}. \tag{5.6.36}$$

Carry out the following steps to prove that problem (5.6.36) always has a solution (by method of analysis):

(a) We have $\eta \leq \|v\|^2$ and the problem is not trivial only when $\eta < \|v\|^2$.

(b) From (a), we must assume $\eta < \|v\|^2$ to avoid triviality. Prove that in this case $T \neq 0$ and that there is a (minimizing) sequence

$$\{u_m\} \subset U \setminus N(T) \tag{5.6.37}$$

such that

$$\lim_{m \to \infty} \|Tu_m - v\|^2 = \eta. \tag{5.6.38}$$

(c) Show that $T' \circ T \in L(U)$ is reducible over $U = N(T) \oplus N(T)^\perp$. That is, $N(T)^\perp$ is invariant under $T' \circ T$.

(d) Show that $T' \circ T$ is positive definite over $N(T)^\perp$. Hence there is some $c_0 > 0$ such that

$$(u, T' \circ Tu) \geq c_0 \|u\|^2, \quad u \in N(T)^\perp. \tag{5.6.39}$$

(e) Show that it suffices to assume $\{u_m\} \subset N(T)^\perp$ in (5.6.38).

(f) Establish the expression

$$\|Tu - v\|^2 = (u, T' \circ Tu) - 2(Tu, v) + \|v\|^2, \quad u \in U. \tag{5.6.40}$$

(g) Use (5.6.39) and the Cauchy–Schwarz inequality to derive from (5.6.40) the lower ("coercive") estimate

$$\|Tu - v\|^2 \geq (c_0 - \varepsilon) \|u\|^2 - \frac{1}{\varepsilon} \|T\|^2 \|v\|^2 + \|v\|^2, \quad u \in N(T)^\perp, \tag{5.6.41}$$

where $\varepsilon > 0$ is an arbitrary constant. (Hint: Use the inequality $2ab \leq a^2 + b^2$.)

(h) Choose $\varepsilon < c_0$ in (5.6.41) to show that any minimizing sequence $\{u_m\}$ in $N(T)^\perp$ must be bounded. Hence it has a convergent subsequence. Show that the limit of this convergent subsequence solves problem (5.6.36). So the existence of a solution to (5.6.36) follows.

6

Complex Quadratic Forms and Self-Adjoint Mappings

In this chapter, we extend our study on real quadratic forms and self-adjoint mappings to the complex situation. We begin by a discussion on the complex version of bilinear forms and the Hermitian structures. We will relate the Hermitian structure of a bilinear form with representing it by a unique self-adjoint mapping. Then, we establish the main spectrum theorem for self-adjoint mappings. We next focus again on the positive definiteness of self-adjoint mappings. We explore the commutativity of self-adjoint mappings and apply it to obtain the main spectrum theorem for normal mappings. We also show how to use self-adjoint mappings to study a mapping between two spaces.

6.1 Complex sesquilinear and associated quadratic forms

Let U be a finite-dimensional vector space over $\mathbb{C}$. Extending the standard Hermitian scalar product over $\mathbb{C}^n$, we may formulate the notion of a complex "bilinear" form as follows.

Definition 6.1 A complex-valued function $f : U \times U \to \mathbb{C}$ is called a *sesquilinear form*, which is also sometimes loosely referred to as a *bilinear form*, if it satisfies for any $u, v, w \in U$ and $a \in \mathbb{C}$ the following conditions:

(1) $f(u+v, w) = f(u, w) + f(v, w)$, $f(u, v+w) = f(u, v) + f(u, w)$.
(2) $f(au, v) = \bar{a} f(u, v)$, $f(u, av) = a f(u, v)$.

As in the real situation, we may consider how to use a matrix to represent a sesquilinear form. To this end, let $\mathcal{B} = \{u_1, \ldots, u_n\}$ be a basis of U. For u and $v \in U$, let $x = (x_1, \ldots, x_n)^t$ and $y = (y_1, \ldots, y_n)^t \in \mathbb{C}^n$ denote the coordinate vectors of u and v with respect to the basis $\mathcal{B}$. Then

$$f(u, v) = f\left(\sum_{i=1}^n x_i u_i, \sum_{j=1}^n y_j u_j \right) = \sum_{i,j=1}^n \bar{x}_i f(u_i, u_j) y_j = \vec{x}^{\,} A y = x^\dagger A y, \quad (6.1.1)$$

185

where $A = (a_{ij}) = (f(u_i, u_j))$ lies in $\mathbb{C}(n,n)$ which is the matrix representation of the sesquilinear form f with respect to $\mathcal{B}$.

Let $\tilde{\mathcal{B}} = \{\tilde{u}_1, \ldots, \tilde{u}_n\}$ be another basis of U so that the coordinate vectors of u and v are $\tilde{x}$ and $\tilde{y} \in \mathbb{C}^n$ with respect to $\tilde{\mathcal{B}}$. Hence, using $\tilde{A} = (\tilde{a}_{ij}) = (f(\tilde{u}_i, \tilde{u}_j)) \in \mathbb{C}(n,n)$ to denote the matrix representation of the sesquilinear form f with respect to $\tilde{\mathcal{B}}$ and $B = (b_{ij}) \in \mathbb{C}(n,n)$ the basis transition matrix from $\mathcal{B}$ into $\tilde{\mathcal{B}}$ so that

$$\tilde{u}_j = \sum_{i=1}^{n} b_{ij} u_i, \quad j = 1, \ldots, n, \tag{6.1.2}$$

we know that the relations $x = B\tilde{x}$ and $y = B\tilde{y}$ are valid. Therefore, we can conclude with

$$f(u,v) = \tilde{x}^\dagger \tilde{A} \tilde{y} = x^\dagger A y = \tilde{x}^\dagger (B^\dagger A B) \tilde{y}. \tag{6.1.3}$$

Consequently, there holds $\tilde{A} = B^\dagger A B$. As in the real situation, we make the definition that two matrices A and $B \in \mathbb{C}(n,n)$ are said to be *Hermitian congruent*, or simply *congruent*, if there is an invertible matrix $C \in \mathbb{C}(n,n)$ such that

$$A = C^\dagger B C. \tag{6.1.4}$$

Hence, we see that the matrix representations of a sesquilinear form over U with respect to different bases of U are Hermitian congruent.

Let $f : U \times U \to \mathbb{C}$ be a sesquilinear form. Define $q : U \to \mathbb{C}$ by setting

$$q(u) = f(u,u), \quad u \in U. \tag{6.1.5}$$

As in the real situation, we may call q the *quadratic form* associated with f. We have the following homogeneity property:

$$q(zu) = |z|^2 q(u), \quad u \in U, \quad z \in \mathbb{C}. \tag{6.1.6}$$

Conversely, we can also show that f is uniquely determined by q. This fact is in sharp contrast with the real situation.

In fact, since

$$f(u+v, u+v) = f(u,u) + f(v,v) + f(u,v) + f(v,u), \quad u,v \in U, \tag{6.1.7}$$

$$f(u+iv, u+iv) = f(u,u) + f(v,v) + if(u,v) - if(v,u), \quad u,v \in U, \tag{6.1.8}$$

we have the following polarization identity relating a sesquilinear form to its induced quadratic form:

$$f(u,v) = \frac{1}{2}(q(u+v) - q(u) - q(v))$$
$$- \frac{i}{2}(q(u+iv) - q(u) - q(v)), \, u,v \in U. \tag{6.1.9}$$

For U, let $\mathcal{B} = \{u_1, \ldots, u_n\}$ be a basis, $u \in U$, and $x \in \mathbb{C}^n$ the coordinate vector of u with respect to $\mathcal{B}$. In view of (6.1.1), we get

$$q(u) = x^{\dagger} A x = x^{\dagger} \left(\frac{1}{2}(A + A^{\dagger}) + \frac{1}{2}(A - A^{\dagger}) \right) x$$

$$= \frac{1}{2} x^{\dagger} (A + A^{\dagger}) x + \frac{1}{2} x^{\dagger} (A - A^{\dagger}) x = \Re\{q(u)\} + i \Im\{q(u)\}. \qquad (6.1.10)$$

In other words, q is real-valued when A is Hermitian, $A = A^{\dagger}$, which may be checked to be equivalent to the condition

$$f(u, v) = \overline{f(v, u)}, \quad u, v \in U. \qquad (6.1.11)$$

In fact, if q is real-valued, then replacing u by iv and v by u in (6.1.9), we have

$$-if(v, u) = \frac{1}{2} \left(q(u + iv) - q(u) - q(v) \right) - \frac{i}{2} \left(q(u + v) - q(u) - q(v) \right). \qquad (6.1.12)$$

Combining (6.1.9) and (6.1.12), and using the condition that q is real-valued, we arrive at (6.1.11).

Thus we are led to the following definition:

Definition 6.2 A sesquilinear form $f : U \times U \to \mathbb{C}$ is said to be *Hermitian* if it satisfies the condition (6.1.11).

For any column vectors x and $y \in \mathbb{C}^n$, the standard Hermitian scalar product is given by

$$(x, y) = x^{\dagger} y. \qquad (6.1.13)$$

Thus $f : \mathbb{C}^n \times \mathbb{C}^n \to \mathbb{C}$ defined by $f(x, y) = (x, y)$ for x and $y \in \mathbb{C}^n$ is a Hermitian sesquilinear form. More generally, with any $n \times n$ Hermitian matrix A, we see that

$$f(x, y) = x^{\dagger} A y = (x, Ay), \quad x, y \in \mathbb{C}^n, \qquad (6.1.14)$$

is also a Hermitian sesquilinear form. Conversely, the relation (6.1.1) indicates that a Hermitian sesquilinear form over an n-dimensional complex vector space is completely represented, with respect to a given basis, by a Hermitian matrix, in terms of the standard Hermitian scalar product over $\mathbb{C}^n$.

Consider a finite-dimensional complex vector space U with a positive definite scalar product $(\cdot, \cdot)$. Given any sesquilinear form $f : U \times U \to \mathbb{C}$, since

$$g(v) = f(u, v), \quad v \in U, \qquad (6.1.15)$$

is a linear functional in v depending on $u \in U$, there is a unique vector (say) $T(u) \in U$ such that

$$f(u,v) = (T(u),v), \quad u,v \in U, \tag{6.1.16}$$

which in fact defines the correspondence T as an element in $L(U)$. Let T' denote the adjoint of T. Then f may also be represented as

$$f(u,v) = (u,T'(v)), \quad u,v \in U. \tag{6.1.17}$$

In other words, a sesquilinear form f over U may completely be represented by a linear mapping T or T' from U into itself, in terms of the positive definite scalar product over U, alternatively through the expression (6.1.16) or (6.1.17).

Applying (6.1.9) to $f(u,v) = (u,T(v))$ $(u,v \in U)$ for any $T \in L(U)$, we obtain the following useful polarization identity for T:

$$(u,T(v)) = \frac{1}{2}((u+v,T(u+v)) - (u,T(u)) - (v,T(v)))$$
$$- \frac{i}{2}((u+iv,T(u+iv)) - (u,T(u)) - (v,T(v))), \quad u,v \in U. \tag{6.1.18}$$

The Hermitian situation is especially interesting for us.

Theorem 6.3 *Let f be a sesquilinear form over a finite-dimensional complex vector space U with a positive definite scalar product $(\cdot,\cdot)$ and represented by the mapping $T \in L(U)$ through (6.1.16) or (6.1.17). Then f is Hermitian if and only if T is self-adjoint, $T = T'$.*

Proof   If f is Hermitian, then

$$f(u,v) = \overline{f(v,u)} = \overline{(T(v),u)} = (u,T(v)), \quad u,v \in U. \tag{6.1.19}$$

In view of (6.1.17) and (6.1.19), we arrive at $T = T'$. The converse is similar. $\square$

Let $\mathcal{B} = \{u_1,\ldots,u_n\}$ be an orthonormal basis of U, $T \in L(U)$ be self-adjoint, and $A = (a_{ij}) \in \mathbb{C}(n,n)$ the matrix representation of T with respect to $\mathcal{B}$. Then $T(u_j) = \sum_{i=1}^{n} a_{ij}u_i$ $(j = 1,\ldots,n)$ so that

$$a_{ij} = (u_i,T(u_j)) = \overline{(T(u_i),u_j)} = \overline{a}_{ji}, \quad i,j = 1,\ldots,n. \tag{6.1.20}$$

Hence $A = A^\dagger$. Of course the converse is true too. Therefore T is self-adjoint if and only if the matrix representation of T with respect to any orthonormal basis is Hermitian. Consequently, a self-adjoint mapping over a complex vector space with a positive definite scalar product is interchangeably referred to as Hermitian as well.

Exercises

6.1.1 Let U be complex vector space with a positive definite scalar product $(\cdot,\cdot)$ and $T \in L(U)$. Show that $T = 0$ if and only if $(u, T(u)) = 0$ for any $u \in U$. Give an example to show that the same may not hold for a real vector space with a positive definite scalar product.

6.1.2 Let U be complex vector space with a positive definite scalar product $(\cdot,\cdot)$ and $T \in L(U)$. Show that T is self-adjoint or Hermitian if and only if $(u, T(u)) = (T(u), u)$ for any $u \in U$. Give an example to show that the same may not hold for a real vector space with a positive definite scalar product.

6.1.3 Let U be a complex vector space with a basis $\mathcal{B} = \{u_1, \ldots, u_n\}$ and $A = (f(u_i, u_j)) \in \mathbb{C}(n, n)$ the matrix representation of f with respect to $\mathcal{B}$. Show that f is Hermitian if and only if A is Hermitian.

6.1.4 Let U be complex vector space with a positive definite scalar product $(\cdot,\cdot)$ and $T \in L(U)$. Show that T can be uniquely decomposed into a sum $T = R + iS$, where R and $S \in L(U)$ are both self-adjoint.

6.1.5 Show that the inverse of an invertible self-adjoint mapping is also self-adjoint.

6.1.6 Show that I_n and $-I_n$ cannot be Hermitian congruent, although they are congruent as elements in $\mathbb{C}(n, n)$.

6.2 Complex self-adjoint mappings

As in the real situation, we now show that complex self-adjoint or Hermitian mappings are completely characterized by their spectra as well.

Theorem 6.4 *Let U be a complex vector space with a positive definite scalar product $(\cdot,\cdot)$ and $T \in L(U)$. If T is self-adjoint, then the following are valid:*

(1) *The eigenvalues of T are all real.*

(2) *Let $\lambda_1, \ldots, \lambda_k$ be all the distinct eigenvalues of T. Then T may be reduced over the direct sum of mutually perpendicular eigenspaces*

$$U = E_{\lambda_1} \oplus \cdots \oplus E_{\lambda_k}. \tag{6.2.1}$$

(3) *There is an orthonormal basis of U consisting of eigenvectors of T.*

Proof Let T be self-adjoint and $\lambda \in \mathbb{C}$ an eigenvalue of T with $u \in U$ an associated eigenvector. Then, using $T(u) = \lambda u$, we have

$$\lambda \|u\|^2 = (u, T(u)) = (T(u), u) = \overline{\lambda} \|u\|^2, \tag{6.2.2}$$

which gives us $\lambda = \overline{\lambda}$ so that $\lambda \in \mathbb{R}$. This establishes (1). Note that this proof does not assume that U is finite dimensional.

To establish (2), we use induction on $\dim(U)$.

If $\dim(U) = 1$, there is nothing to show.

Assume that the statement (2) is valid if $\dim(U) \leq n-1$ for some $n \geq 2$. Consider $\dim(U) = n$, $n \geq 2$.

Let λ_1 be an eigenvalue of T in $\mathbb{C}$. From (1), we know that actually $\lambda_1 \in \mathbb{R}$. Use E_{λ_1} to denote the eigenspace of T associated with λ_1:

$$E_{\lambda_1} = \{u \in U \mid T(u) = \lambda_1 u\} = N(\lambda_1 I - T). \tag{6.2.3}$$

If $E_{\lambda_1} = U$, then $T = \lambda_1 I$ and there is nothing more to show. We now assume $E_{\lambda_1} \neq U$.

It is clear that E_{λ_1} is invariant under T. In fact, T is reducible over the direct sum $U = E_{\lambda_1} \oplus (E_{\lambda_1})^{\perp}$. To see this, we need to establish the invariance $T((E_{\lambda_1})^{\perp}) \subset (E_{\lambda_1})^{\perp}$. Indeed, for any $u \in E_{\lambda_1}$ and $v \in (E_{\lambda_1})^{\perp}$, we have

$$(u, T(v)) = (T(u), v) = \lambda_1(u, v) = 0. \tag{6.2.4}$$

Thus $T(v) \in (E_{\lambda_1})^{\perp}$ and $T(E_{\lambda_1})^{\perp} \subset (E_{\lambda_1})^{\perp}$.

Using the fact that $\dim((E_{\lambda_1})^{\perp}) = \dim(U) - \dim(E_{\lambda_1}) \leq n-1$ and the inductive assumption, we see that T is reduced over a direct sum of mutually perpendicular eigenspaces of T in $(E_{\lambda_1})^{\perp}$:

$$(E_{\lambda_1})^{\perp} = E_{\lambda_2} \oplus \cdots \oplus E_{\lambda_k}, \tag{6.2.5}$$

where $\lambda_2, \ldots, \lambda_k$ are all the distinct eigenvalues of T over the invariant subspace $(E_{\lambda_1})^{\perp}$, which are real.

Finally, we need to show that $\lambda_1, \ldots, \lambda_k$ obtained above are all possible eigenvalues of T. For this purpose, let λ be an eigenvalue of T and u an associated eigenvector. Then there are $u_1 \in E_{\lambda_1}, \ldots, u_k \in E_{\lambda_k}$ such that $u = u_1 + \cdots + u_k$. Hence the relation $T(u) = \lambda u$ gives us $\lambda_1 u_1 + \cdots + \lambda_k u_k = \lambda(u_1 + \cdots + u_k)$. That is,

$$(\lambda_1 - \lambda)u_1 + \cdots + (\lambda_k - \lambda)u_k = 0. \tag{6.2.6}$$

Since $u \neq 0$, there exists some $i = 1, \ldots, k$ such that $u_i \neq 0$. Thus, taking scalar product of the above equation with u_i, we get $(\lambda_i - \lambda)\|u_i\|^2 = 0$, which implies $\lambda = \lambda_i$. Therefore $\lambda_1, \ldots, \lambda_k$ are all the possible eigenvalues of T.

To establish (3), we simply construct an orthonormal basis over each eigenspace E_{λ_i}, obtained in (2), denoted by $\mathcal{B}_{\lambda_i}$, $i = 1, \ldots, k$. Then $\mathcal{B} = \mathcal{B}_{\lambda_1} \cup \cdots \cup \mathcal{B}_{\lambda_k}$ is a desired orthonormal basis of U as stated in (3). $\square$

Let $A = (a_{ij}) \in \mathbb{C}(n,n)$ and consider the mapping $T \in L(\mathbb{C}^n)$ defined by

$$T(x) = Ax, \quad x \in \mathbb{C}^n. \tag{6.2.7}$$

Using the mapping (6.2.7) and Theorem 6.4, we may obtain the following characterization of a Hermitian matrix, which may also be regarded as a matrix
version of Theorem 6.4.

Theorem 6.5 *A matrix $A \in \mathbb{C}(n,n)$ is Hermitian if and only if there is a unitary
matrix $P \in \mathbb{C}(n,n)$ such that*

$$A = P^\dagger D P, \tag{6.2.8}$$

*where $D \in \mathbb{R}(n,n)$ is a real diagonal matrix whose diagonal entries are the
eigenvalues of A.*

Proof   If (6.2.8) holds, it is clear that A is Hermitian, $A = A^\dagger$.

Conversely, assume A is Hermitian. Using $\mathcal{B}_0 = \{e_1, \ldots, e_n\}$ to denote the
standard basis of $\mathbb{C}^n$ equipped with the usual Hermitian positive definite scalar
product $(x,y) = x^\dagger y$ for $x, y \in \mathbb{C}^n$, we see that $\mathcal{B}_0$ is an orthonormal basis of $\mathbb{C}^n$.
With the mapping T defined by (6.2.7), we have $(T(x),y) = (Ax)^\dagger y = x^\dagger A y = (x, T(y))$ for any $x, y \in \mathbb{C}^n$, and

$$T(e_j) = \sum_{i=1}^{n} a_{ij} e_i, \quad j = 1, \ldots, n. \tag{6.2.9}$$

Thus T is self-adjoint or Hermitian. Using Theorem 6.4, there is an orthonor-
mal basis, say $\mathcal{B}$, consisting of eigenvalues of T, say $\lambda_1, \ldots, \lambda_n$, which are
all real. With respect to $\mathcal{B}$, the matrix representation of T is diagonal, $D = \text{diag}\{\lambda_1, \ldots, \lambda_n\}$. Now since the basis transition matrix from $\mathcal{B}_0$ into $\mathcal{B}$ is
unitary, thus (6.2.8) must hold for some unitary matrix P as expected. $\square$

In other words, Theorem 6.5 states that a complex square matrix is diagonalizable through a unitary matrix into a real diagonal matrix if and only if it is
Hermitian.

Practically, the decomposition (6.2.8) for a Hermitian matrix A may also,
more preferably, be established as in the proof of Theorem 5.6 as follows. Find
an orthonormal basis $\{u_1, \ldots, u_n\}$ of $\mathbb{C}^n$, with the standard Hermitian scalar
product, consisting of eigenvectors of A: $Au_i = \lambda_i u_i$, $i = 1, \ldots, n$. Let $Q \in \mathbb{C}(n,n)$
be made of taking $u_1, \ldots, u_n$ as its respective column vectors. Then Q is unitary
and $AQ = QD$, where $D = \text{diag}\{\lambda_1, \ldots, \lambda_n\}$. Thus $P = Q^\dagger$ renders (6.2.8).

Exercises

6.2.1   Let $A \in \mathbb{C}(n,n)$ be Hermitian. Show that $\det(A)$ must be a real number.

6.2.2   (Extension of Problem 6.2.1) Let U be a complex vector space with a positive definite scalar product and $T \in L(U)$. If T is self-adjoint, then the coefficients of the characteristic polynomial of T are all real.

6.2.3   Let U be a complex vector space with a positive definite scalar product and S and $T \in L(U)$ self-adjoint and commutative, $S \circ T = T \circ S$.

(a) Prove the identity

$$\|(S \pm iT)(u)\|^2 = \|S(u)\|^2 + \|T(u)\|^2, \quad u \in U. \qquad (6.2.10)$$

(b) Show that $S \pm iT$ is invertible if either S or T is so. However, the converse is not true.

(This is an extended version of Problem 4.3.4)

6.2.4   Let U be a complex vector space with a positive definite scalar product and V a subspace of U. Show that the mapping $P \in L(U)$ that projects U onto V along $V^{\perp}$ is self-adjoint.

6.2.5   (A strengthened version of Problem 6.2.4) If $P \in L(U)$ is idempotent, $P^2 = P$, show that P projects U onto $R(P)$ along $R(P)^{\perp}$ if and only if $P' = P$.

6.2.6   We rewrite any $A \in \mathbb{C}(n,n)$ into the form $A = B + iC$, where B and $C \in \mathbb{R}(n,n)$. Show that a necessary and sufficient condition for A to be unitary is that $B^t C = C^t B$ and $B^t B + C^t C = I_n$.

6.2.7   Let $U = C[0,1]$ be the vector space of all complex-valued continuous functions in the variable $t \in [0,1]$ equipped with the positive definite scalar product

$$(u,v) = \int_0^1 \overline{u(t)} v(t) \, dt, \quad u,v \in U. \qquad (6.2.11)$$

(a) Show that $T(u)(t) = tu(t)$ for $t \in [0,1]$ defines a self-adjoint mapping in over U.

(b) Show that T does not have an eigenvalue whatsoever.

6.2.8   Let $T \in L(U)$ be self-adjoint where U is a finite-dimensional complex vector space with a positive definite scalar product.

(a) Show that if $T^k = 0$ for some integer $k \geq 1$ then $T = 0$.

(b) (A sharpened version of (a)) Given $u \in U$ show that if $T^k(u) = 0$ for some integer $k \geq 1$ then $T(u) = 0$.

6.3 Positive definiteness

We now consider the notion of positive definiteness in the complex situation. Assume that U is a finite-dimensional complex vector space with a positive definite scalar product $(\cdot,\cdot)$.

6.3.1 Definition and basic characterization

We start with the definition of the positive definiteness of various subjects of our interest in the complex situation.

Definition 6.6 The positive definiteness of a quadratic form, a self-adjoint or Hermitian mapping, or a Hermitian matrix may be defined as follows:

(1) A real-valued quadratic form q over U (hence it is generated from a sesquilinear Hermitian form) is *positive definite* if

$$q(u) > 0, \quad u \in U, \quad u \neq 0. \tag{6.3.1}$$

(2) A self-adjoint or Hermitian mapping $T \in L(U)$ is *positive definite* if

$$(u, T(u)) > 0, \quad u \in U, \quad u \neq 0. \tag{6.3.2}$$

(3) A Hermitian matrix $A \in \mathbb{C}(n,n)$ is *positive definite* if

$$x^\dagger A x > 0, \quad x \in \mathbb{C}^n, \quad x \neq 0. \tag{6.3.3}$$

Let $f : U \times U \to \mathbb{C}$ be a sesquilinear Hermitian form and the quadratic form q is obtained from f through (6.1.5). Then there is a unique self-adjoint or Hermitian mapping $T \in L(U)$ such that

$$q(u) = f(u,u) = (u, T(u)), \quad u \in U. \tag{6.3.4}$$

Thus we see that the positive definiteness of q and that of T are equivalent. Besides, let $\{u_1, \ldots, u_n\}$ be any basis of U and write $u \in U$ as $u = \sum_{i=1}^{n} x_i u_i$, where $x = (x_1, \ldots, x_n)^t \in \mathbb{C}^n$ is the coordinate vector of u. Then, in view of (6.1.1),

$$q(u) = x^\dagger A x, \quad A = (f(u_i, u_j)), \tag{6.3.5}$$

and the real-valuedness of q is seen to be equivalent to A being Hermitian and, thus, the positive definiteness of A and that of q are equivalent. Therefore the positive definiteness of a self-adjoint or Hermitian mapping is central, which is our focus of this section.

Parallel to Theorem 5.8, we have

Theorem 6.7 *That a self-adjoint or Hermitian mapping $T \in L(U)$ is positive definite is equivalent to any of the following statements:*

(1) *All the eigenvalues of T are positive.*
(2) *There is a positive constant, $\lambda_0 > 0$, such that*

$$(u, T(u)) \geq \lambda_0 \|u\|^2, \quad u \in U. \tag{6.3.6}$$

(3) *There is a positive definite self-adjoint or Hermitian mapping $S \in L(U)$ such that $T = S^2$.*
(4) *The Hermitian matrix A defined by*

$$A = (a_{ij}), \quad a_{ij} = (u_i, T(u_j)), \quad i, j = 1, \ldots, n, \tag{6.3.7}$$

with respect to an arbitrary basis $\{u_1, \ldots, u_n\}$ of U, is positive definite.
(5) *The eigenvalues of the Hermitian matrix A defined in (6.3.7) with respect to an arbitrary basis are all positive.*
(6) *The Hermitian matrix A defined in (6.3.7) with respect to an arbitrary basis enjoys the factorization $A = B^2$ for some Hermitian positive definite matrix $B \in \mathbb{C}(n, n)$.*

The proof of Theorem 6.7 is similar to that of Theorem 5.8 and thus left as an exercise. Here we only check that the matrix A defined in (6.3.7) is indeed Hermitian. In fact, this may be seen from the self-adjointness of T through

$$\bar{a}_{ji} = \overline{(u_j, T(u_i))} = \overline{(T(u_j), u_i)} = (u_i, T(u_j)) = a_{ij}, \quad i, j = 1, \ldots, n. \tag{6.3.8}$$

Note also that if $\{u_1, \ldots, u_n\}$ is an orthonormal basis of U, then the quantities a_{ij} $(i, j = 1, \ldots, n)$ in (6.3.7) simply give rise to the matrix representation of T with respect to this basis. That is, $T(u_j) = \sum_{i=1}^{n} a_{ij} u_i$ $(j = 1, \ldots, n)$.

A matrix version of Theorem 6.7 may be stated as follows.

Theorem 6.8 *That a Hermitian matrix $A \in \mathbb{C}(n, n)$ is positive definite is equivalent to any of the following statements:*

(1) *All the eigenvalues of A are positive.*
(2) *There is a unitary matrix $P \in \mathbb{C}(n, n)$ and a diagonal matrix $D \in \mathbb{R}(n, n)$ whose diagonal entries are all positive such that $A = P^\dagger DP$.*
(3) *There is a positive definite Hermitian matrix $B \in \mathbb{C}(n, n)$ such that $A = B^2$.*
(4) *A is Hermitian congruent to the identity matrix. That is, there is a non-singular matrix $B \in \mathbb{C}(n, n)$ such that $A = B^\dagger I_n B = B^\dagger B$.*

The proof is similar to that of Theorem 5.9 and left as an exercise.

Positive semi-definiteness and negative definiteness can be defined and investigated analogously as in the real situation in Section 5.3 and are skipped here.

6.3.2 Determinant characterization of positive definite Hermitian matrices

If $A \in \mathbb{C}(n,n)$ is Hermitian, then $\det(A) \in \mathbb{R}$. Moreover, if A is positive definite, Theorem 6.8 says there is a non-singular matrix $B \in \mathbb{C}(n,n)$ such that $A = B^\dagger B$. Thus $\det(A) = \det(B^\dagger) \det(B) = |\det(B)|^2 > 0$. Such a property suggests that it may be possible to extend Theorem 5.10 to Hermitian matrices as well, as we now do.

Theorem 6.9 *Let* $A = (a_{ij}) \in \mathbb{C}(n,n)$ *be Hermitian. The matrix* A *is positive definite if and only if all its leading principal minors are positive, that is, if and only if*

$$a_{11} > 0, \quad \begin{vmatrix} a_{11} & a_{12} \\ a_{21} & a_{22} \end{vmatrix} > 0, \quad \ldots, \quad \begin{vmatrix} a_{11} & \cdots & a_{1n} \\ \cdots & \cdots & \cdots \\ a_{n1} & \cdots & a_{nn} \end{vmatrix} > 0. \quad (6.3.9)$$

Proof The necessity proof is similar to that for Theorem 5.10 and omitted. The sufficiency proof is also similar but needs some adaptation to meet the delicacy with handling complex numbers. To see how this is done, we may assume that (6.3.9) holds and we show that A is positive definite by an inductive argument. Again, if $n = 1$, there is nothing to show, and assume that the assertion is valid at $n - 1$ (i.e., when $A \in \mathbb{C}(n-1, n-1)$) ($n \geq 2$). It remains to prove that A is positive definite at $n \geq 2$.

As before, we rewrite the Hermitian matrix A in the blocked form

$$A = \begin{pmatrix} A_{n-1} & \alpha \\ \alpha^\dagger & a_{nn} \end{pmatrix}, \quad (6.3.10)$$

where $\alpha = (a_{1n}, \ldots, a_{n-1,n})^t \in \mathbb{C}^{n-1}$. By the inductive assumption at $n - 1$, we know that the Hermitian matrix A_{n-1} is positive definite. Hence, applying Theorem 6.8, we have a non-singular matrix $B \in \mathbb{C}(n-1, n-1)$ such that $A_{n-1} = B^\dagger B$. Thus

$$A = \begin{pmatrix} A_{n-1} & \alpha \\ \alpha^\dagger & a_{nn} \end{pmatrix} = \begin{pmatrix} B^\dagger & 0 \\ 0 & 1 \end{pmatrix} \begin{pmatrix} I_{n-1} & \beta \\ \beta^\dagger & a_{nn} \end{pmatrix} \begin{pmatrix} B & 0 \\ 0 & 1 \end{pmatrix}, \quad (6.3.11)$$

where $\beta = (B^\dagger)^{-1} \alpha \equiv (b_1, \ldots, b_{n-1})^t \in \mathbb{C}^{n-1}$. Since $\det(A) > 0$, we obtain with some suitable row operations the result

$$\frac{\det(A)}{\det(A_{n-1})} = \det\begin{pmatrix} I_{n-1} & \beta \\ \beta^\dagger & a_{nn} \end{pmatrix} = a_{nn} - |b_1|^2 - \cdots - |b_{n-1}|^2 > 0, \quad (6.3.12)$$

with $\det(B^\dagger B) = \det(A_{n-1})$. Taking $x \in \mathbb{C}^n$ and setting

$$y = \begin{pmatrix} B & 0 \\ 0 & 1 \end{pmatrix} x \equiv (y_1, \ldots, y_{n-1}, y_n)^t, \quad (6.3.13)$$

we have

$$x^\dagger A x = y^\dagger \begin{pmatrix} I_{n-1} & \beta \\ \beta^\dagger & a_{nn} \end{pmatrix} y$$

$$= (\bar{y}_1 + \bar{b}_1 \bar{y}_n, \ldots, \bar{y}_{n-1} + \bar{b}_{n-1} \bar{y}_n, \ b_1 \bar{y}_1 + \cdots + b_{n-1} \bar{y}_{n-1} + a_{nn} \bar{y}_n) \begin{pmatrix} y_1 \\ \vdots \\ y_n \end{pmatrix}$$

$$= |y_1 + b_1 y_n|^2 + \cdots + |y_{n-1} + b_{n-1} y_n|^2$$
$$+ (a_{nn} - |b_1|^2 - \cdots - |b_{n-1}|^2)|y_n|^2, \quad (6.3.14)$$

which is positive when $y \neq 0$ or $x \neq 0$ because of the condition (6.3.12).

Thus the proof is complete. $\quad\square$

6.3.3 Characterization by the Cholesky decomposition

In this subsection, we show that the Cholesky decomposition theorem is also valid in the complex situation.

Theorem 6.10 *Let $A = (a_{ij}) \in \mathbb{C}(n,n)$ be Hermitian. The matrix A is positive definite if and only if there is a non-singular lower triangular matrix $L \in \mathbb{C}(n,n)$ such that $A = LL^\dagger$. Moreover, when A is positive definite, there is a unique such L with the property that all the diagonal entries of L are positive.*

Proof   Assume $A = LL^\dagger$ with L being non-singular. Then $y = L^\dagger x \neq 0$ whenever $x \neq 0$. Thus we have $x^\dagger A x = x^\dagger (LL^\dagger) x = y^\dagger y > 0$, which proves that A is positive definite.

Assume now A is positive definite. We show that there is a unique lower triangular matrix $L = (l_{ij}) \in \mathbb{C}(n,n)$ whose diagonal entries are all positive, $l_{11}, \ldots, l_{nn} > 0$, so that $A = LL^\dagger$. We again use induction.

When $n = 1$, we have $A = (a_{11})$ and the unique choice for L is $L = (\sqrt{a_{11}})$ since $a_{11} > 0$ in view of Theorem 6.9.

Assume the conclusion is established at $n-1$ ($n \geq 2$).

We proceed to establish the conclusion at n ($n \geq 2$).

Rewrite A in the blocked form (6.3.10). Applying Theorem 6.9, we know that $A_{n-1} \in \mathbb{C}(n-1, n-1)$ is positive definite. So there is a unique lower triangular matrix $L_1 \in \mathbb{C}(n-1, n-1)$ with positive diagonal entries such that $A_{n-1} = L_1 L_1^\dagger$.

Now consider $L = \mathbb{C}(n, n)$ of the form

$$L = \begin{pmatrix} L_2 & 0 \\ \gamma^\dagger & a \end{pmatrix}, \tag{6.3.15}$$

where $L_2 \in \mathbb{C}(n-1, n-1)$ is a lower triangular matrix, $\gamma \in \mathbb{C}^{n-1}$ is a column vector, and $a \in \mathbb{R}$ is a suitable number. Then, if we set $A = LL^\dagger$, we obtain

$$A = \begin{pmatrix} A_{n-1} & \alpha \\ \alpha^\dagger & a_{nn} \end{pmatrix} = \begin{pmatrix} L_2 & 0 \\ \gamma^\dagger & a \end{pmatrix} \begin{pmatrix} L_2^\dagger & \gamma \\ 0 & a \end{pmatrix}$$

$$= \begin{pmatrix} L_2 L_2^\dagger & L_2 \gamma \\ (L_2 \gamma)^\dagger & \gamma^\dagger \gamma + a^2 \end{pmatrix}. \tag{6.3.16}$$

Therefore we arrive at the relations

$$A_{n-1} = L_2 L_2^\dagger, \quad \alpha = L_2 \gamma, \quad \gamma^\dagger \gamma + a^2 = a_{nn}. \tag{6.3.17}$$

Thus, if we require that all the diagonal entries of L_2 be positive, then the inductive assumption gives us $L_2 = L_1$. So the vector γ is also uniquely determined, $\gamma = L_1^{-1} \alpha$. Thus it only remains to show that the number a may be uniquely determined as well. To this end, we need to show in (6.3.17) that

$$a_{nn} - \gamma^\dagger \gamma > 0. \tag{6.3.18}$$

In fact, inserting $L_1 = B^\dagger$ in (6.3.11), we have $\beta = \gamma$. That is,

$$A = \begin{pmatrix} A_{n-1} & \alpha \\ \alpha^\dagger & a_{nn} \end{pmatrix} = \begin{pmatrix} L_1 & 0 \\ 0 & 1 \end{pmatrix} \begin{pmatrix} I_{n-1} & \gamma \\ \gamma^\dagger & a_{nn} \end{pmatrix} \begin{pmatrix} L_1^\dagger & 0 \\ 0 & 1 \end{pmatrix}. \tag{6.3.19}$$

Hence (6.3.18) follows as a consequence of $\det(A) > 0$. Thus the third equation in (6.3.17) leads to the unique determination of the positive number a as in the real situation:

$$a = \sqrt{a_{nn} - \gamma^\dagger \gamma} > 0. \tag{6.3.20}$$

The inductive proof is now complete. $\qquad\qquad\qquad\qquad\qquad\qquad\qquad\qquad\square$

Exercises

6.3.1   Prove Theorem 6.7.

6.3.2   Prove Theorem 6.8.

6.3.3   Let $A \in \mathbb{C}(n,n)$ be non-singular. Prove that there is a unique lower triangular matrix $L \in \mathbb{C}(n,n)$ with positive diagonal entries so that $AA^\dagger = LL^\dagger$.

6.3.4   Show that if $A_1,\dots,A_k \in \mathbb{C}(n,n)$ are unitary matrices, so is their product $A = A_1 \cdots A_k$.

6.3.5   Show that if $A \in \mathbb{C}(n,n)$ is Hermitian and all its eigenvalues are ± 1 then A is unitary.

6.3.6   Assume that $u \in \mathbb{C}^n$ is a nonzero column vector. Show that there is a unitary matrix $Q \in \mathbb{C}(n,n)$ such that

$$Q^\dagger(uu^\dagger)Q = \text{diag}\{u^\dagger u, 0, \dots, 0\}. \tag{6.3.21}$$

6.3.7   Let $A = (a_{ij}) \in \mathbb{C}(n,n)$ be Hermitian. Show that if A is positive definite then $a_{ii} > 0$ for $i = 1,\dots,n$ and

$$|a_{ij}| < \sqrt{a_{ii}a_{jj}}, \quad i,j = 1,\dots,n. \tag{6.3.22}$$

6.3.8   Is the Hermitian matrix

$$A = \begin{pmatrix} 5 & i & 2-i \\ -i & 4 & 1-i \\ 2+i & 1+i & 3 \end{pmatrix} \tag{6.3.23}$$

positive definite?

6.3.9   Let $A = (a_{ij}) \in \mathbb{C}(n,n)$ be a positive semi-definite Hermitian matrix. Show that $a_{ii} \geq 0$ for $i = 1,\dots,n$ and that

$$\det(A) \leq a_{11} \cdots a_{nn}. \tag{6.3.24}$$

6.3.10   Let $u_1,\dots,u_n$ be n column vectors of $\mathbb{C}^n$, with the standard Hermitian scalar product, which are the n column vectors of a matrix $A \in \mathbb{C}(n,n)$. Establish the inequality

$$|\det(A)| \leq \|u_1\| \cdots \|u_n\|. \tag{6.3.25}$$

What is the Hadamard inequality in the context of a complex matrix?

6.3.11   Let $A \in \mathbb{C}(n,n)$ be non-singular. Show that there exist a unitary matrix $P \in \mathbb{C}(n,n)$ and a positive definite Hermitian matrix $B \in \mathbb{C}(n,n)$ such that $A = PB$. Show also that, if A is real, then the afore-mentioned matrices P and B may also be chosen to be real so that P is orthogonal and B positive definite.

6.3.12   Let $A \in \mathbb{C}(n,n)$ be non-singular. Since $A^\dagger A$ is positive definite, its eigenvalues, say $\lambda_1,\dots,\lambda_n$, are all positive. Show that there exist unitary matrices P and Q in $\mathbb{C}(n,n)$ such that $A = PDQ$, where

$$D = \text{diag}\left\{\sqrt{\lambda_1},\dots,\sqrt{\lambda_n}\right\} \tag{6.3.26}$$

is a diagonal matrix. Show also that the same conclusion is true for some orthogonal matrices P and Q in $\mathbb{R}(n,n)$ when A is real.

6.3.13 Let U be a finite-dimensional complex vector space with a positive definite scalar product $(\cdot,\cdot)$ and $T \in L(U)$ a positive definite mapping.

(a) Establish the *generalized Schwarz inequality*

$$|(u,T(v))| \le \sqrt{(u,T(u))}\sqrt{(v,T(v))}, \quad u,v \in U. \qquad (6.3.27)$$

(b) Show that the equality in (6.3.27) occurs if and only if u,v are linearly dependent.

6.4 Commutative self-adjoint mappings and consequences

In this section, we extend our investigation about commutativity of self-adjoint mappings in the real situation to that in the complex situation. It is not hard to see that the conclusion in this extended situation is the same as in the real situation.

Theorem 6.11 *Let U be a finite-dimensional complex vector space with a positive definite scalar product $(\cdot,\cdot)$ and S and $T \in L(U)$ be self-adjoint or Hermitian. Then there is an orthonormal basis of U consisting of eigenvectors of both S and T if and only if S and T commute, or $S \circ T = T \circ S$.*

The proof of the theorem is the same as that for the real situation and omitted.

We now use Theorem 6.11 to study a slightly larger class of linear mappings commonly referred to as *normal mappings*.

Definition 6.12 Let $T \in L(U)$. We say that T is *normal* if T and T' commute:

$$T \circ T' = T' \circ T. \qquad (6.4.1)$$

Assume T is normal. Decompose T in the form

$$T = \frac{1}{2}(T+T') + \frac{1}{2}(T-T') = P+Q, \qquad (6.4.2)$$

where $P = \frac{1}{2}(T+T') \in L(U)$ is self-adjoint or Hermitian and $Q = \frac{1}{2}(T-T') \in L(U)$ is *anti-self-adjoint* or *anti-Hermitian* since it satisfies $Q' = -Q$. From

$$(u,iQ(v)) = i(u,Q(v)) = -i(Q(u),v) = (iQ(u),v), \quad u,v \in U, \qquad (6.4.3)$$

we see that $iQ \in L(U)$ is self-adjoint or Hermitian. In view of (6.4.1), P and iQ are commutative. Applying Theorem 6.11, we conclude that U has an orthonormal basis consisting of eigenvectors of P and iQ simultaneously, say $\{u_1, \ldots, u_n\}$, associated with the corresponding real eigenvalues, $\{\varepsilon_1, \ldots, \varepsilon_n\}$ and $\{\delta_1, \ldots, \delta_n\}$, respectively. Consequently, we have

$$
\begin{aligned}
T(u_i) &= (P+Q)(u_i) \\
&= (P - i[iQ])(u_i) = (\varepsilon_i + i\omega_i)u_i, \quad i = 1, \ldots, n,
\end{aligned}
\tag{6.4.4}
$$

where $\omega_i = -\delta_i$ $(i = 1, \ldots, n)$. This discussion leads us to the following theorem.

Theorem 6.13 *Let $T \in L(U)$. Then T is normal if and only if there is an orthonormal basis of U consisting of eigenvectors of T.*

Proof Suppose that U has an orthonormal basis consisting of eigenvectors of T, say $\mathcal{B} = \{u_1, \ldots, u_n\}$, with the corresponding eigenvalues $\{\lambda_1, \ldots, \lambda_n\}$. Let $T' \in L(U)$ be represented by the matrix $B = (b_{ij}) \in \mathbb{C}(n,n)$ with respect to $\mathcal{B}$. Then

$$
T'(u_j) = \sum_{i=1}^{n} b_{ij} u_i, \quad j = 1, \ldots, n.
\tag{6.4.5}
$$

Therefore, we have

$$
(T'(u_j), u_i) = \left(\sum_{k=1}^{n} b_{kj} u_k, u_i \right) = \overline{b}_{ij}, \quad i, j = 1, \ldots, n,
\tag{6.4.6}
$$

and

$$
(T'(u_j), u_i) = (u_j, T(u_i)) = (u_j, \lambda_i u_i) = \lambda_i \delta_{ij}, \quad i, j = 1, \ldots, n.
\tag{6.4.7}
$$

Combining (6.4.6) and (6.4.7), we find $b_{ij} = \overline{\lambda}_i \delta_{ij}$ $(i, j = 1, \ldots, n)$. In other words, B is diagonal, $B = \mathrm{diag}\{\overline{\lambda}_1, \ldots, \overline{\lambda}_n\}$, such that $\overline{\lambda}_1, \ldots, \overline{\lambda}_n$ are the eigenvalues of T' with the corresponding eigenvectors $u_1, \ldots, u_n$. In particular,

$$
(T \circ T')(u_i) = \lambda_i \overline{\lambda}_i u_i = \overline{\lambda}_i \lambda_i u_i = (T' \circ T)(u_i), \quad i = 1, \ldots, n,
\tag{6.4.8}
$$

which proves $T \circ T' = T' \circ T$. That is, T is normal.

Conversely, if T is normal, the existence of an orthonormal basis of U consisting of eigenvectors of T is already shown.

The proof is complete. $\square$

To end this section, we consider commutative normal mappings.

Theorem 6.14 *Let S and $T \in L(U)$ be normal. Then there is an orthonormal basis of U consisting of eigenvectors of S and T simultaneously if and only if S and T are commutative, $S \circ T = T \circ S$.*

Proof If $\{u_1,\ldots,u_n\}$ is an orthonormal basis of U so that $S(u_i) = \gamma_i u_i$ and $T(u_i) = \lambda_i u_i$, $\gamma_i, \lambda_i \in \mathbb{C}$, $i = 1,\ldots,n$, then $(S \circ T)(u_i) = \gamma_i \lambda_i u_i = (T \circ S)(u_i)$, $i = 1,\ldots,n$, which establishes the commutativity of S and T: $S \circ T = T \circ S$.

Conversely, assume S and T are commutative normal mappings. Let $\lambda_1,\ldots,\lambda_k$ be all the distinct eigenvalues of T and $E_{\lambda_1},\ldots,E_{\lambda_k}$ the associated eigenspaces that may readily be checked to be mutually perpendicular (left as an exercise in this section). Fix $i = 1,\ldots,k$. For any $u \in E_{\lambda_i}$, we have $T(S(u)) = S(T(u)) = S(\lambda_i u) = \lambda_i S(u)$. Thus $S(u) \in E_{\lambda_i}$. That is, E_{λ_i} is invariant under S. Since S is normal, E_{λ_i} has an orthonormal basis, say $\{u_{i,1},\ldots,u_{i,m_i}\}$, consisting of the eigenvectors of S. Therefore, the set of vectors

$$\{u_{1,1},\ldots,u_{1,m_1},\ldots,u_{k,1},\ldots,u_{k,m_k}\} \tag{6.4.9}$$

is an orthonormal basis of U consisting of the eigenvectors of both S and T. $\quad\square$

An obvious but pretty general example of a normal mapping that is not necessarily self-adjoint or Hermitian is a unitary mapping $T \in L(U)$ since it satisfies $T \circ T' = T' \circ T = I$. Consequently, for a unitary mapping $T \in L(U)$, there is an orthonormal basis of U consisting of eigenvectors of T. Furthermore, since T is isometric, it is clear that all eigenvalues of T are of absolute value 1, as already observed in Section 4.3.

The matrix versions of Definition 6.12 and Theorems 6.13 and 6.14 may be stated as follows.

Definition 6.15 A matrix $A \in \mathbb{C}(n,n)$ is said to be *normal* if it satisfies the property $AA^{\dagger} = A^{\dagger}A$.

Theorem 6.16 *Normal matrices have the following characteristic properties:*

(1) *A matrix $A \in \mathbb{C}(n,n)$ is diagonalizable through a unitary matrix, that is, there is a unitary matrix $P \in \mathbb{C}(n,n)$ and a diagonal matrix $D \in \mathbb{C}(n,n)$ such that $A = P^{\dagger}DP$, if and only if A is normal.*

(2) *Two normal matrices A and $B \in \mathbb{C}(n,n)$ are simultaneously diagonalizable through a unitary matrix, that is, there is a unitary matrix $P \in \mathbb{C}(n,n)$ and two diagonal matrices D_1 and $D_2 \in \mathbb{C}(n,n)$ such that $A = P^{\dagger}D_1P$ and $B = P^{\dagger}D_2P$, if and only if A and B are commutative: $AB = BA$.*

To prove the theorem, we may simply follow the standard way to associate a matrix $A \in \mathbb{C}(n,n)$ with the mapping it generates over $\mathbb{C}^n$ through $x \mapsto Ax$ for $x \in \mathbb{C}^n$ as before and apply Theorems 6.13 and 6.14.

Moreover, if $A \in \mathbb{C}(n,n)$ is unitary, $AA^{\dagger} = A^{\dagger}A = I_n$, then there is a unitary matrix $P \in \mathbb{C}(n,n)$ and a diagonal matrix $D = \text{diag}\{\lambda_1,\ldots,\lambda_n\}$ with $|\lambda_i| = 1$, $i = 1,\ldots,n$, such that $A = P^{\dagger}DP$.

Exercises

6.4.1   Let U be a finite-dimensional complex vector space with a positive definite scalar product $(\cdot, \cdot)$ and $T \in L(U)$. Show that T is normal if and only if T satisfies the identity

$$\|T(u)\| = \|T'(u)\|, \quad u \in U. \tag{6.4.10}$$

6.4.2   Use Problem 6.4.1 and the property $(\lambda I - T)' = \bar{\lambda} I - T'$ ($\lambda \in \mathbb{C}$) to show directly that if T is normal and λ is an eigenvalue of T with an associated eigenvector $u \in U$, then $\bar{\lambda}$ and u are a pair of eigenvalue and eigenvector of T'.

6.4.3   Let U be a finite-dimensional complex vector space with a positive definite scalar product $(\cdot, \cdot)$ and $T \in L(U)$ satisfy the property that, if $\lambda \in \mathbb{C}$ is an eigenvalue of T and $u \in U$ an associated eigenvector, then $\bar{\lambda}$ is an eigenvalue and u an associated eigenvector of T'. Show that if λ and $\mu \in \mathbb{C}$ are two different eigenvalues of T and u and v are the associated eigenvectors, respectively, then $(u, v) = 0$.

6.4.4   Assume that $A \in \mathbb{C}(n, n)$ is triangular and normal. Show that A must be diagonal.

6.4.5   Let U be a finite-dimensional complex vector space with a positive definite scalar product and $T \in L(U)$ be normal.

   (a) Show that T is self-adjoint if and only if all the eigenvalues of T are real.

   (b) Show that T is anti-self-adjoint if and only if all the eigenvalues of T are imaginary.

   (c) Show that T is unitary if and only if all the eigenvalues of T are of absolute value 1.

6.4.6   Let $T \in L(U)$ be a normal mapping where U is a finite-dimensional complex vector space with a positive definite scalar product.

   (a) Show that if $T^k = 0$ for some integer $k \geq 1$ then $T = 0$.

   (b) (A sharpened version of (a)) Given $u \in U$ show that if $T^k(u) = 0$ for some integer $k \geq 1$ then $T(u) = 0$.

   (This is an extended version of Problem 6.2.8.)

6.4.7   If R and $S \in L(U)$ are normal and commutative, show that $T = R \pm iS$ is normal.

6.4.8   Consider the matrix

$$A = \begin{pmatrix} 1 & i \\ i & 1 \end{pmatrix}. \tag{6.4.11}$$

   (a) Show that A is not Hermitian nor unitary but normal.

(b) Find an orthonormal basis of $\mathbb{C}^2$, with the standard Hermitian scalar product, consisting of the eigenvectors of A.

6.4.9 Let U be a finite-dimensional complex vector space with a positive definite scalar product and $T \in L(U)$ be normal. Show that for any integer $k \geq 1$, there is a normal element $S \in L(U)$ such that $T = S^k$. Moreover, if T is unitary, then there is a unitary element $S \in L(U)$ such that $T = S^k$.

6.4.10 Recall that any $c \in \mathbb{C}$ may be rewritten as $c = a\mathrm{e}^{\mathrm{i}\theta}$, where $a, \theta \in \mathbb{R}$ and $a \geq 0$, known as a polar decomposition of c. Let U be a finite-dimensional complex vector space with a positive definite scalar product and $T \in L(U)$. Show that T enjoys a similar *polar decomposition* property such that there are a positive semi-definite element R and a unitary element S, both in $L(U)$, satisfying $T = R \circ S = S \circ R$, if and only if T is normal.

6.4.11 Let U be complex vector space with a positive definite scalar product $(\cdot, \cdot)$. A mapping $T \in L(U)$ is called *hyponormal* if it satisfies

$$(u, (T \circ T' - T' \circ T)(u)) \leq 0, \quad u \in U. \tag{6.4.12}$$

(a) Show that T is normal if any only if T and T' are both hyponormal.
(b) Show that T being hyponormal is equivalent to

$$\|T'(u)\| \leq \|T(u)\|, \quad u \in U. \tag{6.4.13}$$

(c) If T is hyponormal, so is $T + \lambda I$, where $\lambda \in \mathbb{C}$.
(d) Show that if $\lambda \in \mathbb{C}$ is an eigenvalue and u an associated eigenvector of a hyponormal mapping $T \in L(U)$, then $\overline{\lambda}$ and u are an eigenvalue and an associated eigenvector of T'.

6.4.12 Let U be an n-dimensional ($n \geq 2$) complex vector space with a positive definite scalar product and $T \in L(U)$. Show that T is normal if and only if there is a complex-coefficient polynomial $p(t)$ of degree at most $n - 1$ such that $T' = p(T)$.

6.4.13 Let U be a finite-dimensional complex vector space with a positive definite scalar product and $T \in L(U)$. Show that T is normal if and only if T and T' have the same invariant subspaces of U.

6.4.14 Let $\lambda \in \mathbb{C}$ be an eigenvalue of a matrix $A \in \mathbb{C}(n, n)$. Must $\overline{\lambda}$ be an eigenvalue of $A^{\dagger}$?

6.5 Mappings between two spaces via self-adjoint mappings

As in the real situation, we show that self-adjoint or Hermitian mappings may be used to study mappings between two complex vector spaces.

As in Section 5.6, let U and V denote two complex vector spaces of finite dimensions, with positive definite Hermitian scalar products $(\cdot,\cdot)_U$ and $(\cdot,\cdot)_V$, respectively. Given $T \in L(U,V)$ and $v \in V$, it is clear that

$$f(u) = (v, T(u))_V, \quad u \in U, \tag{6.5.1}$$

defines an element f in U' which depends on $v \in V$. So there is a unique element in U depending on v, say $T'(v)$, such that $f(u) = (T'(v), u)_U$. That is,

$$(v, T(u))_V = (T'(v), u)_U, \quad u \in U, v \in V. \tag{6.5.2}$$

Thus we have obtained a well-defined mapping $T' : V \to U$.

It may easily be verified that T' is linear. Thus $T' \in L(V,U)$. As in the real situation, we can consider the composed mappings $T' \circ T \in L(U)$ and $T \circ T' \in L(V)$, which are both self-adjoint or Hermitian.

Besides, from the relations

$$(u, (T' \circ T)(u))_U = (T(u), T(u))_V \geq 0, \quad u \in U, \tag{6.5.3}$$

$$(v, (T \circ T')(v))_V = (T'(v), T'(v))_U \geq 0, \quad v \in V, \tag{6.5.4}$$

it is seen that $T' \circ T \in L(U)$ and $T \circ T' \in L(V)$ are both positive semi-definite.

Use $\|\cdot\|_U$ and $\|\cdot\|_V$ to denote the norms induced from $(\cdot,\cdot)_U$ and $(\cdot,\cdot)_V$, respectively. Then

$$\|T\| = \sup\{\|T(u)\|_V \mid \|u\|_U = 1, u \in U\}. \tag{6.5.5}$$

On the other hand, using the fact that $T' \circ T \in L(U)$ is self-adjoint and positive semi-definite, we know that there is an orthonormal basis $\{u_1,\ldots,u_n\}$ of U consisting of eigenvectors of $T' \circ T$, with the corresponding non-negative eigenvalues, $\sigma_1,\ldots,\sigma_n$, respectively. Hence, for any $u = \sum_{i=1}^n a_i u_i \in U$ with $\|u\|_U^2 = \sum_{i=1}^n |a_i|^2 = 1$, we have

$$\|T(u)\|_V^2 = (T(u), T(u))_V = ((T' \circ T)(u), u)_U = \sum_{i=1}^n \sigma_i |a_i|^2 \leq \sigma_0, \tag{6.5.6}$$

where

$$\sigma_0 = \max_{1 \leq i \leq n} \{\sigma_i\} \geq 0, \tag{6.5.7}$$

which shows $\|T\| \leq \sqrt{\sigma_0}$. Moreover, let $i = 1,\ldots,n$ be such that $\sigma_0 = \sigma_i$. Then (6.5.5) gives us

$$\|T\|^2 \geq \|T(u_i)\|_V^2 = ((T' \circ T)(u_i), u_i)_U = \sigma_i = \sigma_0. \tag{6.5.8}$$

Consequently, as in the real situation, we conclude with

$$\|T\| = \sqrt{\sigma_0}, \quad \text{where } \sigma_0 \geq 0 \text{ is the largest eigenvalue of the mapping } T' \circ T. \tag{6.5.9}$$

Therefore the norm of a linear mapping $T \in L(U,V)$ may be obtained by computing the largest eigenvalue of the induced self-adjoint or Hermitian mapping $T' \circ T \in L(U)$.

In particular, if $T \in L(U)$ is already self-adjoint, then, since $\|T\|$ is simply the radical root of the largest eigenvalue of T^2, we arrive at

$$\|T\| = \max_{1 \le i \le n}\{|\lambda_i|\}, \qquad (6.5.10)$$

where $\lambda_1, \dots, \lambda_n$ are the eigenvalues of T.

Thus, a combination of (6.5.9) and (6.5.10) leads to the formula

$$\|T\|^2 = \|T' \circ T\|, \quad T \in L(U,V). \qquad (6.5.11)$$

Analogously, from (6.5.10), we note that, when $T \in L(U)$ is self-adjoint, the quantity $\|T\|$ may also be evaluated accordingly by

$$\|T\| = \sup\{|(u, T(u))| \,|\, u \in U, \|u\| = 1\}, \qquad (6.5.12)$$

as in the real situation.

We can similarly show how to extend (6.5.12) to evaluate the norm of an arbitrary linear mapping between U and V.

Theorem 6.17 *Let U and V be finite-dimensional complex vector spaces with positive definite Hermitian scalar products $(\cdot,\cdot)_U$ and $(\cdot,\cdot)_V$, respectively. For $T \in L(U,V)$, we have*

$$\|T\| = \sup\{|(v, T(u))_V| \,|\, u \in U, v \in V, \|u\|_U = 1, \|v\|_V = 1\}. \qquad (6.5.13)$$

The proof is identical to that for Theorem 5.14.

As in the real situation, we may use Theorem 6.17 to establish the fact that the norms of a linear mapping and its dual, over two complex vector spaces with positive definite Hermitian scalar products, share the same value.

Theorem 6.18 *Let U and V be finite-dimensional complex vector spaces with positive definite Hermitian scalar products $(\cdot,\cdot)_U$ and $(\cdot,\cdot)_V$, respectively. For $T \in L(U,V)$ and its dual $T' \in L(V,U)$, we have $\|T\| = \|T'\|$. Thus $\|T' \circ T\| = \|T \circ T'\|$ and the largest eigenvalues of the positive semi-definite self-adjoint or Hermitian mappings $T' \circ T \in L(U)$ and $T \circ T' \in L(V)$ are the same.*

Proof The fact that $\|T\| = \|T'\|$ may be deduced from applying (6.5.13) to T' and the relation $(v, T(u))_V = (T'(v), u)_U$ ($u \in U$ and $v \in V$). The conclusion $\|T' \circ T\| = \|T \circ T'\|$ follows from (6.5.11) and that the largest eigenvalues of the positive semi-definite self-adjoint or Hermitian mappings $T' \circ T$ and $T \circ T'$ are the same is a consequence of the eigenvalue characterization of the norm of a self-adjoint mapping stated in (6.5.10) and $\|T' \circ T\| = \|T \circ T'\|$. $\square$

However, as in the real situation, Theorem 6.18 is natural and hardly surprising in view of Theorem 5.16.

We continue to study a mapping $T \in L(U,V)$, where U and V are two complex vector spaces with positive definite Hermitian product $(\cdot,\cdot)_U$ and $(\cdot,\cdot)_V$ and of dimensions n and m, respectively. Let $\sigma_1,\ldots,\sigma_n$ be all the eigenvalues of the positive semi-definite mapping $T' \circ T \in L(U)$, counting multiplicities, among which $\sigma_1,\ldots,\sigma_k$ are positive, say. Use $\{u_1,\ldots,u_k,\ldots,u_n\}$ to denote an orthonormal basis of U consisting of eigenvectors of $T' \circ T$ associated with the eigenvalues $\sigma_1,\ldots,\sigma_k,\ldots,\sigma_n$. Then we have

$$
\begin{aligned}
(T(u_i),T(u_j))_V &= (u_i,(T' \circ T)(u_j))_U \\
&= \sigma_j(u_i,u_j)_U = \sigma_j\delta_{ij}, \quad i,j=1,\ldots,n. \quad (6.5.14)
\end{aligned}
$$

This simple expression indicates that $T(u_i) = 0$ for $i > k$ (if any) and that $\{T(u_1),\ldots,T(u_k)\}$ forms an orthogonal basis of $R(T)$. In particular, $k = r(T)$.

Now set

$$
v_i = \frac{1}{\|T(u_i)\|_V}T(u_i), \quad i=1,\ldots,k. \tag{6.5.15}
$$

Then $\{v_1,\ldots,v_k\}$ is an orthonormal basis for $R(T)$. Taking $i = j = 1,\ldots,k$ in (6.5.14), we see that $\|T(u_i)\|_V = \sqrt{\sigma_i}, i = 1,\ldots,k$. In view of this and (6.5.15), we arrive at

$$
T(u_i) = \sqrt{\sigma_i}v_i, \quad i=1,\ldots,k, \quad T(u_j) = 0, \quad j>k \quad \text{(if any)}. \tag{6.5.16}
$$

In the above construction, the positive numbers $\sqrt{\sigma_1},\ldots,\sqrt{\sigma_k}$ are called the *singular values* of T and the expression (6.5.16) the *singular value decomposition* for T. This result may conveniently be summarized as a theorem.

Theorem 6.19 *Let U and V be finite-dimensional complex vector spaces with positive definite Hermitian scalar products and $T \in L(U,V)$ is of rank $k \geq 1$. Then there are orthonormal bases $\{u_1,\ldots,u_k,\ldots,u_n\}$ and $\{v_1,\ldots,v_k,\ldots,v_m\}$ of U and V, respectively, and positive numbers $\lambda_1,\ldots,\lambda_k$, referred to as the singular values of T, such that*

$$
T(u_i) = \lambda_i v_i, \quad i=1,\ldots,k, \quad T(u_j) = 0, \quad j>k \quad \text{(if any)}. \tag{6.5.17}
$$

In fact the numbers $\lambda_1^2,\ldots,\lambda_k^2$ are all the positive eigenvalues, counting multiplicities, and $u_1,\ldots,u_k$ the associated eigenvectors of the self-adjoint mapping $T' \circ T \in L(U)$.

Let $A \in \mathbb{C}(m,n)$. Theorem 6.19 implies that there are unitary matrices $P \in \mathbb{C}(n,n)$ and $Q \in \mathbb{C}(m,m)$ such that

$$\left\{ \begin{array}{l} AP = Q\Lambda \quad \text{or} \quad A = Q\Lambda P^\dagger, \\[2mm] \Lambda = \begin{pmatrix} D & 0 \\ 0 & 0 \end{pmatrix} \in \mathbb{R}(m,n), \quad D = \text{diag}\{\lambda_1, \ldots, \lambda_k\}, \end{array} \right. \tag{6.5.18}$$

where $k = r(A)$ and $\lambda_1, \ldots, \lambda_k$ are some positive numbers for which $\lambda_1^2, \ldots, \lambda_k^2$ are all the positive eigenvalues of the Hermitian matrix $A^\dagger A$. The numbers $\lambda_1, \ldots, \lambda_k$ are called the *singular values* of the matrix A and the expression (6.5.18) the *singular value decomposition* for the matrix A.

Note that Problem 6.3.12 of Section 6.3 may be regarded as an early and special version of the general singular value decomposition procedure here.

Of course the results above may also be established similarly for mappings between real vector spaces and for real matrices.

Exercises

6.5.1 Verify that for $T \in L(U,V)$ the mapping T' given in (6.5.2) is a well-defined element in $L(V,U)$ although the scalar products $(\cdot,\cdot)_U$ and $(\cdot,\cdot)_V$ of U and V are both sesquilinear.

6.5.2 Consider the matrix

$$A = \begin{pmatrix} 2 & 1-i & 3 \\ 3i & -1 & 3+2i \end{pmatrix}. \tag{6.5.19}$$

(a) Find the eigenvalues of $AA^\dagger$ and $A^\dagger A$ and compare.
(b) Find $\|A\|$.

6.5.3 Apply (6.5.9) and use the fact that the nonzero eigenvalues of $T' \circ T$ and $T \circ T'$ are the same to prove directly that $\|T\| = \|T'\|$ as stated in Theorem 6.18.

6.5.4 If U is a finite-dimensional vector space with a positive definite scalar product and $T \in L(U)$ is normal, show that $\|T^2\| = \|T\|^2$. Can this result be extended to $\|T^m\| = \|T\|^m$ for any positive integer m?

6.5.5 Consider the matrix

$$A = \begin{pmatrix} 1+i & 2 & -1 \\ 2 & 1-i & 1 \end{pmatrix}. \tag{6.5.20}$$

(a) Find the singular values of A.
(b) Find a singular value decomposition of A.

6.5.6 Let $A \in \mathbb{C}(m,n)$. Show that A and $A^\dagger$ have the same singular values.

6.5.7   Let $A \in \mathbb{C}(n,n)$ be invertible. Investigate the relationship between the singular values of A and those of A^{-1}.

6.5.8   Let $A \in \mathbb{C}(n,n)$. Use the singular value decomposition for A to show that A may be rewritten as $A = PB = CQ$, where P and Q are some unitary matrices and B and C some positive semi-definite Hermitian matrices.

6.5.9   Let U be a finite-dimensional complex vector space with a positive definite Hermitian scalar product and $T \in L(U)$ positive semi-definite. Show that the singular values of T are simply the positive eigenvalues of T.

6.5.10 Show that a square full-rank complex matrix is unitary if and only if its singular values are all 1.

6.5.11 Let the singular values of $A \in \mathbb{C}(m,n)$ be

$$\lambda_1 \geq \cdots \geq \lambda_k > 0, \quad k = r(A). \tag{6.5.21}$$

Recall that the Hilbert–Schmidt norm of A, also known as the Frobenius norm of A, is defined by

$$\|A\| = \sqrt{\mathrm{Tr}(A^\dagger A)}. \tag{6.5.22}$$

Establish the result

$$\|A\| = \sqrt{\lambda_1^2 + \cdots + \lambda_k^2}. \tag{6.5.23}$$

6.5.12 Let the singular values of A and $B \in \mathbb{C}(m,n)$ be given by

$$\alpha_1 \geq \cdots \geq \alpha_k > 0, \quad \beta_1 \geq \cdots \geq \beta_l > 0, \quad k = r(A), \quad l = r(B), \tag{6.5.24}$$

respectively. Prove that, if A and B have *simultaneous* singular value decompositions, that is, there is *one* pair of unitary matrices $P \in \mathbb{C}(m,m)$ and $Q \in \mathbb{C}(n,n)$ (rather than *two* pairs) such that

$$A = P\Lambda_1 Q, \quad B = \pm P\Lambda_2 Q, \quad \Lambda_{1,2} = \begin{pmatrix} D_{1,2} & 0 \\ 0 & 0 \end{pmatrix} \in \mathbb{C}(m,n), \tag{6.5.25}$$

where

$$D_1 = \mathrm{diag}\{\alpha_1, \ldots, \alpha_k\}, \quad D_2 = \mathrm{diag}\{\beta_1, \ldots, \beta_l\}, \tag{6.5.26}$$

then

$$|\mathrm{Tr}(A^\dagger B)| = \alpha_1 \beta_1 + \cdots + \alpha_s \beta_s, \quad s = \min\{k,l\}. \tag{6.5.27}$$

Note that, actually, for arbitrary A and $B \in \mathbb{C}(m,n)$, there holds

$$|\mathrm{Tr}(A^\dagger B)| \leq \alpha_1 \beta_1 + \cdots + \alpha_s \beta_s, \quad s = \min\{k,l\}. \tag{6.5.28}$$

This statement is known as the von Neumann trace inequality which will be encountered again later.

6.5.13 Combine Problem 6.5.11 and the von Neumann trace inequality (6.5.28) to reestablish

$$|\text{Tr}(A^\dagger B)| \leq \|A\| \|B\|, \qquad (6.5.29)$$

which simply recovers the Cauchy–Schwarz inequality, although (6.5.28) is stronger than (6.5.29). Explain why.

6.5.14 Let U and V be two finite-dimensional vector spaces over $\mathbb{R}$ or $\mathbb{C}$ equipped with positive definite scalar products. For any $T \in L(U,V)$, show that T and $T' \in L(V,U)$ have the same singular values.

6.5.15 Let U be an n-dimensional vector space over $\mathbb{R}$ or $\mathbb{C}$ equipped with a positive definite scalar product. For $T \in L(U)$, use $\|T\|$ to denote the induced norm of T. If T is invertible and

$$\lambda_1 \geq \cdots \geq \lambda_n > 0 \qquad (6.5.30)$$

are the singular values of T, establish the formula

$$\|T^{-1}\| = \frac{1}{\lambda_n}. \qquad (6.5.31)$$

Note that most of the exercise problems in Chapter 5 may be restated in the context of the complex situation of this chapter and are omitted.

7

Jordan Decomposition

In this chapter, we establish the celebrated Jordan decomposition theorem which allows us to reduce a linear mapping over $\mathbb{C}$ into a canonical form in terms of its eigenspectrum. As a preparation, we first recall some facts regarding factorization of polynomials. Then, we show how to reduce a linear mapping over a set of its invariant subspaces determined by a prime factorization of the characteristic polynomial of the mapping. Next, we reduce a linear mapping over its generalized eigenspaces. Finally, we prove the Jordan decomposition theorem by understanding how a mapping behaves itself over each of its generalized eigenspaces.

7.1 Some useful facts about polynomials

Let $\mathcal{P}$ be the vector space of all polynomials with coefficients in a given field $\mathbb{F}$ and in the variable t. Various technical computations and concepts involving elements in $\mathcal{P}$ may be simplified considerably with the notion "ideal" as we now describe.

Definition 7.1 A non-empty subset $\mathcal{I} \subset \mathcal{P}$ is called an *ideal* of $\mathcal{P}$ if it satisfies the following two conditions:

(1) $f + g \in \mathcal{I}$ for any f and $g \in \mathcal{I}$.
(2) $fg \in \mathcal{I}$ for any $f \in \mathcal{P}$ and $g \in \mathcal{I}$.

Since $\mathbb{F}$ may naturally be viewed as a subset of $\mathcal{P}$, we see that $af \in \mathcal{I}$ for any $a \in \mathbb{F}$ and $f \in \mathcal{I}$. Hence an ideal is also a subspace.

Let $g_1, \ldots, g_k \in \mathcal{P}$. Construct the subset of $\mathcal{P}$ given by

$$\{f_1 g_1 + \cdots + f_k g_k \mid f_1, \ldots, f_k \in \mathcal{P}\}. \tag{7.1.1}$$

It is readily checked that the subset defined in (7.1.1) is an ideal of $\mathcal{P}$. We may say that this deal is generated from $g_1, \ldots, g_k$ and use the notation $\mathcal{I}(g_1, \ldots, g_k)$ to denote it.

There are two trivial ideals: $\mathcal{I} = \{0\}$ and $\mathcal{I} = \mathcal{P}$, and it is obvious that $\{0\} = \mathcal{I}(0)$ and $\mathcal{P} = \mathcal{I}(1)$. That is, both $\{0\}$ and $\mathcal{P}$ are generated from some single elements in $\mathcal{P}$. The following theorem establishes that any ideal in $\mathcal{P}$ may be generated from a single element in $\mathcal{P}$.

Theorem 7.2 *Any ideal in $\mathcal{P}$ is singly generated. More precisely, if $\mathcal{I} \neq \{0\}$ is an ideal of $\mathcal{P}$, then there is an element $g \in \mathcal{P}$ such that $\mathcal{I} = \mathcal{I}(g)$. Moreover, if there is another $h \in \mathcal{P}$ such that $\mathcal{I} = \mathcal{I}(h)$, then g and h are of the same degree. Besides, if the coefficients of the highest-degree terms of g and h coincide, then $g = h$.*

Proof If $\mathcal{I} = \mathcal{I}(g)$ for some $g \in \mathcal{P}$, it is clear that g will have the lowest degree among all elements in $\mathcal{I}$ in view of the definition of $\mathcal{I}(g)$. Such an observation indicates what to look for in our proof. Indeed, since $\mathcal{I} \neq \{0\}$, we may choose g to be an element in $\mathcal{I} \setminus \{0\}$ which is of the lowest degree. Then it is clear that $\mathcal{I}(g) \subset \mathcal{I}$. For any $h \in \mathcal{I} \setminus \{0\}$, we claim $g|h$ (i.e., g divides h). Otherwise, we may rewrite h as

$$h(t) = q(t)g(t) + r(t) \tag{7.1.2}$$

for some q and $r \in \mathcal{P}$ so that the degree of r is lower than that of g. Since $g \in \mathcal{I}$, we have $qg \in \mathcal{I}$. Thus $r = h - qg \in \mathcal{I}$, which contradicts the definition of g. Consequently, $g|h$. So $h \in \mathcal{I}(g)$ as expected, which proves $\mathcal{I} \subset \mathcal{I}(g)$.

If there is another $h \in \mathcal{P}$ such that $\mathcal{I} = \mathcal{I}(h)$, of course g and h must have the same degree since $h|g$ and $g|h$. If the coefficients of the highest-degree terms of g and h coincide, then $g = h$, otherwise $g - h \in \mathcal{I} \setminus \{0\}$ would be of a lower degree, which contradicts the choice of g given earlier.

The theorem is proved. $\qquad\square$

Let $g_1, \ldots, g_k \in \mathcal{P} \setminus \{0\}$. Choose $g \in \mathcal{P}$ such that $\mathcal{I}(g) = \mathcal{I}(g_1, \ldots, g_k)$. Then there are elements $f_1, \ldots, f_k \in \mathcal{P}$ such that

$$g = f_1 g_1 + \cdots + f_k g_k, \tag{7.1.3}$$

which implies that g contains all common divisors of $g_1, \ldots, g_k$. In other words, if $h|g_1, \ldots, h|g_k$, then $h|g$. On the other hand, the definition of $\mathcal{I}(g_1, \ldots, g_k)$ already gives us $g|g_1, \ldots, g|g_k$. So g itself is a common divisor of $g_1, \ldots, g_k$. In view of Theorem 7.2, we see that the coefficient of the highest-degree term of g determines g completely. Thus, we may fix g by taking the coefficient of the highest-degree term of g to be 1. Such a polynomial g is referred to as the *greatest common divisor* of $g_1, \ldots, g_k$ and often denoted as $g = \gcd(g_1, \ldots, g_k)$. Therefore, we see that the notion of an ideal and its generator provides an effective tool for the study of greatest common divisors.

Given $g_1, \ldots, g_k \in \mathcal{P} \setminus \{0\}$, if there does not exist a common divisor of a nontrivial degree (≥ 1) for $g_1, \ldots, g_k$, then we say that $g_1, \ldots, g_k$ are *relatively prime* or *co-prime*. Thus $g_1, \ldots, g_k$ are relatively prime if and only if $\gcd(g_1, \ldots, g_k) = 1$. In this situation, there are elements $f_1, \ldots, f_k \in \mathcal{P}$ such that the identity

$$f_1(t)g_1(t) + \cdots + f_k(t)g_k(t) = 1 \tag{7.1.4}$$

is valid for arbitrary t. This fact will be our starting point in the subsequent development.

A polynomial $p \in \mathcal{P}$ of degree at least 1 is called a *prime polynomial* or an *irreducible polynomial* if p cannot be factored into the product of two polynomials in $\mathcal{P}$ of degrees at least 1. Two polynomials are said to be *equivalent* if one is a scalar multiple of the other.

Exercises

7.1.1   For f and $g \in \mathcal{P}$ show that $f | g$ if and only if $\mathcal{I}(f) \supset \mathcal{I}(g)$.

7.1.2   Consider $\mathcal{I} \subset \mathcal{P}$ over a field $\mathbb{F}$ given by

$$\mathcal{I} = \{f \in \mathcal{P} \mid f(a_1) = \cdots = f(a_k) = 0\}, \tag{7.1.5}$$

where $a_1, \ldots, a_k \in \mathbb{F}$ are distinct.

(a) Show that $\mathcal{I}$ is an ideal.
(b) Find a concrete element $g \in \mathcal{P}$ such that $\mathcal{I} = \mathcal{I}(g)$.
(c) To what extent, is the element g obtained in (b) unique?

7.1.3   Show that if $f \in \mathcal{P}$ and $(t - 1) | f(t^n)$, where $n \geq 1$ is an integer, then $(t^n - 1) | f(t^n)$.

7.1.4   Let f and $g \in \mathcal{P}$, $g \neq 0$, and $n \geq 1$ be an integer. Prove that $f | g$ if and only if $f^n | g^n$.

7.1.5   Let f and $g \in \mathcal{P}$ be nonzero polynomials and $n \geq 1$ an integer. Show that

$$(\gcd(f, g))^n = \gcd(f^n, g^n). \tag{7.1.6}$$

7.1.6   Let U be a finite-dimensional vector space over a field $\mathbb{F}$, $T \in L(U)$, and $\mathcal{P}$ the vector space of all polynomials over $\mathbb{F}$. Show that $\mathcal{I} = \{p \in \mathcal{P} \mid p(T) = 0\}$ is an ideal of $\mathcal{P}$. Let $g \in \mathcal{P}$ be such that $\mathcal{I} = \mathcal{I}(g)$ and the

coefficient of the highest-degree term of g is 1. Is g the minimal polynomial of T which has the minimum degree among all the polynomials
that annihilate T?

7.2 Invariant subspaces of linear mappings

In this section, we show how to use characteristic polynomials and their prime
factorization to resolve or reduce linear mappings into mappings over invariant subspaces. As a preparation, we first establish a factorization theorem for
characteristic polynomials.

Theorem 7.3 *Let U be a finite-dimensional vector space over a field $\mathbb{F}$ and
$T \in L(U)$. Assume that V and W are nontrivial subspaces of U such that $U =
V \oplus W$ and V and W are invariant under T. Use R and S to denote T restricted
to V and W and $p_R(\lambda), p_S(\lambda)$, and $p_T(\lambda)$ the characteristic polynomials of
$R \in L(V), S \in L(W)$, and $T \in L(U)$, respectively. Then $p_T(\lambda) = p_R(\lambda)p_S(\lambda)$.*

Proof   Let $\{v_1, \ldots, v_k\}$ and $\{w_1, \ldots, w_l\}$ be bases of V, W, respectively. Then
$\{v_1, \ldots, v_k, w_1, \ldots, w_l\}$ is a basis of U. Assume that $B \in \mathbb{F}(k, k)$ and $C \in \mathbb{F}(l, l)$
are the matrix representations of R and S, with respect to the bases $\{v_1, \ldots, v_k\}$
and $\{w_1, \ldots, w_l\}$, respectively. Then

$$A = \begin{pmatrix} B & 0 \\ 0 & C \end{pmatrix} \tag{7.2.1}$$

is the matrix representation of T with respect to the basis $\{v_1, \ldots, v_k, w_1, \ldots, w_l\}$.
Consequently, we have

$$p_T(\lambda) = \det(\lambda I - A) = \det \begin{pmatrix} \lambda I_k - B & 0 \\ 0 & \lambda I_l - C \end{pmatrix}$$
$$= \det(\lambda I_k - B)\det(\lambda I_l - C) = p_R(\lambda)p_S(\lambda), \tag{7.2.2}$$

as asserted, and the theorem is proved.   □

We can now demonstrate how to use the factorization of the characteristic
polynomial of a linear mapping to naturally resolve it over invariant subspaces.

Theorem 7.4 *Let U be a finite-dimensional vector space over a field $\mathbb{F}$ and
$T \in L(U)$ and use $p_T(\lambda)$ to denote the characteristic polynomial of T. Factor
$p_T(\lambda)$ into the form*

$$p_T = p_1^{n_1} \cdots p_k^{n_k}, \tag{7.2.3}$$

where $p_1, \ldots, p_k$ are nonequivalent prime polynomials in $\mathcal{P}$, and set

$$g_i = \frac{p_T}{p_i^{n_i}} = p_1^{n_1} \cdots \widehat{p_i^{n_i}} \cdots p_k^{n_k}, \quad i = 1, \ldots, k, \tag{7.2.4}$$

where $\widehat{}$ denotes the factor that is missing. Then we have the following:

(1) *The vector space U has the direct decomposition*

$$U = V_1 \oplus \cdots \oplus V_k, \quad V_i = N(p_i^{n_i}(T)), \quad i = 1, \ldots, k. \tag{7.2.5}$$

(2) *T is invariant over each V_i, $i = 1, \ldots, k$.*

Proof Since $p_T(\lambda)$ is the characteristic polynomial, the Cayley–Hamilton theorem gives us $p_T(T) = 0$. Since $g_1, \ldots, g_k$ are relatively prime, there are polynomials $f_1, \ldots, f_k \in \mathcal{P}$ such that

$$f_1(\lambda)g_1(\lambda) + \cdots + f_k(\lambda)g_k(\lambda) = 1. \tag{7.2.6}$$

Therefore, we have

$$f_1(T)g_1(T) + \cdots + f_k(T)g_k(T) = I. \tag{7.2.7}$$

Thus, given $u \in U$, we can rewrite u as

$$u = u_1 + \cdots + u_k, \quad u_i = f_i(T)g_i(T)u, \quad i = 1, \ldots, k. \tag{7.2.8}$$

Now since $p_i^{n_i}(T)u_i = p_i^{n_i}(T)f_i(T)g_i(T)u = f_i(T)p_T(T)u = 0$, we get $u_i \in N(p_i^{n_i}(T))$ $(i = 1, \ldots, k)$, which proves $U = V_1 + \cdots + V_k$.

For any $i = 1, \ldots, k$, we need to show

$$W_i \equiv V_i \cap \left(\sum_{1 \le j \le k, j \ne i} V_j \right) = \{0\}. \tag{7.2.9}$$

In fact, since $p_i^{n_i}$ and g_i are relatively prime, there are polynomials q_i and r_i in $\mathcal{P}$ such that

$$q_i p_i^{n_i} + r_i g_i = 1. \tag{7.2.10}$$

This gives us the relation

$$q_i(T)p_i^{n_i}(T) + r_i(T)g_i(T) = I. \tag{7.2.11}$$

Note also that the definition of g_i indicates that

$$\sum_{1 \le j \le k, j \ne i} V_j \subset N(g_i(T)). \tag{7.2.12}$$

Thus, let $u \in W_i$. Then, applying (7.2.11) to u and using (7.2.12), we find

$$u = q_i(T)p_i^{n_i}(T)u + r_i(T)g_i(T)u = 0. \tag{7.2.13}$$

Therefore (1) is established.

Let $u \in V_i$. Then $p_i^{n_i}(T)(T(u_i)) = T p_i^{n_i}(T)(u) = 0$. Thus $T(u) \in V_i$ and the invariance of V_i under T is proved, which establishes (2). □

Exercises

7.2.1 Let S and $T \in L(U)$ where U is a finite-dimensional vector space over a field $\mathbb{F}$. Show that if the characteristic polynomials $p_S(\lambda)$ and $p_T(\lambda)$ of S and T are relatively prime then $p_S(T)$ and $p_T(S)$ are both invertible.

7.2.2 Let S and $T \in L(U)$ where U is a finite-dimensional vector space over a field $\mathbb{F}$. Use Problem 7.2.1 to show that if the characteristic polynomials of S and T are relatively prime and $R \in L(U)$ satisfies $R \circ S = T \circ R$ then $R = 0$.

7.2.3 Let U be finite-dimensional vector space over a field $\mathbb{F}$ and $T \in L(U)$. Show that T is idempotent, $T^2 = T$, if and only if

$$r(T) + r(I - T) = \dim(U). \tag{7.2.14}$$

7.2.4 Let U be a finite-dimensional vector space over a field $\mathbb{F}$ and $T \in L(U)$. Prove the following slightly extended version of Theorem 7.4.

Suppose that the characteristic polynomial of T, say $p_T(\lambda)$, has the factorization $p_T(\lambda) = g_1(\lambda)g_2(\lambda)$ over $\mathbb{F}$, where g_1 and g_2 are relatively prime polynomials. Then $U = N(g_1(T)) \oplus N(g_2(T))$ and both $N(g_1(T))$ and $N(g_2(T))$ are invariant under T.

7.2.5 Let $u \in \mathbb{C}^n$ be a nonzero column vector and set $A = uu^{\dagger} \in \mathbb{C}(n,n)$.

  (a) Show that $A^2 = aA$, where $a = u^{\dagger}u$.
  (b) Find a non-singular matrix $B \in \mathbb{C}(n,n)$ so that $A = B^{-1}DB$, where $D \in \mathbb{C}(n,n)$ is diagonal and determine D.
  (c) Describe

$$N(A) = \{x \in \mathbb{C}^n \,|\, Ax = 0\}, \quad N(A - aI_n) = \{x \in \mathbb{C}^n \,|\, (A - aI_n)x = 0\}, \tag{7.2.15}$$

  as two invariant subspaces of the mapping $T \in L(\mathbb{C}^n)$ given by $x \mapsto Ax$ $(x \in \mathbb{C}^n)$ over which T reduces.
  (d) Determine $r(T)$.

7.2.6 Let U be a finite-dimensional vector space and $T_1, \ldots, T_k \in L(U)$ satisfy $T_i^2 = T_i$ $(i = 1, \ldots, k)$ and $T_i \circ T_j = 0$ $(i, j = 1, \ldots, k, i \neq j$, if any$)$. Show that there holds the space decomposition

$$U = R(T_1) \oplus \cdots \oplus R(T_k) \oplus V, \quad V = \bigcap_{i=1}^{k} N(T_i), \tag{7.2.16}$$

which reduces $T_1, \ldots, T_k$ simultaneously.

7.3 Generalized eigenspaces as invariant subspaces

In Section 7.3.1, we will carry out a study of nilpotent mappings, which will be crucial for the understanding of the structure of a general linear mapping in terms of its eigenvalues, to be seen in Section 7.3.2.

7.3.1 Reducibility of nilpotent mappings

Let U be a finite-dimensional vector space over a field $\mathbb{F}$ and $T \in L(U)$ a nilpotent mapping. Recall that T is said to be of degree $m \geq 1$ if m is the smallest integer such that $T^m = 0$. When $m = 1$, then $T = 0$ and the situation is trivial. In the nontrivial case, $m \geq 2$, we know that if $u \in U$ is of period m (that is, $T^m(u) = 0$ but $T^{m-1}(u) \neq 0$), then $u, T(u), \ldots, T^{m-1}(u)$ are linearly independent vectors in U. Thus, in any nontrivial situation, m satisfies $2 \leq m \leq \dim(U)$.

For a nontrivial nilpotent mapping, we have the following key results.

Theorem 7.5 *Let $T \in L(U)$ be a nilpotent mapping of degree $m \geq 2$, where U is finite-dimensional. Then, the following are valid:*

(1) *There are k vectors $u_1, \ldots, u_k$ and k integers $m_1 \geq 2, \ldots, m_k \geq 2$ such that $u_1, \ldots, u_k$ are of periods $m_1, \ldots, m_k$, respectively, such that U has a basis of the form*

$$u_1^0, \ldots, u_{k_0}^0, u_1, T(u_1), \ldots, T^{m_1-1}(u_1), \ldots, u_k, T(u_k), \ldots, T^{m_k-1}(u_k), \tag{7.3.1}$$

where $u_1^0, \ldots, u_{k_0}^0$, if any, are some vectors taken from $N(T)$. Thus, setting

$$\begin{cases} U_0 &= \operatorname{Span}\{u_1^0, \ldots, u_{k_0}^0\}, \\ U_i &= \operatorname{Span}\{u_i, T(u_i), \ldots, T^{m_i-1}(u_i)\}, \quad i = 1, \ldots, k, \end{cases} \tag{7.3.2}$$

we have the decomposition

$$U = U_0 \oplus U_1 \oplus \cdots \oplus U_k, \tag{7.3.3}$$

and that $U_0, U_1, \ldots, U_k$ are invariant under T.

(2) *The degree m of T is given by*

$$m = \max\{m_i \mid i = 1, \ldots, k\}. \tag{7.3.4}$$

(3) *The sum of the integers k_0 and k is the nullity of T:*

$$k_0 + k = n(T). \tag{7.3.5}$$

Proof (1) We proceed inductively with $m \geq 2$.

(i) $m = 2$.

Let $\{v_1, \ldots, v_k\}$ be a basis of $R(T)$. Choose $u_1, \ldots, u_k \in U$ such that $T(u_1) = v_1, \ldots, T(u_k) = v_k$. We assert that

$$u_1, T(u_1), \ldots, u_k, T(u_k) \tag{7.3.6}$$

are linearly independent vectors in U. To see this, consider

$$a_1 u_1 + b_1 T(u_1) + \cdots + a_k u_k + b_k T(u_k) = 0, \quad a_i, b_i \in \mathbb{F}, \quad i = 1, \ldots, k. \tag{7.3.7}$$

Applying T to (7.3.7), we obtain $a_1 v_1 + \cdots + a_k v_k = 0$. Thus $a_1 = \cdots = a_k = 0$. Inserting this result into (7.3.7), we obtain $b_1 v_1 + \cdots + b_k v_k = 0$, which leads to $b_1 = \cdots = b_k = 0$.

It is clear that $v_1, \ldots, v_k \in N(T)$ since $T^2 = 0$. Choose $u_1^0, \ldots, u_{k_0}^0 \in N(T)$, if any, so that $\{u_1^0, \ldots, u_{k_0}^0, v_1, \ldots, v_k\}$ becomes a basis of $N(T)$.

For any $u \in U$, we rewrite $T(u)$ as

$$T(u) = a_1 v_1 + \cdots + a_k v_k = T(a_1 u_1 + \cdots + a_k u_k), \tag{7.3.8}$$

where $a_1, \ldots, a_k \in \mathbb{F}$ are uniquely determined. Therefore we conclude that $u - (a_1 u_1 + \cdots + a_k u_k) \in N(T)$ and that there are unique scalars $c_1, \ldots, c_{k_0}, b_1, \ldots, b_k$ such that

$$u - (a_1 u_1 + \cdots + a_k u_k) = c_1 u_1^0 + \cdots + c_{k_0} u_{k_0}^0 + b_1 v_1 + \cdots + b_k v_k. \tag{7.3.9}$$

Consequently we see that

$$u_1^0, \ldots, u_{k_0}^0, u_1, T(u_1), \ldots, u_k, T(u_k) \tag{7.3.10}$$

form a basis of U, as described.

(ii) $m \geq 3$.

We assume the statement in (1) holds when the degree of a nilpotent mapping is up to $m - 1 \geq 2$.

Let $T \in L(U)$ be of degree $m \geq 3$ and set $V = R(T)$. Since V is invariant under T, we may regard T as an element in $L(V)$. To avoid confusion, we use T_V to denote the restriction of T over V.

It is clear that the degree of T_V is $m - 1 \geq 2$. Applying the inductive assumption to T_V, we see that there are vectors $v_1, \ldots, v_l \in V$ with respective periods $m_1' \geq 2, \ldots, m_l' \geq 2$ and vectors $v_1^0, \ldots, v_{l_0}^0 \in N(T_V)$, if any, such that

$$v_1^0, \ldots, v_{l_0}^0, v_1, T_V(v_1), \ldots, T_V^{m_1'-1}(v_1), \ldots, v_l, T_V(v_l), \ldots, T_V^{m_l'-1}(v_l) \tag{7.3.11}$$

form a basis of V.

Since $V = R(T)$, we can find some $w_1, \ldots, w_{l_0}, u_1, \ldots, u_l \in U$ such that

$$T(w_1) = v_1^0, \ldots, T(w_{l_0}) = v_{l_0}^0, T(u_1) = v_1, \ldots, T(u_l) = v_l. \tag{7.3.12}$$

Hinted by (i), we assert that

$$w_1, v_1^0 \ldots, w_{l_0}, v_{l_0}^0, u_1, T(u_1), \ldots, T^{m_1'}(u_1), \ldots, u_l, T(u_l), \ldots, T^{m_l'}(u_l) \quad (7.3.13)$$

are linearly independent. In fact, consider the relation

$$\sum_{i=1}^{l_0}(a_i w_i + b_i v_i^0) + \sum_{i=1}^{l}\sum_{j=0}^{m_i'} b_{ij} T^j(u_i) = 0, \quad (7.3.14)$$

where a_i's, b_i's, and b_{ij}'s are scalars. Applying T to (7.3.14) and using (7.3.12), we arrive at

$$\sum_{i=1}^{l_0} a_i v_i^0 + \sum_{i=1}^{l}\sum_{j=0}^{m_i'-1} b_{ij} T_V^j(v_i) = 0, \quad (7.3.15)$$

which results in the conclusion

$$a_i = 0, \quad i = 1, \ldots, l_0; \quad b_{ij} = 0, \quad j = 0, 1, \ldots, m_i' - 1, \quad i = 1, \ldots, l. \quad (7.3.16)$$

Substituting (7.3.16) into (7.3.14) we have

$$b_1 v_1^0 + \cdots + b_{l_0} v_{l_0}^0 + b_{1m_1'} T^{m_1'}(u_1) + \cdots + b_{lm_l'} T^{m_l'}(u_l) = 0. \quad (7.3.17)$$

In other words, we get

$$b_1 v_1^0 + \cdots + b_{l_0} v_{l_0}^0 + b_{1m_1'} T_V^{m_1'-1}(v_1) + \cdots + b_{lm_l'} T_V^{m_l'-1}(v_l) = 0. \quad (7.3.18)$$

Since the vectors given in (7.3.11) are linearly independent, we find

$$b_1 = \cdots = b_{l_0} = b_{1m_1'} = \cdots = b_{lm_l'} = 0 \quad (7.3.19)$$

as well, which establishes that the vectors given in (7.3.13) are linearly independent.

It is clear that

$$v_1^0, \ldots, v_{l_0}^0, T^{m_1'}(u_1), \ldots, T^{m_l'}(u_l) \in N(T). \quad (7.3.20)$$

Find $u_1^0, \ldots, u_{k_0}^0 \in N(T)$, if any, such that

$$u_1^0, \ldots, u_{k_0}^0, v_1^0, \ldots, v_{l_0}^0, T^{m_1'}(u_1), \ldots, T^{m_l'}(u_l) \quad (7.3.21)$$

form a basis of $N(T)$.

Take any $u \in U$. We know that $T(u) \in V$ implies that

$$T(u) = \sum_{i=1}^{l_0} b_i v_i^0 + \sum_{i=1}^{l}\sum_{j=0}^{m_i'-1} b_{ij} T_V^j(v_i)$$

$$= \sum_{i=1}^{l_0} b_i T(w_i) + \sum_{i=1}^{l}\sum_{j=1}^{m_i'} b_{ij} T^j(u_i) \quad (7.3.22)$$

for some unique b_i's and b_{ij}'s in $\mathbb{F}$, which gives rise to the result

$$u - \sum_{i=1}^{l_0} b_i w_i - \sum_{i=1}^{l} \sum_{j=0}^{m'_i-1} b_{ij} T^j(u_i) \in N(T). \qquad (7.3.23)$$

Consequently, we see that

$$u - \sum_{i=1}^{l_0} b_i w_i - \sum_{i=1}^{l} \sum_{j=0}^{m'_i-1} b_{ij} T^j(u_i) = \sum_{i=1}^{k_0} a_i u_i^0 + \sum_{i=1}^{l_0} c_i v_i^0 + \sum_{i=1}^{l} d_i T^{m'_i}(u_i) \qquad (7.3.24)$$

for some unique a_i's, c_i's, and d_i's in $\mathbb{F}$. Thus we conclude that the vectors

$$u_1^0, \dots, u_{k_0}^0, w_1, T(w_1), \dots, w_{l_0}, T(w_{l_0}), u_i, T(u_i), \dots, T^{m'_i}(u_i), i = 1, \dots, l, \qquad (7.3.25)$$

form a basis of U so that $u_1^0, \dots, u_{k_0}^0 \in N(T)$ (if any), $w_1, \dots, w_{l_0}$ are of period 2, and $u_1, \dots, u_l$ of periods $m'_1 + 1, \dots, m'_l + 1$, respectively.

Thus (1) is proved.

The statement (2) is obvious.

In (7.3.25), we have seen that $k = l_0 + l$. Hence, from (7.3.21), we have $n(T) = k_0 + k$ as anticipated, which proves (3). □

We are now prepared to consider general linear mappings over complex vector spaces.

7.3.2 Reducibility of a linear mapping via generalized eigenspaces

We now assume that the field we work on is $\mathbb{C}$. In this situation, any prime polynomial must be of degree 1, which allows us to make statements in Theorem 7.4 concrete and explicit.

Theorem 7.6 *Let U be a complex n-dimensional vector space and $T \in L(U)$. For the characteristic polynomial $p_T(\lambda)$ of T, let $\lambda_1, \dots, \lambda_k$ be the distinct roots of $p_T(\lambda)$ so that*

$$p_T(\lambda) = (\lambda - \lambda_1)^{n_1} \cdots (\lambda - \lambda_k)^{n_k}, \quad n_1, \dots, n_k \in \mathbb{N}. \qquad (7.3.26)$$

Then, the following statements are valid:

(1) The vector space U has the decomposition

$$U = V_1 \oplus \cdots \oplus V_k, \quad V_i = N((T - \lambda_i I)^{n_i}), \quad i = 1, \dots, k. \qquad (7.3.27)$$

(2) Each V_i is invariant under T and $T - \lambda_i I$ is nilpotent over V_i, $i = 1, \dots, k$.

(3) *The characteristic polynomial of T restricted to V_i is simply*

$$p_{V_i}(\lambda) = (\lambda - \lambda_i)^{n_i}, \tag{7.3.28}$$

and the dimension of V_i is n_i, $i = 1, \ldots, k$.

Proof   In view of Theorem 7.4, it only remains to establish (3).

From Theorem 7.2, we have the factorization

$$p_T(\lambda) = p_{V_1}(\lambda) \cdots p_{V_k}(\lambda), \tag{7.3.29}$$

where $p_{V_i}(\lambda)$ is the characteristic polynomial of T restricted to V_i ($i = 1, \ldots, k$).

For fixed i, we consider T over V_i. Since $T - \lambda_i I$ is nilpotent on V_i, we can find vectors $u_1^0, \ldots, u_{m_0}^0 \in N(T - \lambda_i I)$, if any, and cyclic vectors $u_1, \ldots, u_l$ of respective periods $m_1 \geq 2, \ldots, m_l \geq 2$, if any, so that

$$\left\{ \begin{array}{l} u_1^0, \ldots, u_{m_0}^0, \\ u_1, (T - \lambda_i I)(u_1), \ldots, (T - \lambda_i)^{m_1 - 1}(u_1), \\ \cdots\cdots\cdots\cdots \\ u_l, (T - \lambda_i I)(u_l), \ldots, (T - \lambda_i)^{m_l - 1}(u_l), \end{array} \right. \tag{7.3.30}$$

form a basis for V_i, in view of Theorem 7.5. In particular, we have

$$d_i \equiv \dim(V_i) = m_0 + m_1 + \cdots + m_l. \tag{7.3.31}$$

With respect to such a basis, the matrix of $T - \lambda_i I$ is seen to take the following boxed diagonal form:

$$\begin{pmatrix} 0 & 0 & \cdots & 0 \\ 0 & S_1 & \cdots & 0 \\ 0 & \cdots & \ddots & 0 \\ 0 & \cdots & 0 & S_l \end{pmatrix}, \tag{7.3.32}$$

where, in the diagonal of the above matrix, 0 is the zero matrix of size $m_0 \times m_0$ and $S_1, \ldots, S_l$ are shift matrices of sizes $m_1 \times m_1, \ldots, m_l \times m_l$, respectively. Therefore the matrix of $T = (T - \lambda_i I) + \lambda_i I$ with respect to the same basis is simply

$$A_i = \begin{pmatrix} 0 & 0 & \cdots & 0 \\ 0 & S_1 & \cdots & 0 \\ 0 & \cdots & \ddots & 0 \\ 0 & \cdots & 0 & S_l \end{pmatrix} + \lambda_i I_{d_i}. \tag{7.3.33}$$

Consequently, it follows immediately that the characteristic polynomial of T restricted to V_i may be computed by

$$p_{V_i}(\lambda) = \det(\lambda I_{d_i} - A_i) = (\lambda - \lambda_i)^{d_i}. \tag{7.3.34}$$

Finally, inserting (7.3.34) into (7.3.29), we obtain

$$p_T(\lambda) = (\lambda - \lambda_1)^{d_1} \cdots (\lambda - \lambda_k)^{d_k}. \tag{7.3.35}$$

Comparing (7.3.35) with (7.3.26), we arrive at $d_i = n_i$, $i = 1, \ldots, k$, as asserted.

$\square$

Note that the factorization expressed in (7.3.26) indicates that, each eigenvalue, λ_i, repeats itself n_i times as a root in the characteristic polynomial of T. For this reason, the integer n_i is called the *algebraic multiplicity* of the eigenvalue λ_i of T.

Given $T \in L(U)$, let λ_i be an eigenvalue of T. For any integer $m \geq 1$, the nonzero vectors in $N((T - \lambda_i I)^m)$ are called the *generalized eigenvectors* and $N((T - \lambda_i I)^m)$ the generalized eigenspace associated to the eigenvalue λ_i. If $m = 1$, generalized eigenvectors and eigenspace are simply eigenvectors and eigenspace, respectively, associated to the eigenvalue λ_i of T, and

$$n(T - \lambda_i I) = \dim(N(T - \lambda_i I)) \tag{7.3.36}$$

is the geometric multiplicity of the eigenvalue λ_i. Since $N(T - \lambda_i I) \subset N((T - \lambda_i)^{n_i})$, we have

$$n(T - \lambda_i I) \leq \dim(N(T - \lambda_i)^{n_i}) = n_i. \tag{7.3.37}$$

In other words, the geometric multiplicity is less than or equal to the algebraic multiplicity, of any eigenvalue, of a linear mapping.

Exercises

7.3.1 Let U be a finite-dimensional vector space over a field $\mathbb{F}$ and $T \in L(U)$. Assume that $\lambda_0 \in \mathbb{F}$ is an eigenvalue of T and consider the eigenspace $E_{\lambda_0} = N(T - \lambda_0 I)$ associated with λ_0. Let $\{u_1, \ldots, u_k\}$ be a basis of E_{λ_0} and extend it to obtain a basis of U, say $\mathcal{B} = \{u_1, \ldots, u_k, v_1, \ldots, v_l\}$. Show that, using the matrix representation of T with respect to the basis $\mathcal{B}$, the characteristic polynomial of T may be shown to take the form

$$p_T(\lambda) = (\lambda - \lambda_0)^k q(\lambda), \tag{7.3.38}$$

where $q(\lambda)$ is a polynomial of degree l with coefficients in $\mathbb{F}$. In particular, use (7.3.38) to infer again, without relying on Theorem 7.6, that the geometric multiplicity does not exceed the algebraic multiplicity, of the eigenvalue λ_0.

7.3.2 Let U be a finite-dimensional vector space over a field $\mathbb{F}$ and $T \in L(U)$. We say that U is a *cyclic vector space* with respect to T if there is a vector $u \in U$ such that the vectors

$$u, T(u), \ldots, T^{n-1}(u) \qquad (7.3.39)$$

form a basis for U, and we call the vector u a *cyclic vector* of T.

(a) Find the matrix representation of T with respect to the basis

$$\{T^{n-1}(u), \ldots, T(u), u\} \qquad (7.3.40)$$

with specifying

$$T(T^{n-1}(u)) = a_{n-1}T^{n-1}(u) + \cdots + a_1 T(u) + a_0 u,$$
$$a_0, a_1, \ldots, a_{n-1} \in \mathbb{F}. \qquad (7.3.41)$$

Note that T is nilpotent of degree n only when $a_0 = a_1 = \cdots = a_{n-1} = 0$.

(b) Find the characteristic polynomial and the minimal polynomial of T in terms of $a_0, a_1, \ldots, a_{n-1}$.

7.3.3 Let U be a n-dimensional vector space ($n \geq 2$) over a field $\mathbb{F}$ and $T \in L(U)$. Assume that T has a cyclic vector of period n. Show that $S \in L(U)$ commutes with T if and only $S = p(T)$ for some polynomial $p(t)$ of degree at most $n - 1$ and with coefficients in $\mathbb{F}$.

7.3.4 Let U be a n-dimensional complex vector space with a positive definite scalar product and $T \in L(U)$ a normal mapping.

(a) Show that if T has a cyclic vector then T has n distinct eigenvalues.

(b) Assume that T has n distinct eigenvalues and $u_1, \ldots, u_n$ are the associated eigenvectors. Show that

$$u = a_1 u_1 + \cdots + a_n u_n \qquad (7.3.42)$$

is a cyclic vector of T if and only if $a_i \neq 0$ for any $i = 1, \ldots, n$.

7.3.5 Let U be an n-dimensional vector space over a field $\mathbb{F}$ and $T \in L(U)$ a degree n nilpotent mapping, $n \geq 2$. Show that there is no $S \in L(U)$ such that $S^2 = T$.

7.4 Jordan decomposition theorem

Let $T \in L(U)$ where U is an n-dimensional vector space over $\mathbb{C}$ and $\lambda_1, \ldots, \lambda_k$ be all the distinct eigenvalues of T, of respective algebraic multiplicities $n_1, \ldots, n_k$, so that the characteristic polynomial of T assumes the form

$$p_T(\lambda) = (\lambda - \lambda_i)^{n_1} \cdots (\lambda - \lambda_k)^{n_k}. \tag{7.4.1}$$

For each $i = 1, \ldots, k$, use V_i to denote the generalized eigenspace associated with the eigenvalue λ_i:

$$V_i = N((T - \lambda_i I)^{n_i}). \tag{7.4.2}$$

Then we have seen the following:

(1) V_i is invariant under T.
(2) There are eigenvectors $u_{i,1}^0, \ldots, u_{i,l_i^0}^0$ of T, if any, associated with the eigenvalue λ_i, and cyclic vectors $u_{i,1}, \ldots, u_{i,l_i}$ of respective periods $m_{i,1} \geq 2, \ldots, m_{i,l_i} \geq 2$, if any, relative to $T - \lambda_i I$, such that V_i has a basis, denoted by $\mathcal{B}_i$, consisting of vectors

$$\begin{cases} u_{i,1}^0, \ldots, u_{i,l_i^0}^0, \\ u_{i,1}, (T - \lambda_i I)(u_{i,1}), \ldots, (T - \lambda_i)^{m_{i,1}-1}(u_{i,1}), \\ \quad \cdots \cdots \cdots \cdots \\ u_{i,l_i}, (T - \lambda_i I)(u_{i,l_i}), \ldots, (T - \lambda_i)^{m_{i,l_i}-1}(u_{i,l_i}). \end{cases} \tag{7.4.3}$$

(3) $T - \lambda_i I$ is nilpotent of degree m_i on V_i where

$$m_i = \max\{1, m_{i,1}, \ldots, m_{i,l_i}\}. \tag{7.4.4}$$

Since $(T - \lambda_i I)^{n_i}$ is null over V_i, we have $m_i \leq n_i$. Therefore, applying T to these vectors, we have

$$\begin{cases} T(u_{i,s}^0) &= \lambda_i u_{i,s}^0, \quad s = 1, \ldots, l_i^0, \\ T(u_{i,1}) &= (T - \lambda_i I)(u_{i,1}) + \lambda_i u_{i,1}, \\ T((T - \lambda_i I)(u_{i,1})) &= (T - \lambda_i I)^2(u_{i,1}) + \lambda_i (T - \lambda_i I)(u_{i,1}), \\ \quad \cdots \cdots \quad \cdots \quad \cdots \cdots \\ T((T - \lambda_i)^{m_{i,1}-1}(u_{i,1})) &= \lambda_i (T - \lambda_i)^{m_{i,1}-1}(u_{i,1}), \\ \quad \cdots \cdots \quad \cdots \quad \cdots \cdots \\ T(u_{i,l_i}) &= (T - \lambda_i I)(u_{i,l_i}) + \lambda_i u_{i,l_i}, \\ T((T - \lambda_i I)(u_{i,l_i})) &= (T - \lambda_i I)^2(u_{i,l_i}) + \lambda_i (T - \lambda_i I)(u_{i,l_i}), \\ \quad \cdots \cdots \quad \cdots \quad \cdots \cdots \\ T((T - \lambda_i)^{m_{i,l_i}-1}(u_{i,l_i})) &= \lambda_i (T - \lambda_i)^{m_{i,l_i}-1}(u_{i,l_i}). \end{cases} \tag{7.4.5}$$

From (7.4.5), we see that, as an element in $L(V_i)$, the matrix representation of T with respect to the basis (7.4.3) is

$$
J_i \equiv \begin{pmatrix} J_{i,0} & 0 & \cdots & 0 \\ 0 & J_{i,1} & \cdots & 0 \\ 0 & 0 & \ddots & 0 \\ 0 & \cdots & 0 & J_{i,l_i} \end{pmatrix},
\tag{7.4.6}
$$

where $J_{i,0} = \lambda_i I_{l_i^0}$, and

$$
J_{i,s} = \begin{pmatrix} \lambda_i & 0 & \cdots & \cdots & 0 \\ 1 & \lambda_i & 0 & \cdots & 0 \\ \vdots & \ddots & \ddots & \ddots & 0 \\ \vdots & \cdots & \ddots & \ddots & \vdots \\ 0 & \cdots & \cdots & 1 & \lambda_i \end{pmatrix}
\tag{7.4.7}
$$

is an $m_{i,s} \times m_{i,s}$ matrix, $s = 1, \ldots, l_i$.

Alternatively, we may also reorder the vectors listed in (7.4.3) to get

$$
\left\{ \begin{aligned} & u_{i,1}^0, \ldots, u_{i,l_i^0}^0, \\ & (T - \lambda_i)^{m_{i,1}-1}(u_{i,1}), \ldots, (T - \lambda_i I)(u_{i,1}), u_{i,1}, \\ & \cdots\cdots\cdots \\ & (T - \lambda_i)^{m_{i,l_i}-1}(u_{i,l_i}), \ldots, (T - \lambda_i I)(u_{i,l_i}), u_{i,l_i}. \end{aligned} \right.
\tag{7.4.8}
$$

With the choice of these reordered basis vectors, the submatrix $J_{i,s}$ instead takes the following updated form:

$$
J_{i,s} = \begin{pmatrix} \lambda_i & 1 & 0 & \cdots & 0 \\ 0 & \lambda_i & 1 & \ddots & \vdots \\ \vdots & \ddots & \ddots & \ddots & 0 \\ \vdots & \cdots & \ddots & \ddots & 1 \\ 0 & \cdots & \cdots & 0 & \lambda_i \end{pmatrix} \in \mathbb{C}(m_{i,s}, m_{i,s}), \quad s = 1, \ldots, l_i. \tag{7.4.9}
$$

The submatrix $J_{i,s}$ given in either (7.4.7) or (7.4.9) is the called a *Jordan block*.

To simplify statement, $J_{i,0} = \lambda_i I_{l_i^0}$ is customarily said to consist of l_i^0 1×1 (degenerate) Jordan blocks.

Consequently, if we choose

$$
\mathcal{B} = \mathcal{B}_1 \cup \cdots \cup \mathcal{B}_k,
\tag{7.4.10}
$$

where $\mathcal{B}_i$ is as given in (7.4.3) or (7.4.8), $i = 1,\ldots,k$, to be a basis of U, then the matrix that represents T with respect to $\mathcal{B}$ is

$$
J = \begin{pmatrix} J_1 & 0 & \cdots & 0 \\ 0 & \ddots & \ddots & \vdots \\ \vdots & \ddots & \ddots & 0 \\ 0 & \cdots & 0 & J_k \end{pmatrix}, \tag{7.4.11}
$$

which is called a *Jordan canonical form* or a *Jordan matrix*.

We may summarize the above discussion into the following theorem, which is the celebrated *Jordan decomposition theorem*.

Theorem 7.7 *Let* U *be an n-dimensional vector space over* $\mathbb{C}$ *and* $T \in L(U)$ *so that its distinct eigenvalues are* $\lambda_1,\ldots,\lambda_k$ *with respective algebraic multiplicities* $n_1,\ldots,n_k$. *Then the following hold:*

(1) *U has the decomposition* $U = V_1 \oplus \cdots \oplus V_k$ *and T is invariant over each of the subspaces $V_1,\ldots,V_k$.*
(2) *For each $i = 1,\ldots,k$, $T - \lambda_i I$ is nilpotent of degree m_i over V_i where $m_i \leq n_i$.*
(3) *For each $i = 1,\ldots,k$, appropriate eigenvectors and generalized eigenvectors may be chosen in V_i as stated in (7.4.3) or (7.4.8) which generate a basis of V_i, say $\mathcal{B}_i$.*
(4) *With respect to the basis $\mathcal{B} = \mathcal{B}_1 \cup \cdots \cup \mathcal{B}_k$ of U, the matrix representation of T assumes the Jordan canonical form (7.4.11).*

The theorem indicates that T nullifies the polynomial

$$
m_T(\lambda) = (\lambda - \lambda_1)^{m_1} \cdots (\lambda - \lambda_k)^{m_k}, \tag{7.4.12}
$$

which is a polynomial of the minimum degree among all nonzero polynomials having T as a root. In fact, to show $m_T(T) = 0$, we rewrite any $u \in U$ in the form

$$
u = u_1 + \cdots + u_k, \quad u_1 \in V_1,\ldots,u_k \in V_k. \tag{7.4.13}
$$

Thus, we have

$$
m_T(T)(u) = (T - \lambda_1 I)^{m_1} \cdots (T - \lambda_k I)^{n_k}(u_1 + \cdots + u_k)
$$
$$
= \sum_{i=1}^{k} \left((T - \lambda_1 I)^{m_1} \cdots [\widehat{(T - \lambda_i I)^{m_i}}] \cdots (T - \lambda_k I)^{n_k} \right) (T - \lambda_i I)^{m_i}(u_i)
$$
$$
= 0, \tag{7.4.14}
$$

which establishes $m_T(T) = 0$ as asserted. Here $\widehat{}$ denotes an item that is missing in the expression. Since $m_T(\lambda)|p_T(\lambda)$, we arrive at $p_T(T) = 0$, as stated in the Cayley–Hamilton theorem.

Given $T \in L(U)$, use $\mathcal{P}$ to denote the vector space of all polynomials with coefficients in $\mathbb{C}$ and consider the subspace of $\mathcal{P}$:

$$\mathcal{A}_T = \{p \in \mathcal{P} \mid p(T) = 0\}. \tag{7.4.15}$$

It is clear that $\mathcal{A}_T$ is an ideal in $\mathcal{P}$. Elements in $\mathcal{A}_T$ are also called *annihilating polynomials* of the mapping T. Let $\mathcal{A}_T$ be generated by some $m(\lambda)$. Then $m(\lambda)$ has the property that it is a minimal-degree polynomial among all nonzero elements in $\mathcal{A}_T$. If we normalize the coefficient of the highest-degree term of $m(\lambda)$ to 1, then m is uniquely determined and is called the *minimal polynomial* of the linear mapping T. It is clear that, given $T \in L(U)$, if $\lambda_1, \ldots, \lambda_k$ are all the distinct eigenvalues of T and $m_1, \ldots, m_k$ are the corresponding degrees of nilpotence of $T - \lambda_1 I, \ldots, T - \lambda_k I$ over the respective generalized eigenspaces (7.4.2), then $m_T(\lambda)$ defined in (7.4.12) is the minimal polynomial of T.

For example, if T is nilpotent of degree k, then $m_T(\lambda) = \lambda^k$; if T is idempotent, that is, $T^2 = T$, and $T \neq 0$, $T \neq I$, then $m_T(\lambda) = \lambda^2 - \lambda$.

We may use minimal polynomials to characterize a *diagonalizable* linear mapping whose matrix representation with respect to a suitable basis is diagonal. In such a situation it is clear that this basis is made of eigenvectors and the diagonal entries of the diagonal matrix are the corresponding eigenvalues of the mapping.

Theorem 7.8 *Let U be n-dimensional vector space over $\mathbb{C}$ and $T \in L(U)$. Then T is diagonalizable if and only if its minimal polynomial has only simple roots.*

Proof Let $m_T(\lambda)$ be the minimal polynomial of T and $\lambda_1, \ldots, \lambda_k$ the distinct eigenvalues of T. If all roots of $m_T(\lambda)$ are simple, then for each eigenvalue λ_i, the degree of $T - \lambda_i I$ over the generalized eigenspace associated to λ_i is 1, or $T = \lambda_i I$, $i = 1, \ldots, k$. Thus all the Jordan blocks given in (7.4.6) are diagonal, which makes J stated in (7.4.11) diagonal.

Conversely, assume T is diagonalizable and $\lambda_1, \ldots, \lambda_k$ are all the distinct eigenvalues of T. Since $U = E_{\lambda_1} \oplus \cdots \oplus E_{\lambda_k}$, we see that $h(\lambda) = (\lambda - \lambda_1) \cdots (\lambda - \lambda_k) \in \mathcal{A}_T$ defined in (7.4.15). We claim that $m_T(\lambda) = h(\lambda)$. To see this, we only need to show that for any element $p \in \mathcal{A}_T$ we have $p(\lambda_i) = 0$, $i = 1, \ldots, k$. Assume otherwise $p(\lambda_1) \neq 0$ (say). Then $p(\lambda)$ and $\lambda - \lambda_1$ are coprime. So there are polynomials f and g such that $f(\lambda)p(\lambda) + g(\lambda)(\lambda - \lambda_1) \equiv 1$. Consequently, $I = g(T)(T - \lambda_1 I)$, which leads to the contradiction $u = 0$ for any $u \in E_{\lambda_1}$. Thus, we see that h is the lowest degree element in $\mathcal{A}_T \setminus \{0\}$. In other words, $m_T = h$. $\square$

As a matrix version of Theorem 7.7, we may state that, any $n \times n$ complex matrix is similar to a Jordan matrix of the form (7.4.11). For matrices, minimal polynomials may be defined analogously and, hence, omitted. Besides, the matrix version of Theorem 7.8 may read as follows: An $n \times n$ matrix is diagonalizable if and only if the roots of its minimal polynomial are all simple.

Exercises

7.4.1 Show that a diagonalizable nilpotent mapping must be trivial, $T = 0$, and a nontrivial idempotent mapping $T \neq I$ is diagonalizable.

7.4.2 Let $\lambda_1, \ldots, \lambda_k$ be all the distinct eigenvalues of a Hermitian mapping T over an n-dimensional vector space U over $\mathbb{C}$. Show that

$$m_T(\lambda) = (\lambda - \lambda_1) \cdots (\lambda - \lambda_k). \tag{7.4.16}$$

7.4.3 Let U be an n-dimensional vector space over $\mathbb{C}$ and $S, T \in L(U)$. In the proof of Theorem 7.8, it is shown that if $S \sim T$, then $m_S(\lambda) = m_T(\lambda)$. Give an example to show that the condition $m_S(\lambda) = m_T(\lambda)$ is not sufficient to ensure $S \sim T$.

7.4.4 Let $A \in \mathbb{C}(n, n)$. Show that $A \sim A^t$.

7.4.5 Let A and $B \in \mathbb{R}(n, n)$. Show that if there is a non-singular element $C \in \mathbb{C}(n, n)$ such that $A = C^{-1} BC$ then there is a non-singular element $K \in \mathbb{R}(n, n)$ such that $A = K^{-1} BK$.

7.4.6 Let A and $B \in \mathbb{C}(n, n)$ be normal. Show that if the characteristic polynomials of A and B coincide then $A \sim B$.

7.4.7 Let $T \in L(\mathbb{C}^n)$ be defined by

$$T(x) = \begin{pmatrix} x_n \\ \vdots \\ x_1 \end{pmatrix}, \quad x = \begin{pmatrix} x_1 \\ \vdots \\ x_n \end{pmatrix} \in \mathbb{C}^n. \tag{7.4.17}$$

(a) Determine all the eigenvalues of T.
(b) Find the minimal polynomial of T.
(c) Does $\mathbb{C}^n$ have a basis consisting of eigenvectors of T?

7.4.8 Consider the matrix

$$A = \begin{pmatrix} a & 0 & 0 \\ 1 & 0 & 1 \\ 0 & 1 & 0 \end{pmatrix}, \tag{7.4.18}$$

where $a \in \mathbb{R}$.

(a) For what value(s) of a can or cannot the matrix A be diagonalized?
(b) Find the Jordan forms of A corresponding to various values of a.

7.4.9 Let $A \in \mathbb{C}(n,n)$ and $k \geq 2$ be an integer such that $A \sim A^k$.

    (a) Show that, if λ is an eigenvalue of A, so is λ^k.

    (b) Show that, if in addition A is non-singular, then each eigenvalue of A is a root of unity. In other words, if $\lambda \in \mathbb{C}$ is an eigenvalue of A, then there is an integer $s \geq 1$ such that $\lambda^s = 1$.

7.4.10 Let $A \in \mathbb{C}(n,n)$ satisfy $A^m = aI_n$ for some integer $m \geq 1$ and nonzero $a \in \mathbb{C}$. Use the information about the minimal polynomial of A to prove that A is diagonalizable.

7.4.11 Let A and $B \in \mathbb{F}(n,n)$ satisfy $A \sim B$. Show that $\mathrm{adj}(A) \sim \mathrm{adj}(B)$.

7.4.12 Show that the $n \times n$ matrices

$$A = \begin{pmatrix} 1 & 1 & \cdots & 1 \\ 1 & 1 & \cdots & 1 \\ \vdots & \vdots & \ddots & \vdots \\ 1 & 1 & \cdots & 1 \end{pmatrix}, \quad B = \begin{pmatrix} n & 0 & \cdots & 0 \\ b_2 & 0 & \cdots & 0 \\ \vdots & \vdots & \ddots & \vdots \\ b_n & 0 & \cdots & 0 \end{pmatrix}, \quad (7.4.19)$$

where $b_2, \ldots, b_n \in \mathbb{R}$, are similar and diagonalizable.

7.4.13 Show that the matrices

$$A = \begin{pmatrix} 2 & 0 & 0 \\ 0 & 0 & 1 \\ 0 & 1 & 0 \end{pmatrix} \quad \text{and} \quad B = \begin{pmatrix} 1 & 0 & 0 \\ 0 & -1 & 0 \\ 0 & -6 & 2 \end{pmatrix} \quad (7.4.20)$$

are similar and find a non-singular element $C \in \mathbb{R}(3,3)$ such that $A = C^{-1}BC$.

7.4.14 Show that if $A \in \mathbb{C}(n,n)$ has a single eigenvalue then A is not diagonalizable unless $A = \lambda I_n$. In particular, a triangular matrix with identical diagonal entries can never be diagonalizable unless it is already diagonal.

7.4.15 Consider $A \in \mathbb{C}(n,n)$ and express its characteristic polynomial $p_A(\lambda)$ as

$$p_A(\lambda) = (\lambda - \lambda_1)^{n_1} \cdots (\lambda - \lambda_k)^{n_k}, \quad (7.4.21)$$

where $\lambda_1, \ldots, \lambda_k \in \mathbb{C}$ are the distinct eigenvalues of A and $n_1, \ldots, n_k$ the respective algebraic multiplicities of these eigenvalues such that $\sum_{i=1}^{k} n_i = n$. Show that A is diagonalizable if and only if $r(\lambda_i I_n - A) = n - n_i$ for $i = 1, \ldots, k$.

7.4.16 Show that the matrix

$$A = \begin{pmatrix} 0 & -1 & 0 & 0 \\ 0 & 0 & 1 & 0 \\ 0 & 0 & 0 & -1 \\ a^4 & 0 & 0 & 0 \end{pmatrix}, \qquad (7.4.22)$$

where $a > 0$, is diagonalizable in $\mathbb{C}(4,4)$ but not in $\mathbb{R}(4,4)$.

7.4.17 Show that there is no matrix in $\mathbb{R}(3,3)$ whose minimal polynomial is $m(\lambda) = \lambda^2 + 3\lambda + 4$.

7.4.18 Let $A \in \mathbb{R}(n,n)$ and satisfy $A^2 + I_n = 0$.

(a) Show that the minimal polynomial of A is simply $\lambda^2 + 1$.
(b) Show that n must be an even integer, $n = 2m$.
(c) Show that there are n linearly independent vectors, $u_1, v_1, \ldots, u_m, v_m$, in $\mathbb{R}^n$ such that

$$Au_i = -v_i, \quad Av_i = u_i, \quad i = 1, \ldots, m. \qquad (7.4.23)$$

(d) Use the vectors $u_1, v_1, \ldots, u_m, v_m$ to construct an invertible matrix $B \in \mathbb{R}(n,n)$ such that

$$A = B \begin{pmatrix} 0 & I_m \\ -I_m & 0 \end{pmatrix} B^{-1}. \qquad (7.4.24)$$

7.4.19 Consider the space $U = \mathcal{P}_5$ and $T = D^2$ (the second derivative map over U).

(a) Show that T is nilpotent and determine its degree of nilpotence.
(b) Show that 0 is the only eigenvalue of T and write down the characteristic polynomial of T.
(c) Determine the generalized eigenspace of T associated with the eigenvalue 0.
(d) Determine the geometric and algebraic multiplicities of the eigenvalue 0.
(e) Using the meaning of T, find two (why two?) linearly independent polynomials p and $q \in U$ having the largest period (what is the period?).
(f) Use p and q to generate a cyclic (Jordan) basis of U.
(g) Use (f) to obtain a Jordan canonical form for the matrix of T.
(h) Find a Jordan canonical form of the map $S = T - 7I$.

8

Selected Topics

In this chapter, we present a few selected subjects that are important in applications as well but are not usually included in a standard linear algebra course. These subjects may serve as supplemental or extracurricular materials. The first subject is the Schur decomposition theorem, the second is about the classification of skew-symmetric bilinear forms, the third is the Perron–Frobenius theorem for positive matrices, and the fourth concerns the Markov or stochastic matrices.

8.1 Schur decomposition

In this section, we establish the *Schur decomposition theorem*, which serves as a useful complement to the Jordan decomposition theorem and renders further insight and fresh view angles into various subjects, such as the spectral structures of normal mappings and self-adjoint mappings, already covered.

Theorem 8.1 *Let U be a finite-dimensional complex vector space with a positive definite scalar product and $T \in L(U)$. There is an orthonormal basis $B = \{u_1, \ldots, u_n\}$ of U such that the matrix representation of T with respect to B is upper triangular. That is,*

$$T(u_j) = \sum_{i=1}^{j} b_{ij} u_i, \quad j = 1, \ldots, n, \tag{8.1.1}$$

for some $b_{ij} \in \mathbb{C}, i = 1, \ldots, j, j = 1, \ldots, n$. In particular, the diagonal entries $b_{11}, \ldots, b_{nn}$ of the upper triangular matrix $B = (b_{ij}) \in \mathbb{C}(n, n)$ are the eigenvalues of T, which are not necessarily distinct.

Proof We prove the theorem by induction on $\dim(U)$.

When $\dim(U) = 1$, there is nothing to show.

Assume that the theorem holds when $\dim(U) = n - 1 \geq 1$.
We proceed to establish the theorem when $\dim(U) = n \geq 2$.
Let w be an eigenvector of T' associated with the eigenvalue λ and consider

$$V = (\text{Span}\{w\})^{\perp}. \tag{8.1.2}$$

We assert that V is invariant under T. In fact, for any $v \in V$, we have

$$(w, T(v)) = (T'(w), v) = (\lambda w, v) = \overline{\lambda}(w, v) = 0, \tag{8.1.3}$$

which verifies $T(v) \in V$. Thus $T \in L(V)$.

Applying the inductive assumption on $T \in L(V)$ since $\dim(V) = n - 1$, we see that there is an orthonormal basis $\{u_1, \ldots, u_{n-1}\}$ of V and scalars $b_{ij} \in \mathbb{C}$, $i = 1, \ldots, j, j = 1, \ldots, n - 1$ such that

$$T(u_j) = \sum_{i=1}^{j} b_{ij} u_i, \quad j = 1, \ldots, n - 1. \tag{8.1.4}$$

Finally, setting $u_n = (1/\|w\|)w$, we conclude that $\{u_1, \ldots, u_{n-1}, u_n\}$ is an orthonormal basis of U with the stated properties. $\qquad\square$

Using the Schur decomposition theorem, Theorem 8.1, the Cayley–Hamilton theorem (over $\mathbb{C}$) can be readily proved.

In fact, since $T(u_1) = b_{11} u_1$, we have $(T - b_{11} I)(u_1) = 0$. For $j - 1 \geq 1$, we assume $(T - b_{11} I) \cdots (T - b_{j-1,j-1} I)(u_k) = 0$ for $k = 1, \ldots, j - 1$. Using the matrix representation of T with respect to $\{u_1, \ldots, u_n\}$, we have

$$(T - b_{jj} I)(u_j) = \sum_{i=1}^{j-1} b_{ij} u_i. \tag{8.1.5}$$

Hence we arrive at the general conclusion

$$(T - b_{11} I) \cdots (T - b_{jj} I)(u_j)$$

$$= (T - b_{11} I) \cdots (T - b_{j-1,j-1} I) \left(\sum_{i=1}^{j-1} b_{ij} u_i \right) = 0 \tag{8.1.6}$$

for $j = 2, \ldots, n$. Hence $(T - b_{11} I) \cdots (T - b_{jj} I)(u_k) = 0$ for $k = 1, \ldots, j$, $j = 1, \ldots, n$. In particular, since the characteristic polynomial of T takes the form

$$p_T(\lambda) = \det(\lambda I - B) = (\lambda - b_{11}) \cdots (\lambda - b_{nn}), \tag{8.1.7}$$

we have

$$p_T(T)(u_i) = (T - b_{11} I) \cdots (T - b_{nn} I)(u_i) = 0, \quad i = 1, \ldots, n. \tag{8.1.8}$$

Consequently, $p_T(T) = 0$, as anticipated.

It is clear that, in Theorem 8.1, if the upper triangular matrix $B \in \mathbb{C}(n,n)$ is diagonal, then the orthonormal basis $\mathcal{B}$ is made of the eigenvectors of T. In this situation T is normal. Likewise, if B is diagonal and real, then T is self-adjoint.

We remark that Theorem 8.1 may also be proved without resorting to the adjoint mapping.

In fact, use the notation of Theorem 8.1 and proceed to the nontrivial situation $\dim(U) = n \geq 2$ directly. Let $\lambda_1 \in \mathbb{C}$ be an eigenvalue of T and u_1 an associated unit eigenvector. Then we have the orthogonal decomposition

$$U = \mathrm{Span}\{u_1\} \oplus V, \quad V = (\mathrm{Span}\{u_1\})^{\perp}. \tag{8.1.9}$$

Let $P \in L(U)$ be the projection of U onto V along $\mathrm{Span}\{u_1\}$ and set $S = P \circ T$. Then S may be viewed as an element in $L(V)$.

Since $\dim(V) = n - 1$, we may apply the inductive assumption to obtain an orthonormal basis, say $\{u_2, \ldots, u_n\}$, of V, such that

$$S(u_j) = \sum_{i=2}^{j} b_{ij} u_i, \quad j = 2, \ldots, n, \tag{8.1.10}$$

for some b_{ij}'s in $\mathbb{C}$. Of course the vectors $u_1, u_2, \ldots, u_n$ now form an orthonormal basis of U. Moreover, in view of (8.1.10) and $R(I - P) = \mathrm{Span}\{u_1\}$, we have

$$\begin{aligned}
T(u_1) &= \lambda_1 u_1 \equiv b_{11} u_1, \\
T(u_j) &= ((I - P) \circ T)(u_j) + (P \circ T)(u_j) \\
&= b_{1j} u_1 + \sum_{i=2}^{j} b_{ij} u_i, \quad b_{1j} \in \mathbb{C}, \quad j = 2, \ldots, n,
\end{aligned} \tag{8.1.11}$$

where the matrix $B = (b_{ij}) \in \mathbb{C}(n,n)$ is clearly upper triangular as described.

Thus Theorem 8.1 is again established.

The matrix version of Theorem 8.1 may be stated as follows.

Theorem 8.2 *For any matrix $A \in \mathbb{C}(n,n)$ there is a unitary matrix $P \in \mathbb{C}(n,n)$ and an upper triangular matrix $B \in \mathbb{C}(n,n)$ such that*

$$A = P^{\dagger} B P. \tag{8.1.12}$$

That is, the matrix A is Hermitian congruent or similar through a unitary matrix to an upper triangular matrix B whose diagonal entries are all the eigenvalues of A.

The proof of Theorem 8.2 may be obtained by applying Theorem 8.1 where we take $U = \mathbb{C}^n$ with the standard Hermitian scalar product and define $T \in L(\mathbb{C}^n)$ by setting $T(u) = Au$ for $u \in \mathbb{C}^n$. In fact, with $\mathcal{B} = \{u_1, \ldots, u_n\}$ being the orthonormal basis of $\mathbb{C}^n$ stated in Theorem 8.1, the unitary matrix P in (8.1.12) is such that the ith column vector of $P^{\dagger}$ is simply u_i, $i = 1, \ldots, n$.

From (8.1.12) we see immediately that A is normal if and only if B is diagonal and that A is Hermitian if and only if B is diagonal and real.

Exercises

8.1.1 Show that the matrix B may be taken to be lower triangular in Theorems 8.1 and 8.2.

8.1.2 Let $A = (a_{ij})$ and $B = (b_{ij}) \in \mathbb{C}(n,n)$ be stated in the relation (8.1.12). Use the fact $\text{Tr}(A^\dagger A) = \text{Tr}(B^\dagger B)$ to infer the identity

$$\sum_{i,j=1}^{n} |a_{ij}|^2 = \sum_{1 \le i \le j \le n} |b_{ij}|^2. \tag{8.1.13}$$

8.1.3 Denote by $\lambda_1, \ldots, \lambda_n$ all the eigenvalues of a matrix $A \in \mathbb{C}(n,n)$. Show that A is normal if and only if it satisfies the equation

$$\text{Tr}(A^\dagger A) = |\lambda_1|^2 + \cdots + |\lambda_n|^2. \tag{8.1.14}$$

8.1.4 Consider the matrix

$$A = \begin{pmatrix} 5 & 7 \\ -2 & -4 \end{pmatrix}.$$

Find a Schur decomposition of A.

8.1.5 Show that if $A \in \mathbb{R}(n,n)$ is normal and all the roots of its characteristic polynomial are real then A must be symmetric.

8.1.6 Let U be a finite-dimensional complex vector space with a positive definite scalar product and S and $T \in L(U)$. If S and T are commutative, show that U has an orthonormal basis, say $\mathcal{B}$, such that, with respect to $\mathcal{B}$, the matrix representations of S and T are both upper triangular.

8.1.7 Consider A and $B \in \mathbb{C}(n,n)$ satisfying $AB = BA$.

(a) Show that there is a unitary matrix $P \in \mathbb{C}(n,n)$ such that

$$A = P^\dagger S P, \quad B = P^\dagger T P,$$

where S and $T \in \mathbb{C}(n,n)$ are upper triangular matrices whose diagonal entries consist of the eigenvalues of A and B, respectively.

(b) Use (a) to show that the eigenvalues of AB are some products of the eigenvalues of A and B.

(c) Explain why some products of the eigenvalues of A and B may not be the eigenvalues of AB.

(d) Give an example to show that the property stated in (a) does not imply $AB = BA$ in general.

8.1.8   Given $A \in \mathbb{C}(n,n)$ $(n \geq 2)$, follow the steps below to carry out an inductive but also constructive proof of Theorem 8.2.

(a) Find an eigenvalue, say λ_1, of A, and a unit eigenvector $u_1 \in \mathbb{C}^n$, taken as a column vector. Let $u_2, \ldots, u_n \in \mathbb{C}^n$ be chosen so that $u_1, u_2, \ldots, u_n$ form an orthonormal basis of $\mathbb{C}^n$. Use $u_1, u_2, \ldots, u_n$ as the first, second,..., and the nth column vectors of a matrix called Q_1. Then Q_1 is unitary. Check that

$$Q_1^\dagger A Q_1 = \begin{pmatrix} \lambda_1 & \alpha \\ 0 & A_{n-1} \end{pmatrix}, \tag{8.1.15}$$

where $A_{n-1} \in \mathbb{C}(n-1, n-1)$ and $\alpha \in \mathbb{C}^{n-1}$ is a row vector.

(b) Apply the inductive assumption to get a unitary element $Q \in \mathbb{C}(n-1, n-1)$ so that $Q^\dagger A_{n-1} Q$ is upper triangular. Show that

$$Q_2 = \begin{pmatrix} 1 & 0 \\ 0 & Q \end{pmatrix} \tag{8.1.16}$$

is a unitary element in $\mathbb{C}(n,n)$ such that

$$(Q_1 Q_2)^\dagger A (Q_1 Q_2) = Q_2^\dagger Q_1^\dagger A Q_1 Q_2 \tag{8.1.17}$$

becomes upper triangular as desired.

8.1.9   Let $T \in L(U)$ where U is a finite-dimensional complex vector space with a positive definite scalar product. Prove that if λ is an eigenvalue of T then $\overline{\lambda}$ is an eigenvalue of T'.

8.2 Classification of skew-symmetric bilinear forms

Let U be a finite-dimensional vector space over a field $\mathbb{F}$. A bilinear form $f : U \times U \to \mathbb{F}$ is called *skew-symmetric* or *anti-symmetric* if it satisfies

$$f(u,v) = -f(v,u), \quad u,v \in U. \tag{8.2.1}$$

Let $\mathcal{B} = \{u_1, \ldots, u_n\}$ be a basis of U. For u and $v \in U$ with coordinate vectors $x = (x_1, \ldots, x_n)^t$ and $y = (y_1, \ldots, y_n)^t \in \mathbb{F}^n$ with respect to $\mathcal{B}$, we can rewrite $f(u,v)$ as

$$f(u,v) = f\left(\sum_{i=1}^n x_i u_i, \sum_{j=1}^n y_j u_j\right) = \sum_{i,j=1}^n x_i f(u_i, u_j) y_j = x^t A y, \tag{8.2.2}$$

where $A = (a_{ij}) = (f(u_i, u_j)) \in \mathbb{F}(n,n)$ is the matrix representation of f with respect to the basis $\mathcal{B}$. Thus, combining (8.2.1) and (8.2.2), we see that A must be skew-symmetric or anti-symmetric, $A = -A^t$, because

$$-x^t Ay = -f(u,v) = f(v,u) = y^t Ax = (y^t Ax)^t = x^t A^t y, \qquad (8.2.3)$$

and x and $y \in \mathbb{F}^n$ are arbitrary.

Let $\tilde{\mathcal{B}} = \{\tilde{u}_1, \dots, \tilde{u}_n\}$ be another basis of U and $\tilde{A} = (\tilde{a}_{ij}) = (f(\tilde{u}_i, \tilde{u}_j)) \in \mathbb{F}(n,n)$ is the matrix representation of f with respect to $\tilde{\mathcal{B}}$. If the transition matrix between $\mathcal{B}$ and $\tilde{\mathcal{B}}$ is $B = (b_{ij}) \in \mathbb{F}(n,n)$ so that

$$\tilde{u}_j = \sum_{i=1}^{n} b_{ij} u_i, \quad j = 1, \dots, n, \qquad (8.2.4)$$

then we know that $\tilde{A}$ and A are congruent through B, $\tilde{A} = B^t AB$, as discussed in Chapter 5.

In this section, we study the canonical forms of skew-symmetric forms through a classification of skew-symmetric matrices by congruent relations.

Theorem 8.3 *Let $A \in \mathbb{F}(n,n)$ be skew-symmetric. Then there is some non-singular matrix $C \in \mathbb{F}(n,n)$ satisfying $\det(C) = \pm 1$ so that A is congruent through C to a matrix $\Gamma \in \mathbb{F}(n,n)$ of the following canonical form:*

$$\Gamma \equiv \begin{pmatrix} \alpha_1 & \cdots & 0 & \cdots & \cdots & 0 \\ \vdots & \ddots & \vdots & \vdots & \vdots & \vdots \\ \vdots & \cdots & \alpha_k & \cdots & & \vdots \\ 0 & \cdots & \cdots & 0 & \cdots & \vdots \\ \vdots & \vdots & \vdots & \vdots & \ddots & \vdots \\ 0 & \cdots & \cdots & \cdots & \cdots & 0 \end{pmatrix} = CAC^t, \qquad (8.2.5)$$

where

$$\alpha_i = \begin{pmatrix} 0 & a_i \\ -a_i & 0 \end{pmatrix}, \quad i = 1, \dots, k, \qquad (8.2.6)$$

are some 2×2 skew-symmetric matrices given in terms of k (if any) nonzero scalars $a_1, \dots, a_k \in \mathbb{F}$.

Proof In the trivial case $A = 0$, there is nothing to show.

We now make induction on n.

If $n = 1$, then $A = 0$ and there is nothing to show. If $n = 2$ and but $A \neq 0$, then

$$A = \begin{pmatrix} 0 & a \\ -a & 0 \end{pmatrix}, \quad a \neq 0. \qquad (8.2.7)$$

Hence there is nothing to show either.

Assume the statement of the theorem is valid for any $n \leq l$ for some $l \geq 2$.

Let $n = l + 1$. Assume the nontrivial case $A = (a_{ij}) \neq 0$. So there is some $a_{ij} \neq 0$ for $i, j = 1, \ldots, n$. Let i be the smallest among $\{1, \ldots, n\}$ such that $a_{ij} \neq 0$ for some $j = 1, \ldots, n$. If $i = 1$ and $j = 2$, then $a_{12} \neq 0$. If $i = 1$ and $j > 2$, we let E_1 be the elementary matrix obtained from interchanging the second and jth rows of I_n. Then $\det(E_1) = -1$ and the entry at the first row and second column of $E_1 A E_1^t$ is nonzero. If $i > 1$, then the first row of A is a zero row. Let E_1 be the elementary matrix obtained from interchanging the first and ith rows of I_n. Then $\det(E_1) = -1$ and the first row of $E_1 A E_1^t = (b_{ij})$ is nonzero. So there is some $2 \leq j \leq n$ such that $b_{1j} \neq 0$. Let E_2 be the elementary matrix obtained from interchanging the second and jth rows of $E_1 A E_1^t$. Then, in $E_2 E_1 A E_1^t E_2^t = (c_{ij})$, we have $c_{12} \equiv a_1 \neq 0$. We also have $\det(E_2) = \pm 1$ depending on whether $j = 2$ or $j \neq 2$.

Thus we may summarize that there is a matrix E with $\det(E) = \pm 1$ such that

$$EAE^t = \begin{pmatrix} \alpha_1 & \beta \\ -\beta^t & A_{n-2} \end{pmatrix}, \quad \alpha_1 = \begin{pmatrix} 0 & a_1 \\ -a_1 & 0 \end{pmatrix}, \tag{8.2.8}$$

where $\beta \in \mathbb{F}(2, n-2)$ and $A_{n-2} \in \mathbb{F}(n-2, n-2)$ is skew-symmetric.

Consider $P \in \mathbb{F}(n, n)$ of the form

$$P = \begin{pmatrix} I_2 & 0 \\ \gamma & I_{n-2} \end{pmatrix}, \tag{8.2.9}$$

where $\gamma \in \mathbb{F}(2, n-2)$ is to be determined. Then $\det(P) = 1$ and

$$P^t = \begin{pmatrix} I_2 & \gamma \\ 0 & I_{n-2} \end{pmatrix}. \tag{8.2.10}$$

Consequently, we have

$$PEAE^t P^t = \begin{pmatrix} I_2 & 0 \\ \gamma & I_{n-2} \end{pmatrix} \begin{pmatrix} \alpha_1 & \beta \\ -\beta^t & A_{n-2} \end{pmatrix} \begin{pmatrix} I_2 & \gamma \\ 0 & I_{n-2} \end{pmatrix}$$
$$= \begin{pmatrix} \alpha_1 & \alpha_1 \gamma + \beta \\ \gamma \alpha_1 - \beta^t & \gamma \alpha_1 \gamma - \beta^t \gamma + \gamma \beta + A_{n-2} \end{pmatrix}. \tag{8.2.11}$$

To proceed, we choose γ to satisfy

$$\alpha_1 \gamma + \beta = 0. \tag{8.2.12}$$

This is possible to do since $\alpha_1 \in \mathbb{F}(2, 2)$ is invertible. In other words, we are led to the unique choice

$$\gamma = -\alpha_1^{-1} \beta, \quad \alpha_1^{-1} = \begin{pmatrix} 0 & -a_1^{-1} \\ a_1 & 0 \end{pmatrix}. \tag{8.2.13}$$

Thus, in view of (8.2.12), we see that (8.2.11) becomes

$$PEAE^tP^t = \begin{pmatrix} \alpha_1 & 0 \\ 0 & G \end{pmatrix}, \tag{8.2.14}$$

where

$$G = \gamma^t \alpha_1 \gamma - \beta^t \gamma + \gamma^t \beta + A_{n-2} \tag{8.2.15}$$

is a skew-symmetric element in $\mathbb{F}(n-2, n-2)$.

Using the inductive assumption, we can find an element $D \in \mathbb{F}(n-2, n-2)$ satisfying $\det(D) = \pm 1$ such that DGD^t is of the desired canonical form stated in the theorem.

Now let

$$Q = \begin{pmatrix} I_2 & 0 \\ 0 & D \end{pmatrix}. \tag{8.2.16}$$

Then $\det(Q) = \pm 1$ and

$$QPEAE^tP^tQ^t = \begin{pmatrix} \alpha_1 & 0 \\ 0 & DGD^t \end{pmatrix} = \Gamma, \tag{8.2.17}$$

where the matrix $\Gamma \in \mathbb{F}(n, n)$ is as given in (8.2.5).

Taking $C = QPE$, we see that the theorem is established. $\square$

Furthermore, applying a sequence of elementary matrices to the left and right of the canonical matrix Γ given in (8.2.5) realizing suitable row and column interchanges, we see that we can use a non-singular matrix of determinant ± 1 to congruently reduce Γ into another canonical form,

$$\tilde{\Gamma} = \begin{pmatrix} 0 & D_k & 0 \\ -D_k & 0 & 0 \\ 0 & 0 & 0 \end{pmatrix} \in \mathbb{F}(n, n), \tag{8.2.18}$$

where $D_k \in \mathbb{F}(k, k)$ is the diagonal matrix

$$D_k = \mathrm{diag}\{a_1, \ldots, a_k\}. \tag{8.2.19}$$

As a by-product of the above discussion, we infer that the rank of a skew-symmetric matrix is $2k$, an even number.

We now consider the situation when $\mathbb{F} = \mathbb{R}$. By row and column operations if necessary, we may assume without loss of generality that $a_1, \ldots, a_k > 0$ in (8.2.6). With this assumption, define

$$R = \begin{pmatrix} \beta_1 & \cdots & 0 & 0 \\ \vdots & \ddots & \vdots & \vdots \\ 0 & \cdots & \beta_k & 0 \\ 0 & \cdots & 0 & I_{n-2k} \end{pmatrix}, \quad \beta_i = \begin{pmatrix} \frac{1}{\sqrt{a_i}} & 0 \\ 0 & \frac{1}{\sqrt{a_i}} \end{pmatrix}, \quad i = 1,\ldots,k.$$

(8.2.20)

Then

$$R\Gamma R^t = \Omega \equiv \begin{pmatrix} J_2 & \cdots & 0 & 0 \\ \vdots & \ddots & \vdots & \vdots \\ 0 & \cdots & J_2 & 0 \\ 0 & \cdots & \cdots & 0 \end{pmatrix}, \tag{8.2.21}$$

where

$$J_2 = \begin{pmatrix} 0 & 1 \\ -1 & 0 \end{pmatrix} \tag{8.2.22}$$

appears k times in (8.2.21).

Of course, as before, we may also further congruently reduce Ω in (8.2.21) into another canonical form,

$$\tilde{\Omega} = \begin{pmatrix} J_{2k} & 0 \\ 0 & 0 \end{pmatrix}, \tag{8.2.23}$$

where $J_{2k} \in \mathbb{R}(2k, 2k)$ is given by

$$J_{2k} = \begin{pmatrix} 0 & I_k \\ -I_k & 0 \end{pmatrix}. \tag{8.2.24}$$

As in the case of scalar products, for a skew-symmetric bilinear form $f : U \times U \to \mathbb{F}$, define

$$S^\perp = \{u \in U \mid f(u,v) = 0, v \in S\} \tag{8.2.25}$$

for a non-empty subset S of U. Then $S^\perp$ is a subspace of U.

Set

$$U_0 = U^\perp = \{u \in U \mid f(u,v) = 0, v \in U\}. \tag{8.2.26}$$

We call f nondegenerate if $U_0 = \{0\}$. In other words, a nondegenerate skew-symmetric bilinear form f is characterized by the fact that when $u \in U$ and $f(u,v) = 0$ for all $v \in U$ then $u = 0$. Thus f is called degenerate if $U_0 \neq \{0\}$ or $\dim(U_0) \geq 1$. It is not hard to show that f is degenerate if and only if the rank of the matrix representation of f with respect to any basis of U is smaller than $\dim(U)$.

If f is degenerate over U, then the canonical form (8.2.5) indicates that we may obtain a basis

$$\{u_1,\ldots,u_{n_0},v_1,\ldots,v_k,w_1,\ldots,w_k\} \tag{8.2.27}$$

of U so that $\{u_1,\ldots,u_{n_0}\}$ is a basis of U_0 with $n_0 = \dim(U_0)$ and that

$$\begin{cases} f(v_i,w_i) = a_i \neq 0, & i = 1,\ldots,k, \\ f(v_i,v_j) = f(w_i,w_j) = 0, & i,j = 1,\ldots,k, \\ f(v_i,w_j) = 0, & i \neq j, \quad i,j = 1,\ldots,k. \end{cases} \tag{8.2.28}$$

Of particular interest is when f is nondegenerate over U. In such a situation, $\dim(U)$ must be an even number, $2k$.

Definition 8.4 Let U be a vector space of $2k$ dimensions. A skew-symmetric bilinear form $f : U \times U \to \mathbb{F}$ is called *symplectic* if f is nondegenerate. An even dimensional vector space U equipped with a symplectic form is called a *symplectic vector space*. A basis $\{v_1,\ldots,v_k,w_1,\ldots,w_k\}$ of a $2k$-dimensional symplectic vector space U equipped with the symplectic form f is called *symplectic* if

$$\begin{cases} f(v_i,w_i) = 1, & i = 1,\ldots,k, \\ f(v_i,v_j) = f(w_i,w_j) = 0, & i,j = 1,\ldots,k, \\ f(v_i,w_j) = 0, & i \neq j, \quad i,j = 1,\ldots,k. \end{cases} \tag{8.2.29}$$

A symplectic basis is also called a *Darboux basis*.

Therefore, we have seen that a real symplectic vector space always has a symplectic basis.

Definition 8.5 Let U be a symplectic vector space equipped with a symplectic form f. A subspace V of U is called *isotropic* if $f(u,v) = 0$ for any $u,v \in V$. If $\dim(U) = 2k$ then any k-dimensional isotropic subspace of U is called a *Lagrangian subspace*.

Let U be a symplectic vector space with a symplectic basis given as in (8.2.29). Then we see that both

$$V = \text{Span}\{v_1,\ldots,v_k\} \quad \text{and} \quad W = \text{Span}\{w_1,\ldots,w_k\} \tag{8.2.30}$$

are Lagrangian subspaces of U.

If U is a finite-dimensional complex vector space, we may consider a skew-symmetric sesquilinear form f from $U \times U$ into $\mathbb{C}$. Such a form satisfies

$$f(u,v) = -\overline{f(v,u)}, \quad u,v \in U, \tag{8.2.31}$$

and is called *Hermitian skew-symmetric* or *skew-Hermitian*. Therefore, the form

$$g(u,v) = if(u,v), \quad u,v \in U, \tag{8.2.32}$$

is Hermitian. It is clear that the matrix representation, say $A \in \mathbb{C}(n,n)$, with respect to any basis of U of a Hermitian skew-symmetric form is anti-Hermitian or skew-Hermitian, $A^\dagger = -A$. Hence iA is Hermitian. Applying the knowledge about Hermitian forms and Hermitian matrices studied in Chapter 6, it is not hard to come up with a complete understanding of skew-Hermitian forms and matrices, in the same spirit of Theorem 8.3. We leave this as an exercise.

Exercises

8.2.1   Let f be a bilinear form over a vector space U. Show that f is skew-symmetric if and only if $f(u,u) = 0$ for any $u \in U$.

8.2.2   Let f be a skew-symmetric bilinear form over U and $S \subset U$ a non-empty subset. Show that $S^\perp$ defined in (8.2.25) is a subspace of U.

8.2.3   Let U be a finite-dimensional vector space and f a skew-symmetric bilinear form over U. Define U_0 by (8.2.26) and use A to denote the matrix representation of f with respect to any given basis of U. Show that

$$\dim(U_0) = \dim(U) - r(A). \tag{8.2.33}$$

8.2.4   Let U be a symplectic vector space equipped with a symplectic form f. If V is a subspace of U, $V^\perp$ is called the *symplectic complement* of V in U. Prove

(a) $\dim(V) + \dim(V^\perp) = \dim(U)$.
(b) $(V^\perp)^\perp = V$.

8.2.5   Let U be a symplectic vector space equipped with a symplectic form f. If V is a space of U, we can consider the restriction of f over V. Prove that f is symplectic over V if and only if $V \cap V^\perp = \{0\}$.

8.2.6   Show that a subspace V of a symplectic vector space U is isotropic if and only if $V \subset V^\perp$.

8.2.7   Show that a subspace V of a symplectic vector space U is Lagrangian if and only if $V = V^\perp$.

8.2.8   For $A \in \mathbb{C}(n,n)$, show that A is skew-Hermitian if and only if there is a non-singular element $C \in \mathbb{C}(n,n)$ such that $CAC^\dagger = iD$, where D is an $n \times n$ real diagonal matrix.

8.3 Perron–Frobenius theorem for positive matrices

Let $A = (a_{ij}) \in \mathbb{R}(n,n)$. We say that A is *positive (non-negative)* if $a_{ij} > 0$ ($a_{ij} \geq 0$) for all $i, j = 1, \ldots, n$. Likewise we say that a vector $u = (a_i) \in \mathbb{R}^n$ is *positive (non-negative)* if $a_i > 0$ ($a_i \geq 0$) for all $i = 1, \ldots, n$. More generally, for A and $B \in \mathbb{R}(n,n)$, we write $A > B$ ($A \geq B$) if $(A - B) > 0$ (($A - B) \geq 0$); for u and $v \in \mathbb{R}^n$, we write $u > v$ ($u \geq v$) if $(u - v) > 0$ (($u - v) \geq 0$).

The *Perron–Frobenius theorem* concerns the existence and properties of a positive eigenvector, associated to a positive eigenvalue, of a positive matrix and may be stated as follows.

Theorem 8.6 *Let $A = (a_{ij}) \in \mathbb{R}(n,n)$ be a positive matrix. Then there is a positive eigenvalue, r, of A, called the dominant eigenvalue, satisfying the following properties:*

(1) *There is a positive eigenvector, say u, associated with r, such that any other non-negative eigenvectors of A associated with r must be positive multiples of u.*
(2) *r is a simple root of the characteristic polynomial of A.*
(3) *If $\lambda \in \mathbb{C}$ is any other eigenvalue of A, then*

$$|\lambda| < r. \tag{8.3.1}$$

Furthermore, any non-negative eigenvector of A must be associated with the dominant eigenvalue r.

Proof We equip $\mathbb{R}^n$ with the norm

$$\|x\| = \max\{|x_i| \,|\, i = 1, \ldots, n\}, \quad x = \begin{pmatrix} x_1 \\ \vdots \\ x_n \end{pmatrix} \in \mathbb{R}^n, \tag{8.3.2}$$

and consider the subset

$$S = \{x \in \mathbb{R}^n \,|\, \|x\| = 1, x \geq 0\} \tag{8.3.3}$$

of $\mathbb{R}^n$. Then define

$$\Lambda = \{\lambda \in \mathbb{R} \,|\, \lambda \geq 0, Ax \geq \lambda x \text{ for some } x \in S\}. \tag{8.3.4}$$

We can show that Λ is an interval in $\mathbb{R}$ of the form $[0, r]$ for some $r > 0$.

In fact, take a test vector, say $y = (1, \ldots, 1)^t \in \mathbb{R}^n$. Since A is positive, it is clear that $Ay \geq \lambda y$ if λ satisfies

$$0 < \lambda \leq \min\left\{ \sum_{j=1}^{n} a_{ij} \,\bigg|\, i = 1, \ldots, n \right\}. \tag{8.3.5}$$

Moreover, the definition of Λ implies immediately that if $\lambda \in \Lambda$ then $[0, \lambda] \subset \Lambda$. Thus Λ is connected.

Let $\lambda \in \Lambda$. Then there is some $x \in S$ such that $Ax \geq \lambda x$. Therefore, since $\|x\| = 1$, we have

$$\lambda = \|\lambda x\| \leq \|Ax\| = \max \left\{ \left| \sum_{j=1}^{n} a_{ij}x_j \right| \,\Big|\, i = 1, \ldots, n \right\}$$

$$\leq \max \left\{ \sum_{j=1}^{n} a_{ij} \,\Big|\, i = 1, \ldots, n \right\}, \tag{8.3.6}$$

which establishes the boundedness of Λ.

Now set

$$r = \sup \left\{ \lambda \in \Lambda \right\}. \tag{8.3.7}$$

Then, we have seen that r satisfies $0 < r < \infty$. Hence, there is a sequence $\{\lambda_k\} \subset \Lambda$ such that

$$r = \lim_{k \to \infty} \lambda_k. \tag{8.3.8}$$

On the other hand, the assumption that $\{\lambda_k\} \subset \Lambda$ indicates that there is a sequence $\{x^{(k)}\} \subset S$ such that

$$Ax^{(k)} \geq \lambda_k x^{(k)}, \quad k = 1, 2, \ldots. \tag{8.3.9}$$

Using the compactness of S, we may assume that there is subsequence of $\{x^{(k)}\}$ which we still denote by $\{x^{(k)}\}$ without loss of generality such that it converges to some element, say $u = (a_1, \ldots, a_n)^t$, in S, as $k \to \infty$. Letting $k \to \infty$ in (8.3.9), we arrive at

$$Au \geq ru. \tag{8.3.10}$$

In particular, this proves $r \in \Lambda$. Thus indeed $\Lambda = [0, r]$.

We next show that equality must hold in (8.3.10).

Suppose otherwise there is some $i_0 = 1, \ldots, n$ such that

$$\sum_{j=1}^{n} a_{i_0 j} a_j > r a_{i_0}; \quad \sum_{j=1}^{n} a_{ij} a_j \geq r a_i \quad \text{for } i \in \{1, \ldots, n\} \setminus \{i_0\}. \tag{8.3.11}$$

Let $z = (z_i) = u + s e_{i_0}$ ($s > 0$). Since $z_i = a_i$ for $i \neq i_0$, we conclude from the second inequality in (8.3.11) that

$$\sum_{j=1}^{n} a_{ij} z_j > r z_i \quad \text{for } i \in \{1, \ldots, n\} \setminus \{i_0\}, \quad s > 0. \tag{8.3.12}$$

However, the first inequality in (8.3.11) gives us

$$\sum_{j=1}^{n} a_{i_0 j} a_j > r(a_{i_0} + s) \quad \text{when } s > 0 \text{ is sufficiently small,} \tag{8.3.13}$$

which leads to

$$\sum_{j=1}^{n} a_{i_0 j} z_j > r z_{i_0} \quad \text{when } s > 0 \text{ is sufficiently small.} \tag{8.3.14}$$

Thus, $Az > rz$. Set $v = z/\|z\|$. Then, $v \in S$ and $Av > rv$. Hence we may choose $\varepsilon > 0$ small such that $Av > (r + \varepsilon)v$. So $r + \varepsilon \in \Lambda$ which contradicts the definition of r made in (8.3.7).

Therefore, $Au = ru$ and r is a positive eigenvalue of A.

The positivity of $u = (a_i)$ follows easily since $u \in S$ and $A > 0$ so that $ru = Au$ leads to

$$\lambda a_i = \sum_{j=1}^{n} a_{ij} a_j > 0, \quad i = 1, \ldots, n, \tag{8.3.15}$$

because $u \neq 0$. That is, a non-negative eigenvector of A associated to a positive eigenvalue must be positive.

Let v be any non-negative eigenvector of A associated to r. We show that there is a positive number a such that $v = au$. For this purpose, we construct the vector

$$u_s = u - sv, \quad s > 0. \tag{8.3.16}$$

Of course, u_s is a non-negative eigenvector of A associated to r, when $s > 0$ is small. Set

$$s_0 = \sup\{s \mid u_s \geq 0\}. \tag{8.3.17}$$

Then $s_0 > 0$ and $u_{s_0} \geq 0$ but $u_{s_0} \not> 0$. If $u_{s_0} \neq 0$ then $u_{s_0} > 0$ since u_{s_0} is an eigenvector of A associated to r which contradicts the definition of s_0. Therefore, we must have $u_{s_0} = 0$, which gives us the result $v = (s_0^{-1})u$ as desired.

In order to show that r is the simple root of the characteristic polynomial of A, we need to prove that there is only one Jordan block associated with the eigenvalue r and that this Jordan block can only be 1×1. To do so, we first show that the geometric multiplicity of r is 1. Then, we show that there is no generalized eigenvector. That is, the equation

$$(A - rI_n)v = u, \quad v \in \mathbb{C}^n, \tag{8.3.18}$$

has no solution.

Suppose otherwise that the dimension of the eigenspace E_r is greater than 1. Then, there is some $v \in E_r$, which is not a scalar multiple of u. Write v as

$v = v_1 + iv_2$ with v_1 and $v_2 \in \mathbb{R}^n$. Then $Av_1 = rv_1$ and $Av_2 = rv_2$. We assert that one of the sets of vectors $\{u, v_1\}$ and $\{u, v_2\}$ must be linearly independent over $\mathbb{R}$. Otherwise there are a_1 and $a_2 \in \mathbb{R}$ such that $v_1 = a_1 u$ and $v_2 = a_2 u$ which imply $v = (a_1 + ia_2)u$, a contradiction. Use w to denote either v_1 or v_2 which is linearly independent from u over $\mathbb{R}$. Then we know that $\pm w$ can never be non-negative. Thus there are components of w which have different signs. Now consider the vector

$$u_s = u + sw, \quad s > 0. \tag{8.3.19}$$

It is clear that $u_s > 0$ when $s > 0$ is sufficiently small since $u > 0$. So there is some $s_0 > 0$ such that $u_{s_0} \geq 0$ but a component of u_{s_0} is zero. However, since $u_{s_0} \neq 0$ due to the presence of a positive component in w, we arrive at a contradiction because u_{s_0} is seen to be an eigenvector of A associated to r.

We next show that (8.3.18) has no solution. Since $u \in \mathbb{R}^n$, we need only to consider $v \in \mathbb{R}^n$ in (8.3.18). We proceed as follows.

Consider

$$w_s = v + su, \quad s > 0. \tag{8.3.20}$$

Since $Au = ru$, we see that w_s also satisfies (8.3.18). Take $s > 0$ sufficient large so that $w_s > 0$. Hence $(A - rI_n)w_s = u > 0$ or

$$Aw_s > rw_s. \tag{8.3.21}$$

Thus, if $\delta > 0$ is sufficiently small, we have $Aw_s > (r + \delta)w_s$. Rescaling w_s if necessary, we may assume $w_s \in S$. This indicates $r + \delta \in \Lambda$ which violates the definition of r stated in (8.3.7).

So the assertion is r is a simple root of the characteristic polynomial of A follows.

Moreover, let $\lambda \in \mathbb{C}$ be an eigenvalue of A which is not r and $v = (b_i) \in \mathbb{C}^n$ an associated eigenvector. From $Av = \lambda v$, we have

$$|\lambda||b_i| \leq \sum_{j=1}^{n} a_{ij}|b_j|, \quad i = 1, \ldots, n. \tag{8.3.22}$$

Rescaling if necessary, we may also assume

$$w = \begin{pmatrix} |b_1| \\ \vdots \\ |b_n| \end{pmatrix} \in S. \tag{8.3.23}$$

Since (8.3.22) implies $Aw \geq |\lambda|w$, we see in view of the definition of Λ that $|\lambda| \in \Lambda$. Using (8.3.7), we have $|\lambda| \leq r$.

If $|\lambda| < r$, there is nothing more to do. If $|\lambda| = r$, the discussion just made in the earlier part of the proof shows that w is a non-negative eigenvector of A associated to r and thus equality holds in (8.3.22) for all $i = 1, \ldots, n$. Therefore, since $A > 0$, the complex numbers $b_1, \ldots, b_n$ must share the same phase angle, $\theta \in \mathbb{R}$, so that

$$b_i = |b_i|e^{i\theta}, \quad i = 1, \ldots, n. \tag{8.3.24}$$

On the other hand, since w is a non-negative eigenvector of A associated to r, there is a number $a > 0$ such that

$$w = au. \tag{8.3.25}$$

Combining (8.3.23)–(8.3.25), we have $v = ae^{i\theta}u$. In particular, $\lambda = r$.

Finally let v be an arbitrary eigenvector of A which is non-negative and associated to an eigenvalue $\lambda \in \mathbb{C}$. Since $A^t > 0$ and the characteristic polynomial of A^t is the same as that of A, we conclude that r is also the dominant eigenvalue of A^t. Now use w to denote a positive eigenvector of A^t associated to r. Then, $A^t w = rw$. Thus

$$\lambda v^t w = (Av)^t w = v^t A^t w = rv^t w. \tag{8.3.26}$$

However, in view of the fact that $v \geq 0, v \neq 0$, and $w > 0$, we have $v^t w > 0$. So it follows from (8.3.26) that $\lambda = r$.

The proof of the theorem is complete. $\qquad\qquad\qquad\qquad\qquad\qquad\square$

The dominant eigenvalue r of the distinguished characteristics of a positive matrix A stated in Theorem 8.6 is also called the Perron or Perron–Frobenius eigenvalue of the matrix A.

For an interesting historical account of the Perron–Frobenius theory and a discussion of its many applications and generalizations and attributions of the proofs including the one presented here, see Bellman [12].

Exercises

8.3.1   Let $A = (a_{ij}) \in \mathbb{R}(n,n)$ be a positive matrix and $r > 0$ the Perron–Frobenius eigenvalue of A. Show that r satisfies the estimate

$$\min_{1 \leq i \leq n} \left(\sum_{j=1}^{n} a_{ij} \right) \leq r \leq \max_{1 \leq i \leq n} \left(\sum_{j=1}^{n} a_{ij} \right). \tag{8.3.27}$$

8.3.2   Consider the matrix

$$A = \begin{pmatrix} 1 & 2 & 3 \\ 2 & 3 & 1 \\ 3 & 2 & 1 \end{pmatrix}. \qquad (8.3.28)$$

(a) Use (8.3.27) to find the dominant eigenvalue of A.

(b) Check to see that $u = (1,1,1)^t$ is a positive eigenvector of A. Use u and Theorem 8.6 to find the dominant eigenvalue of A and confirm that this is exactly what obtained in part (a).

(c) Compute all the eigenvalues of A directly and confirm the result obtained in part (a) or (b).

8.3.3   Let A and $B \in \mathbb{R}(n,n)$ be positive matrices and use r_A and r_B to denote the dominant eigenvalues of A and B, respectively. Show that $r_A \leq r_B$ if $A \leq B$.

8.3.4   Let $A \in \mathbb{R}(n,n)$ be a positive matrix and $u = (a_i) \in \mathbb{R}^n$ a non-negative eigenvector of A. Show that the dominant eigenvalue r of A may be computed by the formula

$$r = \frac{1}{\sum_{i=1}^{n} a_i} \left(\sum_{j,k=1}^{n} a_{jk} a_k \right). \qquad (8.3.29)$$

8.4 Markov matrices

In this section, we discuss a type of non-negative matrices known as the Markov or stochastic matrices, which are important in many areas of applications.

Definition 8.7 Let $A = (a_{ij}) \in \mathbb{R}(n,n)$ so that $a_{ij} \geq 0$ for $i,j = 1,\ldots,n$. If A satisfies

$$\sum_{j=1}^{n} a_{ij} = 1, \quad i = 1,\ldots,n, \qquad (8.4.1)$$

that is, the components of each row vector in A sum up to 1, then A is called a *Markov* or *stochastic* matrix. If there is an integer $m \geq 1$ such that A^m is a positive matrix, then A is called a *regular Markov* or *regular stochastic* matrix.

A few immediate consequences follow directly from the definition of a Markov matrix and are stated as below.

Theorem 8.8 *Let $A = (a_{ij}) \in \mathbb{R}(n,n)$ be a Markov matrix. Then* 1 *is an eigenvalue of A which enjoys the following properties:*

(1) *The vector $u = (1,\dots,1)^t \in \mathbb{R}^n$ is an eigenvector of A associated to the eigenvalue* 1.
(2) *Any eigenvalue $\lambda \in \mathbb{C}$ of A satisfies*

$$|\lambda| \le 1. \tag{8.4.2}$$

Proof Using (8.4.1), the fact that $u = (1,\dots,1)^t$ satisfies $Au = u$ may be checked directly.

Now let $\lambda \in \mathbb{C}$ be any eigenvalue of A and $v = (b_i) \in \mathbb{C}^n$ an associated eigenvector. Then there is some $i_0 = 1,\dots,n$ such that

$$|b_{i_0}| = \max\{|b_i| \,|\, i = 1,\dots,n\}. \tag{8.4.3}$$

Thus, from the relation $Av = \lambda v$, we have

$$\lambda b_{i_0} = \sum_{j=1}^{n} a_{i_0 j} b_j, \tag{8.4.4}$$

which in view of (8.4.1) gives us

$$|\lambda||b_{i_0}| \le \sum_{j=1}^{n} a_{i_0 j}|b_j| \le |b_{i_0}| \sum_{j=1}^{n} a_{i_0 j} = |b_{i_0}|. \tag{8.4.5}$$

Since $|b_{i_0}| > 0$, we see that the bound (8.4.2) follows. $\qquad\square$

In fact, for a non-negative element $A \in \mathbb{R}(n,n)$, it is clear that A being a Markov matrix is equivalent to the vector $u = (1,\dots,1)^t$ being an eigenvector of A associated to the eigenvalue 1. This simple fact establishes that the product of any number of the Markov matrices in $\mathbb{R}(n,n)$ is also a Markov matrix.

Let $A \in \mathbb{R}(n,n)$ be a Markov matrix. It will be interesting to identify a certain condition under which the eigenvalue 1 of A becomes dominant as in the Perron–Frobenius theorem. Below is such a result.

Theorem 8.9 *If $A \in \mathbb{R}(n,n)$ is a regular Markov matrix, then the eigenvalue* 1 *of A is the dominant eigenvalue of A which satisfies the following properties:*

(1) *The absolute value of any other eigenvalue $\lambda \in \mathbb{C}$ of A is less than* 1. *That is, $|\lambda| < 1$.*
(2) 1 *is a simple root of the characteristic polynomial of A.*

Proof Let $m \ge 1$ be an integer such that $A^m > 0$. Since A^m is a Markov matrix, we see that 1 is the dominant eigenvalue of A^m. On the other hand, if $\lambda \in \mathbb{C}$ is any eigenvalue of A other than 1, since λ^m is an eigenvalue of A^m other than 1, we have in view of Theorem 8.6 that $|\lambda^m| < 1$, which proves $|\lambda| < 1$.

We now show that 1 is a simple root of the characteristic polynomial of A. If $m = 1$, the conclusion follows from Theorem 8.6. So we may assume $m \ge 2$.

Recall that there is an invertible matrix $C \in \mathbb{C}(n,n)$ such that

$$A = CBC^{-1}, \tag{8.4.6}$$

where $B \in \mathbb{C}(n,n)$ takes the block or boxed diagonal form

$$B = \text{diag}\{J_1, \cdots, J_k\} \tag{8.4.7}$$

in which each J_i $(i = 1, \ldots, k)$ is either a diagonal matrix of the form λI or a Jordan block of the form

$$
J = \begin{pmatrix}
\lambda & 1 & 0 & \cdots & 0 \\
0 & \lambda & 1 & \ddots & \vdots \\
\vdots & \ddots & \ddots & \ddots & 0 \\
\vdots & \cdots & \ddots & \ddots & 1 \\
0 & \cdots & \cdots & 0 & \lambda
\end{pmatrix}, \tag{8.4.8}
$$

where we use λ to denote a generic eigenvalue of A. In either case, we may rewrite J as the sum of two matrices, in the form

$$J = \lambda I + P, \tag{8.4.9}$$

where for some integer $l \geq 1$ (with l being the degree of nilpotence of P) we have $P^l = 0$. Hence, in view of the binomial expansion formula, we have

$$
\begin{aligned}
J^m &= (\lambda I + P)^m \\
&= \sum_{s=0}^{m} \frac{m!}{s!(m-s)!} \lambda^{m-s} P^s \\
&= \sum_{s=0}^{l-1} \frac{m!}{s!(m-s)!} \lambda^{m-s} P^s \\
&= \lambda^m I + \lambda^{m-1} m P + \cdots + \lambda^{m-l+1} \frac{m(m-1)\cdots(m-[l-2])}{(l-1)!} P^{l-1}, \quad (8.4.10)
\end{aligned}
$$

which is an upper triangular matrix with the diagonal entries all equal to λ^m.

From (8.4.6) and (8.4.7), we have

$$A^m = C\text{diag}\{J_1^m, \ldots, J_k^m\}C^{-1}. \tag{8.4.11}$$

However, using the condition $A^m > 0$, Theorem 8.6, and (8.4.10), we conclude that there exists exactly one Jordan block among the Jordan blocks $J_1, \ldots, J_k$ of A with $\lambda = 1$ and that such a Jordan block can only be 1×1.

The proof of the theorem is thus complete. $\qquad\square$

Since in (8.4.10) there are l terms, we see that $J^m \to 0$ as $m \to \infty$ when $|\lambda| < 1$. This observation leads to the following important convergence theorem for the Markov matrices.

Theorem 8.10 *If $A \in \mathbb{R}(n,n)$ is a regular Markov matrix, then*

$$\lim_{m \to \infty} A^m = K, \tag{8.4.12}$$

where K is a positive Markov matrix with n identical row vectors, $v_1^t = \cdots = v_n^t = v^t$, where $v = (b_1, \ldots, b_n)^t \in \mathbb{R}^n$ is the unique positive vector satisfying

$$A^t v = v, \quad \sum_{i=1}^{n} b_i = 1. \tag{8.4.13}$$

Proof If A is a regular Markov matrix, then there is an integer $m \geq 1$ such that $A^m > 0$. Thus $(A^t)^m = (A^m)^t > 0$. Since A^t and A have the same characteristic polynomial, A^t has 1 as its dominant eigenvalue as A does. Let $v = (b_i)$ be an eigenvector of A^t associated to the eigenvalue 1. Then v is also an eigenvector of $(A^t)^m$ associated to 1. Since $(A^t)^m > 0$ and its eigenvalue 1 as the dominant eigenvalue of $(A^t)^m$ is simple, we see in view of Theorem 8.6 that we may choose $v \in \mathbb{R}^n$ so that either $v > 0$ or $v < 0$. We now choose $v > 0$ and normalize it so that its components sum to 1. That is, v satisfies (8.4.13)

Using Theorem 8.9, let $C \in \mathbb{C}(n,n)$ be invertible such that

$$A = C\text{diag}\{1, J_1, \ldots, J_k\}C^{-1}. \tag{8.4.14}$$

Rewriting (8.4.14) as $AC = C\text{diag}\{1, J_1, \cdots, J_k\}$, we see that the first column vector of C is an eigenvector of A associated to the eigenvalue 1 which is simple by Theorem 8.9. Hence we may choose this first column vector of C to be $u = a(1, \ldots, 1)^t$ for some $a \in \mathbb{C}$, $a \neq 0$.

On the other hand, rewrite $D = C^{-1}$ and express (8.4.14) as

$$DA = \text{diag}\{1, J_1, \ldots, J_k\}D. \tag{8.4.15}$$

We see that the first row vector of D, say w^t for some $w \in \mathbb{C}^n$, satisfies $w^t A = w^t$. Hence $A^t w = w$. Since 1 is a simple root of the characteristic polynomial of A^t, we conclude that there is some $b \in \mathbb{C}$, $b \neq 0$, such that $w = bv$ where v satisfies (8.4.13).

Since $D = C^{-1}$, we have $w^t u = 1$ which leads to

$$abv^t(1, \ldots, 1)^t = ab \sum_{i=1}^{n} b_i = ab = 1. \tag{8.4.16}$$

Finally, for any integer $m \geq 1$, (8.4.14) gives us

$$A^m = C\text{diag}\{1, J_1^m, \ldots, J_k^m\}D. \tag{8.4.17}$$

Since each of the Jordan blocks $J_1, \ldots, J_k$ is of the form (8.4.9) for some $\lambda \in \mathbb{C}$ satisfying $|\lambda| < 1$, we have seen that

$$J_1^m, \ldots, J_k^m \to 0 \quad \text{as } m \to \infty. \tag{8.4.18}$$

Therefore, taking $m \to \infty$ in (8.4.17), we arrive at

$$\lim_{m \to \infty} A^m = C\mathrm{diag}\{1, 0, \ldots, 0\}D. \tag{8.4.19}$$

Inserting the results that the first column vector of C is $u = a(1, \ldots, 1)^t$, the first row vector of D is $w^t = bv^t$ where v satisfies (8.4.13), and $ab = 1$ into (8.4.19), we obtain

$$\lim_{m \to \infty} A^m = \begin{pmatrix} b_1 & \cdots & b_n \\ \vdots & \cdots & \vdots \\ b_1 & \cdots & b_n \end{pmatrix} = K, \tag{8.4.20}$$

as asserted.   □

For a Markov matrix $A \in \mathbb{R}(n, n)$, the power of A, A^m, may or may not approach a limiting matrix as $m \to \infty$. If the limit of A^m as $m \to \infty$ exists and is some $K \in \mathbb{R}(n, n)$, then it is not hard to show that K is also a Markov matrix, which is called the *stable matrix* of A and A is said to be a *stable Markov matrix*. Theorem 8.10 says that a regular Markov matrix A is stable and, at the same time, gives us a constructive method to find the stable matrix of A.

Exercises

8.4.1   Let $A \in \mathbb{R}(n, n)$ be a stable Markov matrix and $K \in \mathbb{R}(n, n)$ the stable matrix of A. Show that K is also a Markov matrix and satisfies $AK = KA = K$.

8.4.2   Show that the matrix

$$A = \begin{pmatrix} 0 & 1 \\ 1 & 0 \end{pmatrix} \tag{8.4.21}$$

is a Markov matrix which is not regular. Is A stable?

8.4.3   Consider the Markov matrix

$$A = \frac{1}{2} \begin{pmatrix} 1 & 1 & 0 \\ \frac{1}{2} & 1 & \frac{1}{2} \\ 0 & 1 & 1 \end{pmatrix}. \tag{8.4.22}$$

(a) Check that A is regular by showing that $A^2 > 0$.
(b) Find the stable matrix of A.

(c) Use induction to establish the formula

$$A^m = \frac{1}{2^m}\begin{pmatrix} \frac{1}{2}+2^{m-2} & 2^{m-1} & -\frac{1}{2}+2^{m-2} \\ 2^{m-2} & 2^{m-1} & 2^{m-2} \\ -\frac{1}{2}+2^{m-2} & 2^{m-1} & \frac{1}{2}+2^{m-2} \end{pmatrix}, \quad m = 1,2,\dots.$$

$$(8.4.23)$$

Take $m \to \infty$ in (8.4.23) and verify your result obtained in (b).

8.4.4 Let $A \in \mathbb{R}(n,n)$ be a Markov matrix. If A^t is also a Markov matrix, A is said to be a *doubly Markov matrix*. Show that the stable matrix K of a regular doubly Markov matrix is simply given by

$$K = \frac{1}{n}\begin{pmatrix} 1 & \cdots & 1 \\ \vdots & \cdots & \vdots \\ 1 & \cdots & 1 \end{pmatrix}. \qquad (8.4.24)$$

8.4.5 Let $A_1,\dots,A_k \in \mathbb{R}(n,n)$ be k doubly Markov matrices. Show that their product $A = A_1 \cdots A_k$ is also a doubly Markov matrix.

9

Excursion: Quantum Mechanics in a Nutshell

The content of this chapter may serve as, yet, another supplemental topic to meet the needs and interests beyond those of a usual course curriculum. Here we shall present an over-simplified, but hopefully totally transparent, description of some of the fundamental ideas and concepts of quantum mechanics, using a pure linear algebra formalism.

9.1 Quantum mechanics and linear algebra

Linear algebra is essential for understanding quantum mechanics because it provides the mathematical framework for representing quantum states, operators, and observables. Here is a brief explanation why and how it is so.

State representation: In quantum mechanics, the state of a physical system is represented by a vector in a complex vector space known as a Hilbert space, whose simplest form is the space $\mathbb{C}^n$. Linear algebra provides the language and tools for describing these quantum states. The state vector of a quantum system is typically represented using the Dirac notation, in which kets ($|\psi\rangle$) represent state vectors and bras ($\langle\psi|$) represent their conjugates, which are easily conceptualized as column vectors in $\mathbb{C}^n$ and their Hermitian conjugates.

Operators and observables: Physical observables in quantum mechanics, such as position, momentum, and spin, are represented by self-adjoint linear operators or mappings acting on state vectors, making linear algebra a necessary tool in analyzing their properties and computing the outcomes of their measurements.

Superposition: Superposition, a fundamental functional principle of quantum mechanics, refers to the ability of quantum systems to exist in multiple states simultaneously which are represented using linear combinations of state vectors in a state space. This feature makes linear algebra the starting point for the formalism of quantum mechanics.

Quantum dynamics: Linear algebra may well be used to understand the time evolution of quantum systems governed either by the Schrödinger equation or the Heisenberg equation of motion. Time evolution operators, such as unitary evolution operators, are represented by linear transformations on the state space, and their properties are analyzed using techniques from linear algebra. As a by-product, the equivalence of the Schrödinger picture and the Heisenberg picture, a foundational result in quantum mechanics, can be established rigorously and completely, due to the convenience of working on a finite-dimensional setting.

Quantum measurement: When a quantum system is measured, its state "collapses" to one of its possible outcomes, that is, the eigenstates, according to the rules of quantum mechanics. With linear algebra, this measurement process is well described which enables one to calculate the probabilities of different measurement outcomes based on the initial state of the system and the measurement operators.

Uncertainty principle: The principle states that it is impossible to obtain measurements of two noncommutative observables with simultaneously high precision so that a high precision achieved for one will be compromised for another. Within linear algebra, this principle can well be appreciated as an application of the Cauchy–Schwarz inequality.

Therefore, in this chapter, the power and insight of linear algebra will be explored that will help us understand some fundamental concepts of quantum mechanics. Besides, this self-contained study will also provide a preparation for the materials in Chapter 13 on quantum information and quantum entanglement.

9.2 Vectors in $\mathbb{C}^n$ and Dirac bracket

Consider the vector space $\mathbb{C}^n$, consisting of column vectors, and use $\{e_1, \ldots, e_n\}$ to denote the standard basis of $\mathbb{C}^n$. For $u, v \in \mathbb{C}^n$ with

$$u = \begin{pmatrix} a_1 \\ \vdots \\ a_n \end{pmatrix} = \sum_{i=1}^{n} a_i e_i, \quad v = \begin{pmatrix} b_1 \\ \vdots \\ b_n \end{pmatrix} = \sum_{i=1}^{n} b_i e_i, \tag{9.2.1}$$

recall that the Hermitian scalar product is given by

$$(u, v) = u^\dagger v = \sum_{i=1}^{n} \bar{a}_i b_i, \tag{9.2.2}$$

so that $\{e_1, \ldots, e_n\}$ is a unitary basis, satisfying $(e_i, e_j) = \delta_{ij}, \ i, j = 1, \ldots, n$.

In quantum mechanics, it is customary to rewrite the scalar product (9.2.2) in a *bracket* form, $\langle u|v\rangle$. Then it was Dirac who suggested to view $\langle u|v\rangle$ as the scalar pairing of a "bra" vector $\langle u|$ and a "ket" vector $|v\rangle$, representing the row vector $u^\dagger$ and the column vector v. Thus we may use $|e_1\rangle,\ldots,|e_n\rangle$ to denote the standard basis vectors of $\mathbb{C}^n$ and represent the vector u in $\mathbb{C}^n$ as

$$|u\rangle = \sum_{i=1}^{n} a_i|e_i\rangle, \tag{9.2.3}$$

therefore the bra-counterpart of $|u\rangle$ is simply given as

$$\langle u| = (|u\rangle)^\dagger = \sum_{i=1}^{n} \bar{a}_i\langle e_i|. \tag{9.2.4}$$

As a consequence, the orthonormal condition regarding the basis $\{e_1,\ldots,e_n\}$ becomes

$$\langle e_i|e_j\rangle = \delta_{ij}, \quad i,j=1,\ldots,n, \tag{9.2.5}$$

and the Hermitian scalar product of the vectors $|u\rangle$ and $|v\rangle$ assumes the form

$$\langle u|v\rangle = \sum_{i=1}^{n} \bar{a}_i b_i = \overline{\langle v|u\rangle}. \tag{9.2.6}$$

For the vector $|u\rangle$ given in (9.2.3), we find that

$$a_i = \langle e_i|u\rangle, \quad i=1,\ldots,n. \tag{9.2.7}$$

Now rewriting $|u\rangle$ as

$$|u\rangle = \sum_{i=1}^{n} |e_i\rangle a_i, \tag{9.2.8}$$

and inserting (9.2.7) into (9.2.8), we obtain

$$|u\rangle = \sum_{i=1}^{n} |e_i\rangle\langle e_i|u\rangle \equiv \left(\sum_{i=1}^{n} |e_i\rangle\langle e_i|\right)|u\rangle, \tag{9.2.9}$$

which suggests that the strange-looking "quantity," $\sum_{i=1}^{n}|e_i\rangle\langle e_i|$, should naturally be identified as the identity mapping or matrix,

$$\sum_{i=1}^{n} |e_i\rangle\langle e_i| = I, \tag{9.2.10}$$

which readily follows from the associativity property of matrix multiplication. Similarly, we have

$$\langle u| = \sum_{i=1}^{n} \overline{\langle e_i|u\rangle}\langle e_i| = \sum_{i=1}^{n} \langle u|e_i\rangle\langle e_i| = \langle u|\left(\sum_{i=1}^{n} |e_i\rangle\langle e_i|\right). \tag{9.2.11}$$

Thus (9.2.10) can be applied to both bra and ket vectors symmetrically and what it expresses is simply the fact that $|e_1\rangle, \ldots, |e_n\rangle$ form an orthonormal basis of $\mathbb{C}^n$.

We may reexamine some familiar linear mappings under the new notation.

Let $|u\rangle$ be a unit vector in $\mathbb{C}^n$. Use $P_{|u\rangle}$ to denote the mapping that projects $\mathbb{C}^n$ onto $\mathrm{Span}\{|u\rangle\}$ along $(\mathrm{Span}\{|u\rangle\})^\perp$. Then we have

$$P_{|u\rangle}|v\rangle = \langle u|v\rangle |u\rangle, \quad |v\rangle \in \mathbb{C}^n. \tag{9.2.12}$$

Placing the scalar number $\langle u|v\rangle$ to the right-hand side of the above expression, we see that the mapping $P_{|u\rangle}$ can be rewritten as

$$P_{|u\rangle} = |u\rangle\langle u|. \tag{9.2.13}$$

Let $\{|u_1\rangle, \ldots, |u_n\rangle\}$ be any orthonormal basis of $\mathbb{C}^n$. Then $P_{|u_i\rangle}$ projects $\mathbb{C}^n$ onto $\mathrm{Span}\{|u_i\rangle\}$ along

$$\mathrm{Span}\{|u_1\rangle, \ldots, \widehat{|u_i\rangle}, \ldots, |u_n\rangle\}. \tag{9.2.14}$$

It is clear that

$$I = \sum_{i=1}^n P_{|u_i\rangle}, \tag{9.2.15}$$

since $a_i = \langle u_i|u\rangle$, which generalizes the result (9.2.10).

If T is an arbitrary Hermitian operator with eigenvalues $\lambda_1, \ldots, \lambda_n$ and the associated orthonormal eigenvectors $u_1, \ldots, u_n$ (which form a basis of $\mathbb{C}^n$), then we have the representation

$$T = \sum_{i=1}^n \lambda_i |u_i\rangle\langle u_i|, \tag{9.2.16}$$

as may be checked easily. Besides, in $\langle u|T|v\rangle$, we may interpret T as applied either to the ket vector $|v\rangle$, from the left, or to the bra vector $\langle u|$, from the right, which will not cause any ambiguity.

To summarize, we may realize $\mathbb{C}^n$ by column vectors as ket vectors

$$|u\rangle = \begin{pmatrix} a_1 \\ \vdots \\ a_n \end{pmatrix}, \quad |v\rangle = \begin{pmatrix} b_1 \\ \vdots \\ b_n \end{pmatrix} \in \mathbb{C}^n, \tag{9.2.17}$$

so that the bra vectors are represented as row vectors

$$\langle u| = (|u\rangle)^\dagger = (\bar{a}_1, \ldots, \bar{a}_n), \quad \langle v| = (|v\rangle)^\dagger = (\bar{b}_1, \ldots, \bar{b}_n). \tag{9.2.18}$$

Consequently,

$$\langle u|v\rangle = (|u\rangle)^\dagger |v\rangle = \sum_{i=1}^{n} \bar{a}_i b_i, \tag{9.2.19}$$

$$|u\rangle\langle v| = |u\rangle(|v\rangle)^\dagger = \begin{pmatrix} a_1 \\ \vdots \\ a_n \end{pmatrix} (\bar{b}_1, \ldots, \bar{b}_n)$$

$$= \begin{pmatrix} a_1\bar{b}_1 & \cdots & a_1\bar{b}_n \\ \cdots & \cdots & \cdots \\ a_n\bar{b}_1 & \cdots & a_n\bar{b}_n \end{pmatrix} = (a_i\bar{b}_j). \tag{9.2.20}$$

In particular, we have the representation

$$\sum_{i=1}^{n} |e_i\rangle\langle e_i| = \begin{pmatrix} 1 & 0 & \cdots & 0 \\ 0 & 1 & \cdots & 0 \\ 0 & 0 & \ddots & 0 \\ 0 & 0 & \cdots & 1 \end{pmatrix}. \tag{9.2.21}$$

Finally, if $A \in \mathbb{C}(n,n)$ is a Hermitian matrix (or equivalently viewed as a self-adjoint mapping over $\mathbb{C}^n$), it is clear to explain A in $\langle u|A|v\rangle$ as to be applied on the ket vector $|v\rangle$ from the left or on the row vector $\langle u|$ from the right, without ambiguity, since

$$\langle u|A|v\rangle = \langle u|(A|v\rangle) = ((\langle u|A)|v\rangle = (A|u\rangle)^\dagger |v\rangle. \tag{9.2.22}$$

Exercises

9.2.1   Consider the orthonormal basis of $\mathbb{C}^2$ consisting of the ket vectors

$$|u_1\rangle = \frac{1}{2}\begin{pmatrix} 1+i \\ 1-i \end{pmatrix}, \quad |u_2\rangle = \frac{1}{2}\begin{pmatrix} 1+i \\ -1+i \end{pmatrix}. \tag{9.2.23}$$

Verify the identity

$$|u_1\rangle\langle u_1| + |u_2\rangle\langle u_2| = I_2. \tag{9.2.24}$$

9.2.2   Consider the Hermitian matrix

$$A = \begin{pmatrix} -1 & 3+i \\ 3-i & 2 \end{pmatrix} \tag{9.2.25}$$

in $\mathbb{C}(2,2)$.

(a) Find an orthonormal basis of $\mathbb{C}^2$ consisting of eigenvectors, say $|u_1\rangle$ and $|u_2\rangle$, associated with the eigenvalues, say λ_1 and λ_2, of A.

(b) Find the matrices that represent the orthogonal projections

$$P_{|u_1\rangle} = |u_1\rangle\langle u_1|, \quad P_{|u_2\rangle} = |u_2\rangle\langle u_2|. \tag{9.2.26}$$

(c) Verify the identity

$$A = \lambda_1 P_{|u_1\rangle} + \lambda_2 P_{|u_2\rangle} = \lambda_1 |u_1\rangle\langle u_1| + \lambda_2 |u_2\rangle\langle u_2|. \tag{9.2.27}$$

9.3 Quantum-mechanical postulates

In physics literature, quantum mechanics may be formulated in terms of two objects referred to as *states* and *observables*. A state, also called a *wave function*, contains statistical information of a certain physical observable, which often describes one of some measurable quantities such as energy, momenta, and position coordinates of a mechanical system, such as those encountered in describing the motion of a hypothetical particle. Mathematically, states are vectors in a complex vector space with a positive definite Hermitian scalar product, called the *state space*, and observables are Hermitian mappings over the state space given. In this section, we present an over-simplified formalism of quantum mechanics using the space $\mathbb{C}^n$ as the state space and Hermitian matrices in $\mathbb{C}(n,n)$ as observables.

To proceed, we state a number of axioms, in our context, called the quantum mechanical postulates, based on which quantum mechanics is built.

- *State postulate.* The state of a mechanical system, hereby formally referred to as a "particle," is described by a unit ket vector $|\phi\rangle$ in $\mathbb{C}^n$.
- *Observable postulate.* A physically measurable quantity of the particle such as energy, momenta, and so on, called an observable, is represented by a Hermitian matrix $A \in \mathbb{C}(n,n)$, so that its expected value for the particle in the state $|\phi\rangle$, denoted by $\langle A \rangle$, is given by

$$\langle A \rangle = \langle \phi | A | \phi \rangle. \tag{9.3.1}$$

- *Measurement postulate.* Let $A \in \mathbb{C}(n,n)$ be an observable with eigenvalues $\lambda_1, \ldots, \lambda_n$, which are known to be all real, and the corresponding eigenvectors of A be denoted by $|u_1\rangle, \ldots, |u_n\rangle$, which form an orthonormal basis of $\mathbb{C}^n$. As a *random variable*, the measurement X_A of the observable A, when the particle lies in the state $|\phi\rangle$, can only result in a reading among the eigenvalues of A, and obeys the probability distribution

$$P(\{X_A = \lambda\}) = \begin{cases} \sum_{\lambda_i = \lambda} |\phi_i|^2, & \text{if } \lambda \text{ is an eigenvalue,} \\ 0, & \text{if } \lambda \text{ is not an eigenvalue,} \end{cases} \tag{9.3.2}$$

where $\phi_1, \ldots, \phi_n \in \mathbb{C}$ are the coordinates of $|\phi\rangle$ with respect to the basis $\{u_1, \ldots, u_n\}$, that is,

$$|\phi\rangle = \sum_{i=1}^{n} \phi_i |u_i\rangle. \tag{9.3.3}$$

The measurement postulate (9.3.2) may be viewed as directly motivated from the expected (or expectation) value formula

$$\langle A \rangle = \sum_{i=1}^{n} \lambda_i |\phi_i|^2, \tag{9.3.4}$$

which can be obtained by substituting (9.3.3) into (9.3.1).

In the context of the measurement postulate, we have the following quantum mechanical interpretation of an eigenstate.

Theorem 9.1 *The particle lies in an eigenstate of the observable A if and only if all measurements of A render the same value that must be some eigenvalue of A.*

Proof   Suppose that λ is an eigenvalue of A. Use E_λ to denote the corresponding eigenspace. If the particle lies in a state $|\phi\rangle \in E_\lambda$, then $\phi_i = \langle u_i | \phi \rangle = 0$ when $\lambda_i \neq \lambda$, where $\lambda_1, \ldots, \lambda_n$ are all the possible eigenvalues of A and $|u_1\rangle, \ldots, |u_n\rangle$ the corresponding eigenvectors that form an orthonormal basis of $\mathbb{C}^n$. Therefore

$$P(\{X_A = \lambda_i\}) = 0 \quad \text{when } \lambda_i \neq \lambda, \tag{9.3.5}$$

which leads to $P(\{X_A = \lambda\}) = \langle \phi | \phi \rangle = 1$.

Conversely, if $P(\{X_A = \lambda\}) = 1$ holds for some λ when the particle lies in the state $|\phi\rangle$, then according to the measurement postulate (9.3.2), we have $\lambda = \lambda_i$ for some $i = 1, \ldots, n$. Hence

$$0 = 1 - P(\{X_A = \lambda_i\}) = \sum_{\lambda_j \neq \lambda_i} |\phi_j|^2, \tag{9.3.6}$$

which follows that $\phi_j = 0$ when $j \neq i$. Thus $|\phi\rangle \in E_{\lambda_i}$ as claimed.   □

We are now at a position to consider how a state evolves itself with respect to time t.

- *Time evolution postulate.* A state vector $|\phi\rangle$ follows time evolution according to the law

$$i\hbar \frac{\mathrm{d}}{\mathrm{d}t} |\phi\rangle = H |\phi\rangle, \tag{9.3.7}$$

where $H \in \mathbb{C}(n, n)$ is a Hermitian matrix called the *Hamiltonian* of the system and $\hbar > 0$ a universal constant called the *Planck constant*. The equation (9.3.7) is the matrix version of the celebrated *Schrödinger equation*.

In classical mechanics, the Hamiltonian H of a system measures the total energy which may be written as the sum of the kinetic energy K and potential energy V,

$$H = K + V. \qquad (9.3.8)$$

If the system consists of a single particle of mass $m > 0$, then K may be expressed in terms of the momentum P of the particle through the relation

$$K = \frac{1}{2m} P^2. \qquad (9.3.9)$$

In quantum mechanics in our context here, both P and V are taken to be Hermitian matrices.

Let $|\phi(0)\rangle = |\phi_0\rangle$ be the initial state of the time-dependent state vector $|\phi(t)\rangle$. Solving (9.3.7), we obtain

$$|\phi(t)\rangle = U(t)|\phi_0\rangle, \quad U(t) = \mathrm{e}^{-\frac{\mathrm{i}}{\hbar} tH}. \qquad (9.3.10)$$

Since H is Hermitian, $-\frac{\mathrm{i}}{\hbar} H$ is anti-Hermitian. Hence $U(t)$ is unitary,

$$U(t)U^\dagger(t) = I, \qquad (9.3.11)$$

which ensures the conservation of the normality of the state vector $|\phi(t)\rangle$. That is,

$$\langle \phi(t)|\phi(t)\rangle = \langle \phi_0|\phi_0\rangle = 1. \qquad (9.3.12)$$

Assume that the eigenvalues of H, say $\lambda_1, \ldots, \lambda_n$ are all positive. Let $u_1, \ldots, u_n$ be the associated eigenstates of H which form an orthonormal basis of $\mathbb{C}^n$. With the expansion

$$|\phi_0\rangle = \sum_{i=1}^n \phi_{0,i}|u_i\rangle, \qquad (9.3.13)$$

we may rewrite the state vector $|\phi(t)\rangle$ as

$$|\phi(t)\rangle = \sum_{i=1}^n \phi_{0,i}\mathrm{e}^{-\frac{\mathrm{i}}{\hbar} tH}|u_i\rangle = \sum_{i=1}^n \mathrm{e}^{-\frac{\mathrm{i}}{\hbar}\lambda_i t}\phi_{0,i}|u_i\rangle = \sum_{i=1}^n \mathrm{e}^{-\mathrm{i}\omega_i t}\phi_{0,i}|u_i\rangle, \qquad (9.3.14)$$

where

$$\omega_i = \frac{\lambda_i}{\hbar}, \quad i = 1, \ldots, n, \qquad (9.3.15)$$

are *angular frequencies* of the *eigenmodes*

$$\mathrm{e}^{-\mathrm{i}\omega_i t}\phi_{0,i}|u_i\rangle, \quad i = 1, \ldots, n. \qquad (9.3.16)$$

In other words, the state vector or wave function $|\phi(t)\rangle$ is a superposition of n eigenmodes with the associated angular frequencies determined by the eigenvalues of the Hamiltonian through (9.3.15).

As an observable, the Hamiltonian H measures the total energy. Thus the eigenvalues $\lambda_1, \ldots, \lambda_n$ of H are the possible energy values of the system. For this reason, we may use E to denote a generic energy value, among $\lambda_1, \ldots, \lambda_n$. Correspondingly, we use ω to denote a generic angular frequency, among $\omega_1, \ldots, \omega_n$. Therefore we arrive at the generic relation

$$E = \hbar\omega, \tag{9.3.17}$$

known as the *Einstein formula*, arising originally in the work of Einstein toward an understanding of the *photoelectric effect*, which later became one of the two basic equations in the *wave-particle duality hypothesis* of de Broglie and laid the very foundation of quantum mechanics. Roughly speaking, the formula (9.3.17) indicates that a particle of energy E behaves like a wave of angular frequency ω and that a wave of angular frequency ω also behaves like a particle of energy E.

Let A be an observable and assume that the system lies in the state $|\phi(t)\rangle$ which is governed by the Schrödinger equation (9.3.7). We investigate whether the expectation value $\langle A \rangle(t) = \langle \phi(t)|A|\phi(t)\rangle$ is conserved or time-independent. For this purpose, we compute

$$
\begin{aligned}
\frac{\mathrm{d}}{\mathrm{d}t}\langle A\rangle(t) &= \frac{\mathrm{d}}{\mathrm{d}t}\langle\phi(t)|A|\phi(t)\rangle \\
&= \left(\frac{\mathrm{d}}{\mathrm{d}t}\langle\phi(t)|\right)A|\phi(t)\rangle + \langle\phi(t)|A\left(\frac{\mathrm{d}}{\mathrm{d}t}|\phi(t)\rangle\right) \\
&= \left(\frac{\mathrm{d}}{\mathrm{d}t}|\phi(t)\rangle\right)^{\dagger}A|\phi(t)\rangle + \langle\phi(t)|A\left(\frac{\mathrm{d}}{\mathrm{d}t}|\phi(t)\rangle\right) \\
&= \left(-\frac{\mathrm{i}}{\hbar}H|\phi(t)\rangle\right)^{\dagger}A|\phi(t)\rangle + \langle\phi(t)|A\left(-\frac{\mathrm{i}}{\hbar}H|\phi(t)\rangle\right) \\
&= \left\langle\phi(t)\left|\frac{\mathrm{i}}{\hbar}[H,A]\right|\phi(t)\right\rangle,
\end{aligned}
\tag{9.3.18}
$$

where

$$[H,A] = HA - AH \tag{9.3.19}$$

is the *commutator* of H and A which measures the noncommutativity of H and A. Hence we see that $\langle A \rangle$ is time-independent if A commutes with H:

$$[H,A] = 0. \tag{9.3.20}$$

In particular, the average energy $\langle H \rangle$ is always conserved. Furthermore, if H is related through the momentum P and potential V through (9.3.8) and (9.3.9) so that P commutes with V, then $[H, P] = 0$ and the average momentum $\langle P \rangle$ is also conserved. These are the quantum mechanical extensions of laws of conservation for energy and momentum.

Exercises

9.3.1   Consider the Hermitian matrix

$$
A = \begin{pmatrix} 5 & i & 0 \\ -i & 3 & 1-i \\ 0 & 1+i & 5 \end{pmatrix} \tag{9.3.21}
$$

as an observable of a system.

(a) Find the eigenvalues of A as possible readings when measuring the observable A.

(b) Find the corresponding unit eigenvectors of A, say $|u_1\rangle, |u_2\rangle$, and $|u_3\rangle$, which form an orthonormal basis of the state space $\mathbb{C}^3$.

(c) Assume that the state the system occupies is given by the vector

$$
|\phi\rangle = \frac{1}{\sqrt{15}} \begin{pmatrix} i \\ 2+i \\ 3 \end{pmatrix}. \tag{9.3.22}
$$

Resolve $|\phi\rangle$ in terms of $|u_1\rangle, |u_2\rangle$, and $|u_3\rangle$.

(d) If the system stays in the state $|\phi\rangle$, determine the probability distribution function of the random variable X_A, which is the random value read each time when a measurement about A is made.

(e) If the system stays in the state $|\phi\rangle$, evaluate the expected value, $\langle A \rangle$, of X_A.

9.3.2   Consider the Schrödinger equation

$$
i\hbar \frac{d}{dt} |\phi\rangle = H|\phi\rangle, \quad H = \begin{pmatrix} 4 & i \\ -i & 4 \end{pmatrix}. \tag{9.3.23}
$$

(a) Find the orthonormal eigenstates of H and use them to construct the solution $|\phi(t)\rangle$ of (9.3.23) satisfying the initial condition

$$
|\phi(0)\rangle = |\phi_0\rangle = \frac{1}{\sqrt{2}} \begin{pmatrix} 1 \\ -1 \end{pmatrix}. \tag{9.3.24}
$$

(b) Consider a perturbation of the Hamiltonian H given as

$$H_\varepsilon = H + \varepsilon\sigma_1, \quad \sigma_1 = \begin{pmatrix} 0 & 1 \\ 1 & 0 \end{pmatrix}, \quad \varepsilon \in \mathbb{R}, \qquad (9.3.25)$$

where σ_1 is known as one of the *Pauli matrices*. Show that the commutator of H and H_ε is

$$[H, H_\varepsilon] = 2\mathrm{i}\varepsilon\sigma_3, \quad \sigma_3 = \begin{pmatrix} 1 & 0 \\ 0 & -1 \end{pmatrix}, \qquad (9.3.26)$$

where σ_3 is another Pauli matrix, and use it and (9.3.18) to evaluate the rate of change of the time-dependent expected value $\langle H_\varepsilon \rangle(t) = \langle \phi(t) | H_\varepsilon | \phi(t) \rangle$.

(c) Establish the formula

$$\langle \phi(t) | H_\varepsilon | \phi(t) \rangle = \langle \phi_0 | H | \phi_0 \rangle + \varepsilon \langle \phi(t) | \sigma_1 | \phi(t) \rangle \qquad (9.3.27)$$

and use it to verify the result regarding $\frac{\mathrm{d}}{\mathrm{d}t}\langle H_\varepsilon \rangle(t)$ obtained in (b) through the commutator identity (9.3.18).

9.4 Noncommutativity and uncertainty principle

Let $\{|u_i\rangle\}$ be an orthonormal set of eigenstates of a Hermitian matrix $A \in \mathbb{C}(n,n)$ with the associated real eigenvalues $\{\lambda_i\}$ and use X_A to denote the random variable of measurement of A. If the system lies in the state $|\phi\rangle$ with $\phi_i = \langle u_i | \phi \rangle$ $(i = 1, 2, \cdots)$, the distribution function of X_A is as given in (9.3.2). Thus the *variance* of X_A can be calculated according to the formula

$$\begin{aligned}
\sigma_A^2 &= \sum_{i=1}^{n} (\lambda_i - \langle A \rangle)^2 |\phi_i|^2 \\
&= (|(A - \langle A \rangle I)|\phi\rangle)^\dagger (A - \langle A \rangle I)|\phi\rangle \\
&= \langle \phi | (A - \langle A \rangle I)|(A - \langle A \rangle I)|\phi\rangle, \qquad (9.4.1)
\end{aligned}$$

which is given in a form free of the choice of the basis $\{|u_i\rangle\}$. Thus, with (9.4.1), if A and B are two observables, then the Schwartz inequality implies that

$$\sigma_A^2 \sigma_B^2 \geq |\langle \phi | (A - \langle A \rangle I)|(B - \langle B \rangle I)|\phi\rangle|^2 \equiv |c|^2, \qquad (9.4.2)$$

where the complex number c is given by

$$
\begin{aligned}
c &= \langle \phi | (A - \langle A \rangle I) | (B - \langle B \rangle I) | \phi \rangle \\
&= \langle \phi | (A - \langle A \rangle I)(B - \langle B \rangle I) | \phi \rangle \\
&= \langle \phi | AB | \phi \rangle - \langle B \rangle \langle \phi | A | \phi \rangle - \langle A \rangle \langle \phi | B | \phi \rangle + \langle A \rangle \langle B \rangle \langle \phi | \phi \rangle \\
&= \langle AB \rangle - \langle A \rangle \langle B \rangle.
\end{aligned}
\tag{9.4.3}
$$

Interchanging A and B, we have

$$
\begin{aligned}
\overline{c} &= \langle \phi | (B - \langle B \rangle I) | (A - \langle A \rangle I) | \phi \rangle \\
&= \langle BA \rangle - \langle A \rangle \langle B \rangle.
\end{aligned}
\tag{9.4.4}
$$

Therefore, we obtain

$$
\Im(c) = \frac{1}{2i}(c - \overline{c}) = \frac{1}{2i} \langle [A, B] \rangle.
\tag{9.4.5}
$$

Inserting (9.4.5) into (9.4.2), we arrive at the inequality

$$
\sigma_A^2 \sigma_B^2 \geq \left(\frac{1}{2i} \langle [A, B] \rangle \right)^2,
\tag{9.4.6}
$$

which roughly says that if two observables are noncommutative, we cannot achieve simultaneous high-precision measurements for them. To put the statement in another way, if we know one observable with high precision, we do not know the other observable at the same time with high precision. This fact, in particular, the inequality (9.4.6), in quantum mechanics, is known as the *Heisenberg uncertainty principle*.

On the other hand, when A and B commute, we know that A and B share the same eigenstates that may form an orthonormal basis of $\mathbb{C}^n$. Let ϕ be a commonly shared eigenstate so that

$$
A | \phi \rangle = \lambda_A | \phi \rangle, \quad B | \phi \rangle = \lambda_B | \phi \rangle, \quad \lambda_A, \lambda_B \in \mathbb{R}.
\tag{9.4.7}
$$

If the system lies in the state $| \phi \rangle$, then, with simultaneous full certainty, the measured values of the observables A and B are λ_A and λ_B, respectively.

Let $k \geq 1$ be an integer and define the kth moment of an observable A in the state $| \phi \rangle$ by

$$
\langle A^k \rangle = \langle \phi | A^k | \phi \rangle.
\tag{9.4.8}
$$

Thus, in (9.4.3), when we set $B = A$, we see that the variance σ_A^2 of A in the state $| \phi \rangle$ may be computed using the formula

$$
\sigma_A^2 = \langle A^2 \rangle - \langle A \rangle^2.
\tag{9.4.9}
$$

In probability theory, the radical root of the variance, $\sigma_A = \sqrt{\sigma_A^2}$, is called *standard deviation*. In quantum mechanics, σ_A is also called *uncertainty*, which measures the randomness of the observed values of the observable A.

It will be instructive to identify those states that will render the maximum uncertainty. To simplify our discussion, we shall assume that $A \in \mathbb{C}(n,n)$ has n distinct eigenvalues $\lambda_1, \ldots, \lambda_n$. As before, we use $|u_1\rangle, \ldots, |u_n\rangle$ to denote the corresponding eigenstates of A which form an orthonormal basis of the state space $\mathbb{C}^n$. In order to emphasize the dependence of the uncertainty on the underlying state $|\phi\rangle$, we use $\sigma_{A,|\phi\rangle}^2$ to denote the associated variance. We are to solve the problem

$$\max\{\sigma_{A,|\phi\rangle}^2 \mid \langle\phi|\phi\rangle = 1\}. \tag{9.4.10}$$

For this purpose, we write any normalized state vector $|\phi\rangle$ as

$$|\phi\rangle = \sum_{i=1}^{n} \phi_i|u_i\rangle. \tag{9.4.11}$$

Then $\langle A^2\rangle$ and $\langle A\rangle$ are given by

$$\langle A^2\rangle = \sum_{i=1}^{n} \lambda_i^2|\phi_i|^2, \quad \langle A\rangle = \sum_{i=1}^{n} \lambda_i|\phi_i|^2. \tag{9.4.12}$$

Hence, we have

$$\sigma_{A,|\phi\rangle}^2 = \sum_{i=1}^{n} \lambda_i^2|\phi_i|^2 - \left(\sum_{i=1}^{n} \lambda_i|\phi_i|^2\right)^2. \tag{9.4.13}$$

To ease computation, we replace $|\phi_i|$ by $x_i \in \mathbb{R}$ $(i = 1, \ldots, n)$ and consider instead the constrained maximization problem

$$\begin{cases} \max\left\{\sum_{i=1}^{n} \lambda_i^2 x_i^2 - \left(\sum_{i=1}^{n} \lambda_i x_i^2\right)^2\right\}, \\ \sum_{i=1}^{n} x_i^2 = 1. \end{cases} \tag{9.4.14}$$

Thus, using calculus, the maximum points are to be sought among the solutions of the equations

$$x_i\left(\lambda_i^2 - 2\lambda_i\left[\sum_{j=1}^{n} \lambda_j x_j^2\right] - \xi\right) = 0, \quad i = 1, \ldots, n, \tag{9.4.15}$$

where $\xi \in \mathbb{R}$ is a Lagrange multiplier. Multiplying this equation by x_i and summing over $i = 1, \ldots, n$, we find

$$\xi = \langle A^2\rangle - 2\langle A\rangle^2. \tag{9.4.16}$$

Consequently (9.4.15) takes the form

$$x_i \left(\lambda_i^2 - 2\langle A \rangle \lambda_i - [\langle A^2 \rangle - 2\langle A \rangle^2] \right) = 0, \quad i = 1, \ldots, n. \tag{9.4.17}$$

On the other hand, since the quadratic equation

$$\lambda^2 - 2\langle A \rangle \lambda - (\langle A^2 \rangle - 2\langle A \rangle^2) = 0 \tag{9.4.18}$$

has two real roots

$$\lambda = \langle A \rangle \pm \sqrt{\langle A^2 \rangle - \langle A \rangle^2} = \langle A \rangle \pm \sigma_A \tag{9.4.19}$$

in the nontrivial situation $\sigma_A > 0$, we see that there are at least $n - 2$ values of $i = 1, \ldots, n$ such that

$$\lambda_i^2 - 2\langle A \rangle \lambda_i - (\langle A^2 \rangle - 2\langle A \rangle^2) \neq 0, \tag{9.4.20}$$

which leads us to conclude in view of (9.4.17) that $x_i = 0$ at those values of i. For definiteness, we assume $x_i = 0$ when $i \neq 1, 2$. Hence (9.4.14) is reduced into

$$\begin{cases} \max \left\{ \lambda_1^2 x_1^2 + \lambda_2^2 x_2^2 - \left(\lambda_1 x_1^2 + \lambda_2 x_2^2 \right)^2 \right\}, \\ x_1^2 + x_2^2 = 1. \end{cases} \tag{9.4.21}$$

Using the constraint in (9.4.21), we may simplify the objective function of the problem (9.4.21) into the form

$$(\lambda_1 - \lambda_2)^2 (x_1^2 - x_1^4), \tag{9.4.22}$$

which may be further maximized to give us the solution

$$|x_1| = |x_2| = \frac{1}{\sqrt{2}}. \tag{9.4.23}$$

In this case, it is straightforward to check that λ_1 and λ_2 are indeed the two roots of the equation (9.4.19). In particular,

$$\sigma_{A,|\phi\rangle}^2 = \frac{1}{4} (\lambda_1 - \lambda_2)^2. \tag{9.4.24}$$

Consequently, if we use $\lambda_{\min}$ and $\lambda_{\max}$ to denote the smallest and largest eigenvalues of A, then we see in view of (9.4.24) that the *maximum uncertainty* is given by

$$\sigma_{A,\max} = \frac{1}{2} (\lambda_{\max} - \lambda_{\min}), \tag{9.4.25}$$

which is achieved when the system occupies the *maximum uncertainty state*

$$|\phi_{\max}\rangle = a|u_{\lambda_{\max}}\rangle + b|u_{\lambda_{\min}}\rangle, \quad a, b \in \mathbb{C}, \quad |a| = |b| = \frac{1}{\sqrt{2}}. \tag{9.4.26}$$

In other words, we have the result

$$\sigma_{A,|\phi_{max}\rangle} = \sigma_{A,max}.$$ (9.4.27)

Exercises

9.4.1   Let A and $B \in \mathbb{C}(2,2)$ be two observables given by

$$A = \begin{pmatrix} 1 & i \\ -i & 2 \end{pmatrix}, \quad B = \begin{pmatrix} -2 & 1-i \\ 1+i & 3 \end{pmatrix}.$$ (9.4.28)

Evaluate the quantities σ_A^2, σ_B^2, and $\langle [A,B] \rangle$ in the state

$$|\phi\rangle = \frac{1}{\sqrt{5}} \begin{pmatrix} -1 \\ 2 \end{pmatrix},$$ (9.4.29)

and use them to check the uncertainty principle (9.4.6).

9.4.2   Consider an observable $A \in \mathbb{C}(n,n)$ which has n distinct eigenvalues $\lambda_1,\ldots,\lambda_n$. Let $|u_1\rangle,\ldots,|u_n\rangle$ be the corresponding eigenstates of A which form an orthonormal basis of $\mathbb{C}^n$. Consider the *uniform state* given by

$$|\phi\rangle = \frac{1}{\sqrt{n}} \sum_{i=1}^{n} |u_i\rangle.$$ (9.4.30)

(a) Compute the uncertainty of A in the state $|\phi\rangle$.
(b) Compare your result with the maximum uncertainty given in the expression (9.4.25).

9.4.3   Let $A \in \mathbb{C}(3,3)$ be an observable given by

$$A = \begin{pmatrix} 1 & 2+i & 0 \\ 2-i & -3 & 0 \\ 0 & 0 & 5 \end{pmatrix}.$$ (9.4.31)

(a) Compute the maximum uncertainty of A.
(b) Find all maximum uncertainty states of A.
(c) Let $|\phi\rangle$ be the uniform state defined in Problem 9.4.1. Find the uncertainty of A in the state $|\phi\rangle$ and compare it with the maximum uncertainty of A found in (a).

9.5 Heisenberg picture for quantum mechanics

So far our description of quantum mechanics has been based on the Schrödinger equation which governs the evolution of a state vector. Such a description of quantum mechanics is also called a *Schrödinger picture*. Here we study another important description of quantum mechanics called the *Heisenberg picture* within which the state vector is time-independent but observables evolve with respect to time following a dynamical equation similar to that seen in classical mechanics.

We start from the Schrödinger equation (9.3.7) defined by the Hamiltonian H. Let $|\phi(t)\rangle$ be a state vector and $|\phi(0)\rangle = |\phi_0\rangle$. Then

$$|\phi(t)\rangle = e^{-\frac{i}{\hbar}tH}|\phi_0\rangle. \tag{9.5.1}$$

Thus, for any observable A, its expected value in the state $|\phi(t)\rangle$ is given by

$$\langle A\rangle(t) = \langle \phi(t)|A|\phi(t)\rangle = \langle \phi_0|e^{\frac{i}{\hbar}tH}Ae^{-\frac{i}{\hbar}tH}|\phi_0\rangle. \tag{9.5.2}$$

On the other hand, if we use time-independent state vector, $|\phi_0\rangle$, and replace A with a correctly formulated time-dependent version, $A(t)$, we are to have the same mechanical conclusion. In particular, the expected value of A at time t in the state $|\phi(t)\rangle$ must be equal to the expected value of $A(t)$ in the state $|\phi_0\rangle$. That is,

$$\langle \phi(t)|A|\phi(t)\rangle = \langle \phi_0|A(t)|\phi_0\rangle. \tag{9.5.3}$$

Comparing (9.5.2) and (9.5.3), we obtain

$$\langle \phi_0|e^{\frac{i}{\hbar}tH}Ae^{-\frac{i}{\hbar}tH} - A(t)|\phi_0\rangle = 0. \tag{9.5.4}$$

Therefore, using the arbitrariness of $|\phi_0\rangle$, we arrive at the relation

$$A(t) = e^{\frac{i}{\hbar}tH}Ae^{-\frac{i}{\hbar}tH}, \tag{9.5.5}$$

which indicates how an observable should evolve itself with respect to time. Differentiating (9.5.5), we are led to the equation

$$\frac{d}{dt}A(t) = \frac{i}{\hbar}[H,A(t)], \tag{9.5.6}$$

which spells out the dynamical law that a time-dependent observable must follow and is known as the *Heisenberg equation*.

We next show that (9.5.6) implies (9.3.7) as well.

As a preparation, we establish the following *Gronwall inequality* which is useful in the study of differential equations: If $f(t)$ and $g(t)$ are continuous non-negative functions in $t \geq 0$ and satisfy

$$f(t) \leq a + \int_0^t f(\tau)g(\tau)\,d\tau, \quad t \geq 0, \tag{9.5.7}$$

for some constant $a \geq 0$, then

$$f(t) \leq a \exp\left\{ \int_0^t g(\tau) \, d\tau \right\}, \quad t \geq 0. \tag{9.5.8}$$

To prove it, we modify (9.5.7) into

$$f(t) < a + \varepsilon + \int_0^t f(\tau) g(\tau) \, d\tau, \quad t \geq 0, \tag{9.5.9}$$

and set

$$h(t) = a + \varepsilon + \int_0^t f(\tau) g(\tau) \, d\tau, \quad t \geq 0, \tag{9.5.10}$$

where $\varepsilon > 0$. Therefore $h(t)$ is positive-valued and differentiable with $h'(t) = f(t)g(t)$ for $t > 0$ and $h(0) = a + \varepsilon$. Multiplying (9.5.9) by $g(t)$, we have

$$\frac{h'(t)}{h(t)} \leq g(t), \quad t \geq 0. \tag{9.5.11}$$

Integrating (9.5.11), we obtain

$$h(t) \leq h(0) \exp\left\{ \int_0^t g(\tau) \, d\tau \right\}, \quad t \geq 0. \tag{9.5.12}$$

However, (9.5.10) and (9.5.11) indicate that $f(t) < h(t)$ $(t \geq 0)$. Hence we may use (9.5.12) to get

$$f(t) < (a + \varepsilon) \exp\left\{ \int_0^t g(\tau) \, d\tau \right\}, \quad t \geq 0. \tag{9.5.13}$$

Finally, since $\varepsilon > 0$ is arbitrary, we see that (9.5.8) follows.

We now turn our attention back to the Heisenberg equation (9.5.6). Suppose that $B(t)$ is another solution of the equation. Then $C(t) = A(t) - B(t)$ satisfies

$$\frac{d}{dt} C(t) = \frac{i}{\hbar}[H, C(t)], \quad C(0) = A(0) - B(0). \tag{9.5.14}$$

Therefore, we have

$$\|C'(t)\| \leq \frac{2}{\hbar} \|H\| \|C(t)\|. \tag{9.5.15}$$

On the other hand, we may use the triangle inequality to get

$$\|C(t+h)\| - \|C(t)\| \leq \|C(t+h) - C(t)\|, \quad h > 0, \tag{9.5.16}$$

which allows us to conclude with

$$\frac{d}{dt} \|C(t)\| \leq \|C'(t)\|. \tag{9.5.17}$$

Inserting (9.5.17) into (9.5.15) and integrating, we find

$$\|C(t)\| \leq \|C(0)\| + \frac{2}{\hbar} \int_0^t \|H\| \|C(\tau)\| \, d\tau, \quad t \geq 0. \tag{9.5.18}$$

Consequently, it follows from applying the Gronwall inequality that

$$\|C(t)\| \leq \|C(0)\| e^{\frac{2}{\hbar} \|H\| t}, \quad t \geq 0. \tag{9.5.19}$$

The same argument may be carried out in the domain $t \leq 0$ with the time flipping $t \mapsto -t$. Thus, in summary, we obtain the collective conclusion

$$\|C(t)\| \leq \|C(0)\| e^{\frac{2}{\hbar} \|H\| |t|}, \quad t \in \mathbb{R}. \tag{9.5.20}$$

In particular, if $C(0) = 0$, then $C(t) \equiv 0$, which implies that the solution to the initial value problem of the Heisenberg equation (9.5.6) is unique. Hence, if $A(0) = A$, then the solution is uniquely given by (9.5.5). As a consequence, if A commutes with H, $A(t) = A$ for all time t. In other words, if an observable commutes with the Hamiltonian initially, it stays commutative with the Hamiltonian and remains in fact constant for all time.

We can now derive the Schrödinger equation (9.3.7) from the Heisenberg equation (9.5.6).

In fact, let $A(t)$ be the unique solution of (9.5.6) evolving from its initial state A. Then $A(t)$ is given by (9.5.5). Let $|\phi(t)\rangle$ denote the state vector that evolves with respect to t from its initial state vector $|\phi_0\rangle$ so that it gives rise to the same expected value as that evaluated using the Heisenberg equation through $A(t)$. Then (9.5.3) holds. We hope to examine, to what extent, the relation (9.5.1) must be valid. To this end, we assume that

$$\langle u|A|u \rangle = \langle v|A|v \rangle \tag{9.5.21}$$

holds for any Hermitian matrix $A \in \mathbb{C}(n, n)$ for some $|u\rangle$ and $|v\rangle \in \mathbb{C}^n$ and we investigate how $|u\rangle$ and $|v\rangle$ are related.

If $|v\rangle = 0$, then $\langle u|A|u\rangle = 0$ for any Hermitian matrix A, which implies $|u\rangle = 0$ as well. Thus, in the following, we only consider the nontrivial situation $|u\rangle \neq 0$ and $|v\rangle \neq 0$.

If $v \neq 0$, we set $V = \text{Span}\{|v\rangle\}$ and $W = V^\perp$. Choose a Hermitian matrix A so that $|x\rangle \mapsto A|x\rangle$ ($|x\rangle \in \mathbb{C}^n$) defines the projection of $\mathbb{C}^n$ onto V along W. Write $|u\rangle = a|v\rangle + |w\rangle$ where $a \in \mathbb{C}$ and $|w\rangle \in W$. Then $A|u\rangle = a|v\rangle$. Thus

$$\langle u|A|u \rangle = |a|^2 \langle v|v \rangle = |a|^2 \langle v|A|v \rangle, \tag{9.5.22}$$

which leads to

$$|a| = 1 \quad \text{or} \quad a = e^{i\theta}, \quad \theta \in \mathbb{R}. \tag{9.5.23}$$

Moreover, let $A \in \mathbb{C}(n,n)$ be a Hermitian matrix so that $|x\rangle \mapsto A|x\rangle$ $(|x\rangle \in \mathbb{C}^n)$ defines the projection of $\mathbb{C}^n$ onto W along V. Then $A|v\rangle = 0$ and $A|w\rangle = |w\rangle$. Inserting these into (9.5.21), we find

$$0 = \langle v|A|v\rangle = \langle u|A|u\rangle = \langle w|w\rangle, \tag{9.5.24}$$

which gives us the result $w = 0$. In other words, that (9.5.21) holds for any Hermitian matrix $A \in \mathbb{C}(n,n)$ implies that $|u\rangle$ and $|v\rangle$ differ from each other by a phase factor a satisfying (9.5.23). That is,

$$|u\rangle = e^{i\theta}|v\rangle, \quad \theta \in \mathbb{R}. \tag{9.5.25}$$

Now inserting (9.5.5) into (9.5.3), we obtain

$$\langle \phi(t)|A|\phi(t)\rangle = \langle \psi(t)|A|\psi(t)\rangle, \tag{9.5.26}$$

where

$$|\psi(t)\rangle = e^{-\frac{i}{\hbar}tH}|\phi_0\rangle. \tag{9.5.27}$$

Consequently, in view of the conclusion (9.5.25), we arrive at the relation

$$|\phi(t)\rangle = e^{i\theta(t)}e^{-\frac{i}{\hbar}tH}|\phi_0\rangle, \tag{9.5.28}$$

where $\theta(t)$ is a real-valued function of t, which simply cancels itself out in (9.5.26) and may well be taken to be zero. In other words, we are prompted to conclude that the state vector should follow the law of evolution given simply by (9.5.1) or by (9.5.28) with setting $\theta(t) \equiv 0$, which is the unique solution of the Schrödinger equation (9.3.7) subject to the initial condition $|\phi(0)\rangle = |\phi_0\rangle$. Thus the Schrödinger equation inevitably comes into being as anticipated.

Exercises

9.5.1   Consider the Heisenberg equation (9.5.6) subject to the initial condition $A(0) = A_0$. An integration of (9.5.6) gives

$$A(t) = A_0 + \frac{i}{\hbar} \int_0^t [H, A(\tau)]\, d\tau. \tag{9.5.29}$$

(a) From (9.5.29) derive the result

$$[H, A(t)] = [H, A_0] + \frac{i}{\hbar} \int_0^t [H, [H, A(\tau)]]\, d\tau. \tag{9.5.30}$$

(b) Use (9.5.30) and the Gronwall inequality to come up with an alternative proof that H and $A(t)$ commute at all time if and only if they do so initially.

9.5.2 Consider the Hamiltonian H and an observable A given by

$$H = \begin{pmatrix} 2 & i \\ -i & 2 \end{pmatrix}, \quad A = \begin{pmatrix} 1 & 1-i \\ 1+i & 1 \end{pmatrix}. \tag{9.5.31}$$

(a) Solve the Schrödinger equation (9.3.7) to obtain the time-dependent state $|\phi(t)\rangle$ evolving from the initial state

$$|\phi_0\rangle = \frac{1}{\sqrt{2}} \begin{pmatrix} 1 \\ i \end{pmatrix}. \tag{9.5.32}$$

(b) Evaluate the expected value of the observable A assuming that the system lies in the state $|\phi(t)\rangle$.

(c) Solve the Heisenberg equation (9.5.6) with the initial condition $A(0) = A$ and use it to evaluate the expected value of the same observable within the Heisenberg picture. That is, compute the quantity $\langle\phi_0|A(t)|\phi_0\rangle$. Compare the result with that obtained in (b) and explain.

(d) Verify the relation

$$\langle\phi(0)|A|\phi(0)\rangle = \langle\phi_0|A(0)|\phi_0\rangle = \langle\phi_0|A|\phi_0\rangle, \tag{9.5.33}$$

at the initial time.

10

Excursion: Problems in Machine Learning

A rich and important area for the applications of linear algebra is *machine learning*. In machine learning, one aims to achieve optimized or *learned* understanding of various kinds of real-world phenomena from data collected or observed, without real comprehension of the functioning mechanisms of such phenomena. These functioning mechanisms are often impossible or unpractical to grasp anyway. In this chapter, we present several introductory and fundamental problems in supervised machine learning including linear regression, data classification, and logistic regression and the associated mathematical and computational methods.

10.1 Supervised learning and linear regression

Generally speaking, supervised learning and unsupervised learning are two primary categories of machine learning algorithms, differing primarily in the presence or absence of labeled training data. In supervised learning, the algorithm is trained on a labeled data set, where each input example is associated with an output label. The goal is to learn a mapping from inputs to outputs based on the training data, such that the algorithm can predict the correct output for new, unseen future inputs. Supervised learning can further be divided into two main types: regression and classification. In a regression task, the algorithm learns to predict a continuous numerical value or a range of values for the output variable, given an input variable; in a classification task, the output variable is categorical or discrete and the algorithm learns to classify inputs into one of several classes or categories. In *unsupervised learning*, the algorithm is given an unlabeled data set and tasked with finding patterns or structure in the data without explicit guidance. Unlike supervised learning, there are no predefined output labels to guide the learning process. Unsupervised learning tasks include clustering and dimensionality reduction. In the former context, the task is to group similar data points together into clusters based on some metric, and, in the latter context, the task is to reduce the number of features or dimensions in the data set while preserving its essential structure.

Specifically, *supervised learning* is a subject in machine learning which in its simplest form is concerned with learning about a function $y = h(x)$ that makes prediction through its output value $y \in \mathbb{R}$ from its input data x which is often taken to be in $\mathbb{R}^m$, representing the space of m-feature inputs. In other words, input data consist of a collection of readings expressed as m real numbers or features. In such a process, the function h is hypothesized to assume a specific form depending on some adjustable parameters, say $\alpha_1, \ldots, \alpha_k$, called *learning parameters*, *learning features*, or simply *parameter features*, so that the hypothesized function h reads

$$y = h(x; \alpha_1, \ldots, \alpha_k) \equiv h(x; \alpha). \tag{10.1.1}$$

Within the hypothesized form of the function h given in (10.1.1), we now collect n sample observations, grouped in the forms of data pairs, say (input, output), denoted by

$$(A_1, b_1), \ldots, (A_n, b_n), \quad A_j = \begin{pmatrix} a_{1j} \\ \vdots \\ a_{mj} \end{pmatrix} \in \mathbb{R}^m, \quad b_j \in \mathbb{R}, \quad j = 1, \ldots, n.$$

$$\tag{10.1.2}$$

The data set (10.1.2) is often referred to as *training data*. Hence a_{ij} represents the ith featured input in the jth sample. Then the task of supervised learning is to determine the adjustable parameters $\alpha_1, \ldots, \alpha_k$ so that, roughly speaking, the errors,

$$z_j = h(A_j; \alpha) - b_j, \quad j = 1, \ldots, n, \quad \alpha = (\alpha_1, \ldots, \alpha_k), \tag{10.1.3}$$

occurring at the training data points are kept jointly as small as possible, so that the predictions made by the hypothesized function at the input data points are as close as possible to the actual outputs given by the training data set. The product of this learning process is the function (10.1.1) with α so determined. That is, the function (10.1.1) provides an output y as its prediction at any input data vector x for the problem under consideration.

In practice, it is convenient to consider the total squared sum of errors

$$E = \sum_{j=1}^{n} z_j^2 = \sum_{j=1}^{n} (h(A_j; \alpha) - b_j)^2, \tag{10.1.4}$$

as a quantity to be minimized with respect to $\alpha \in \mathbb{R}^k$. This is a general formalism of the least-squares optimization.

At this juncture, it will be enlightening to present a statistical interpretation of the concept of the least-squares optimization in the context of machine learning as well. For this purpose, regard the training data given in (10.1.2) as the observed values of two random variables, X and Y, respectively, such that the errors (10.1.3) are realized as the observed values of the random variable

$$Z = h(X; \alpha) - Y. \tag{10.1.5}$$

On the other hand, as error, it is natural to assume that Z follows a zero-mean normal distribution with the probability density function

$$p(Z = z) = p(z) = \frac{1}{\sqrt{2\pi}\sigma} e^{-\frac{z^2}{2\sigma^2}}, \quad -\infty < z < \infty, \tag{10.1.6}$$

where $\sigma > 0$ is the standard deviation of the distribution. Thus, since the events $\{Z = z_1\}, \ldots, \{Z = z_n\}$ are independent, we have

$$p(Z = z_1, \ldots, Z = z_n) = \prod_{j=1}^{n} p(Z = z_j)$$

$$= \frac{1}{(2\pi)^{\frac{n}{2}} \sigma^n} e^{-\frac{1}{2\sigma^2} \sum_{j=1}^{n} z_j^2}$$

$$= \frac{1}{(2\pi)^{\frac{n}{2}} \sigma^n} e^{-\frac{1}{2\sigma^2} \sum_{j=1}^{n} (h(A_j; \alpha) - b_j)^2}, \tag{10.1.7}$$

in view of (10.1.4). Consequently, applying the *maximum likelihood* consideration for the occurrence of the event $\{Z = z_1, \ldots, Z = z_n\}$, we are prompted to maximize the right-hand side of (10.1.7). In other words, we arrive at the least-squares optimization given by the quantity (10.1.4) as stated earlier.

In the simplest situation, the hypothesized function is *affine linear* of the form

$$y = \sum_{i=1}^{m} \theta_i x_i + \theta_0 = \theta^t x + \theta_0 = x^t \theta + \theta_0; \quad \theta = \begin{pmatrix} \theta_1 \\ \vdots \\ \theta_m \end{pmatrix}, x = \begin{pmatrix} x_1 \\ \vdots \\ x_m \end{pmatrix} \in \mathbb{R}^m, \tag{10.1.8}$$

where $\theta_0 \in \mathbb{R}$ is called *bias* which gives a perception of the value of the output y in absence of the input x and $\theta_1, \ldots, \theta_m$ are called *learning weights*. Jointly $\theta_0, \theta_1, \ldots, \theta_m$ are adjustable learning parameters in the current context. The optimization problem in such a setting is called a linear regression problem and a solution to the problem a linear regression solution or fit.

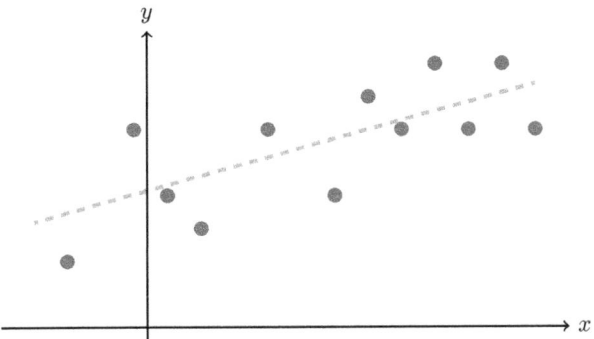

Figure 10.1 Linear regression by the least-squares optimization based on a given training data set.

Figure 10.1 is an illustrative plot of such a linear regression solution by the least-squares optimization, displayed as a straight line passing through the data set, with a minimized sum of squared errors. The bias exhibits itself as the y-intercept of the regression line.

For conciseness and by custom, we may expand the vectors θ and x in (10.1.8) to start from the zeroth components with $x_0 = 1$ so that without loss of generality we may focus on (10.1.8) with $\theta_0 = 0$ in our subsequent study. In other words, it suffices to consider the zero-bias situation:

$$y = x^t \theta; \quad \theta = \begin{pmatrix} \theta_1 \\ \vdots \\ \theta_m \end{pmatrix}, x = \begin{pmatrix} x_1 \\ \vdots \\ x_m \end{pmatrix} \in \mathbb{R}^m, \quad m \geq 2. \qquad (10.1.9)$$

In view of the training data (10.1.2) and the linear-function hypothesis (10.1.9), we see that the total squared sum of errors (10.1.4) becomes

$$E = \sum_{j=1}^{n} (A_j^t \theta - b_j)^2$$
$$= \|A^t \theta - b\|^2, \qquad (10.1.10)$$

where

$$A = (A_1, \ldots, A_n) = \begin{pmatrix} a_{11} & \cdots & a_{1n} \\ \vdots & \ddots & \vdots \\ a_{m1} & \cdots & a_{mn} \end{pmatrix}, \quad b = \begin{pmatrix} b_1 \\ \vdots \\ b_n \end{pmatrix}. \qquad (10.1.11)$$

By Section 5.6, we know that the problem

$$\eta = \inf \left\{ \|A^t \theta - b\|^2 \mid \theta \in \mathbb{R}^m \right\}, \qquad (10.1.12)$$

referred to as the linear regression problem, is always solvable, so that its solution is obtained by solving the normal equation

$$AA^t\theta = Ab \qquad (10.1.13)$$

in our current context. That is, the supervised learning problem with any training data can always be solved under the linear regression hypothesis.

If $r(A) = m$, then $AA^t \in \mathbb{R}(m,m)$ is invertible. Under such an assumption, the unique solution of (10.1.13) is given by the formula

$$\theta = (AA^t)^{-1}(Ab). \qquad (10.1.14)$$

Thus we see from (10.1.9) and (10.1.14) that the training data set (10.1.2) provides us the function

$$y = x^t(AA^t)^{-1}(Ab) = (Ab)^t(AA^t)^{-1}x \qquad (10.1.15)$$

as the predicted output over the input data vector x.

Note also that the linear regression idea here may directly be extended further to accommodate a broader range of applications. For example, for $x \in \mathbb{R}^k$, consider a map $\phi : \mathbb{R}^k \to \mathbb{R}^m$, or

$$\phi(x) = \begin{pmatrix} \phi_1(x) \\ \vdots \\ \phi_m(x) \end{pmatrix}, \quad x \in \mathbb{R}^k, \qquad (10.1.16)$$

in which the real-valued functions $\phi_1(x), \ldots, \phi_m(x)$ are called *basis functions* which may be arbitrarily chosen otherwise. With basis functions, the linear hypothesized function (10.1.9) is modified into

$$y = (\phi(x))^t\theta, \qquad (10.1.17)$$

and the input data matrix (10.1.11) is replaced by

$$\Phi = (\phi(A_1), \ldots, \phi(A_n)); \quad A_j = \begin{pmatrix} a_{1j} \\ \vdots \\ a_{kj} \end{pmatrix}, \quad j = 1, \ldots, n. \qquad (10.1.18)$$

In Figure 10.2, we display an example of a training data set that obviously may not be well described by a linear function fit. Instead, it clearly favors a cubic polynomial fit. As a result, we use

$$y = \theta_0 + \theta_1 x + \theta_2 x^2 + \theta_3 x^3 \qquad (10.1.19)$$

to model the problem in which $\theta_0, \theta_1, \theta_2$, and θ_3 are learning parameters.

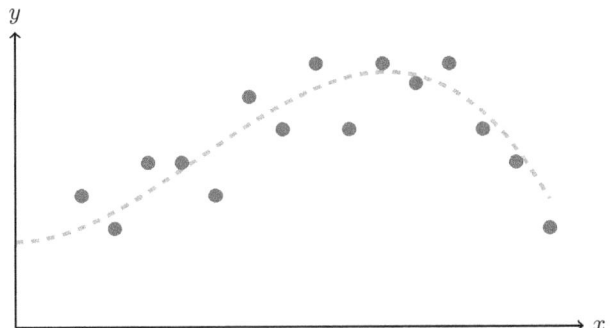

Figure 10.2 The method of linear regression may be extended to fit a nonlinearly distributed training data set.

In this situation, the simplest feature space is two-dimensional where the two features are the slope and the y-intercept of the regression line. In order to describe the more complicated dependence behavior of the output variable on the input one exhibited by the data set, we enlarge the feature space into a four-dimensional one to capture the fine details shown by data with a cubic regression curve.

With the formalism (10.1.16)–(10.1.18), the optimization problem (10.1.12) is replaced by

$$\eta = \inf\left\{ \|\Phi^t\theta - b\|^2 \mid \theta \in \mathbb{R}^m \right\}, \tag{10.1.20}$$

so that the equation (10.1.13) becomes

$$\Phi\Phi^t\theta = \Phi b. \tag{10.1.21}$$

In applications, the positive semi-definite matrix $\Phi\Phi^t \in \mathbb{R}(m,m)$ may fail to be invertible in general, which complicates the problem. In order to tackle this difficulty, we may consider the following λ-regularized problem:

$$\inf\left\{ \|\Phi^t\theta - b\|^2 + \lambda\|\theta\|^2 \mid \theta \in \mathbb{R}^m \right\}, \tag{10.1.22}$$

with $\lambda > 0$ an adjustable parameter. For the problem (10.1.22), the associated normal equation is

$$(\Phi\Phi^t + \lambda I_m)\theta = \Phi b, \tag{10.1.23}$$

with the solution

$$\theta = (\Phi\Phi^t + \lambda I_m)^{-1}\Phi b. \tag{10.1.24}$$

Substituting (10.1.24) into (10.1.17), we obtain

$$y = (\phi(x))^t(\Phi\Phi^t + \lambda I_m)^{-1}\Phi b. \tag{10.1.25}$$

In order to simplify (10.1.25) further, we note that there holds the identity

$$(\Phi\Phi^t + \lambda I_m)^{-1}\Phi = \Phi(\Phi^t\Phi + \lambda I_n)^{-1}. \qquad (10.1.26)$$

(See Problem 10.1.4.) Inserting (10.1.26) into (10.1.25), we have

$$y = (\phi(x))^t\Phi(\Phi^t\Phi + \lambda I_n)^{-1}b. \qquad (10.1.27)$$

Now of particular interest is the $n \times n$ matrix $\Phi^t\Phi$ in (10.1.27):

$$\Phi^t\Phi = ((\phi(A_i))^t\phi(A_j)) = ((\phi(A_i), \phi(A_j))). \qquad (10.1.28)$$

In other words, the (i,j)th entry of $\Phi^t\Phi$ is simply the scalar product of the feature vectors $\phi(A_i)$ and $\phi(A_j)$,

$$(\Phi^t\Phi)_{ij} = ((\phi(A_i), \phi(A_j))), \quad i,j = 1,\ldots,n. \qquad (10.1.29)$$

In machine learning practice, one follows the idea of *Mercer's theorem* to further represent (10.1.29) by a symmetric positive semi-definite function $K(x,x')$ $(x,x' \in \mathbb{R}^k)$, called the *kernel function*, satisfying

(1) $K(x,x') = K(x',x)$.
(2) The matrix $(K(A_i,A_j))$ is positive semi-definite for any $A_1,\ldots,A_n \in \mathbb{R}^k$,

such that

$$(\phi(A_i), \phi(A_j)) = K(A_i,A_j), \quad i,j = 1,\ldots,n, \qquad (10.1.30)$$
$$(\phi(x))^t\Phi = ((\phi(x))^t\phi(A_1),\ldots,(\phi(x))^t\phi(A_n))$$
$$= (K(x,A_1),\ldots,K(x,A_n)). \qquad (10.1.31)$$

As a consequence, if we use K to denote the matrix $(K(A_i,A_j))$, also called the *Gram matrix* or *kernel matrix*, generated by the kernel function $K(x,x')$, then (10.1.27)–(10.1.31) give us the solution

$$y = (K(x,A_1),\ldots,K(x,A_n))(K + \lambda I_n)^{-1}b, \qquad (10.1.32)$$

explicitly expressed in terms of the kernel function evaluated over the training data. This widely used formalism is known as the *kernel method* in machine learning practice. It is interesting and important to note that this method enables one to *by-pass* the feature map ϕ completely, which is related to the kernel function implicitly by

$$K(x,x') = (\phi(x))^t\phi(x') = (\phi(x), \phi(x')), \quad x,x' \in \mathbb{R}^k. \qquad (10.1.33)$$

Some useful examples of kernel functions include the *polynomial kernel* of degree *d*, the *sigmoidal kernel*, and the *radial basis kernel* or *Gaussian kernel* given by

$$K(x,x') = (x^t x' + a)^d, \quad a \in \mathbb{R}, \quad a \geq 0, \quad d = 1, 2, \ldots, \quad (10.1.34)$$

$$K(x,x') = \tanh(x^t x' + a), \quad a \in \mathbb{R}, \quad (10.1.35)$$

$$K(x,x') = \exp\left(-\frac{\|x - x'\|^2}{2\sigma^2}\right), \quad \sigma > 0, \quad (10.1.36)$$

respectively, for *x* and $x' \in \mathbb{R}^k$.

Exercises

10.1.1 For $A \in \mathbb{R}(m,n)$, show that the square matrix AA^t is non-singular if and only if $r(A) = m$.

10.1.2 Consider a one-feature input problem and set

$$\phi_1(x) = 1, \phi_2(x) = x, \ldots, \phi_m(x) = x^{m-1}, \quad m \geq 2.$$

Show that the input data matrix (10.1.18) is of rank *m* if and only if the input data set has *m* distinct input sample points.

10.1.3 Derive the normal equation (10.1.23).

10.1.4 Verify (10.1.26) by establishing the matrix identity

$$(AA^t + \lambda I_m)^{-1} A = A(A^t A + \lambda I_n)^{-1}, \quad A \in \mathbb{R}(m,n), \quad \lambda > 0. \quad (10.1.37)$$

10.1.5 Show that the kernel function defined by (10.1.33) is positive semi-definite for any map ϕ.

10.1.6 For $d = 2$ and $k = 2$, find a map $\phi : \mathbb{R}^2 \to \mathbb{R}^m$ to fulfill (10.1.33) and (10.1.34). In particular, determine a suitable *m* to make it work.

10.1.7 With (10.1.33), establish the inequality

$$(K(x,x'))^2 \leq K(x,x)K(x',x'), \quad x,x' \in \mathbb{R}^k. \quad (10.1.38)$$

10.2 Solution to the linear regression problem by iterations

In this section, we present some practical methods that solve the optimization problem (10.1.12). For this purpose, we use the Euclidean space scalar product to expand the right-hand side of (10.1.10) to get

$$\|A^t \theta - b\|^2 = (\theta, AA^t \theta) - 2(\theta, Ab) + (b,b). \quad (10.2.1)$$

Hence, with

$$B = AA^t \in \mathbb{R}(m,m), \quad c = Ab \in \mathbb{R}^m, \qquad (10.2.2)$$

it suffices to minimize the function

$$F(\theta) = \frac{1}{2}(\theta, B\theta) - (\theta, c), \quad \theta \in \mathbb{R}^m. \qquad (10.2.3)$$

First, using

$$\nabla_\theta F(\theta) = \begin{pmatrix} \frac{\partial F}{\partial \theta_1} \\ \vdots \\ \frac{\partial F}{\partial \theta_m} \end{pmatrix}, \quad \frac{dF(\theta + \tau\xi)}{d\tau}\bigg|_{\tau=0} = \xi^t(\nabla_\theta F(\theta)), \quad \xi \in \mathbb{R}^m,$$

$$\qquad (10.2.4)$$

where τ is a real parameter, we have

$$\nabla_\theta F(\theta) = B\theta - c. \qquad (10.2.5)$$

Thus a critical point of the function (10.2.3) solves the equation

$$B\theta = c, \qquad (10.2.6)$$

which recovers the normal equation (10.1.13) in view of (10.2.2).

Conversely, we show that any solution to (10.2.6) minimizes the function (10.2.3). In fact, since B in (10.2.2) is positive semi-definite, we have for any $\xi \in \mathbb{R}^m$ the result

$$\begin{aligned} 0 \le (\xi - \theta, B(\xi - \theta)) &= (\xi, B\xi) - 2(\xi, B\theta) + (\theta, B\theta) \\ &= ((\xi, B\xi) - 2(\xi, B\theta)) - ((\theta, B\theta) - 2(\theta, B\theta)) \\ &= ((\xi, B\xi) - 2(\xi, c)) - ((\theta, B\theta) - 2(\theta, c)) \\ &= 2F(\xi) - 2F(\theta), \end{aligned} \qquad (10.2.7)$$

which establishes that a solution to (10.2.6) minimizes (10.2.3). In other words, we have shown that θ minimizes (10.2.3) if and only if it solves (10.2.6).

Next, we look for a solution to (10.2.6) when the matrix B in (10.2.2) is positive definite. This condition is equivalent to the condition $r(A) = m$.

In order to solve (10.2.6), we may consider the *gradient descent flow* in $\mathbb{R}^m$ defined by the equation

$$\frac{d\theta}{d\tau} = -\nabla_\theta F(\theta), \quad \tau > 0; \quad \theta(0) = \theta_0, \qquad (10.2.8)$$

starting from an arbitrary initial state vector $\theta_0 \in \mathbb{R}^m$ at $\tau = 0$, where τ is viewed as a time variable. Then we have

$$\frac{dF(\theta)}{d\tau} = -\|\nabla_\theta F(\theta)\|^2, \quad \tau > 0, \qquad (10.2.9)$$

so that the limit

$$\lim_{\tau \to \infty} \theta(\tau) \equiv \theta_\infty \qquad (10.2.10)$$

exists and is the unique critical point of $F(\theta)$ or the solution to (10.2.6).

Mathematically, we may solve (10.2.6) using the Fourier expansion as we have done in Chapter 4 without assuming the positive definiteness of the matrix B. Indeed, since B is positive semi-definite, we know that $\mathbb{R}^m$ has an orthonormal basis, say $\{u_1, \ldots, u_m\}$, consisting of eigenvectors of B associated with non-negative eigenvalues, say $\lambda_1, \ldots, \lambda_m$, among which the first l eigenvalues $\lambda_1, \ldots, \lambda_l$ are positive ($l \geq 1$ in a nontrivial situation). For any $\theta \in \mathbb{R}^m$, we have

$$\theta = a_1 u_1 + \cdots + a_m u_m, \qquad (10.2.11)$$

where $a_i = (u_i, \theta)$ $(i = 1, \ldots, m)$ are the Fourier coefficients of θ with respect to the orthonormal basis $\{u_1, \ldots, u_m\}$. On the other hand, since $c = Ab$ given in (10.2.2) lies in Span$\{u_1, \ldots, u_l\}$, we have

$$c = c_1 u_1 + \cdots + c_l u_l, \qquad (10.2.12)$$

where $c_i = (u_i, c)$ $(i = 1, \ldots, l)$ are the nontrivial Fourier coefficients of c. Inserting (10.2.11) and (10.2.12) into (10.2.6), we obtain the solution (10.2.11) with

$$a_i = \frac{c_i}{\lambda_i} = \frac{(u_i, c)}{\lambda_i}, \quad i = 1, \ldots, l. \qquad (10.2.13)$$

We now return to consider the case when B is positive definite.

Motivated by the gradient flow convergence result, (10.2.10), we consider the discretized version of the dynamical system (10.2.8) where we replace the derivative on the left-hand side of the equation by the following difference quotient:

$$\frac{\theta(\tau_{k+1}) - \theta(\tau_k)}{\delta} \approx \frac{\theta_{k+1} - \theta_k}{\delta}, \quad k = 1, 2, \ldots, \qquad (10.2.14)$$

where $\delta > 0$ is the step size, $\delta = \tau_{k+1} - \tau_k$, and $\theta_k \in \mathbb{R}^m$ approximates $\theta(\tau_k) \in \mathbb{R}^m$ which should not be confused with the coordinate notation in Section 10.1. Thus, in view of (10.2.8) and (10.2.14), we arrive at the computational scheme

$$\theta_{k+1} = \theta_k - \delta(B\theta_k - c), \quad k = 0, 1, 2, \ldots. \qquad (10.2.15)$$

This scheme is rather crude so that it is unreasonable to expect the sequence $\{\theta_k\}$ so defined to converge to the solution of (10.2.6) without some further elaboration. Indeed, as a matter of fact, the sequence $\{F(\theta_k)\}$ may fail to decrease along the sequence (10.2.15) if the step size δ, also called *learning rate*, is not appropriately chosen. To tackle such an issue, we proceed as in [49] as follows.

Inserting (10.2.15) into (10.2.3), we get

$$F(\theta_{k+1}) = \frac{1}{2}(\theta_k - \delta\xi_k, B(\theta_k - \delta\xi_k)) - (\theta_k - \delta\xi_k, c), \quad \xi_k \equiv B\theta_k - c.$$
(10.2.16)

If $\xi_k = 0$, we see that θ_k solves (10.2.6) already and we stop the process. So we may assume $\xi_k \neq 0$ and express (10.2.16) as

$$F(\theta_{k+1}) = F(\theta_k) - \delta(\xi_k, \xi_k) + \frac{\delta^2}{2}(\xi_k, B\xi_k).$$
(10.2.17)

Since the quantity $(\xi_k, B\xi_k)$ is positive, we see that (10.2.17) is minimized at

$$\delta = \delta_k = \frac{(\xi_k, \xi_k)}{(\xi_k, B\xi_k)}.$$
(10.2.18)

With (10.2.18), (10.2.17) has the value

$$F(\theta_{k+1}) = F(\theta_k) - \frac{\delta_k}{2}(\xi_k, \xi_k),$$
(10.2.19)

which truly descends from $F(\theta_k)$ since $\delta_k > 0$ as given in (10.2.18).

Inserting $\xi = \theta_k$ into (10.2.7), using the notation $\varepsilon_k = \theta_k - \theta$, applying (10.2.19), we have

$$(\varepsilon_{k+1}, B\varepsilon_{k+1}) = 2F(\theta_{k+1}) - 2F(\theta)$$
$$= 2\left(F(\theta_k) - \frac{\delta_k}{2}(\xi_k, \xi_k)\right) - 2F(\theta)$$
$$= (\varepsilon_k, B\varepsilon_k) - \delta_k(\xi_k, \xi_k).$$
(10.2.20)

On the other hand, note that $B\varepsilon_k = B(\theta_k - \theta) = B\theta_k - c = \xi_k$. Thus

$$(\varepsilon_k, B\varepsilon_k) = (B^{-1}B\varepsilon_k, B\varepsilon_k) = (\xi_k, B^{-1}\xi_k).$$
(10.2.21)

Hence, if $\varepsilon_k \neq 0$ or equivalently $\xi_k \neq 0$, we may combine (10.2.20) and (10.2.21) and insert (10.2.18) to obtain

$$(\varepsilon_{k+1}, B\varepsilon_{k+1}) = (\varepsilon_k, B\varepsilon_k)\left(1 - \frac{\delta_k(\xi_k, \xi_k)}{(\varepsilon_k, B\varepsilon_k)}\right)$$
$$= (\varepsilon_k, B\varepsilon_k)\left(1 - \frac{(\xi_k, \xi_k)}{(\xi_k, B\xi_k)}\frac{(\xi_k, \xi_k)}{(\xi_k, B^{-1}\xi_k)}\right).$$
(10.2.22)

Order the positive eigenvalues $\lambda_1, \ldots, \lambda_m$ of B such that $\lambda_1 \leq \cdots \leq \lambda_m$. Then $\lambda_1^{-1}, \ldots, \lambda_m^{-1}$ are the eigenvalues of B^{-1}. Therefore

$$\lambda_1 \leq \frac{(\xi, B\xi)}{(\xi, \xi)} \leq \lambda_m, \quad \frac{1}{\lambda_m} \leq \frac{(\xi, B^{-1}\xi)}{(\xi, \xi)} \leq \frac{1}{\lambda_1}, \quad \xi \in \mathbb{R}^m, \quad \xi \neq 0. \quad (10.2.23)$$

In particular, we have

$$\frac{(\xi, \xi)}{(\xi, B\xi)} \frac{(\xi, \xi)}{(\xi, B^{-1}\xi)} \geq \frac{\lambda_1}{\lambda_m}, \quad \xi \in \mathbb{R}^m, \quad \xi \neq 0. \quad (10.2.24)$$

Inserting (10.2.24) into (10.2.22), we arrive at

$$(\varepsilon_{k+1}, B\varepsilon_{k+1}) \leq (\varepsilon_k, B\varepsilon_k) \left(1 - \frac{\lambda_1}{\lambda_m} \right), \quad k = 0, 1, 2, \ldots, \quad (10.2.25)$$

which then leads to

$$(\varepsilon_k, B\varepsilon_k) \leq (\varepsilon_0, B\varepsilon_0) \left(1 - \frac{\lambda_1}{\lambda_m} \right)^k, \quad k = 1, 2, \ldots. \quad (10.2.26)$$

As a consequence, applying (10.2.23) again, we get the following error estimate:

$$\|\varepsilon_k\|^2 \leq \|\varepsilon_0\|^2 \frac{\lambda_m}{\lambda_1} \left(1 - \frac{\lambda_1}{\lambda_m} \right)^k, \quad k = 1, 2, \ldots. \quad (10.2.27)$$

In summary, we can state

Theorem 10.1 *Let $B \in \mathbb{R}(m, m)$ be positive definite and ρ the ratio of the smallest and greatest eigenvalues of B. For any $c \in \mathbb{R}^m$, the unique solution θ to the equation (10.2.6) may be obtained as the limit of the sequence $\{\theta_k\}$ constructed by the iterative scheme*

$$\theta_{k+1} = \theta_k - \delta_k (B\theta_k - c), \quad \delta_k = \frac{(B\theta_k - c, B\theta_k - c)}{(B\theta_k - c, B(B\theta_k - c))}, \quad k = 0, 1, 2, \ldots, \quad (10.2.28)$$

starting from an arbitrary initial state vector $\theta_0 \in \mathbb{R}^m$, whenever $B\theta_k - c \neq 0$, otherwise we terminate the scheme and obtain the solution in finite steps already. Moreover, there holds the error estimate

$$\|\theta_k - \theta\|^2 \leq \|\theta_0 - \theta\|^2 \frac{(1 - \rho)^k}{\rho}, \quad k = 1, 2, \ldots, \quad (10.2.29)$$

establishing a power-law type convergence result.

For a non-singular matrix $A \in \mathbb{R}(m, m)$ (say), the quantity

$$\kappa(A) = \|A\| \, \|A^{-1}\|, \quad (10.2.30)$$

given as the product of the induced norms of A and A^{-1}, is often referred to as the *condition number* of A. In the current context of the positive definite matrix B with the ordered eigenvalues $0 < \lambda_1 \leq \cdots \leq \lambda_m$ as specified earlier, we have

$$\|B\| = \lambda_m, \quad \|B^{-1}\| = \frac{1}{\lambda_1}. \tag{10.2.31}$$

Therefore

$$\kappa(B) = \frac{\lambda_m}{\lambda_1} = \frac{1}{\rho} \geq 1. \tag{10.2.32}$$

In view of (10.2.29) and (10.2.32), we conclude that a lower condition number of the problem ensures a faster convergence for the iterative scheme (10.2.28).

We next pursue to obtain a convergent iterative scheme of the form (10.2.15) with a *uniform* step size δ. That is, the step size is independent of the step count k. Such a property enables a much simplified computational scheme for implementation, of course.

In fact, consider the mapping $T : \mathbb{R}^m \to \mathbb{R}^m$ defined by

$$T(u) = u - \delta(Bu - c), \quad u \in \mathbb{R}^m. \tag{10.2.33}$$

Then

$$\|T(u) - T(v)\| = \|(I - \delta B)(u - v)\|$$
$$\leq \|I - \delta B\| \, \|u - v\|, \quad u, v \in \mathbb{R}^m, \tag{10.2.34}$$

where

$$\|I - \delta B\| = \max\{|1 - \delta\lambda_1|, \ldots, |1 - \delta\lambda_m|\}. \tag{10.2.35}$$

If the eigenvalues of B are ordered as before, $\lambda_1 \leq \cdots \leq \lambda_m$, we choose δ small such that $\delta\lambda_m < 1$. Then $\|I - \delta B\| = 1 - \delta\lambda_1 < 1$. Hence T defined in (10.2.33) is contractive. Consequently, the iterative sequence $\theta_{k+1} = T(\theta_k)$ or (10.2.15) starting from an arbitrary initial state vector $\theta_0 \in \mathbb{R}^m$ converges to the unique solution θ of the equation (10.2.6) realized as the unique fixed point of T, satisfying $T(\theta) = \theta$. As a result, we have

$$\theta_{k+1} - \theta = T(\theta_k) - T(\theta) = (I - \delta B)(\theta_k - \theta). \tag{10.2.36}$$

Since $I - \delta B$ is non-singular, we see that $\theta_k \neq \theta$ implies $\theta_{k+1} \neq \theta$. In other words, if $\theta_0 \neq \theta$, the iteration $\theta_{k+1} = T(\theta_k)$ will never terminate in finite steps. Nevertheless, the relation (10.2.36) yields the error estimate

$$\|\theta_k - \theta\| \leq \|I - \delta B\|^k \|\theta_0 - \theta\|, \quad k = 1, 2, \ldots, \tag{10.2.37}$$

which enjoys the same power-law type convergence rate as that stated in (10.2.29) for the scheme (10.2.28) with a then step-dependent step size, $\delta = \delta_k$.

Exercises

10.2.1 Verify (10.2.5).

10.2.2 Compute the *Hessian*

$$H_F(\theta) \equiv \left(\frac{\partial^2 F}{\partial \theta_i \partial \theta_j} \right) \tag{10.2.38}$$

of the function (10.2.3) and obtain $H_F(\theta) = B$.

10.2.3 Derive (10.2.9) and prove that the limit in (10.2.10) uniquely exists and solves (10.2.6) if the matrix B is positive definite. Show also that, in this situation, the limit in (10.2.10) may be achieved exponentially fast.

10.2.4 Show that the vector c given in (10.2.2) enjoys the expression (10.2.12) for any $b \in \mathbb{R}^n$.

10.2.5 Give an example to show that the iterative scheme (10.2.28) could terminate in finite steps.

10.3 Data classification by separation theorem

Let A and B be two disjoint non-empty subsets of $\mathbb{R}^m$ viewed as two m-feature data sets. A basic problem in data classification is to separate the data sets A and B by a hyperplane of $\mathbb{R}^m$ of the form

$$P_b = \left\{ x = (x_1, \dots, x_m) \in \mathbb{R}^m \,|\, (u, x) = a_1 x_1 + \cdots + a_m x_m = b \right\}, \tag{10.3.1}$$

where $u = (a_1, \dots, a_m) \in \mathbb{R}^m$ is a nonzero vector and $b \in \mathbb{R}$, such that

$$(u, x) > b, \quad x \in A; \quad (u, x) < b, \quad x \in B. \tag{10.3.2}$$

In other words, the hyperplane P_b divides the space $\mathbb{R}^m$ into two half spaces bordered by P_b which contain A and B, respectively. Such a hyperplane P_b, if it exists, is referred to as a *linear classifier* or *separating hyperplane* for the data sets. It is easily observed that a linear classifier may not exist when A and B are not well separated in the sense that their convex hulls overlap. Here recall that some $K \subset \mathbb{R}^m$ is said to be convex if K contains all convex combinations of any finite subset of K and that a convex combination of the vectors $u_1, \dots, u_k$ is a sum of the form $\lambda_1 u_1 + \cdots + \lambda_k u_k$ with $\lambda_1, \dots, \lambda_k \geq 0$ and $\lambda_1 + \cdots + \lambda_k = 1$. The *convex hull* H of a non-empty set $A \subset \mathbb{R}^m$ (say) is defined as

$$H_A = \text{the set of all convex combinations of all finite subets of } A. \tag{10.3.3}$$

It is clear that H_A is the smallest convex subset of $\mathbb{R}^m$ containing A. That is, if K is a convex set containing A, then $H_A \subset K$. Besides, if A is convex, then $H_A = A$, and if A is bounded and closed, or compact, so is H_A.

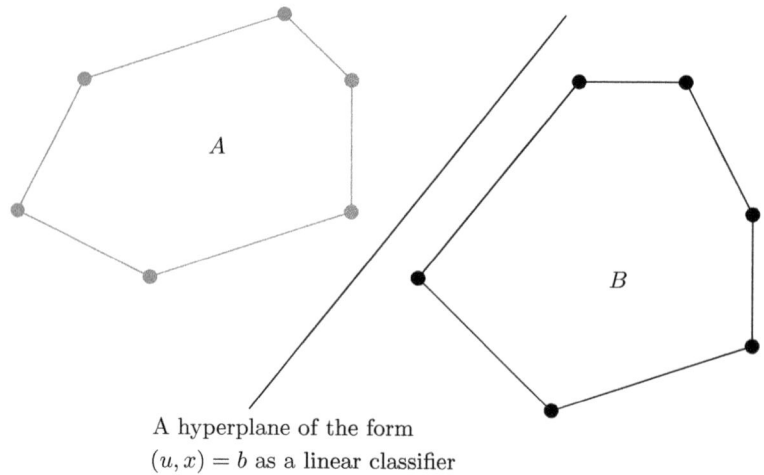

A hyperplane of the form
$(u,x) = b$ as a linear classifier

Figure 10.3 Separation of two non-overlapping compact convex data sets by a hyperplane.

Thus, by taking their convex hulls if necessary, we assume that A and B are compact and convex in our subsequent study. Note that, since data sets in applications are finite sets, the compactness assumption is universally met.

Figure 10.3 illustrates two non-overlapping convex data sets separated by a hyperplane.

Geometrically, when A and B are compact and convex, we may look for some $v \in A$ and $w \in B$ such that

$$\|v - w\| = \text{dist}(A, B) \equiv \inf\{\|x - y\| \mid x \in A, y \in B\}. \tag{10.3.4}$$

Now set $u = v - w$. Then $u \neq 0$ if $A \cap B = \emptyset$. With so constructed vector u and a suitable choice of $b \in \mathbb{R}$, we can show that the hyperplane (10.3.1) fulfills (10.3.2) as a linear classifier. In the following, we realize such a construction.

Motivated by (10.3.4), we consider

$$K_{AB} = \{z \in \mathbb{R}^m \mid z = x - y, x \in A, y \in B\}. \tag{10.3.5}$$

Then K_{AB} is also a compact and convex set. With this construction, we have

$$\text{dist}(A, B) = \inf\{\|z\| \mid z \in K_{AB}\}. \tag{10.3.6}$$

Since K_{AB} is compact, we see that

$$\text{dist}(A, B) \equiv \eta > 0. \tag{10.3.7}$$

Otherwise, if $\eta = 0$, then there is a sequence $\{z_k\}$ in K_{AB} such that $\|z_k\| \to 0$ as $k \to \infty$. This implies $0 \in K_{AB}$ which is in contradiction with the assumption $A \cap B = \emptyset$.

Let $\{z_k\}$ be a minimizing sequence of the problem (10.3.6). By compactness, there is some $u \in K_{AB}$ which is a limiting vector of the sequence $\{z_k\}$ satisfying

$$\|u\| \leq \|u - z_k\| + \|z_k\|, \tag{10.3.8}$$

such that

$$\|u\| \leq \liminf_{k \to \infty} \|u - z_k\| + \lim_{k \to \infty} \|z_k\| = \eta, \tag{10.3.9}$$

which establishes $\|u\| = \eta$ since $u \in K_{AB}$. In particular, we see again that $\eta > 0$ since $u \neq 0$. In fact, we can also show that the convexity of K_{AB} implies that such a vector u is unique. To see this fact, let $\tilde{u} \in K_{AB}$ such that $\|\tilde{u}\| = \eta$. Then

$$\|u + \tilde{u}\| \geq 2\eta, \quad \text{since } \frac{1}{2}(u + \tilde{u}) \in K_{AB}. \tag{10.3.10}$$

On the other hand, in view of the scalar product, we have the identity

$$\|u - \tilde{u}\|^2 + \|u + \tilde{u}\|^2 = 2\|u\|^2 + 2\|\tilde{u}\|^2. \tag{10.3.11}$$

As a consequence of (10.3.10), (10.3.11), and $\|u\| = \|\tilde{u}\| = \eta$, we have $\|u - \tilde{u}\| \leq 0$ or $u = \tilde{u}$ as anticipated.

With this unique $u \in K_{AB}$ and any vector $z \in K_{AB}$, we have

$$\begin{aligned} \eta^2 &\leq \|(1 - \lambda)u + \lambda z\|^2 = \|u + \lambda(z - u)\|^2 \\ &= \|u\|^2 + 2\lambda(u, z - u) + \lambda^2 \|z - u\|^2 \\ &= \eta^2 + 2\lambda(u, z - u) + \lambda^2 \|z - u\|^2, \quad \lambda \in [0, 1]. \end{aligned} \tag{10.3.12}$$

Simplifying (10.3.12), we get

$$0 \leq (u, z) - \eta^2 + \frac{\lambda}{2}\|z - u\|^2, \quad \lambda \in (0, 1]. \tag{10.3.13}$$

Since $\lambda \in (0, 1]$ is arbitrary, we see that (10.3.13) actually implies that $(u, z) \geq \eta^2$ for any $z \in K_{AB}$, or equivalently,

$$(u, x) \geq \eta^2 + (u, y), \quad \text{for any pair of vectors } x \in A, y \in B. \tag{10.3.14}$$

As a consequence of (10.3.14), we see that the quantities

$$b_A \equiv \inf\{(u, x) \mid x \in A\}, \quad b_B \equiv \sup\{(u, x) \mid x \in B\}, \tag{10.3.15}$$

satisfy

$$b_A \geq \eta^2 + b_B. \tag{10.3.16}$$

Moreover, choose $v \in A$ and $w \in B$ such that $u = v - w$. Then

$$\eta^2 = \|u\|^2 = (u,v) - (u,w) \quad \text{or} \quad (u,v) = \eta^2 + (u,w) \qquad (10.3.17)$$

gives us

$$b_A \leq \eta^2 + b_B. \qquad (10.3.18)$$

Combining (10.3.16) and (10.3.18), we obtain the following exact gap width:

$$b_A - b_B = \eta^2, \qquad (10.3.19)$$

between b_A and b_B. This result is important because it allows us to infer from (10.3.17) again that

$$b_A \leq (u,v) = \eta^2 + b_B = b_A, \qquad (10.3.20)$$

which renders the exact results $(u,v) = b_A$ and $(u,w) = b_B$. In other words, we conclude that there hold the inequalities

$$(u,x) \geq b_A, \quad x \in A; \quad (u,x) \leq b_B, \quad x \in B, \qquad (10.3.21)$$

which are saturated by some $v \in A$ and $w \in B$ satisfying $u = v - w$. In particular, for any $b \in (b_B, b_A)$, we have

$$(u,x) \geq b_A > b, \quad x \in A; \quad (u,x) \leq b_B < b, \quad x \in B, \qquad (10.3.22)$$

which establishes (10.3.2) as a by-product.

Geometrically, the vector u is a normal vector to the plane P_b given in (10.3.1) and $u^\perp = P_0$. The line

$$L = \text{Span}\{u\}, \qquad (10.3.23)$$

which is perpendicular to all separating hyperplanes, is called the *separating axis*. Now, with

$$\mathbb{R}^m = L \oplus P_0, \qquad (10.3.24)$$

we see that the projection of $\mathbb{R}^m$ onto L along P_0, or simply the orthogonal projection onto L, is given by

$$x \mapsto \frac{(u,x)}{(u,u)} u = (u_0, x) u_0, \quad x \in \mathbb{R}^m, \qquad (10.3.25)$$

where $u_0 \equiv \frac{u}{\eta}$ is the normalized (unit) basis vector of L such that in terms of the coordinate the orthogonal projection of x onto the axis L reads

$$(u_0, x) = \frac{1}{\eta}(u,x). \qquad (10.3.26)$$

In view of (10.3.15) and (10.3.26), we find

$$(u_0, x) \geq \frac{b_A}{\eta}, \quad x \in A; \quad (u_0, x) \leq \frac{b_B}{\eta}, \quad x \in B, \tag{10.3.27}$$

separating the projections of the data sets A and B on the separating axis, where the bounds are attained at $v \in A$ and $w \in B$ fulfilling $\|v - w\| = \eta$, realizing the distance between A and B.

Summarizing the study, we state the following *data classification* or *separation theorem*.

Theorem 10.2 *Let A and $B \subset \mathbb{R}^m$ be two non-overlapping compact and convex data sets.*

(1) *There are $v \in A$ and $w \in B$ to realize the quantity (10.3.4) so that $u = v - w$ is a unique nonzero vector.*

(2) *For the so-constructed vector u, the quantities b_A and b_B defined in (10.3.15) satisfies the equality*

$$b_A - b_B = \|u\|^2, \tag{10.3.28}$$

and the inequalities

$$(u, x) \geq b_A, x \in A, (u, v) = b_A; (u, x) \leq b_B, x \in B, (u, w) = b_B. \tag{10.3.29}$$

(3) *For any $b \in (b_B, b_A)$, the hyperplane (10.3.1) fulfills (10.3.2), hence providing a linear classifier or separating hyperplane for the data sets A and B.*

(4) *The projections of the data sets A and B onto the separating axis (10.3.23) fall into the intervals*

$$\left[\frac{b_A}{\|u\|}, \infty \right), \quad \left(-\infty, \frac{b_B}{\|u\|} \right], \tag{10.3.30}$$

respectively.

(5) *The separating hyperplane P_b given in (10.3.1) with*

$$b = b_0 \equiv \frac{1}{2}(b_A + b_B) \tag{10.3.31}$$

is equal distant from the hyperplanes with $b = b_A$ and $b = b_B$ satisfying

$$dist(P_{b_0}, P_{b_A}) = dist(P_{b_0}, P_{b_B}) = \frac{1}{2}\|u\| = \frac{1}{2}dist(A, B), \tag{10.3.32}$$

which bisects the gap between the data sets given by

$$G = \{x \in \mathbb{R}^m \mid b_B \leq (u, x) \leq b_A\}. \tag{10.3.33}$$

Proof We only need to prove (5). To this end, we note that $z = \frac{1}{2}(v + w)$ is half-way between $v \in A$ and $w \in B$ satisfying

$$\|v - z\| = \|w - z\| = \frac{1}{2}\|v - w\| = \frac{1}{2}\|u\|, \quad (u, z) = \frac{1}{2}(b_A + b_B) = b_0, \quad (10.3.34)$$

by the facts that $(u, v) = b_A$ and $(u, w) = b_B$. Thus P_{b_0} is the hyperplane that passes through z and bisects the gap G given in (10.3.33) as desired.   □

Theorem 10.2 is a specialized but also a strengthened version of the *Minkowski hyperplane separation theorem* which states that two disjoint convex and open subsets in $\mathbb{R}^m$, or in a normed real vector space of infinite dimension in general, may be separated by a hyperplane represented by a continuous functional, based on the celebrated Hahn–Banach theorem [49, 67, 83].

The hyperplane given in (5) of Theorem 10.2 maximizes its distance from both data sets A and B and is often referred to as the *maximum-margin separating hyperplane* or *maximum-margin linear classifier*. The gap G given by (10.3.33) is also called *margin*. The vectors $v \in A$ and $w \in B$ stated in (1) of Theorem 10.2 are called *support vectors* in the context of *support-vector machine algorithms* in machine learning which deal with data sets consisting of finitely many vectors instead of convex data sets necessarily having infinitely many vectors.

A simple but widely used algorithm in supervised machine learning based on the idea of a linear classifier is called *perceptron* which utilizes the binary-valued function

$$h(x) = \begin{cases} 1, & (u, x) \geq b, \\ 0, & (u, x) < b, \end{cases} \quad x \in \mathbb{R}^m, \quad\quad (10.3.35)$$

to indicate whether or not a certain action is taken, where $u \in \mathbb{R}^m$ is referred to as the weight vector and $b \in \mathbb{R}$ the bias. In this context, the goal of learning is to find the optimal weight vector and bias level based on training data collected to guide future decision for actions executed by an activator, as shown in Figure 10.4.

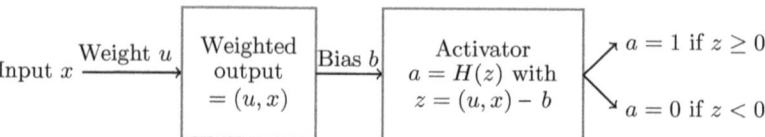

Figure 10.4 A schematic illustration of the working mechanism of a perceptron in terms of the Heaviside step function.

Exercises

10.3.1 Show that the distance from the plane P_b given in (10.3.1) to the origin of $\mathbb{R}^m$ is $\frac{|b|}{\|u\|}$.

10.3.2 Show that the convex hull H_A of a finite subset A in $\mathbb{R}^m$ is compact.

10.3.3 For the convex hull H_A of a subset A in $\mathbb{R}^m$, show that H_A is compact if A is compact, which extends Problem 10.3.2.

10.3.4 Show that the set K_{AB} defined by (10.3.5) is compact and convex if A and B are compact and convex.

10.3.5 Show that, for two compact subsets A and B in $\mathbb{R}^m$, there holds $\text{dist}(A,B) > 0$ if and if $A \cap B = \emptyset$.

10.3.6 Give an example to show that the support vectors v and w that realize a unique vector u as stated in Theorem 10.2 may not be unique in general.

10.3.7 Let $K \subset \mathbb{R}^m$. Show that the ε-neighborhood of K, namely,

$$K_\varepsilon = \{x \in \mathbb{R}^m \,|\, \text{dist}(x,K) < \varepsilon\}, \tag{10.3.36}$$

where $\text{dist}(x,K) = \inf\{\|x-y\| \,|\, y \in K\}$, is the same as the set

$$\{x \in \mathbb{R}^m \,|\, \|x-y\| < \varepsilon \text{ for some } y \in K\}. \tag{10.3.37}$$

Use this result to show that K_ε is convex if K is convex.

10.3.8 Let A and $B \subset \mathbb{R}^m$ be convex, compact, and disjoint. With the notation in Problem 10.3.7, show that A_ε and B_ε are open, convex, and disjoint when $\varepsilon > 0$ is sufficiently small. Use this result and the Minkowski separation theorem to establish the existence of a hyperplane (10.3.1) that separates A and B as described in (10.3.2).

10.3.9 Give an example to show that, if one of the two compact subsets A and B in $\mathbb{R}^m$ is not convex, the minimization problem (10.3.6) still has a solution but the solution may not be unique.

10.4 Logistic regression

Consider two data sets A and B in $\mathbb{R}^m$ which are separated by a hyperplane as described in (10.3.1) and (10.3.2). In practice, the data set A may be associated to the presence of a certain event and B the absence of such an event. In correspondence, we may use another variable y to denote the presence or absence of this event so that $y = 1$ indicates the occurrence of the event and $y = 0$ otherwise. In other words, we have come up with the following training data:

$$y = 1, \quad x \in A; \quad y = 0, \quad x \in B. \tag{10.4.1}$$

In applications, the occurrence of an event may not be certain but rather prob-abilistic. Thus, we may assign a probability $p \in (0,1)$ for the occurrence of the event such that we have the probability distribution for the presence or absence of the event to be, say, $f(1) \equiv P(\{y = 1\}) = p$ and $f(0) \equiv P(\{y = 0\}) = 1 - p$. That is,

$$f(y) = p^y(1-p)^{1-y}, \quad y = 0, 1, \tag{10.4.2}$$

collectively as a function of the variable y. This is known as the *Bernoulli* or *binary distribution*.

Let $x \in \mathbb{R}^m$ be a newly observed feature vector. Our goal now is to use the training data (10.4.1) to estimate the probability p for the occurrence of the event as given in (10.4.2).

To this goal, we note that if x falls near the classifier P_b we are more uncertain whether the event really happens or not, or whether $y = 1$ or $y = 0$. However, if x falls well above P_b or $(u, x) \gg b$, we are more certain that the event occurs. In the same spirit, if x falls well below P_b or $(u, x) \ll b$, we are more certain that the event will not occur. This description naturally prompts us to model the problem with a hypothesized function $h(\tau)$ with

$$\tau = (u, x) - b, \tag{10.4.3}$$

which satisfies the properties

(1) $h(\tau)$ increases for $\tau \in (-\infty, \infty)$.
(2) $h(0) = \frac{1}{2}$.
(3) $h(-\infty) = 0$ and $h(\infty) = 1$.

A function of these properties, as depicted in Figure 10.5, is referred to as a *sigmoid function*, after the S-shapedness of the curve it represents. In machine learning, a popular choice of the function h is the *logistic function*

$$h(\tau) = \frac{1}{1 + e^{-\tau}}, \tag{10.4.4}$$

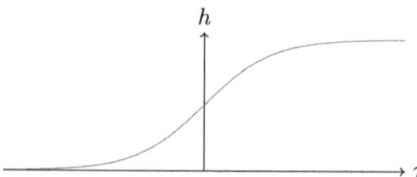

Figure 10.5 The properties of a hypothesized probability distribution function realized by a sigmoid function.

which is a solution of the classical logistic equation modeling population growth and will be used in our subsequent study.

In terms of (10.4.3) and (10.4.4), we can rewrite the probability p in (10.4.2) as

$$p = p(x; u, b) = \frac{1}{1 + e^{-((u,x)-b)}}. \tag{10.4.5}$$

This expression clearly imposes the parameter b as a bias as in Section 10.1. Thus, as done in Section 10.1, we may expand the feature space if necessary to normalize the bias to zero. With such a normalization, (10.4.5) assumes the simplified form

$$p = p(x; u) = \frac{1}{1 + e^{-(u,x)}}. \tag{10.4.6}$$

The expression (10.4.6) predicts the probability of the event in the Bernoulli distribution as a function of the feature vector $x \in \mathbb{R}^m$. In other words, if we use X to denote the random feature vector and Y the random variable in the Bernoulli distribution, then the conditional probability of the event $\{Y = y\}$ given $\{X = x\}$ reads

$$
\begin{aligned}
f(y|x; u) &= p(x; u)^y (1 - p(x; u))^{1-y} \\
&= \frac{e^{-(1-y)(u,x)}}{1 + e^{-(u,x)}}, \quad x \in \mathbb{R}^m, \quad y = 0, 1,
\end{aligned} \tag{10.4.7}
$$

in view of (10.4.2) and (10.4.6).

With the above preparation, we are ready to launch a learning task based on the training data (10.4.1). Specifically, we may rewrite (10.4.1) in the format of (10.1.2) as

$$(A_1, b_1), \ldots, (A_n, b_n), \quad A_j = \begin{pmatrix} a_{1j} \\ \vdots \\ a_{mj} \end{pmatrix} \in \mathbb{R}^m, \quad b_j = 0, 1, \quad j = 1, \ldots, n.$$

$$\tag{10.4.8}$$

As a consequence of (10.4.7), we obtain from the sample set (10.4.8) the likelihood function

$$L(u) = \prod_{j=1}^{n} f(b_j | A_j; u) = \prod_{j=1}^{n} \frac{e^{-(1-b_j)(A_j,u)}}{1 + e^{-(A_j,u)}}. \tag{10.4.9}$$

The parameter vector $u \in \mathbb{R}^m$ that maximizes the likelihood function (10.4.9), or equivalently the logarithmic likelihood function

$$l(u) = \ln L(u) = -\sum_{j=1}^{n} (1 - b_j)(A_j, u) - \sum_{j=1}^{n} \ln \left(1 + e^{-(A_j,u)} \right) \tag{10.4.10}$$

gives us the learned probability function (10.4.6) based on the training data, thereby accomplishing the *logistic regression* task for the problem. This is by nature a nonlinear regression problem.

Exercises

10.4.1 Consider the initial value problem of the logistic differential equation

$$\frac{dh}{d\tau} = rh\left(1 - \frac{h}{C}\right), \quad h(0) = h_0, \tag{10.4.11}$$

where the positive constants r, C, and h_0 are growth rate, carrying capacity, and initial population, of the population of a certain species in evolution, respectively. Find the solution of this problem and determine r, C, and h_0 to recover the logistic regression function (10.4.4).

10.4.2 For the sigmoid function

$$S(\tau) = \frac{1}{(1 + e^{-\tau})^\alpha}, \quad \alpha \geq 1, \tag{10.4.12}$$

derive a first-order differential equation that governs such a function as a solution of the equation and generalizes the logistic equation.

10.4.3 For the logistic function $h(\tau)$ defined in (10.4.4), show that for any $q \in (\frac{1}{2}, 1)$ there holds

$$q < h(\tau) < 1 - q, \quad -\ln\left(\frac{1-q}{q}\right) < \tau < \ln\left(\frac{1-q}{q}\right). \tag{10.4.13}$$

As a consequence, the function $h_a(\tau) = h(a\tau)$ $(a > 0)$ has the property

$$q < h_a(\tau) < 1 - q, \quad -\frac{1}{a}\ln\left(\frac{1-q}{q}\right) < \tau < \frac{1}{a}\ln\left(\frac{1-q}{q}\right), \tag{10.4.14}$$

indicating that a larger scale factor a gives rise to a sharper transition of the function $h_a(\tau)$ between its two limiting values 0 and 1.

10.4.4 In the situation of a Bernoulli distribution (10.4.2), the *odds* is defined to be the ratio of p over $1 - p$ which is the expected ratio of the number of the occurrences of the event of probability p against the number of nonoccurrences of the event in a repeated collection of independent observations. Show that the model (10.4.4) arises from taking τ to measure the logarithm of the odds. That is,

$$\ln\frac{p}{1-p} = \tau. \tag{10.4.15}$$

10.4.5 Use the rational sigmoid function

$$R(\tau) = \frac{1}{2}\left(1 + \frac{\tau}{\sqrt{1+\tau^2}}\right), \quad -\infty < \tau < \infty, \qquad (10.4.16)$$

to model the probability p in the Bernoulli distribution such that (10.4.6) is replaced by $p(x;u) = R((u,x))$ and obtain the associated logarithmic likelihood function for the training data (10.4.8).

10.5 Solution and nonsolution to the logistic regression problem

In this section, we are to obtain a parameter vector $u \in \mathbb{R}^m$ that maximizes the logarithmic likelihood function (10.4.10). For convenience, we consider the equivalent minimization problem

$$\eta \equiv \inf\{F(u) \mid u \in \mathbb{R}^m\}, \qquad (10.5.1)$$

where

$$F(u) = -l(u)$$

$$= \sum_{j=1}^{n}(1-b_j)(A_j,u) + \sum_{j=1}^{n}\ln\left(1+e^{-(A_j,u)}\right), u = \begin{pmatrix} a_1 \\ \vdots \\ a_m \end{pmatrix} \in \mathbb{R}^m.$$

$$(10.5.2)$$

Using the property

$$s + \ln(1+e^{-s}) = \ln(1+e^s) > 0, \quad s \in \mathbb{R}, \qquad (10.5.3)$$

we see that the quantity η in (10.5.1) is well defined and $\eta \geq 0$. Besides, with (10.4.8), we have

$$\frac{\partial F}{\partial a_i} = \sum_{j=1}^{n}(1-b_j)a_{ij} - \sum_{j=1}^{n}\frac{a_{ij}e^{-(A_j,u)}}{1+e^{-(A_j,u)}}, \quad i=1,\ldots,m, \qquad (10.5.4)$$

$$\frac{\partial^2 F}{\partial a_i \partial a_k} = \sum_{j=1}^{n}\frac{a_{ij}a_{kj}e^{-(A_j,u)}}{(1+e^{-(A_j,u)})^2}, \quad i,k=1,\ldots,m. \qquad (10.5.5)$$

On the other hand, for given $u \in \mathbb{R}^m$, we can show that the Hessian matrix

$$H_F(u) = \left(\frac{\partial^2 F}{\partial a_i \partial a_k}\right) \qquad (10.5.6)$$

is positive semi-definite. For this purpose, let $\xi \in \mathbb{R}^m$ be a testing vector. Then (10.5.5) gives us the result

$$\xi^t H_F(u)\xi = \sum_{j=1}^n \frac{e^{-(A_j,u)}}{(1+e^{-(A_j,u)})^2}((A_j,\xi))^2, \qquad (10.5.7)$$

which is seen to be positive semi-definite. Thus, if the function F has a critical point, that is, a vector $u_0 \in \mathbb{R}^m$ satisfying

$$\nabla_u F(u)|_{u=u_0} = \begin{pmatrix} \frac{\partial F}{\partial a_1} \\ \vdots \\ \frac{\partial F}{\partial a_m} \end{pmatrix}_{u=u_0} = 0, \qquad (10.5.8)$$

then u_0 is a global minimizer of F which solves the problem (10.5.1).

In fact, let u_0 be a critical point satisfying (10.5.8) and $v \in \mathbb{R}^m$ be arbitrary. Set

$$u(\tau) = u_0 + \tau(v - u_0), \quad \tau \in [0,1]. \qquad (10.5.9)$$

Then, with $g(\tau) = F(u(\tau))$, we have

$$g(1) = g(0) + g'(0) + \frac{1}{2}g''(\tau), \qquad (10.5.10)$$

for some $\tau \in [0,1]$. However, in view of the chain rule for derivatives, we have

$$g'(\tau) = (\nabla_u F(u))^t_{u=u(\tau)}(v - u_0), \qquad (10.5.11)$$
$$g''(\tau) = (v - u_0)^t (H_F(u))_{u=u(\tau)}(v - u_0). \qquad (10.5.12)$$

From (10.5.8) and (10.5.11), we have $g'(0) = 0$; from the positive semi-definiteness of (10.5.6) and (10.5.12), we have $g''(\tau) \geq 0$. Substituting these results into (10.5.10), we obtain $F(v) = g(1) \geq g(0) = F(u_0)$ as asserted. In particular, $F(u_0) = \eta$. That is, u_0 solves (10.5.1).

Therefore, we conclude that whether (10.5.1) has a solution is equivalent to whether the function $F(u)$ given in (10.5.2) has a critical point. From (10.5.4) and (10.5.8), we see that this latter question amounts to solving the system of nonlinear equations

$$\sum_{j=1}^n (1-b_j)a_{ij} - \sum_{j=1}^n \frac{a_{ij}e^{-(A_j,u)}}{1+e^{-(A_j,u)}} = 0, \quad i = 1,\ldots,m, \quad u \in \mathbb{R}^m, \qquad (10.5.13)$$

whose matrix form reads

$$A\sigma(u) = A\beta, \qquad (10.5.14)$$

where

$$A = (a_{ij}) = (A_1, \ldots, A_n), \quad \sigma(u) = \begin{pmatrix} \sigma_1(u) \\ \vdots \\ \sigma_n(u) \end{pmatrix}, \quad \beta = \begin{pmatrix} \beta_1 \\ \vdots \\ \beta_n \end{pmatrix}, \quad (10.5.15)$$

with

$$\sigma_j(u) = \frac{e^{-(A_j, u)}}{1 + e^{-(A_j, u)}}, \quad \beta_j = 1 - b_j, \quad j = 1, \ldots, n. \quad (10.5.16)$$

Since $\sigma_j \in (0, 1)$ and $\beta_j = 0, 1$ for all $j = 1, \ldots, n$, it is clear that the equation (10.5.14) may fail to be consistent in general situations. In other words, a solution to this equation may not exist without imposing further conditions in the setup of the problem. In other words, some training data may render the logistic regression problem *unsolvable*.

For example, if $m = n$ and the feature matrix A given in (10.5.15) is non-singular, then (10.5.14) becomes $\sigma(u) = \beta$ which is impossible since $\beta_j = 0, 1$ and $\sigma_j(u) \in (0, 1)$ for $j = 1, \ldots, n$. Recall that the set of non-singular matrices in $\mathbb{R}(m, m)$ is open and dense in $\mathbb{R}(m, m)$. Hence in the $m = n$ situation the logistic regression problem (10.5.1) is generically unsolvable.

If the data matrix A in (10.5.15) is of rank m, then $\text{Span}\{A_1, \ldots, A_n\} = \mathbb{R}^m$. Therefore $\xi^t H_F(u)\xi > 0$ for any nonzero $\xi \in \mathbb{R}^m$. Thus, if $u_0 \in \mathbb{R}^m$ is a critical point of $F(u)$, then (10.5.10)–(10.5.12) indicate that $F(u_0) < F(v)$ for any $v \in \mathbb{R}^m$ with $v \neq u_0$. In other words, if a critical point exists now, it is unique. For $n \geq m$, since the subset of matrices in $\mathbb{R}(m, n)$ being of rank m is open and dense in $\mathbb{R}(m, n)$, we conclude that data matrices permitting a unique solution or no solution to the problem (10.5.1) are generic in $\mathbb{R}(m, n)$ when $n \geq m$.

In view of the notation (10.5.15) and (10.5.4), we have

$$\nabla_u F(u) = A\beta - A\sigma(u). \quad (10.5.17)$$

Thus, as in Section 10.2, we see that the idea of the gradient descent leads us to the iterative scheme

$$u_{k+1} = u_k - \delta(\nabla_u F(u))_{u=u_k} = u_k - \delta A(\beta - \sigma(u_k)), \quad k = 1, 2, \ldots, \quad (10.5.18)$$

in the space $\mathbb{R}^m$, where $\delta > 0$ is the learning rate and u_1 an initial state. Of course, the sequence $\{u_k\}$ may not converge because (10.5.14) may not have a solution. If it does converge, then it converges to a solution to (10.5.14).

Conversely, assume that (10.5.14) has a solution, say u_0. Then $F(u_0) = \eta$ and we show that the sequence $\{F(u_k)\}$ monotonically converges to $F(u_0)$ when the learning rate δ in the gradient descent scheme (10.5.18) is suitably chosen.

To such a goal, we take two trial vectors $u, v \in \mathbb{R}^m$ and consider

$$w(\tau) = u + \tau(v - u), \quad \tau \in [0, 1]. \tag{10.5.19}$$

Then, similar to the formulation consisting of (10.5.9)–(10.5.12), we now have

$$F(v) = F(u) + (\nabla_u F(u))^t (v - u) + \frac{1}{2}(v - u)^t H_F(w(\tau))(v - u) \tag{10.5.20}$$

for some $\tau \in [0, 1]$. From (10.5.5), we see that there is some constant $K_0 > 0$ such that it bounds all the non-negative eigenvalues of $H_F(w(\tau))$. As a consequence, we have

$$0 \le (v - u)^t H_F(w(\tau))(v - u) \le K_0 \|v - u\|^2. \tag{10.5.21}$$

Inserting (10.5.21) into (10.5.20), we have

$$F(v) \le F(u) + (\nabla_u F(u))^t (v - u) + \frac{K_0}{2} \|v - u\|^2, \quad u, v \in \mathbb{R}^m. \tag{10.5.22}$$

With (10.5.22), we can estimate the step size δ in (10.5.18) such that the iteration enables a descending sequence as expected. To do so, we let u and v be related by (10.5.18). That is,

$$v = u - \delta \nabla_u F(u), \tag{10.5.23}$$

with $\delta > 0$ to be determined. Using (10.5.23) or

$$\nabla_u F(u) = \frac{1}{\delta}(u - v) \tag{10.5.24}$$

in (10.5.22), we get

$$F(v) \le F(u) - q(\delta) \|\nabla_u F(u)\|^2, \quad q(\delta) = \delta - \frac{K_0 \delta^2}{2}. \tag{10.5.25}$$

The maximum of $q(\delta)$ is at

$$\delta_0 = \frac{1}{K_0}, \tag{10.5.26}$$

so that $q(\delta_0) = \frac{1}{2K_0}$, with which the one-step descent (10.5.23) achieves the *greatest possible drop* of the objective function from the value $F(u)$ provided that $\nabla_u F(u) \ne 0$, although a drop is ensured with any δ below $\frac{2}{K_0}$ with which $q(\delta) > 0$. Therefore, we arrive at the choice

$$0 < \delta \le \delta_0 = \frac{1}{K_0} \tag{10.5.27}$$

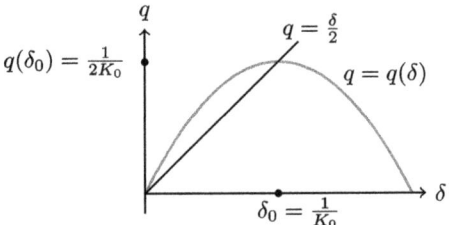

Figure 10.6 The quadratic quantity $q(\delta)$ defined in (10.5.25) stays above the line of slope $\frac{1}{2}$ in the interval $0 < \delta < \delta_0 = \frac{1}{K_0}$.

for the step size. This condition will be observed subsequently which leads to $q(\delta) \geq \frac{\delta}{2}$ as depicted in Figure 10.6 such that (10.5.25) reads

$$F(v) \leq F(u) - \frac{\delta}{2}\|\nabla_u F(u)\|^2$$
$$= F(u) - \frac{1}{2\delta}\|v - u\|^2, \qquad (10.5.28)$$

in view of (10.5.23) or (10.5.24) again. In particular, we obtain the desired monotonicity $F(v) \leq F(u)$.

We next compare $F(u)$ where $u \in \mathbb{R}^m$ is arbitrary with the minimum value of the objective function at u_0 which is assumed to exist. To this end, substituting v by u_0 in (10.5.20) and using the positive semi-definiteness of the Hessian matrix $H_F(w)$ as seen in (10.5.7), we find

$$F(u_0) \geq F(u) + (\nabla_u F(u))^t (u_0 - u). \qquad (10.5.29)$$

Applying (10.5.29) in (10.5.28), we have

$$F(v) \leq F(u_0) + (\nabla_u F(u))^t (u - u_0) - \frac{1}{2\delta}\|v - u\|^2$$
$$= F(u_0) - \frac{1}{2\delta}\left(2(v - u)^t(u - u_0) + \|v - u\|^2\right)$$
$$= F(u_0) + \frac{1}{2\delta}\left(\|u - u_0\|^2 - \|v - u_0\|^2\right), \qquad (10.5.30)$$

where we have used (10.5.24) again.

Then we set $u = u_{k-1}$ and $v = u_k$ to be the two adjacent vectors in the iterative scheme with $k = 2, 3, \ldots$. As a consequence, we have

$$F(u_1) \geq F(u_2) \geq \cdots \geq F(u_k) \geq \cdots, \qquad (10.5.31)$$

and

$$F(u_2) \leq F(u_0) + \frac{1}{2\delta} \left(\|u_1 - u_0\|^2 - \|u_2 - u_0\|^2 \right), \qquad (10.5.32)$$

$$F(u_3) \leq F(u_0) + \frac{1}{2\delta} \left(\|u_2 - u_0\|^2 - \|u_3 - u_0\|^2 \right), \qquad (10.5.33)$$

$$\vdots$$

$$F(u_k) \leq F(u_0) + \frac{1}{2\delta} \left(\|u_{k-1} - u_0\|^2 - \|u_k - u_0\|^2 \right). \qquad (10.5.34)$$

Adding (10.5.32)–(10.5.34) and using (10.5.31), we get

$$F(u_k) \leq F(u_0) + \frac{\|u_1 - u_0\|^2}{2\delta(k-1)}, \quad k = 2, 3, \ldots, \qquad (10.5.35)$$

thereby establishing the monotone convergence and the convergence rate in terms of the values of the objective function for the gradient descent method (10.5.18).

The convergence of the iterative sequence $\{u_k\}$ itself, however, is usually not ensured without imposing some further restrictive conditions on $F(u)$. A simple but convenient such condition is to assume that the lowest eigenvalue of the Hessian defined in general by (10.5.6) is bounded away uniformly by a positive constant λ_0, say. As a result, there holds

$$\xi^t H_F(u)\xi \geq \lambda_0 \|\xi\|^2, \quad \xi \in \mathbb{R}^m. \qquad (10.5.36)$$

A function over $\mathbb{R}^m$ satisfying (10.5.36) is called *strongly convex* or *elliptic*. Hence, if $F(u)$ is strongly convex satisfying (10.5.36), we set $v = u_k$ and $u = u_0$ in (10.5.20) where $\{u_k\}$ is the gradient descent sequence and u_0 is the unique minimizer of the optimization problem, respectively, under discussion, and get

$$F(u_k) \geq F(u_0) + \frac{\lambda_0}{2} \|u_k - u_0\|^2, \qquad (10.5.37)$$

in view of (10.5.36). Combining (10.5.35) and (10.5.37), we obtain

$$\|u_k - u_0\|^2 \leq \frac{\|u_1 - u_0\|^2}{\delta \lambda_0 (k-1)}, \quad k = 2, 3, \ldots. \qquad (10.5.38)$$

In summary of our discussion, we state the following theorem.

Theorem 10.3 *Let $F : \mathbb{R}^m \to \mathbb{R}$ be a twice continuously differentiable function so that its Hessian matrix H_F is positive semi-definite. If $u_0 \in \mathbb{R}^m$ is a critical point of F, then it is a global minimizer of F, satisfying $F(u_0) \leq F(u)$ for any $u \in \mathbb{R}^m$. Furthermore, if the eigenvalues of H_F are bounded from above by constant $K_0 > 0$, then the gradient descent iteration $u_{k+1} = u_k - \delta (\nabla_u F(u))_{u=u_k}$*

*gives rise to a monotone decreasing sequence of the values of the func-
tion F, {F(u_k)}, starting from an arbitrary initial vector $u_1 \in \mathbb{R}^m$, such that
$F(u_k) \to F(u_0)$ as $k \to \infty$, provided that $\delta \in (0, \frac{1}{K_0}]$. Besides, there also holds
the convergence estimate*

$$F(u_0) \leq F(u_k) \leq F(u_0) + \frac{\|u_1 - u_0\|^2}{2\delta(k-1)}, \quad k = 2, 3, \ldots. \tag{10.5.39}$$

*Moreover, if the eigenvalues of H_F are also bounded from below by a constant
$\lambda_0 > 0$ so that F is strongly convex, then the critical point u_0 of F is unique
and the iterative sequence $\{u_k\}$ converges to u_0 following the error estimate
(10.5.38).*

In contrast, the objective function F for logistic regression given in (10.5.2)
is not strongly convex. To overcome this difficulty, we may consider the
following *regularized objective function*:

$$F_\varepsilon(u) = F(u) + \frac{\varepsilon}{2}(u, u), \tag{10.5.40}$$

where ε is a positive constant. This regularized objective function satisfies
all the conditions stated in Theorem 10.3. The method of using a regularized objective function in optimization is referred to as *regularization*, which
is a convenient and popular method in machine learning practice, although
there lacks a suitable statistical interpretation in this context. The ε-augmented
term in (10.5.40) is also called a *penalty term* and the regularized regression
approach *ridge regression* which makes the underlying learning parameters
more manageable [4].

For a comprehensive coverage of the subject of convex optimization, see
[16], for example.

We have now discussed several fundamental machine learning problems.
These problems are all one-layer learning problems. In applications, algorithms involving multi-layer learning, or *deep learning algorithms*, are often
formulated and implemented.

For a systematic presentation of the topics of machine learning under the
general subject of artificial intelligence, see [66].

Exercises

10.5.1 Establish (10.5.7).

10.5.2 Verify (10.5.12).

10.5.3 Show directly that, when $m = n = 1$, the function (10.5.2) in the nontrivial situation,

$$F(u) = a(1-b)u + \ln(1 + e^{-au}), \quad a, u \in \mathbb{R}, \quad a \neq 0, \quad b = 0, 1, \tag{10.5.41}$$

has no global minimizer solving (10.5.1).

10.5.4 When $m = n = 1$, show that the iterative scheme (10.5.18) in the nontrivial situation will never converge regardless how small the learning rate δ is taken.

10.5.5 For the function $F(u)$ defined in (10.5.41), show that a solution to the negative gradient flow equation

$$\frac{du}{d\tau} = -F'(u) \tag{10.5.42}$$

does not have a finite limit as $\tau \to \infty$.

10.5.6 Show that, with the regularization (10.5.40), the constants K_0 and λ_0, given in (10.5.21) and (10.5.36), respectively, are both shifted upward by ε.

10.5.7 Apply Theorem 10.3 to obtain the convergence of a gradient descent sequence to solve the linear regression problem (10.1.12) and consider when a regularization approach needs to be called upon.

11

Excursion: Problems in Data Analysis

In this chapter, we present two important and related problems in data analysis: the low-rank approximation and principal component analysis (PCA), both based on singular value decomposition. First, we consider the low-rank approximation problem for mappings between two vector spaces. Next, we focus on the low-rank approximation problem for matrices in both induced norm and the Frobenius norm, which are of independent interest for applications. Then, we consider PCA. These results are also useful in machine learning. Furthermore, as an extension of the ideas and methods, we present a study of some related matrix nearness problems.

11.1 Low-rank approximation problem for general mappings

Besides its obvious purely mathematical interest, the low-rank approximation problem is important in data analysis for several reasons explained below.

Dimensionality reduction: In many data analysis tasks, data sets may be high-dimensional, containing a large number of features or variables. High dimensionality can lead to computational complexity, overfitting, and difficulty in visualization. Low-rank approximation techniques reduce the dimensionality of the data by approximating it with a lower-dimensional subspace while preserving most of its variance. This simplifies the data representation, makes it easier to analyze and visualize, and reduces computational complexity.

Compression: Low-rank approximation can be used for data compression, where the goal is to represent the data using fewer bits or storage space while minimizing information loss. By approximating the original data matrix with a lower-rank approximation, redundant or irrelevant information can be discarded, resulting in a more compact representation of the data. This is particularly useful for large-scale data sets, where storage and computational resources may be limited.

Noise reduction: Low-rank approximation techniques can help reduce noise and filter out irrelevant information from the data. Since low-rank approximations capture the dominant patterns and structures in the data, they tend to suppress noise and amplify signal, leading to cleaner and more reliable data representations. This is especially beneficial in applications where the data is noisy or corrupted, such as that of sensor data, imaging, and signal processing.

Feature extraction: Low-rank approximation methods extract the most important features or components from the data, which can be used for subsequent analysis or modeling tasks. For example, in image processing, low-rank approximations can extract principal components representing dominant patterns or structures in the image, which can then be used for tasks such as object recognition, image classification, and image retrieval.

For convenience, we begin with a brief review of the construction of singular value decomposition in our context.

Let U, V be two vector spaces over $\mathbb{R}$ or $\mathbb{C}$ equipped with positive definite scalar products, both denoted by the same notation $(\,,\,)$. Assume $\dim(U) = n \geq 1, \dim(V) = m \geq 1$, and $T \in L(U, V)$. Recall that, in view of the construction of singular value decomposition, we may choose an orthonormal basis of U consisting of eigenvectors of $T' \circ T \in L(U)$, say $\{u_1, \ldots, u_k, \ldots, u_n\}$, associated with the eigenvalues

$$\sigma_1 \geq \cdots \geq \sigma_k > 0; \quad \sigma_i = 0, \quad k < i \leq n \quad \text{if any,} \qquad (11.1.1)$$

so that

$$T(u_1) = \sqrt{\sigma_1} v_1, \ldots, T(u_k) = \sqrt{\sigma_k} v_k; \quad T(u_i) = 0, \quad k < i \leq n \quad \text{if any,} \qquad (11.1.2)$$

and $\{v_1, \ldots, v_k\}$ is an orthonormal basis of $R(T)$, which is a subset of an orthonormal basis of V, say $\{v_1, \ldots, v_k, \ldots, v_m\}$. In particular, $r(T) = k$.

In order to make our discussion more transparent and motivated, we consider the simpler case $r(T) = n$ first and study the following low-rank optimization problem:

$$\eta_n \equiv \inf\{\|T - S\| \mid S \in L(U, V), r(S) \leq n - 1\}, \qquad (11.1.3)$$

where $\|\cdot\|$ is the induced norm, also called the spectral norm, or operator norm in functional analysis, which for a mapping $K \in L(U, V)$ is given by

$$\|K\| = \sup\{\|Ku\| \mid u \in U, \|u\| = 1\}. \qquad (11.1.4)$$

To find an estimate for the quantity η_n, we note from (11.1.2) with $k = n$ that we may take the rank $(n-1)$ mapping $T_{n-1} \in L(U,V)$ defined by

$$T_{n-1}(u_1) = \sqrt{\sigma_1}v_1, \dots, T_{n-1}(u_{n-1}) = \sqrt{\sigma_{n-1}}v_{n-1}; \quad T_{n-1}(u_n) = 0. \quad (11.1.5)$$

Hence we have

$$(T - T_{n-1})(u_i) = 0, \quad i = 1,\dots,n-1; \quad (T - T_{n-1})(u_n) = \sqrt{\sigma_n}v_n. \quad (11.1.6)$$

Consequently, for any $u \in U$, $\|u\| = 1$, we see that, with

$$u = \sum_{i=1}^{n} a_i u_i, \quad (a_i) \in \mathbb{C}^n, \quad \sum_{i=1}^{n} |a_i|^2 = 1, \quad (11.1.7)$$

we have

$$\|(T - T_{n-1})(u)\| = \sqrt{\sigma_n}|a_n| \leq \sqrt{\sigma_n}, \quad (11.1.8)$$

which establishes $\|T - T_{n-1}\| \leq \sqrt{\sigma_n}$ by (11.1.4). Besides, we also have the result $\|(T - T_{n-1})(u_n)\| = \sqrt{\sigma_n}$. Hence $\|T - T_{n-1}\| \geq \sqrt{\sigma_n}$ by (11.1.4). So

$$\|T - T_{n-1}\| = \sqrt{\sigma_n}. \quad (11.1.9)$$

We next show that T_{n-1} solves the problem (11.1.3) such that $\eta_n = \sqrt{\sigma_n}$.

In fact, let $S \in L(U,V)$ satisfy $r(S) \leq n-1$. Then $n(S) \geq 1$ by the rank-nullity equation, $r(S) + n(S) = n$. Hence, choose $u \in N(S)$, $\|u\| = 1$. We have in view of (11.1.7) the result

$$\|T - S\| \geq \|(T - S)(u)\| = \|T(u)\|$$
$$= \left(\sum_{i=1}^{n} |a_i|^2 \sigma_i \right)^{\frac{1}{2}} \geq \sqrt{\sigma_n}, \quad (11.1.10)$$

applying (11.1.1) and (11.1.2) with $k = n$. From (11.1.9) and (11.1.10), we see that T_{n-1} indeed solves the problem (11.1.3).

We then turn to the general case, for an integer l satisfying $1 \leq l \leq k = r(T) \leq n$, and consider the following l-dependent optimization problem:

$$\eta_l \equiv \inf\{\|T - S\| \mid S \in L(U,V), r(S) \leq l - 1\}. \quad (11.1.11)$$

To avoid triviality, we assume $l \geq 2$ in our subsequent discussion. Suggested by (11.1.5), we define

$$T_{l-1}(u_1) = \sqrt{\sigma_1}v_1, \dots, T_{l-1}(u_{l-1}) = \sqrt{\sigma_{l-1}}v_{l-1};$$
$$T_{l-1}(u_i) = 0, \quad i = l,\dots,n. \quad (11.1.12)$$

Hence, replacing (11.1.6), we have

$$
\begin{aligned}
(T - T_{l-1})(u_i) &= 0, \quad i = 1, \ldots, l-1; \\
(T - T_{l-1})(u_i) &= \sqrt{\sigma_i} v_i, \quad i = l, \ldots, k; \\
(T - T_{l-1})(u_i) &= 0 \text{ otherwise.}
\end{aligned}
\tag{11.1.13}
$$

As a consequence of (11.1.7) and (11.1.13), we have

$$
\|(T - T_{l-1})(u)\| = \left(\sum_{i=l}^{k} |a_i|^2 \sigma_i \right)^{\frac{1}{2}} \le \sqrt{\sigma_l},
\tag{11.1.14}
$$

or $\|T - T_{l-1}\| \le \sqrt{\sigma_l}$. Besides, we also have $\|(T - T_{l-1})(u_l)\| = \sqrt{\sigma_l}$. So $\|T - T_{l-1}\| \ge \sqrt{\sigma_l}$. These establish

$$
\|T - T_{l-1}\| = \sqrt{\sigma_l},
\tag{11.1.15}
$$

which generalizes (11.1.9).

We now show that T_{l-1} solves (11.1.11). To this end, we take $S \in L(U, V)$ satisfying $r(S) \le l - 1$. Hence $n(S) \ge n - (l - 1)$ such that

$$
W \equiv N(S) \cap \mathrm{Span}\{u_1, \ldots, u_l\} \ne \{0\}.
\tag{11.1.16}
$$

In fact, from the general dimensionality equation

$$
\dim\left(N(S) + \mathrm{Span}\{u_1, \ldots, u_l\}\right) = n(S) + l - \dim(W),
\tag{11.1.17}
$$

we obtain the lower bound

$$
\begin{aligned}
\dim(W) &= n(S) + l - \dim\left(N(S) + \mathrm{Span}\{u_1, \ldots, u_l\}\right) \\
&\ge n - (l-1) + l - n = 1.
\end{aligned}
\tag{11.1.18}
$$

Choosing $u \in W$ with $\|u\| = 1$, we have

$$
S(u) = 0; \quad u = \sum_{i=1}^{l} a_i u_i, \quad (a_i) \in \mathbb{C}^l, \quad \sum_{i=1}^{l} |a_i|^2 = 1.
\tag{11.1.19}
$$

With (11.1.19), we obtain

$$
\begin{aligned}
\|T - S\| &\ge \|(T - S)(u)\| = \|T(u)\| \\
&= \left(\sum_{i=1}^{l} |a_i|^2 \sigma_i \right)^{\frac{1}{2}} \ge \sqrt{\sigma_l},
\end{aligned}
\tag{11.1.20}
$$

similar to (11.1.10). Consequently, we have shown that $\eta_l = \sqrt{\sigma_l}$ in (11.1.11) and T_{l-1} defined in (11.1.12) solves (11.1.11).

In summary, we state:

Theorem 11.1 *Let U and V be finite-dimensional vector spaces over* $\mathbb{R}$ *or* $\mathbb{C}$ *equipped with positive definite scalar products. For any* $T \in L(U,V)$ *with* $r(T) = k \geq 1$, *use* $\sigma_1 \geq \cdots \geq \sigma_k > 0$ *to denote the ordered eigenvalues of the positive semi-definite mapping* $T' \circ T \in L(U)$ *and* $\{u_1, \ldots, u_k, \ldots, u_n\}$ *a corresponding set of eigenvectors of* $T' \circ T$, *which form an orthonormal basis of* U, *associated with the respective positive and zero eigenvalues (if any). Then, for any* $l \leq k$, *the minimization problem (11.1.11) has a solution* T_{l-1} *given by (11.1.12) and the minimum distance* η_l *defined in (11.1.11) is simply* $\sqrt{\sigma_l}$.

This is the *low-rank approximation theorem* obtained conveniently in terms of the induced or spectral norm associated with the scalar products of the spaces involved.

Exercises

11.1.1 Let U be a vector space with a positive definite scalar product and $\dim(U) = n \geq 1$. Assume that $T \in L(U)$ is invertible such that the positive eigenvalues of $T' \circ T$ are $\sigma_1, \ldots, \sigma_n$, in descending order. Show that if $S \in L(U)$ satisfies

$$\|T - S\| < \sqrt{\sigma_n}, \tag{11.1.21}$$

where $\|\cdot\|$ denotes the induced norm equipped for $L(U)$, then S is also invertible. This results provides an alternative proof that, if $r(S) < n$, then $\|T - S\| \geq \sqrt{\sigma_n}$.

11.1.2 For $T \in L(U,V)$ with $r(T) = n = \dim(U)$ and $\dim(V) = m$ (hence $m \geq n$), let the singular values of T be $\lambda_1 \geq \cdots \geq \lambda_n > 0$, which gives rise to a singular value decomposition for T: $T(u_i) = \lambda_i v_i, i = 1, \ldots, n$, where $\{u_1, \ldots, u_n\}$ and $\{v_1, \ldots, v_n, \ldots, v_m\}$ are orthonormal bases of U and V, respectively. For any $j = 1, \ldots, n$, define a rank $(n-1)$ mapping $S \in L(U,V)$ by setting

$$S(u_j) = 0; \quad S(u_i) = \lambda_i v_i, \quad i \neq j, \quad 1 \leq i \leq n. \tag{11.1.22}$$

Find j that minimizes $\|T - S\|$ for S defined by (11.1.22).

11.1.3 With the notation of Problem 11.1.2, assume $r(T) = k \geq 2$. For any $l \leq k$, formulate and solve the $r(S) = l - 1$ version of Problem 11.1.2.

11.2 Low-rank approximation for matrices

For applications, it will be interesting to elaborate on the corresponding *matrix version of the low-rank approximation theorem.*

Let $A \in \mathbb{C}(m,n)$. Assume $r(A) = k \geq 1$ and use $u_1, \ldots, u_k, \ldots, u_n$ to denote the eigenvectors of the positive semi-definite matrix $A^\dagger A \in \mathbb{C}(n,n)$ associated with positive eigenvalues $\sigma_1 \geq \cdots \geq \sigma_k$, in descending order, and zero eigenvalue (if any). With the induced matrix norm

$$\|K\| = \sup\{\|Ku\| \,|\, u \in \mathbb{C}^n, \|u\| = 1\}, \quad K \in \mathbb{C}(m,n), \tag{11.2.1}$$

consider the optimization problem for $l \leq k$:

$$\delta_l \equiv \inf\{\|A - B\| \,|\, B \in \mathbb{C}(m,n), r(B) \leq l - 1\}. \tag{11.2.2}$$

Obviously, in view of the earlier study, we should make the identification

$$U = \mathbb{C}^n, \quad V = \mathbb{C}^m, \quad T(u) = Au, \quad S(u) = Bu, \quad u \in \mathbb{C}^n. \tag{11.2.3}$$

In view of (11.1.12), if the mapping T_{l-1} is represented by some $\mathcal{A}_{l-1} \in \mathbb{C}(m,n)$ such that

$$T_{l-1}(u) = \mathcal{A}_{l-1}u, \quad u \in \mathbb{C}^n, \tag{11.2.4}$$

then we have

$$\mathcal{A}_{l-1}u_1 = \sqrt{\sigma_1}v_1, \ldots, \mathcal{A}_{l-1}u_{l-1} = \sqrt{\sigma_{l-1}}v_{l-1};$$
$$\mathcal{A}_{l-1}u_i = 0, \quad i = l, \ldots, n, \tag{11.2.5}$$

by (11.1.12), so that $\mathcal{A}_k = A$. In matrix form, these relations may be suppressed as

$$\mathcal{A}_{l-1}(u_1, \ldots, u_{l-1}, u_l, \ldots, u_n) = (v_1, \ldots, v_{l-1}, v_l, \ldots, v_m) \begin{pmatrix} D_{l-1} & 0 \\ 0 & 0 \end{pmatrix}, \tag{11.2.6}$$

where $v_l, \ldots, v_m$ are arbitrarily picked unit vectors in $\mathbb{C}^m$ such that the set of vectors $\{v_1, \ldots, v_{l-1}, v_l, \ldots, v_m\}$ forms an orthonormal basis for $\mathbb{C}^m$ and the second matrix factor on the right-hand side of (11.2.6) lies in $\mathbb{C}(m,n)$ in which $D_{l-1} = \text{diag}\{\sqrt{\sigma_1}, \ldots, \sqrt{\sigma_{l-1}}\}$. Resolving (11.2.6), we get

$$\mathcal{A}_{l-1} = (v_1, \ldots, v_{l-1}, v_l, \ldots, v_m) \begin{pmatrix} D_{l-1} & 0 \\ 0 & 0 \end{pmatrix} (u_1, \ldots, u_{l-1}, u_l, \ldots, u_n)^\dagger$$
$$= \sqrt{\sigma_1}v_1 u_1^\dagger + \cdots + \sqrt{\sigma_{l-1}}v_{l-1}u_{l-1}^\dagger. \tag{11.2.7}$$

This matrix, $\mathcal{A}_{l-1}$, solves (11.2.2) with $\delta_l = \sqrt{\sigma_l}$.

Although the induced matrix norm (11.2.1) is handy for theoretical formalism, it is more computationally convenient in applications to use the Hilbert–Schmidt norm, also popularly referred to as the *Frobenius norm*, over the space $\mathbb{C}(m,n)$, associated with the Hilbert–Schmidt scalar product

$$(A,B) = \mathrm{Tr}(A^\dagger B), \quad A,B \in \mathbb{C}(m,n). \tag{11.2.8}$$

Hence, from (11.2.8), we have

$$\|A\|^2 = \mathrm{Tr}(A^\dagger A) = \sum_{i=1}^{m} \sum_{j=1}^{n} |a_{ij}|^2, \quad A = (a_{ij}) \in \mathbb{C}(m,n). \tag{11.2.9}$$

The obvious advantage of (11.2.9) is that it is the sum of the squares of the norms of n column vectors in $\mathbb{C}^m$ or the sum of the squares of the norms of m row vectors in $\mathbb{C}^n$ of the matrix A. For our discussion, we regard the matrix A as being comprised of n column vectors $A_1, \ldots, A_n \in \mathbb{C}^m$, such that

$$\|A\|^2 = \sum_{j=1}^{n} \|A_j\|^2, \quad A = (A_1, \ldots, A_n) = (A_j) \in \mathbb{C}(m,n), \tag{11.2.10}$$

where $\|u\|^2$ for $u \in \mathbb{C}^m$ is associated with the usual Hermitian scalar product $(u,v) = u^\dagger v$ for $u,v \in \mathbb{C}^m$, so that (11.2.8) enjoys the decomposition

$$(A,B) = \mathrm{Tr}(A^\dagger B) = ((A_1, \ldots, A_n), (B_1, \ldots, B_n))$$

$$= \sum_{j=1}^{n} (A_j, B_j) = \sum_{j=1}^{n} A_j^\dagger B_j, \quad A = (A_j), B = (B_j) \in \mathbb{C}(m,n), \tag{11.2.11}$$

for which the accompanying geometric features may effectively be utilized for application purposes.

First, it is clear that since $\mathcal{A}_l$ may be rewritten as

$$\mathcal{A}_l = P \begin{pmatrix} D_l & 0 \\ 0 & 0 \end{pmatrix} Q, \tag{11.2.12}$$

where $P \in \mathbb{C}(m,m)$ and $Q \in \mathbb{C}(n,n)$ are unitary matrices, we see that (11.2.9) leads to

$$\|\mathcal{A}_l\|^2 = \mathrm{Tr}(D_l^2) = \sigma_1 + \cdots + \sigma_l. \tag{11.2.13}$$

Hence, with $A = \mathcal{A}_k$, we have

$$\|A - \mathcal{A}_{l-1}\|^2 = \sigma_l + \cdots + \sigma_k. \tag{11.2.14}$$

We now show that the right-hand side of (11.2.14) gives rise to the quantity δ_l in (11.2.2) under the Frobenius norm such that $\mathcal{A}_{l-1}$ is again the associated matrix minimizer for the problem (11.2.2).

To see this, let A and $B \in \mathbb{C}(m,n)$ be of the ranks r and $s \geq 1$ and have singular values

$$\alpha_1 \geq \cdots \geq \alpha_r > 0, \quad \beta_1 \geq \cdots \geq \beta_s > 0, \qquad (11.2.15)$$

respectively, listed in descending orders. Then we have

$$|\mathrm{Tr}(A^\dagger B)| \leq \sum_{i=1}^{d} \alpha_i \beta_i, \quad d = \min\{r,s\}. \qquad (11.2.16)$$

This result is known as the *von Neumann trace inequality*. Thus, under the Frobenius matrix norm and observing (11.2.13) and (11.2.16), we have

$$\|A - B\|^2 = \|A\|^2 + \|B\|^2 - 2\Re(A,B)$$

$$\geq \sum_{i=1}^{r} \alpha_i^2 + \sum_{i=1}^{s} \beta_i^2 - 2 \sum_{i=1}^{d} \alpha_i \beta_i. \qquad (11.2.17)$$

Now specializing (11.2.17) to the problem (11.2.2), we have $r = r(A) = k$, $s = r(B) \leq l - 1$, and $d = s$, and

$$\alpha_i = \sqrt{\sigma_i}, \quad i = 1, \dots, k, \qquad (11.2.18)$$

such that (11.2.17) renders us the lower bound

$$\|A - B\|^2 \geq \sum_{i=1}^{s} (\sqrt{\sigma_i} - \beta_i)^2 + \sum_{i=s+1}^{k} \sigma_i$$

$$\geq \sigma_l + \cdots + \sigma_k, \qquad (11.2.19)$$

exactly given as on the right-hand side of (11.2.14), as desired.

Summarizing this study, we have

Theorem 11.2 *For any $A \in \mathbb{C}(m,n)$ with $r(A) = k \geq 1$, let $\{u_1, \dots, u_k, \dots, u_n\}$ be an orthonormal basis of $\mathbb{C}^n$ consisting of eigenvectors of the positive semi-definite matrix $A^\dagger A \in \mathbb{C}(n,n)$ associated with the positive eigenvalues $\sigma_1, \dots, \sigma_k$ of $A^\dagger A$, in descending order, and the zero eigenvalue (if any). For any $l \leq k$, the low-rank minimization problem (11.2.2) with the induced matrix norm (11.2.1) or the Frobenius norm (11.2.9) has a solution given by (11.2.7) so that $\delta_l = \sqrt{\sigma_l}$ in the former situation and $\delta_l = \sqrt{\sigma_l + \cdots + \sigma_k}$ in the latter situation.*

Let $\| \cdot \|$ be a norm on the space $\mathbb{C}(m,n)$. It is called unitarily invariant, if for any unitary matrices $P \in \mathbb{C}(m,m)$ and $Q \in \mathbb{C}(n,n)$, and any $K \in \mathbb{C}(m,n)$, there holds $\|PKQ\| = \|K\|$. Theorem 11.2 is in fact valid under any unitarily invariant matrix norm. This general statement is known as the *Schmidt–Eckart–Young–Mirsky Theorem* [54], developed into its present maturity from the study initiated by Schmidt [68], and the further work pursued by Eckart and Young [24], Mirsky [59], and the investigation of other people over a span of many years.

We next elaborate on a proof of the von Neumann trace inequality (11.2.16) following Mirsky [60].

With the notation given in (11.2.15), choose unitary matrices P_1 and $P_2 \in \mathbb{C}(m,m)$ and Q_1 and $Q_2 \in \mathbb{C}(n,n)$ such that

$$A = P_1 \Lambda_1 Q_1, \quad B = P_2 \Lambda_2 Q_2, \quad \Lambda_{1,2} = \begin{pmatrix} D_{1,2} & 0 \\ 0 & 0 \end{pmatrix} \in \mathbb{C}(m,n), \quad (11.2.20)$$

where, for convenience, we set

$$D_1 = \mathrm{diag}\{\alpha_1, \ldots, \alpha_k\}, \quad D_2 = \mathrm{diag}\{\beta_1, \ldots, \beta_k\}, \quad k = \max\{r, s\}, \quad (11.2.21)$$

such that $\alpha_1 \geq \cdots \geq \alpha_r > 0$, $\beta_1 \geq \cdots \geq \beta_s > 0$, and the other α_i's and β_i's are taken to be zero, if any. Then we have

$$\mathrm{Tr}(A^\dagger B) = \mathrm{Tr}(\Lambda_1^\dagger P \Lambda_2 Q), \quad P = P_1^\dagger P_2, \quad Q = Q_2 Q_1^\dagger$$

$$= \sum_{i,j=1}^{k} \alpha_i \beta_j u_{ij} v_{ji}, \quad P = (u_{ij}), \quad Q = (v_{ij}), \quad (11.2.22)$$

where $P \in \mathbb{C}(m,m)$ and $Q \in \mathbb{C}(n,n)$ are unitary matrices. Consequently, we get

$$|\mathrm{Tr}(A^\dagger B)| \leq \sum_{i,j=1}^{k} \alpha_i \beta_j w_{ij}; \quad w_{ij} = |u_{ij} v_{ji}|, \quad i, j = 1, \ldots, k. \quad (11.2.23)$$

On the other hand, from

$$\sum_{i=1}^{k} w_{ij} \leq \frac{1}{2} \left(\sum_{i=1}^{k} |u_{ij}|^2 + \sum_{i=1}^{k} |v_{ji}|^2 \right) \leq 1, \quad j = 1, \ldots, k, \quad (11.2.24)$$

$$\sum_{j=1}^{k} w_{ij} \leq \frac{1}{2} \left(\sum_{j=1}^{k} |u_{ij}|^2 + \sum_{j=1}^{k} |v_{ji}|^2 \right) \leq 1, \quad i = 1, \ldots, k, \quad (11.2.25)$$

we see that the matrix $W = (w_{ij}) \in \mathbb{R}(k,k)$ is *doubly substochastic*. That is, its entries are all non-negative so that the components of each of its row and column vectors add up to a value less than or equal to 1. See [55] for numerous applications of such matrices. The following lemma due to Mirsky [60] is a key step in the proof of (11.2.16).

Lemma 11.3 *Let* $\alpha_1 \geq \cdots \geq \alpha_k \geq 0$ *and* $\beta_1 \geq \cdots \geq \beta_k \geq 0$ *be two descending sequences of numbers and* $W = (w_{ij}) \in \mathbb{R}(k,k)$ *a doubly substochastic matrix. Then*

$$\sum_{i,j=1}^{k} \alpha_i \beta_j w_{ij} \leq \sum_{i=1}^{k} \alpha_i \beta_i. \tag{11.2.26}$$

Proof   First, note that there holds the *Mirsky identity* [60] (cf. Problem 11.2.8)

$$\sum_{1 \leq r,s \leq k} c_{rs} \sum_{r \leq i \leq k} a_i \sum_{s \leq j \leq k} b_j = \sum_{1 \leq i,j \leq k} a_i b_j \sum_{1 \leq r \leq i, 1 \leq s \leq j} c_{rs}, \tag{11.2.27}$$

involving clustered sums for $a_1, \ldots, a_k, b_1, \ldots, b_k \in \mathbb{C}$, and $C = (c_{rs}) \in \mathbb{C}(k,k)$.

Next, represent $\alpha_1, \ldots, \alpha_k$ and $\beta_1, \ldots, \beta_k$ stated in the lemma by the unique non-negative numbers $a_1, \ldots, a_k$ and $b_1, \ldots, b_k$ with the formulas

$$\alpha_r = \sum_{r \leq i \leq k} a_i, \quad \beta_r = \sum_{r \leq i \leq k} b_i, \quad r = 1, \ldots, k. \tag{11.2.28}$$

Then we have

$$\sum_{r=1}^{k} \alpha_r \beta_r - \sum_{r,s=1}^{k} w_{rs} \alpha_r \beta_s = \sum_{r,s=1}^{k} (\delta_{rs} - w_{rs}) \alpha_r \beta_s$$

$$= \sum_{1 \leq r,s \leq k} (\delta_{rs} - w_{rs}) \sum_{r \leq i \leq k} a_i \sum_{s \leq j \leq k} b_j$$

$$= \sum_{1 \leq i,j \leq k} a_i b_j \sum_{1 \leq r \leq i, 1 \leq s \leq j} (\delta_{rs} - w_{rs}), \tag{11.2.29}$$

where we have used the identity (11.2.27) with $c_{rs} = \delta_{rs} - w_{rs}$. Now it suffices to show that the right-hand side of (11.2.29) is non-negative. To this end, we first assume the condition $i \leq j$ in the inner sum to get

$$\sum_{1 \leq r \leq i, 1 \leq s \leq j} (\delta_{rs} - w_{rs}) = i - \sum_{r=1}^{i} \sum_{s=1}^{j} w_{rs} \geq i - i = 0, \tag{11.2.30}$$

by virtue of (11.2.25). We then assume $i \geq j$ in the inner sum to get

$$\sum_{1 \leq r \leq i, 1 \leq s \leq j} (\delta_{rs} - w_{rs}) = j - \sum_{s=1}^{j} \sum_{r=1}^{i} w_{rs} \geq j - j = 0, \tag{11.2.31}$$

by virtue of (11.2.24). This completes the proof of the lemma. □

From (11.2.23) and (11.2.26), we see that the von Neumann trace inequality (11.2.16) follows.

Exercises

11.2.1 Show that, for A and $B \in \mathbb{C}(m,n)$, their Hilbert–Schmidt scalar product (11.2.8) is simply the sum of the Hermitian scalar products of the row vectors of A and B, in respective pairs.

11.2.2 Show that the induced matrix norm (11.2.1) on $\mathbb{C}(m,n)$ is unitarily invariant.

11.2.3 Show that the Hilbert–Schmidt norm (11.2.9) on $\mathbb{C}(m,n)$ is unitarily invariant.

11.2.4 Let the singular values of A and $B \in \mathbb{C}(m,n)$ be given by (11.2.15). Prove that, if A and B have *simultaneous* singular value decompositions, that is, there is *one* pair of unitary matrices $P \in \mathbb{C}(m,m)$ and $Q \in \mathbb{C}(n,n)$ (rather than *two* pairs) such that

$$A = P\Lambda_1 Q, \quad B = \pm P\Lambda_2 Q, \quad \Lambda_{1,2} = \begin{pmatrix} D_{1,2} & 0 \\ 0 & 0 \end{pmatrix} \in \mathbb{C}(m,n), \tag{11.2.32}$$

where D_1 and D_2 are as given in (11.2.21), then the von Neumann trace inequality (11.2.16) is saturated with equality,

$$\pm \operatorname{Tr}(A^\dagger B) = \alpha_1 \beta_1 + \cdots + \alpha_d \beta_d, \quad d = \min\{r,s\}. \tag{11.2.33}$$

11.2.5 Verify (11.2.22).

11.2.6 Let the matrix W in Lemma 11.3 be *doubly stochastic*. That is, W satisfies

$$w_{ij} \geq 0, \quad \sum_{i=1}^{k} w_{ij} = 1, \quad \sum_{j=1}^{k} w_{ij} = 1, \quad i,j = 1,\ldots,k. \tag{11.2.34}$$

Assuming (11.2.34) and $k = 2$, obtain a direct proof of the inequality (11.2.26).

11.2.7 Let $A = (a_{ij}) \in \mathbb{R}(n,n)$ be a doubly substochastic matrix. Show that there is a doubly stochastic matrix $B = (b_{ij})$ satisfying $a_{ij} \leq b_{ij}$, $i,j = 1,\ldots,n$. (This result is due to von Neumann. See [55].)

11.2.8 Establish the identity (11.2.27).

11.2.9 Consider the matrix

$$A = \begin{pmatrix} 0 & 1 & 0 & 0 \\ 0 & 0 & 2 & 0 \\ 0 & 0 & 0 & 3 \\ 0 & 0 & 0 & 0 \end{pmatrix}. \tag{11.2.35}$$

Obtain a singular value decomposition for A and use it to construct the optimal rank-1, -2, and -3 matrix approximations of A. Find also the optimal distances between A and its low rank approximations in both the induced and Frobenius norms.

11.3 Principal component analysis

We now consider the subject of *principal component analysis*, popularly known as PCA, which is an important tool in data science for dimensionality reduction, feature extraction, noise reduction, visualization, and data preprocessing. Its applications span various domains, including machine learning, pattern recognition, image processing, signal processing, and exploratory data analysis, making it a versatile and widely used technique.

Let $v \in \mathbb{C}^m$ be a unit vector. For any $x \in \mathbb{C}^m$, its projection along v is given by

$$P_v(x) = (v, x)v = v(v, x) = (vv^\dagger)x. \tag{11.3.1}$$

That is, the projection is defined by the matrix $vv^\dagger \in \mathbb{C}(m, m)$. More generally, let S be an l-dimensional subspace of $\mathbb{C}^m$ with $v_1, \dots, v_l$ being an orthonormal basis. Then the projection of $x \in \mathbb{C}^m$ onto S is given by

$$P_S(x) = \sum_{i=1}^{l} (v_i, x)v_i = \left(\sum_{i=1}^{l} v_i v_i^\dagger \right) x = VV^\dagger x, \tag{11.3.2}$$

where

$$V = (v_1, \dots, v_l) \in \mathbb{C}(m, l) \tag{11.3.3}$$

satisfies $V^\dagger V = I_l$. That is, $VV^\dagger$ is the matrix representation of the projection mapping $P_S : \mathbb{C}^m \to \mathbb{C}^m$ with respect to the standard basis of $\mathbb{C}^m$, called the *projection matrix* in this context.

For greater generality, let $A_1, \dots, A_n \in \mathbb{C}^m$ be a collection of data vectors. The main task of PCA is to find a vector $A_0 \in \mathbb{C}^m$, called a reference point, and an l-dimensional subspace $S \subset \mathbb{C}^m$, determined by a suitable orthonormal basis $\{v_1, \dots, v_l\}$, such that the sum of the squared distances from the data points $A_1, \dots, A_n$ to their respective projections onto S, relative to the reference point A_0, is minimized. Mathematically, this amounts to solving the minimization problem

$$\eta \equiv \inf \left\{ \sum_{j=1}^{n} \|(A_j - A_0) - VV^\dagger(A_j - A_0)\|^2 \,\Big|\, A_0 \in \mathbb{C}^m, V \in \mathbb{C}(m, l), V^\dagger V = I_l \right\}, \tag{11.3.4}$$

where V is given by (11.3.3).

It will be convenient to reformulate the problem (11.3.4) in terms of the Frobenius norm. For this purpose, we form the matrices

$$A = (A_1, \ldots, A_n), \quad \mathcal{A}_0 = (A_0, \ldots, A_0), \tag{11.3.5}$$

in $\mathbb{C}(m, n)$. Thus $(A_j - A_0) - VV^\dagger(A_j - A_0)$ is the jth column vector of the matrix $(A - \mathcal{A}_0) - VV^\dagger(A - \mathcal{A}_0) = (I - VV^\dagger)(A - \mathcal{A}_0) \in \mathbb{C}(m, n)$ and (11.3.4) now reads

$$\eta = \inf\{\|(I - VV^\dagger)(A - \mathcal{A}_0)\|^2\}, \tag{11.3.6}$$

under the Frobenius norm over $\mathbb{C}(m, n)$.

We first show that a choice for A_0 is the *mean data vector*

$$A_0 = \underline{A} = \frac{1}{n}(A_1 + \cdots + A_n), \tag{11.3.7}$$

also called the center of the data. In fact, we may rewrite A_0 as $A_0 = \underline{A} + x$, where $x \in \mathbb{C}^m$ is arbitrary. For convenience, set

$$\underline{\mathcal{A}} = (\underline{A}, \ldots, \underline{A}), \quad X = (x, \ldots, x) \in \mathbb{C}(m, n). \tag{11.3.8}$$

Then $\mathcal{A}_0 = \underline{\mathcal{A}} + X$. Thus, in terms of the Hilbert–Schmidt scalar product over $\mathbb{C}(m, n)$ and the matrix $B = I - VV^\dagger \in \mathbb{C}(m, m)$, we have

$$\|B(A - \mathcal{A}_0)\|^2 = \|BA\|^2 - 2\Re(BA, B\mathcal{A}_0) + \|B\mathcal{A}_0\|^2$$
$$= \|BA\|^2 - 2\Re(BA, B\underline{\mathcal{A}}) + \|B\underline{\mathcal{A}}\|^2 + \|BX\|^2, \tag{11.3.9}$$

where we have used the result

$$(B\underline{\mathcal{A}}, BX) = (B(\underline{A}, \ldots, \underline{A}), B(x, \ldots, x))$$
$$= n(B\underline{A}, Bx) = \sum_{j=1}^{n}(BA_j, Bx) = (BA, BX), \tag{11.3.10}$$

by the property that the Hilbert–Schmidt scalar product of two matrices is simply the sum of the Hermitian products of their column vectors in respective pairs. Consequently, (11.3.9) gives us the lower bound

$$\|B(A - \mathcal{A}_0)\|^2 \geq \|BA\|^2 - 2\Re(BA, B\underline{\mathcal{A}}) + \|B\underline{\mathcal{A}}\|^2, \tag{11.3.11}$$

which is saturated when $\|BX\|^2 = n\|Bx\|^2 = 0$. Since $B = I - VV^\dagger$ contains the unknown matrix V, we can simply choose $x = 0$ or (11.3.7) to partially solve (11.3.4).

After the determination of A_0 with the solution (11.3.7), we return to the determination of the matrix (11.3.3) in the problem (11.3.4).

To proceed, let $V = W \in \mathbb{C}(m, l)$ be a testing matrix satisfying $W^\dagger W = I_l$ in the problem (11.3.6) and set $\mathcal{A} = A - \mathcal{A}_0 = A - \underline{\mathcal{A}}$ in (11.3.6). As a data matrix, the mean data vector of $\mathcal{A}$ is zero. It is clear that $r(\mathcal{A}) \leq r(A)$. Such a data matrix is also called a *centralized data matrix*. Then, with the Frobenius norm and the Hilbert–Schmidt scalar product, we have

$$\begin{aligned} \|\mathcal{A} - WW^{\dagger}\mathcal{A}\|^2 &= \|\mathcal{A}\|^2 - 2\Re(\mathcal{A}, WW^{\dagger}\mathcal{A}) + \|WW^{\dagger}\mathcal{A}\|^2 \\ &= \|\mathcal{A}\|^2 - \|W^{\dagger}\mathcal{A}\|^2 \end{aligned} \tag{11.3.12}$$

for the quantity on the left-hand side of (11.3.6) to be estimated. Hence, to get the optimal lower bound for (11.3.12), it is equivalent to get the optimal upper bound for $\|W^{\dagger}\mathcal{A}\|^2$, which is what we pursue next.

Following the singular value decomposition for the matrix $\mathcal{A}$ whose mean data vector is normalized to zero, use $\{u_1, \ldots, u_k, \ldots, u_n\}$ to denote an orthonormal basis of $\mathbb{C}^n$ consisting of the eigenvectors of $\mathcal{A}^{\dagger}\mathcal{A} \in \mathbb{C}(n,n)$ associated with the positive eigenvalues $\sigma_1 \geq \cdots \geq \sigma_k > 0$ and zero eigenvalue (if any), where $k = r(\mathcal{A})$. Let $\{v_1, \ldots, v_m\}$ be an orthonormal basis of $\mathbb{C}^m$ satisfying

$$\mathcal{A}u_i = \sqrt{\sigma_i}v_i, \quad i = 1, \ldots, k; \quad \mathcal{A}u_j = 0, \quad j > k, \quad \text{if any.} \tag{11.3.13}$$

Then we have

$$\mathcal{A} = P\Lambda Q^{\dagger}, \quad P = (v_1, \ldots, v_m) \in \mathbb{C}(m,m), \quad Q = (u_1, \ldots, u_n) \in \mathbb{C}(n,n),$$

$$\Lambda = \begin{pmatrix} D & 0 \\ 0 & 0 \end{pmatrix} \in \mathbb{C}(m,n), \quad D = \text{diag}\{\sqrt{\sigma_1}, \ldots, \sqrt{\sigma_k}\}. \tag{11.3.14}$$

Inserting (11.3.14), we obtain

$$\begin{aligned} \|W^{\dagger}\mathcal{A}\|^2 &= \text{Tr}([W^{\dagger}\mathcal{A}]^{\dagger}[W^{\dagger}\mathcal{A}]) \\ &= \text{Tr}(W^{\dagger}\mathcal{A}\mathcal{A}^{\dagger}W) \\ &= \text{Tr}(W^{\dagger}P\Lambda\Lambda^{\dagger}P^{\dagger}W) \\ &= \text{Tr}(Y^{\dagger}\Lambda\Lambda^{\dagger}Y), \quad Y = P^{\dagger}W \in \mathbb{C}(m,l). \end{aligned} \tag{11.3.15}$$

Rewrite Y in (11.3.15) as $Y = (y_1, \ldots, y_l)$. Then $y_1, \ldots, y_l$ are orthonormal vectors in $\mathbb{C}^m$ since $Y^{\dagger}Y = I_l$. Expand $\{y_1, \ldots, y_l\}$ to get an orthonormal basis of $\mathbb{C}^m$, say $\{y_1, \ldots, y_l, \ldots, y_m\}$ and denote the m by m unitary matrix $(y_1, \ldots, y_l, \ldots, y_m)$ by (y_{ij}). Then we have, in particular,

$$\sum_{i=1}^{m} |y_{ij}|^2 = 1, j = 1, \ldots, l; \quad \sum_{j=1}^{l} |y_{ij}|^2 \leq \sum_{j=m}^{l} |y_{ij}|^2 = 1, i = 1, \ldots, m. \tag{11.3.16}$$

On the other hand, the right-hand side of (11.3.15) may be calculated to render

$$\text{Tr}(Y^{\dagger}\Lambda\Lambda^{\dagger}Y) = \|\Lambda^{\dagger}Y\|^2$$

$$= \sigma_1 \sum_{j=1}^{l} |y_{1j}|^2 + \sigma_2 \sum_{j=1}^{l} |y_{2j}|^2 + \cdots + \sigma_k \sum_{j=1}^{l} |y_{kj}|^2, \tag{11.3.17}$$

since

$$\Lambda^\dagger Y = \begin{pmatrix} \sqrt{\sigma_1}y_{11} & \cdots & \sqrt{\sigma_1}y_{1l} \\ \vdots & \cdots & \vdots \\ \sqrt{\sigma_k}y_{k1} & \cdots & \sqrt{\sigma_k}y_{kl} \\ 0 & \cdots & 0 \\ \vdots & \cdots & \vdots \\ 0 & \cdots & 0 \end{pmatrix}, \tag{11.3.18}$$

which lies in $\mathbb{C}(n,l)$. In view of (11.3.16) and (11.3.17), we see that the right-hand side of (11.3.17) may be maximized by setting

$$y_{11} = 1, y_{i1} = 0, i = 2,\ldots,m; \quad y_{22} = 1, y_{i2} = 0, i = 1,\ldots,m, i \neq 2;$$
$$\ldots\ldots; \quad y_{ll} = 1, y_{il} = 0, i = 1,\ldots,m, i \neq l, \tag{11.3.19}$$

since $\sigma_1 \geq \cdots \geq \sigma_k$, which gives rise to the maximum

$$\max\left\{ \operatorname{Tr}(Y^\dagger \Lambda \Lambda^\dagger Y) \right\} = \sum_{i=1}^{l} \sigma_i. \tag{11.3.20}$$

In fact, the first term on the right-hand side of (11.3.17) may be maximized by taking $y_{11} = 1$, which leads to $y_{i1} = 0$ when $i \neq 1$ by (11.3.16). Since y_1 is orthogonal to $y_2,\ldots,y_l$, we get $y_{1j} = 0$ for $j \neq 1$. With $y_{21} = 0$, we see that the second term of the right-hand side of (11.3.17) may be maximized by taking $y_{22} = 1$, which leads to $y_{i2} = 0$ when $i \neq 2$ by (11.3.16). By orthogonality again, we get $y_{2j} = 0$ for $j \neq 2$. This procedure may be executed successively up to the lth term to yield $y_{ll} = 1$, $y_{il} = 0$ when $i \neq l$, and $y_{lj} = 0$ when $j \neq l$. Thus we obtain (11.3.19) and (11.3.20).

Indeed, by Lemma 11.3, in view of (11.3.16), we have the general upper bound

$$\sum_{i=1}^{k}\sum_{j=1}^{l} \sigma_i |y_{ij}|^2 \leq \sum_{i=1}^{l} \sigma_i, \tag{11.3.21}$$

which may be attained by taking Y that satisfies the condition (11.3.19). Therefore, with (11.3.19), we arrive at

$$Y = \begin{pmatrix} I_l \\ 0 \end{pmatrix}. \tag{11.3.22}$$

Inserting (11.3.22) into (11.3.15), we find

$$W = PY = (v_1,\ldots,v_l) = V, \tag{11.3.23}$$

as given in (11.3.3).

We summarize our study on the minimization problem arising in PCA on a complex data matrix $A \in \mathbb{C}(m,n)$ as follows:

Theorem 11.4 *Let the column vectors of the data matrix $A \in \mathbb{C}(m,n)$ be denoted by $A_1,\ldots,A_n$, realizing n data vectors. Relative to a given reference data point $A_0 \in \mathbb{C}^m$, orthogonally project the n data vectors onto an l-dimensional subspace represented by l orthonormal vectors, $v_1,\ldots,v_l$ in $\mathbb{C}^m$. Then among all such reference data points and l-dimensional subspaces, there is a pair of data reference point and an l-dimensional subspace which give rise to a minimized sum of squared distances from the n data points to their orthogonally projected images, with respect to the reference data point. In fact, the optimal choice of the reference data point is the mean data vector, or the center of data, defined by (11.3.7), and the optimal orthonormal vectors $v_1,\ldots,v_l$ are as given in the singular value decomposition (11.3.13) of the mean data vector shifted data matrix $\mathcal{A}$ with zero mean data vector, or the centralized data matrix associated with A, whose singular values are $\sqrt{\sigma_1} \geq \cdots \geq \sqrt{\sigma_k} > 0$, where $k = r(\mathcal{A})$, rendering the minimum sum of squared distances defined in (11.3.4) to be*

$$\eta = \sigma_{l+1} + \cdots + \sigma_k. \tag{11.3.24}$$

Proof It only remains to establish (11.3.24). In fact, from (11.3.15), (11.3.20), and the fact that $\|\mathcal{A}\|^2 = \sigma_1 + \cdots + \sigma_k$, we see that (11.3.24) follows.   □

Note that, in practice and for simplicity, we may choose not to consider a data matrix, say $A = (A_1,\ldots,A_n)$, subject to an optimal column shift, A_0, as stated in (11.3.4), so that the problem (11.3.4) assumes the reduced form

$$\eta \equiv \inf \left\{ \sum_{j=1}^{n} \|A_j - VV^\dagger A_j\|^2 \,\middle|\, V \in \mathbb{C}(m,l), V^\dagger V = I_l \right\}, \tag{11.3.25}$$

with setting $A_0 = 0$ in (11.3.4). In this case, we have $\mathcal{A} = A$ in Theorem 11.4 and all the construction stated there is valid for the situation here.

The set of orthonormal vectors $v_1, v_2, \ldots, v_l$ in Theorem 11.4 are called the principal components of the data matrix $\mathcal{A}$ or A. Specifically, $v_1, v_2, \ldots$, and v_l are called the first, the second,..., and the lth principal components of $\mathcal{A}$ or A, respectively.

Let the singular value decomposition of $A \in \mathbb{C}(m,n)$, say, be given by

$$Au_i = \sqrt{\sigma_i} v_i, \quad i = 1,\ldots,k, \quad k = r(A);$$
$$Au_i = 0, \quad k < i \leq n \text{ (if any)}, \tag{11.3.26}$$

where $\{u_1,\ldots,u_k,\ldots,u_n\}$ and $\{v_1,\ldots,v_k,\ldots,v_m\}$ are orthonormal bases of $\mathbb{C}^n$ and $\mathbb{C}^m$, respectively, and $\sqrt{\sigma_1} \geq \cdots \geq \sqrt{\sigma_k} > 0$ are the singular values of A. Since the matrix form of (11.3.26) reads $A = P\Lambda Q^\dagger$ where $P = (v_1,\ldots,v_m)$, $Q = (u_1,\ldots,u_n)$, and Λ is the usual singular value matrix, the vectors $v_1,\ldots,v_m$ and $u_1,\ldots,u_n$ are also referred to as the *left* and *right singular vectors*, respectively. The study of these two sections indicates that all the features in the decomposition are made full use of: The right and left singular vectors are used to construct the low rank approximation of a matrix, the left singular vectors are used to obtain the principal component representation of a data matrix, and the singular values provide measurements for various optimal values for the approximation errors in specific situations.

Exercises

11.3.1 Show that the map P_S defined by (11.3.2) is indeed a projection by verifying $(VV^\dagger)^2 = VV^\dagger$ for the matrix V defined in (11.3.3).

11.3.2 Let u and v be two column vectors in $\mathbb{R}^n$. Show that the Frobenius norm of the matrix $uu^t - vv^t \in \mathbb{R}(n,n)$ is given by

$$\|uu^t - vv^t\|^2 = \|u\|^4 + \|v\|^4 - 2(u^t v)^2. \tag{11.3.27}$$

In particular, if u and v are unit vectors so that $u^t v = u \cdot v = \cos\theta$, then we have $\|uu^t - vv^t\|^2 = 2\sin^2\theta$.

11.3.3 Let $A \in \mathbb{C}(m,n)$ and $\mathcal{A}$ be its associated centralized matrix. Prove that $r(\mathcal{A}) \leq r(A)$ and obtain an example satisfying $r(\mathcal{A}) < r(A)$.

11.3.4 Establish (11.3.21) by using Lemma 11.3.

11.4 Example illustrating principal component analysis

As a concrete example illustrating PCA, consider the data matrix

$$A = \begin{pmatrix} 3 & 2 & 2 \\ 2 & 3 & -2 \end{pmatrix}, \tag{11.4.1}$$

with data points plotted in Figure 11.1. The least-squares approximation gives us the line $y = \frac{8}{17}x$, and the sum of the least-square errors at the data points is $\frac{225}{17} \approx 13.2353$. On the other hand, the singular values of A are $\lambda_1 = \sqrt{\sigma_1} = 5$ and $\lambda_2 = \sqrt{\sigma_2} = 3$, with the corresponding left and right singular vectors

$$v_1 = \frac{1}{\sqrt{2}} \begin{pmatrix} 1 \\ 1 \end{pmatrix}, \ v_2 = \frac{1}{\sqrt{2}} \begin{pmatrix} 1 \\ -1 \end{pmatrix}, \ u_1 = \frac{1}{\sqrt{2}} \begin{pmatrix} 1 \\ 1 \\ 0 \end{pmatrix}, \ u_2 = \frac{1}{\sqrt{18}} \begin{pmatrix} 1 \\ -1 \\ 4 \end{pmatrix}.$$

(11.4.2)

Using the first principal component v_1, namely, the first left singular vector, we obtain the projection mapping

$$P_{v_1}(x) = (v_1 v_1^t)x = \frac{1}{2} \begin{pmatrix} 1 & 1 \\ 1 & 1 \end{pmatrix} x, \quad x \in \mathbb{R}^2,$$

(11.4.3)

from $\mathbb{R}^2$ onto the subspace spanned by the vector v_1. As a consequence, we have

$$P_{v_1} \begin{pmatrix} 3 \\ 2 \end{pmatrix} = P_{v_1} \begin{pmatrix} 2 \\ 3 \end{pmatrix} = \frac{5}{2} \begin{pmatrix} 1 \\ 1 \end{pmatrix}, \quad P_{v_1} \begin{pmatrix} 2 \\ -2 \end{pmatrix} = 0,$$

(11.4.4)

as the orthogonally projected images of the respective data points to the line spanned by v_1. Therefore, the sum of the squared distances between the data points and their projected images is $\eta = 9$, which verifies the formula (11.3.24) since $\sigma_2 = 9$. This quantity significantly improves that for the least-squares approximation.

Moreover, the centralized data matrix associated with (11.4.1) reads

$$A = \frac{1}{3} \begin{pmatrix} 2 & -1 & -1 \\ 3 & 6 & -9 \end{pmatrix}.$$

(11.4.5)

The positive eigenvalues of $A^t A$ are

$$\sigma_1 = \frac{1}{3}(22 + \sqrt{409}), \quad \sigma_2 = \frac{1}{3}(22 - \sqrt{409}),$$

(11.4.6)

such that σ_2 gives us the optimal value of the sum of the squared distances between the data points and their orthogonally projected images onto the line obtained from shifting the line spanned by the first left singular vector v_1 of the centralized data matrix (11.4.5) to pass through the data center, as shown in Figure 11.1, with the approximate value

$$\eta = \sigma_2 \approx 0.5921,$$

(11.4.7)

which is significantly improved upon the uncentralized data matrix results, $\eta = 9$, found earlier.

Exercises

Consider the data matrix

$$A = \begin{pmatrix} -1 & 2 & 3 \\ 2 & 1 & -2 \end{pmatrix}.$$

(11.4.8)

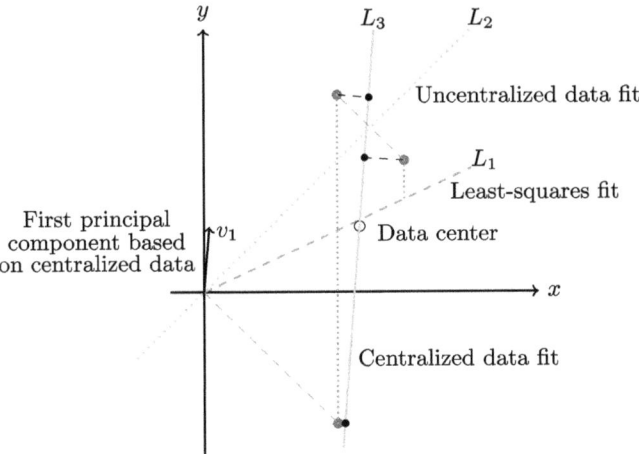

Figure 11.1 Optimal data fits based on principal component analysis versus that based on the least-squares optimization. Three large solid dots represent three data points listed in (11.4.1) as the column vectors of the matrix A. The empty circle indicates the data center of these data points. The lines L_1, L_2, and L_3 are fitting lines representing the least-squares approximation and the one-dimensional principal component fitting lines based on the uncentralized data matrix (11.4.1) and the centralized one, (11.4.5), drawn as dashed, dotted, and solid lines, respectively. In the least-squares approach, the distances are measured from the data points to their respective images on the fitting line L_1, projected along the vertical y-axis direction, shown as dotted line segments; in the principal component approximations, the distances are measured from the data points to their respective images on the fitting lines L_2 and L_3, projected along directions orthogonal to the fittings lines or the first principal components, shown accordingly as dashed line segments. Naturally and as expected, a steady trend of improvement of the optimal approximations is observed in that the principal component methods always provide better approximations and the best result is given by the principal component optimization based on the centralized data matrix which significantly improves that based on the uncentralized data matrix. Also drawn in the figure is the first principal component vector v_1 based on the centralized data matrix $\mathcal{A}$, given in (11.4.5).

11.4.1 Form the centralized data matrix $\mathcal{A}$ associated with A and obtain the singular value decomposition for $\mathcal{A}$.

11.4.2 Use Problem 11.4.1 to find the first principal component of the centralized data matrix and the associated projection matrix.

11.4.3 Find the first principal component of the uncentralized data matrix given by (11.4.8) and the associated projection matrix.

11.4.4   Obtain the sums of the squared distances from the data points to their orthogonally projected images onto the fitting lines, based on the first principal components calculated in Problem 11.4.2 and Problem 11.4.3, respectively, and compare your findings, along those presented in Figure 11.1.

11.5 Some other nearness problems

The low-rank approximation problems studied in Sections 11.1 and 11.2 belong to a broad family of problems called *matrix nearness problems* [39]. In this section, we consider the nearness problems involving symmetric matrices and orthogonal matrices, which find wide applications in data analysis as well, including facial recognition problems.

For greater generality, we again begin with a discussion of the nearness problem involving self-adjoint mappings over a finite-dimensional complex vector space U equipped with a positive-definite scalar product $(\cdot, \cdot)$, given by

$$\eta = \inf\{\|T - S\| \mid S \in L(U), S' = S\}, \tag{11.5.1}$$

where $T \in L(U)$ is arbitrarily picked and $\|\cdot\|$ denotes the induced norm of the mapping involved associated with the scalar product.

We first show that the usual symmetric or self-adjoint part of T, defined by

$$T_{\mathrm{s}} = \frac{1}{2}(T + T'), \tag{11.5.2}$$

solves the problem (11.5.1). In fact, for any self-adjoint element $S \in L(U)$, namely, $S' = S$, we have by the triangle inequality the estimate

$$\|T - T_{\mathrm{s}}\| = \frac{1}{2}\left\|T - T' - S + S'\right\|$$
$$\leq \frac{1}{2}\|T - S\| + \frac{1}{2}\|(T - S)'\| = \|T - S\|, \tag{11.5.3}$$

which proves that (11.5.2) solves (11.5.1). A by-product of this calculation is that the minimal distance η given in (11.5.1) is the norm of the anti-symmetric anti-self-adjoint part, T_{a}, of T:

$$\eta = \|T_{\mathrm{a}}\|, \quad T_{\mathrm{a}} = \frac{1}{2}(T - T'). \tag{11.5.4}$$

A natural question asks whether (11.5.2) is the unique solution to (11.5.1). To investigate this question, we consider a concrete real matrix version of the problem with

$$A = \begin{pmatrix} a & 0 & 0 \\ 0 & 0 & b \\ 0 & -b & 0 \end{pmatrix}, \quad A_{\mathrm{s}} = \frac{1}{2}(A+A^t) = \begin{pmatrix} a & 0 & 0 \\ 0 & 0 & 0 \\ 0 & 0 & 0 \end{pmatrix}, \quad a,b \in \mathbb{R}.$$
(11.5.5)

It is clear that $\|A - A_{\mathrm{s}}\| = |b|$. On the other hand, with

$$B = \begin{pmatrix} a+\delta & 0 & 0 \\ 0 & 0 & 0 \\ 0 & 0 & 0 \end{pmatrix}, \quad \delta \in \mathbb{R},$$
(11.5.6)

we have $\|A - B\| = |b|$ when $|\delta| \leq |b|$. As a consequence, non-uniqueness of a minimizer of the problem occurs when $|b| > 0$.

The observed issue of non-uniqueness prompts us to consider some modification of the problem. Thus, in order to have sufficient flexibility in handling the formulation of the problem, we now specialize our consideration to matrices. In this situation, it turns out that the use of the Frobenius norm or the Hilbert–Schmidt scalar product resolves the non-uniqueness issue, as stated below.

Theorem 11.5 *Let $A \in \mathbb{C}(n,n)$ and consider the minimization problem*

$$\eta = \inf\{\|A - B\| \mid B \in \mathbb{C}(n,n), B^{\dagger} = B\},$$
(11.5.7)

involving Hermitian matrices. If the Frobenius norm is assumed in (11.5.7), then the Hermitian part of A, namely,

$$A_{\mathrm{h}} = \frac{1}{2}(A+A^{\dagger})$$
(11.5.8)

is the unique solution of (11.5.7), so that the minimal distance η defined in (11.5.7) is given by the Frobenius norm of the anti-Hermitian part of A:

$$\eta = \|A_{\mathrm{a}}\|, \quad A_{\mathrm{a}} = \frac{1}{2}(A - A^{\dagger}).$$
(11.5.9)

Proof The key to this result is the fact that Hermitian and anti-Hermitian elements in $\mathbb{C}(n,n)$, say H and K, satisfying $H^{\dagger} = H$ and $K^{\dagger} = -K$, enjoy the Pythagoras identity with respect to the Frobenius norm, in consequence of their Hilbert–Schmidt scalar product:

$$(H,K) = \mathrm{Tr}(H^{\dagger}K) = \mathrm{Tr}(KH) = -\mathrm{Tr}(K^{\dagger}H) = -(K,H) = -\overline{(H,K)}, \quad (11.5.10)$$

resulting in $\Re\{(H,K)\} = 0$ such that

$$\|H + K\|^2 = \|H\|^2 + \|K\|^2. \qquad (11.5.11)$$

With this property in mind, we have for any A and $B \in \mathbb{C}(n,n)$ with $B^\dagger = B$ the decomposition

$$\|A - B\|^2 = \left\| A - \frac{1}{2}A^\dagger + \frac{1}{2}A^\dagger - B \right\|^2$$

$$= \left\| \frac{1}{2}(A + A^\dagger) - B + \frac{1}{2}(A - A^\dagger) \right\|^2$$

$$= \|A_h - B\|^2 + \|A_a\|^2. \qquad (11.5.12)$$

Hence $\|A - B\|^2 \geq \|A_a\|^2$ and this lower bound is attained if and only if $B = A_h$. Thus the theorem follows. $\qquad\qquad\qquad\qquad\qquad\qquad\qquad\qquad\qquad\qquad\square$

Given $A \in \mathbb{C}(n,n)$, we next consider the nearness problem involving unitary matrices:

$$\eta = \inf\{\|A - B\| \,|\, B \in \mathbb{C}(n,n), B^\dagger B = I\}. \qquad (11.5.13)$$

To proceed, we first consider (11.5.13) under the Frobenius norm. Assume $r(A) = k$ and let $\lambda_1 \geq \cdots \geq \lambda_k > 0$ be the singular values of A and represent A in terms of its singular value decomposition of the form

$$A = P\Lambda Q, \quad \Lambda = \mathrm{diag}\{\lambda_1, \ldots, \lambda_k, \ldots, \lambda_n\}, \quad \lambda_i = 0 \text{ for } i > k \text{ if any}, \quad (11.5.14)$$

where P and $Q \in \mathbb{C}(n,n)$ are unitary matrices. We have

$$\|A - B\| = \|P\Lambda Q - B\| = \|\Lambda - P^\dagger B Q^\dagger\| \equiv \|\Lambda - C\|, \qquad (11.5.15)$$

where $C = (c_{ij}) = P^\dagger B Q^\dagger$ is unitary. Therefore, we have

$$\|\Lambda - C\|^2 = \sum_{i=1}^n |\lambda_i - c_{ii}|^2 + \sum_{i,j=1, i\neq j}^n |c_{ij}|^2$$

$$= \sum_{i=1}^n (\lambda_i^2 - 2\lambda_i \Re\{c_{ii}\}) + \sum_{i,j=1}^n |c_{ij}|^2$$

$$= \sum_{i=1}^n \lambda_i^2 + \mathrm{Tr}(C^\dagger C) - 2\sum_{i=1}^n \lambda_i \Re\{c_{ii}\}$$

$$= \sum_{i=1}^n \lambda_i^2 + n - 2\sum_{i=1}^n \lambda_i \Re\{c_{ii}\}. \qquad (11.5.16)$$

On the other hand, since the row and column vectors of C are all of unit norms in $\mathbb{C}^n$, we have $|c_{ii}| \leq 1$, such that $\Re\{c_{ii}\} \in [-1, 1]$ for $i = 1, \ldots, n$. Thus, to minimize the right-hand side of (11.5.16), it suffices to choose $\Re\{c_{ii}\} = 1$ or $c_{ii} = 1$

for all i. It is clear that this choice is unique if $r(A) = n$ such that $\lambda_1, \ldots, \lambda_n > 0$, otherwise it is not unique. In other words, such unique determination for c_{ii} can only be implemented up to $i = k = r(A)$. So, with the choice $c_{ii} = 1$ for $i = 1, \ldots, n$, we obtain $c_{ij} = 0$ for $i \neq j, i, j = 1, \ldots, n$. Therefore, we arrive at the result $C = I$, which gives us the solution

$$B = PQ. \tag{11.5.17}$$

It should be noted that the unitary matrices P and Q here are not unique, but the resulting matrix B, which is the product of P and Q, is unique. That is, the solution (11.5.17) is unique. To see why, we proceed as follows.

For simplicity, assume that the first k singular values of A are identical,

$$\lambda_1 = \cdots \lambda_k \equiv \lambda_0. \tag{11.5.18}$$

Then we have

$$\Lambda = \begin{pmatrix} \lambda_0 I_k & 0 \\ 0 & \Lambda_2 \end{pmatrix}, \tag{11.5.19}$$

where the diagonal matrix $\Lambda_2 \in \mathbb{R}(n-k, n-k)$ collects the rest $n-k$ singular values of A. Moreover, we rewrite (11.5.14) as

$$A(u_1, \ldots, u_k, \ldots, u_n) = (v_1, \ldots, v_k, \ldots, v_n)\Lambda, \quad \text{or} \quad AR = P\Lambda, \tag{11.5.20}$$

where

$$P = (v_1, \ldots, v_k, \ldots, v_n), \quad R = (u_1, \ldots, u_k, \ldots, u_n), \quad R = Q^\dagger. \tag{11.5.21}$$

Then

$$Au_1 = \lambda_0 v_1, \quad \ldots, \quad Au_k = \lambda_0 v_k; \quad Au_i = \lambda_i v_i, \quad k < i \leq n \quad \text{(if any)}. \tag{11.5.22}$$

Thus, setting

$$P_1 = (v_1, \ldots, v_k) \in \mathbb{C}(n, k), \quad P_2 = (v_{k+1}, \ldots, v_n) \in \mathbb{C}(n, n-k), \tag{11.5.23}$$

$$R_1 = (u_1, \ldots, u_k) \in \mathbb{C}(n, k), \quad R_2 = (u_{k+1}, \ldots, u_n) \in \mathbb{C}(n, n-k), \tag{11.5.24}$$

we have

$$\begin{aligned} AR &= A(R_1, R_2) \\ &= (AR_1, AR_2) \\ &= (\lambda_0 P_1, P_2 \Lambda_2) \\ &= (P_1, P_2) \begin{pmatrix} \lambda_0 I_k & 0 \\ 0 & \Lambda_2 \end{pmatrix}. \end{aligned} \tag{11.5.25}$$

Now let $\tilde{P}$ and $\tilde{Q}$ be another pair of unitary matrices to render the solution (11.5.17) such that possible ambiguities may arise from the first k singular values, $\lambda_1 = \cdots = \lambda_k = \lambda_0$, since we are free with choosing an orthonormal basis of the eigenspace $E_{\lambda_0^2}$ of the matrix $A^{\dagger}A$. In other words, we may pick another orthonormal basis, say

$$\{\tilde{u}_1, \ldots, \tilde{u}_k\}, \tag{11.5.26}$$

for $E_{\lambda_0^2}$, to get the modified relations

$$A\tilde{u}_1 = \lambda_0\tilde{v}_1, \quad \ldots, \quad A\tilde{u}_k = \lambda_0\tilde{v}_k. \tag{11.5.27}$$

Thus, we obtain

$$A(\tilde{u}_1, \ldots, \tilde{u}_k, \ldots, \tilde{u}_n) = (\tilde{v}_1, \ldots, \tilde{v}_k, \ldots, \tilde{v}_n)\Lambda, \tag{11.5.28}$$

where $\tilde{u}_i = u_i$ and $\tilde{v}_i = v_i$ for $k < i \leq n$ (if any). Hence we have

$$
\begin{aligned}
A\tilde{R} &= A(\tilde{R}_1, \tilde{R}_2) \\
&= (\tilde{P}_1, \tilde{P}_2)\begin{pmatrix} \lambda_0 I_k & 0 \\ 0 & \Lambda_2 \end{pmatrix},
\end{aligned}
\tag{11.5.29}
$$

where

$$\tilde{P}_1 = (\tilde{v}_1, \ldots, \tilde{v}_k) \in \mathbb{C}(n,k), \quad \tilde{P}_2 = (\tilde{v}_{k+1}, \ldots, \tilde{v}_n) = P_2 \in \mathbb{C}(n,n-k), \tag{11.5.30}$$

$$\tilde{R}_1 = (\tilde{u}_1, \ldots, \tilde{u}_k) \in \mathbb{C}(n,k), \quad \tilde{R}_2 = (\tilde{u}_{k+1}, \ldots, \tilde{u}_n) = R_2 \in \mathbb{C}(n,n-k). \tag{11.5.31}$$

On the other hand, let $\Gamma = (\gamma_{ij}) \in \mathbb{C}(k,k)$ be a unitary matrix such that

$$\tilde{u}_j = \sum_{i=1}^{k} \gamma_{ij} u_i, \quad j = 1, \ldots, k. \tag{11.5.32}$$

Thus, we have $\tilde{R}_1 = R_1\Gamma$, which leads to

$$\lambda_0\tilde{P}_1 = A\tilde{R}_1 = AR_1\Gamma = \lambda_0 P_1\Gamma. \tag{11.5.33}$$

Since $\lambda_0 > 0$, we draw the conclusion $\tilde{P}_1 = P_1\Gamma$, which is natural to anticipate.

In summary, we get the relations

$$\tilde{P} = (\tilde{P}_1, \tilde{P}_2) = (P_1\Gamma, P_2) = (P_1, P_2)\begin{pmatrix} \Gamma & 0 \\ 0 & I_{n-k} \end{pmatrix} = P\begin{pmatrix} \Gamma & 0 \\ 0 & I_{n-k} \end{pmatrix}, \tag{11.5.34}$$

$$\tilde{R} = (\tilde{R}_1, \tilde{R}_2) = (R_1\Gamma, R_2) = (R_1, R_2)\begin{pmatrix} \Gamma & 0 \\ 0 & I_{n-k} \end{pmatrix} = R\begin{pmatrix} \Gamma & 0 \\ 0 & I_{n-k} \end{pmatrix}. \tag{11.5.35}$$

Consequently, with $Q = R^\dagger$ in (11.5.17), and $\tilde{Q} = \tilde{R}^\dagger$, we have

$$\tilde{P}\tilde{Q} = P\begin{pmatrix} \Gamma & 0 \\ 0 & I_{n-k} \end{pmatrix}\begin{pmatrix} \Gamma^\dagger & 0 \\ 0 & I_{n-k} \end{pmatrix}Q = PQ. \qquad (11.5.36)$$

This result confirms that the solution (11.5.17) is unique.
Besides, the right-hand side of (11.5.16) reads

$$\|\Lambda - C\|^2 = \sum_{i=1}^{k}\lambda_i^2 + n - 2\sum_{i=1}^{k}\lambda_i = (n-k) + \sum_{i=1}^{k}(\lambda_i - 1)^2, \qquad (11.5.37)$$

which determines the minimal distance η given in (11.5.13).

We then consider the problem (11.5.13) under the induced norm. With regard to (11.5.15) under the induced matrix norm, we use the standard basis $\{e_1, \ldots, e_n\}$ to get

$$\|(\Lambda - C)e_j\|^2 = |\lambda_j - c_{jj}|^2 + \sum_{i=1, i\neq j}^{n}|c_{ij}|^2$$

$$= \lambda_j^2 - 2\lambda_j\Re\{c_{jj}\} + \sum_{i=1}^{n}|c_{ij}|^2$$

$$= \lambda_j^2 - 2\lambda_j\Re\{c_{jj}\} + 1$$

$$\geq (\lambda_j - 1)^2, \quad j = 1, \ldots, n, \qquad (11.5.38)$$

since $\Re\{c_{jj}\} \in [-1, 1]$. As a consequence of (11.5.38), we obtain the lower bound

$$\inf\{\|\Lambda - C\|^2 \,|\, C \in \mathbb{C}(n,n), C^\dagger C = I\} \geq \max\{(\lambda_i - 1)^2 \,|\, i = 1, \ldots, n\}, \qquad (11.5.39)$$

in view of the definition of the induced norm. On the other hand, since $\Lambda - I$ is diagonal, we see that any unit vector $u = \sum_{i=1}^{n}a_ie_i$ satisfies

$$\|(\Lambda - I)u\|^2 = \sum_{i=1}^{n}(\lambda_i - 1)^2|a_i|^2, \quad \sum_{i=1}^{n}|a_i|^2 = 1. \qquad (11.5.40)$$

From (11.5.40) we readily deduce

$$\|\Lambda - I\| = \max\{|\lambda_i - 1| \,|\, i = 1, \ldots, n\}. \qquad (11.5.41)$$

Consequently, in view of (11.5.39) and (11.5.41), we see that the problem (11.5.13) under the induced matrix norm is still solved by (11.5.17) but the quantity η there is now given by

$$\eta = \|\Lambda - I\| = \max\{|\lambda_i - 1| \,|\, i = 1, \ldots, n\}. \qquad (11.5.42)$$

We give an example to show that the problem (11.5.13) under the induced norm may not have a unique solution. In fact, consider

$$A = \Lambda = \operatorname{diag}\{3,1\}. \tag{11.5.43}$$

Then a solution to (11.5.13) (with $n = 2$) is $B = I$ because $\eta = 2$. However, any member in the family of the unitary matrices

$$C_\theta = \operatorname{diag}\{1, e^{i\theta}\}, \quad \theta \in \mathbb{R}, \tag{11.5.44}$$

also satisfies $\|A - C_\theta\| = 2$. Hence non-uniqueness of a minimizer of (11.5.13) occurs.

Thus we can state:

Theorem 11.6 *For any given $A \in \mathbb{C}(n,n)$, consider the nearness problem (11.5.13) involving unitary matrices. Let the singular value decomposition of A be given by (11.5.14). Under both the Frobenius and induced norms, a solution to the problem (11.5.13) is given by the same formula (11.5.17), such that the minimal distance η in (11.5.13) enjoys the expression*

$$\eta = \begin{cases} \sqrt{\sum_{i=1}^{n}(\lambda_i - 1)^2}, & \text{in the Frobenius norm case,} \\ \max\{|\lambda_i - 1| \,|\, i = 1,\ldots,n\}, & \text{in the induced norm case,} \end{cases} \tag{11.5.45}$$

respectively. In the Frobenius norm case, the solution (11.5.17) is unique when $r(A) = n$, but not unique when $r(A) < n$. In the induced norm case, the solution (11.5.17) fails to be unique, regardless of the rank of A.

Similarly, we can solve the real matrix version of the problem (11.5.13): Given $A \in \mathbb{R}(n,n)$, find a minimizer to the problem

$$\eta = \inf\{\|A - B\| \,|\, B \in \mathbb{R}(n,n), B^t B = I\}, \tag{11.5.46}$$

involving orthogonal matrices.

The methods of this section may be applied to study other matrix nearness problems. For example, given $A \in \mathbb{C}(n,n)$, consider

$$\eta = \inf\{\|A - B\| \,|\, B \in \mathbb{C}(n,n), B^\dagger = B, B \text{ is positive semi-definite}\}. \tag{11.5.47}$$

First, we consider the problem under the Frobenius norm. Express any $A \in \mathbb{C}(n,n)$ as $A = A_h + A_a$ following (11.5.8) and (11.5.9), we have

$$\|A - B\|^2 = \|A_h - B\|^2 + \|A_a\|^2, \quad B \in \mathbb{C}(n,n), \quad B^\dagger = B. \tag{11.5.48}$$

Let the spectral decomposition of A_h be

$$A_h = PDP^\dagger, \quad D = \operatorname{diag}\{\alpha_1,\ldots,\alpha_l,\ldots,\alpha_n\} \in \mathbb{R}(n,n), \tag{11.5.49}$$

where $P \in \mathbb{C}(n,n)$ is unitary and $\alpha_1, \ldots, \alpha_l$ are all the non-negative eigenvalues of A_h (counting multiplicities). Thus, setting $P^\dagger B P = C = (c_{ij})$ in (11.5.48), we have

$$\|A_h - B\|^2 = \|D - C\|^2$$
$$= \sum_{i=1}^{n} |\alpha_i - c_{ii}|^2 + \sum_{i,j=1,i\neq j}^{n} |c_{ij}|^2. \tag{11.5.50}$$

Since B is positive semi-definite, we have $c_{ii} \geq 0$ for $i = 1, \ldots, n$. Thus, in order to minimize the right-hand side of (11.5.50), we choose

$$c_{11} = \alpha_1, \ldots, c_{ll} = \alpha_l; \quad c_{ij} = 0, \quad \text{for other } i, j = 1, \ldots, n. \tag{11.5.51}$$

Consequently, we obtain a solution to (11.5.47) to be

$$B = P\,\mathrm{diag}\{\alpha_1, \ldots, \alpha_l, 0, \ldots, 0\}P^\dagger \equiv PD^+P^\dagger, \tag{11.5.52}$$

which is exactly the non-negative part of A_h in view of (11.5.49). Thus, in this case, we have

$$\|D - C\|^2 = \sum_{i>l}^{n} \alpha_i^2 \quad \text{(if any)}, \tag{11.5.53}$$

which renders the value of the quantity η in (11.5.47) to be

$$\eta = \sqrt{\sum_{i>l}^{n} \alpha_i^2 + \|A_a\|^2}. \tag{11.5.54}$$

Next, we consider the problem under the induced norm. In this situation, we no longer have (11.5.48) available. Thus, we compromise a bit to assume that A itself is Hermitian to consider (11.5.47) again. Thus, we replace (11.5.49) by

$$A = PDP^\dagger, \quad D = \mathrm{diag}\{\alpha_1, \ldots, \alpha_l, \ldots, \alpha_n\} \in \mathbb{R}(n,n), \tag{11.5.55}$$

keeping all other assumptions and notations. Then, since the diagonal entries of $C = (c_{ij})$ are non-negative, we have

$$\|(D-C)e_j\|^2 = (\alpha_j - c_{jj})^2 + \sum_{i=1,i\neq j}^{n} |c_{ij}|^2$$
$$\geq (\alpha_j - c_{jj})^2, \quad j = 1, \ldots, n, \tag{11.5.56}$$

which implies

$$\|D - C\|^2 \geq \max\{\alpha_j^2 \mid \alpha_j < 0 \text{ or } j > l \text{ (if any)}\}. \tag{11.5.57}$$

It is clear that, in order to saturate the lower bound in (11.5.57), it suffices to assume that C satisfies (11.5.51) again, which gives us the minimal distance

$$\eta = \|D - C\| = \max\{|\alpha_i| \mid \alpha_i < 0 \text{ or } i > l \text{ (if any)}\}, \tag{11.5.58}$$

in view of (11.5.57). This solves the nearness problem for a Hermitian matrix, involving positive semi-definite matrices, under the induced norm.

Exercises

11.5.1 Analogous to (11.5.1), consider the problem

$$\eta = \inf\{\|T - R\| \mid R \in L(U), R' = -R\}, \qquad (11.5.59)$$

where $\|\cdot\|$ is the induced norm. Show that the anti-symmetric or anti-self-adjoint part of T given in (11.5.4) solves (11.5.59) and that the quantity η in (11.5.59) is simply $\|T_S\|$, where T_S is defined by (11.5.2).

11.5.2 Give an example to show that the problem (11.5.59) may not have a unique solution.

11.5.3 Prove that, given $A \in \mathbb{C}(n,n)$, the minimization problem

$$\eta = \inf\{\|A - B\| \mid B \in \mathbb{C}(n,n), B^\dagger = -B\}, \qquad (11.5.60)$$

under the Frobenius norm has a unique solution which is given by the anti-Hermitian part of A given in (11.5.9) and the quantity η in (11.5.60) is simply the Frobenius norm of the symmetric part of A. That is, $\eta = \|A_h\|$, where A_h is defined in (11.5.8).

11.5.4 Let H and K be a pair of Hermitian and anti-Hermitian matrices given by

$$H = \begin{pmatrix} 0 & 1 \\ 1 & 0 \end{pmatrix}, \quad K = \begin{pmatrix} 0 & i \\ i & 0 \end{pmatrix}. \qquad (11.5.61)$$

Verify that H and K are not perpendicular with respect to the Hilbert–Schmidt scalar product over $\mathbb{C}(2,2)$, but they fulfill the Pythagoras identity in terms of the Frobenius norm.

11.5.5 Let H and K be a pair of symmetric and anti-symmetric matrices in $\mathbb{R}(n,n)$. Show that H and K are perpendicular with respect to the Hilbert–Schmidt scalar product. In other words, the usual direct sum

$$\mathbb{R}(n,n) = \mathbb{R}_S(n,n) \oplus \mathbb{R}_A(n,n) \qquad (11.5.62)$$

into the subspaces of symmetric and anti-symmetric matrices achieves an orthogonal decomposition of $\mathbb{R}(n,n)$ with respect to the Hilbert–Schmidt scalar product. (However, Problem 11.5.4 indicates that the same conclusion about the orthogonal decomposition is not valid in the complex situation when symmetry is replaced by Hermitian symmetry.)

11.5.6 Establish an existence and uniqueness theorem for the problem (11.5.46) in analogy to Theorem 11.6.

11.5.7 For the matrix

$$A = \begin{pmatrix} 3 & 2 & 1 \\ 1 & 3 & 2 \\ 2 & 1 & 3 \end{pmatrix} \in \mathbb{R}(3,3), \qquad (11.5.63)$$

obtain its unique best orthogonal matrix approximation that solves the problem (11.5.46) with $n = 3$.

12

Excursion: Multilinear Algebra

In this chapter, we extend our study of linear algebraic structures to multilinear ones which have broad and profound applications beyond those covered by linear structures. We first give some remarks on the rich applications of multilinear algebra and consider multilinear forms in a general setting as a starting point which directly generates bilinear forms already studied. We next specialize our discussion to consider tensors and their classifications. We then elaborate on symmetric and anti-symmetric tensors and investigate their properties and characterizations. Finally, we discuss exterior algebras and the Hodge dual correspondence.

12.1 Applications of multilinear algebra

Multilinear algebra, also known as tensor calculus, finds extensive applications in pure mathematics, physics, and engineering due to its ability to describe and analyze multidimensional structures and transformations. Here we explain why and how multilinear algebra is used in these fields.

Pure mathematics

Algebraic structures: Multilinear algebra provides a framework for studying and understanding various algebraic structures such as vector spaces, modules, algebras, and tensors. It allows mathematicians to generalize concepts from linear algebra to higher-dimensional spaces and to define operations and properties that are applicable to multidimensional objects.

Multilinear maps: Multilinear algebra studies multilinear maps, which are mappings that are linear in each of their arguments. Multilinear maps are important in algebraic topology, differential geometry, and differential topology, where they are used to define and study concepts such as wedge products, exterior derivatives, covariant derivatives, differential forms, cohomology, and topological invariants.

Tensor products: Multilinear algebra introduces the concept of tensor products, which provide a way to combine vector spaces and modules to create new spaces. Tensor products are used in algebraic geometry, representation theory, and functional analysis to define and study tensor fields, tensor products of modules, and symmetric and anti-symmetric powers of vector spaces.

Physics

Quantum mechanics: Multilinear algebra plays a crucial role in quantum mechanics, where it is used to describe the state spaces, operators, and observables of quantum systems. Quantum states are represented as vectors in complex Hilbert spaces, and operators are represented as matrices or linear transformations. Techniques such as tensor products and inner products are used to model composite systems, entanglement, and quantum measurements. Some of these constructions will be encountered in Chapter 13.

General relativity: In general relativity, multilinear algebra is used to describe the geometry and curvature of spacetime. Tensors are used to represent physical quantities such as the metric tensor, curvature tensor, and stress-energy tensor. The Einstein field equations, which describe the gravitational field as the curvature of spacetime, are formulated using tensors and differential geometry.

Fluid dynamics: Multilinear algebra is used in fluid dynamics to describe the kinematics and dynamics of fluids. Tensors are used to represent velocity gradients, stress tensors, and fluid flow equations. Techniques such as tensor calculus and multilinear maps are used to derive and solve the Navier–Stokes equations, which describe the motion of viscous fluids.

Electromagnetism: In electromagnetism, multilinear algebra is used to describe electromagnetic fields and interactions. Tensors are used to represent the electromagnetic field tensor, the Maxwell equations, and electromagnetic waves. Techniques such as tensor products and contractions are used to compute electric and magnetic fields, electromagnetic potentials, and electromagnetic forces. In particular, as will be seen in Section 12.5, the Hodge dual operation renders the celebrated electromagnetic duality an expected mathematical structure of the unified electromagnetic field tensor under consideration.

Quantum field theory: Tensors play a central role in quantum field theory by providing a mathematical framework for describing the properties, dynamics, and interactions of quantum fields. They are used to represent physical particles, the action densities governing physical laws, symmetries, quantum states, scattering amplitudes, conservation laws, and other quantities.

Engineering

Solid mechanics: Multilinear algebra is used to describe the stress and strain tensors, which represent the state of deformation and stress in materials. Tensors are used to characterize the mechanical behavior of materials under different loading conditions and to derive constitutive equations governing their response to applied forces.

Continuum mechanics: In continuum mechanics, tensors are used to describe physical quantities such as displacement, velocity, acceleration, and deformation gradients. Multilinear algebra techniques are used to derive the balance equations of mass, momentum, and energy, as well as to solve boundary value problems for solid and fluid mechanics.

Structural analysis: Multilinear algebra is used in structural analysis to model and analyze the behavior of complex structures such as beams, plates, and shells. Tensors are used to represent stiffness matrices, load vectors, and displacement fields, and multilinear algebra techniques are used to solve eigenvalue problems, boundary value problems, and optimization problems in structural design.

In short, multilinear algebra is a powerful mathematical tool that finds diverse applications in pure mathematics, physics, and engineering. Its ability to describe and analyze multidimensional structures and transformations makes it indispensable for understanding complex systems and phenomena in these fields.

12.2 From bilinear forms to multilinear forms

Let U be a vector space over a field $\mathbb{F}$. Recall that a function $f : U \times U \to \mathbb{F}$ is called a *bilinear form* or simply a *2-form* if for any $u, v, w \in U$ and $a \in \mathbb{F}$ it satisfies the following:

(1) $f(u+v, w) = f(u, w) + f(v, w)$, $f(u, v+w) = f(u, v) + f(u, w)$.
(2) $f(au, v) = f(u, av) = af(u, v)$.

This concept can directly be carried over to the more general case $f : U \times V \to \mathbb{F}$, where U and V are a pair of vector spaces over $\mathbb{F}$, so that f is linear in each of its two arguments. In other words, if U' and V' are the dual spaces of U and V, respectively, also called the spaces of *covectors*, then for fixed $u \in U$ and $v \in V$, we have $f(\cdot, v) \in U'$ and $f(u, \cdot) \in V'$. Thus, if U and V are of finite dimensions and $\{u_1, \ldots, u_m\}$ and $\{v_1, \ldots, v_n\}$ are the bases of U and V, respectively, then for $u = \sum_{i=1}^{m} x_i u_i \in U$ and $v = \sum_{j=1}^{n} y_j v_j \in V$, we have

$$f(u,v) = \sum_{i=1}^{m} \sum_{j=1}^{n} x_i f(u_i, v_j) y_j$$

$$= x^t A y, \quad x = \begin{pmatrix} x_1 \\ \vdots \\ x_m \end{pmatrix} \in \mathbb{F}^m, \quad y = \begin{pmatrix} y_1 \\ \vdots \\ y_n \end{pmatrix} \in \mathbb{F}^n, \quad (12.2.1)$$

where $A = (a_{ij}) = (f(u_i, v_j)) \in \mathbb{F}(m,n)$ is the matrix representation of the 2-form f with respect to the given bases. A special situation we have encountered is when V is taken to be U'. Then the pairing

$$\langle \cdot, \cdot \rangle : U \times U' \to \mathbb{F}, \quad \langle u, u' \rangle, \quad u \in U, \quad u' \in U' \quad (12.2.2)$$

defines a bilinear form from $U \times U'$ into $\mathbb{F}$. Motivated by these examples, we naturally consider a *multilinear form*, or *m*-form,

$$f : U_1 \times \cdots \times U_m \to \mathbb{F}, \quad (12.2.3)$$

where $U_1, \ldots, U_m$ are m vector spaces over $\mathbb{F}$ so that f is linear in each of its m arguments. Under the operations of addition of m-forms and scalar multiplication of m-forms by scalars, the set of m-forms becomes a vector space over $\mathbb{F}$. Moreover, let

$$g : V_1 \times \cdots \times V_n \to \mathbb{F} \quad (12.2.4)$$

be an n-form. Then we can define an $(m+n)$-form,

$$h = f \otimes g : U_1 \times \cdots \times U_m \times V_1 \times \cdots \times V_n \to \mathbb{F}, \quad (12.2.5)$$

by setting

$$h(u_1, \ldots, u_m, v_1, \ldots, v_n) = f(u_1, \ldots, u_m) g(v_1, \ldots, v_n), \quad (12.2.6)$$

where $u_i \in U_i, i = 1, \ldots, m, v_j \in V_j, j = 1, \ldots, n$. An application of such a construction is that we may pick $u'_i \in U'_i$ for each $i = 1, \ldots, m$ and form an m-form by setting

$$f = u'_1 \otimes \cdots \otimes u'_m : U_1 \times \cdots \times U_m \to \mathbb{F}. \quad (12.2.7)$$

It is useful and important to observe that in finite-dimensional situations all multilinear forms may be constructed this way. More precisely, we have

Theorem 12.1 *Let* $U_1, \ldots, U_m$ *be m vector spaces over a field* $\mathbb{F}$ *of respective dimensions* $n_1, \ldots, n_m$ *and* $\mathcal{M}$ *the vector space of all m-forms over* $\mathbb{F}$. *Then* $\dim(\mathcal{M}) = n_1, \ldots, n_m$. *Moreover, if*

$$\{u'_{1,1}, \ldots, u'_{1,n_1}\}, \ldots, \{u'_{m,1}, \ldots, u'_{m,n_m}\} \quad (12.2.8)$$

are the bases of the dual spaces $U_1', \ldots, U_m'$ of $U_1, \ldots, U_m$, respectively, then the set of m-forms

$$\{u_{1,j_1}' \otimes \cdots \otimes u_{m,j_m}' \mid j_1 = 1, \ldots, n_1, \ldots, j_m = 1, \ldots, n_m\} \qquad (12.2.9)$$

forms a basis of $\mathcal{M}$.

Proof Let $\{u_{i,1}, \ldots, u_{i,n_i}\}$ be a basis of U_i and $\{u_{i,1}', \ldots, u_{i,n_i}'\}$ a basis of U_i' dual to $\{u_{i,1}, \ldots, u_{i,n_i}\}$, $i = 1, \ldots, m$. Then the vectors

$$(u_{1,j_1}, \ldots, u_{m,j_m}), \quad j_1 = 1, \ldots, n_1, \quad j_m = 1, \ldots, n_m, \qquad (12.2.10)$$

in the product space $U_1 \times \cdots \times U_m$ will serve as appropriate testing vectors for our construction of $\mathcal{M}$. In fact, for any $f \in \mathcal{M}$, set

$$f(u_{1,j_1}, \ldots, u_{m,j_m}) = a_{1,j_1;\ldots;m,j_m}, \quad j_1 = 1, \ldots, n_1, \quad j_m = 1, \ldots, n_m. \qquad (12.2.11)$$

Then define $g \in \mathcal{M}$ by

$$g = \sum_{j_1, \ldots, j_m} a_{1,j_1;\ldots;m,j_m} u_{1,j_1}' \otimes \cdots \otimes u_{m,j_m}'. \qquad (12.2.12)$$

In view of (12.2.11) and (12.2.12), we have

$$f(u_{1,j_1}, \ldots, u_{m,j_m}) = g(u_{1,j_1}, \ldots, u_{m,j_m}), \ j_1 = 1, \ldots, n_1, \ j_m = 1, \ldots, n_m. \qquad (12.2.13)$$

Since both f and g are multilinear, we have $f = g$ over the full product space $U_1 \times \cdots \times U_m$. Thus, we have shown that $\mathcal{M}$ is spanned by the m-forms listed in (12.2.9). To see that these m-forms are linearly independent, set

$$\sum_{j_1, \ldots, j_m} c_{1,j_1;\ldots;m,j_m} u_{1,j_1}' \otimes \cdots \otimes u_{m,j_m}' = 0, \qquad (12.2.14)$$

where the coefficients c's are scalars. Applying the equation (12.2.14) to the vector $(u_{1,j_1}, \ldots, u_{m,j_m})$, we get

$$c_{1,j_1;\ldots;m,j_m} = 0, \quad j_1 = 1, \ldots, n_1, \quad j_m = 1, \ldots, n_m, \qquad (12.2.15)$$

which accomplishes the proof. $\qquad\qquad\qquad\qquad\qquad\qquad\qquad\qquad\qquad\square$

If we denote the vector space $\mathcal{M}$ of m-forms over $U_1 \times \cdots \times U_m$ specifically by $\mathcal{M}(U_1 \times \cdots \times U_m)$, then Theorem 12.1 gives us the relation

$$\mathcal{M}(U_1 \times \cdots \times U_m) = U_1' \otimes \cdots \otimes U_m' = \mathcal{M}(U_1) \otimes \cdots \otimes \mathcal{M}(U_m), \qquad (12.2.16)$$

in a self-evident manner. As a consequence, we have

$$\mathcal{M}(U_1 \times \cdots \times U_k \times V_1 \times \cdots \times V_l) = \mathcal{M}(U_1 \times \cdots \times U_k) \otimes \mathcal{M}(V_1 \times \cdots \times V_l), \qquad (12.2.17)$$

where $U_1, \ldots, U_k, V_1, \ldots, V_l$ are finite-dimensional vector spaces over $\mathbb{F}$.

As an implication, we also have

$$\mathcal{M}(U_1' \times \cdots \times U_m') = U_1 \otimes \cdots \otimes U_m, \qquad (12.2.18)$$

as a special realization of the discussion.

Exercises

12.2.1 For the functionals f and g over $U = \mathbb{R}^3$ given by

$$f(x_1, x_2, x_3) = \sum_{i=1}^{3} a_i x_i, \quad g(x_1, x_2, x_3) = \sum_{i=1}^{3} b_i x_i, \qquad (12.2.19)$$

where $x_i, a_i, b_i \in \mathbb{R}$, $i = 1, 2, 3$, find the matrix that represents the bilinear form $f \otimes g$ under the standard basis of U.

12.2.2 Let $U = \mathbb{R}^2$ and $V = \mathbb{R}^3$ and consider the bilinear form $h : U \times V \to \mathbb{R}$ defined by

$$h(u, v) = u^t \begin{pmatrix} -2 & 3 & 1 \\ 1 & -2 & 0 \end{pmatrix} v, \quad u \in U, \quad v \in V. \qquad (12.2.20)$$

Find some f_1 and $f_2 \in U'$, g_1, g_2, and $g_3 \in V'$, and $h_{ij} \in \mathbb{R}$ ($i = 1, 2, j = 1, 2, 3$) such that

$$h = \sum_{i=1,2, j=1,2,3} h_{ij} (f_i \otimes g_j). \qquad (12.2.21)$$

12.2.3 Let U be a vector space and f and $g \in U'$. Explain that, in general, $f \otimes g \neq g \otimes f$ and find a necessary and sufficient condition to ensure $f \otimes g = g \otimes f$ for any f and $g \in U'$.

12.2.4 Establish (12.2.17).

12.2.5 Let $U_1, \ldots, U_m$ be finite-dimensional vector spaces over a field $\mathbb{F}$ and $U_1', \ldots, U_m'$ their dual spaces, respectively. Show that the pairing

$$\langle u_1 \otimes \cdots \otimes u_m, u_1' \otimes \cdots \otimes u_m' \rangle \equiv \langle u_1, u_1' \rangle \cdots \langle u_m, u_m' \rangle,$$

$$u_i \in U_i, \quad u_i' \in U_i', \quad i = 1, \ldots, m, \qquad (12.2.22)$$

realizes the dual space relation $(U_1 \otimes \cdots \otimes U_m)' = U_1' \otimes \cdots \otimes U_m'$, which generates the dual basis of $(U_1 \otimes \cdots \otimes U_m)'$ from those of $U_1, \ldots, U_m$ correspondingly.

12.3 Tensors, tensor spaces, and tensor types

The most important application of the multilinear form (12.2.3) is when all the vector spaces $U_1, \ldots, U_m$ are a single vector space, U, which is of finite dimension, which will be our focus now. That is, we shall begin to consider multilinear forms of the type

$$T : \underbrace{U \times \cdots \times U}_{m} \to \mathbb{F}, \tag{12.3.1}$$

which is called a *tensor* of order, rank, or degree m. The vector space of all such tensors, also referred to as *covariant tensors* or simply *cotensors*, of order m is denoted $\mathcal{T}^m$. Thus, in the context of the notation (12.2.16), we have

$$\mathcal{T}^m = \mathcal{M}(\underbrace{U \times \cdots \times U}_{m}) = \underbrace{U' \otimes \cdots \otimes U'}_{m}. \tag{12.3.2}$$

So $\mathcal{T}^1 = U'$ and the elements in the dual vector space U' are covariant vectors or covectors.

Let $\{u_1, \ldots, u_n\}$ be a basis of U and $\{u'_1, \ldots, u'_n\}$ the dual basis of U'. It will be convenient to adopt the notation

$$u'_i = u^i, \quad i = 1, \ldots, n, \tag{12.3.3}$$

in the context of the rest of this chapter. Hence we have

$$\langle u_i, u^j \rangle = \langle u^j, u_i \rangle = \delta^i_j = \begin{cases} 1, & i = j, \\ 0, & i \neq j, \end{cases} \quad i, j = 1, \ldots, n, \tag{12.3.4}$$

and in view of (12.3.2), we may represent T in (12.3.1) as

$$T = \sum_{i_1, \ldots, i_m} T_{i_1 \ldots i_m} u^{i_1} \otimes \cdots \otimes u^{i_m}, \quad T_{i_1 \ldots i_m} \in \mathbb{F}, \quad i_1, \ldots, i_m = 1, \ldots, n. \tag{12.3.5}$$

It is clear that

$$T(u_{i_1}, \ldots, u_{i_m}) = T_{i_1 \ldots i_m}, \quad i_1, \ldots, i_m = 1, \ldots, n. \tag{12.3.6}$$

Let $\{\tilde{u}_1, \ldots, \tilde{u}_n\}$ be another basis of U and $\{\tilde{u}^1, \ldots, \tilde{u}^n\}$ the corresponding dual basis of U'. Then there is an invertible matrix $A = (a^i_j) \in \mathbb{F}(n, n)$ such that

$$\tilde{u}_j = \sum_{i=1}^{n} a^i_j u_i, \quad j = 1, \ldots, n. \tag{12.3.7}$$

Thus, from (12.3.6), we have

$$\tilde{T}_{i_1 \ldots i_m} = T(\tilde{u}_{i_1}, \ldots, \tilde{u}_{i_m}) = T\left(\sum_{k_1=1}^{n} a^{k_1}_{i_1} u_{k_1}, \ldots, \sum_{k_m=1}^{n} a^{k_m}_{i_m} u_{k_m} \right)$$

$$= \sum_{k_1=1}^{n} \cdots \sum_{k_m=1}^{n} a^{k_1}_{i_1} \cdots a^{k_m}_{i_m} T_{k_1 \ldots k_m}. \tag{12.3.8}$$

Besides, if the dual basis transforms itself accordingly following the rule:

$$\tilde{u}^j = \sum_{i=1}^{n} b^j_i u^i, \quad j = 1, \ldots, n, \tag{12.3.9}$$

where $B = (b_i^j) \in \mathbb{F}(n,n)$, then (12.3.4) leads to

$$\delta_i^j = \langle \tilde{u}_i, \tilde{u}^j \rangle = \left\langle \sum_{k=1}^{n} a_i^k u_k, \sum_{l=1}^{n} b_l^j u^l \right\rangle = \sum_{k,l=1}^{n} a_i^k b_l^j \delta_k^l = \sum_{k=1}^{n} a_i^k b_k^j, \quad (12.3.10)$$

indicating the fact that $B = A^{-1}$.

We now consider a similar construction of another type of tensor of order m based on the m-form

$$T : \underbrace{U' \times \cdots \times U'}_{m} \to \mathbb{F} \quad (12.3.11)$$

and denote the vector space of all such tensors as $\mathcal{T}_m$. In contrast, these tensors are often referred to as *contravariant tensors*. Since $(U')' = U$, we have

$$\mathcal{T}_m = \mathcal{M}(\underbrace{U' \times \cdots \times U'}_{m}) = \underbrace{U \otimes \cdots \otimes U}_{m}, \quad (12.3.12)$$

in a natural manner so that $\mathcal{T}_1 = U$ and the vectors in U are also referred to as *contravariant vectors*, and analogous to (12.3.5), any $T \in \mathcal{T}_m$ may be expressed as

$$T = \sum_{i_1,\ldots,i_m} T^{i_1 \ldots i_m} u_{i_1} \otimes \cdots \otimes u_{i_m}, \quad T^{i_1 \ldots i_m} \in \mathbb{F}, \quad i_1,\ldots,i_m = 1,\ldots,n.$$
$$(12.3.13)$$

Thus, similar to (12.3.8), we have

$$\tilde{T}^{i_1 \ldots i_m} = \sum_{k_1=1}^{n} \cdots \sum_{k_m=1}^{n} b_{k_1}^{i_1} \cdots b_{k_m}^{i_m} T^{k_1 \ldots k_m}, \quad (12.3.14)$$

under the change of the dual basis given by (12.3.9).

Finally, we consider a mixed-type multilinear form extending those given in (12.3.1) and (12.3.11):

$$T : \underbrace{U' \times \cdots \times U'}_{p} \times \underbrace{U \times \cdots \times U}_{q} \to \mathbb{F} \quad (12.3.15)$$

and denote the vector space of all such forms as $\mathcal{T}_p^q$. With $p+q = m$, an element in $\mathcal{T}_p^q$ is called an order m tensor of the type (p,q). Then we have

$$\mathcal{T}_p^q = \mathcal{M}(\underbrace{U' \times \cdots \times U'}_{p} \times \underbrace{U \times \cdots \times U}_{q}) = \underbrace{U \otimes \cdots \otimes U}_{p} \otimes \underbrace{U' \otimes \cdots \otimes U'}_{q},$$
$$(12.3.16)$$

so that any $T \in \mathcal{T}_p^q$ may be expressed as

$$T = \sum_{i_1,\ldots,i_p,j_1,\ldots,j_q} T^{i_1 \ldots i_p}_{j_1 \ldots j_q} u_{i_1} \otimes \cdots \otimes u_{i_p} \otimes u^{j_1} \otimes \cdots \otimes u^{j_q}, \quad (12.3.17)$$

where $T^{i_1 \ldots i_p}_{j_1 \ldots j_q} \in \mathbb{F}, i_1, \ldots, i_p, j_1, \ldots, j_q = 1, \ldots, n$. Consequently, subject to the changes of bases given by (12.3.7) and (12.3.9), we see that the components of the tenor of the type (p, q) observe the following transition formula:

$$\tilde{T}^{i_1 \ldots i_p}_{j_1 \ldots j_q} = \sum_{k_1=1}^{n} \cdots \sum_{k_p=1}^{n} \sum_{l_1=1}^{n} \cdots \sum_{l_q=1}^{n} b^{i_1}_{k_1} \cdots b^{i_p}_{k_p} a^{l_1}_{j_1} \cdots a^{l_q}_{j_q} T^{k_1 \ldots k_p}_{l_1 \ldots l_q}, \quad (12.3.18)$$

which generalizes (12.3.8) and (12.3.14) for tensors of the types $(0, m)$ and $(m, 0)$, respectively. Thus, $\mathcal{T}^m = \mathcal{T}^m_0$, $\mathcal{T}_m = \mathcal{T}^0_m$, and $\mathcal{T}^q_p$ is the space of mixed tensors with q and p covariant and contravariant tensor indices, respectively.

As a result of Theorem 12.1, we have

$$\dim(\mathcal{T}^q_p) = n^{p+q}. \quad (12.3.19)$$

Let $P \in \mathcal{T}_p$ and $Q \in \mathcal{T}^q$. With

$$P = \sum_{i_1, \ldots, i_p} P^{i_1 \ldots i_p} u_{i_1} \otimes \cdots \otimes u_{i_p}, \quad Q = \sum_{j_1, \ldots, j_q} Q_{j_1 \ldots j_q} u^{j_1} \otimes \cdots \otimes u^{j_q}, \quad (12.3.20)$$

we see that $T = P \otimes Q$ as shown in (12.3.17) is given by

$$T^{i_1 \ldots i_p}_{j_1 \ldots j_q} = P^{i_1 \ldots i_p} Q_{j_1 \ldots j_q}, \quad i_1, \ldots, i_p, j_1, \ldots, j_q = 1, \ldots, n. \quad (12.3.21)$$

A useful operation is called *contraction* which may be carried out by "absorbing" correspondingly indexed vectors and covectors in a mixed tensor by the usual vector-covector pairing. For example, if we use $c_1(T)$ to denote the contraction of the first indexed vectors and covectors in $T \in \mathcal{T}^q_p$ where $p, q \geq 1$, we have

$$c_1(T) = \sum_{i_1, \ldots, i_p} \sum_{j_1, \ldots, j_q} T^{i_1 \ldots i_p}_{j_1 \ldots j_q} \langle u_{i_1}, u^{j_1} \rangle u_{i_2} \otimes \cdots \otimes u_{i_p} \otimes u^{j_2} \otimes \cdots \otimes u^{j_q}$$

$$= \sum_{i_2, \ldots, i_p} \sum_{j_2, \ldots, j_q} \sum_{k=1}^{n} T^{k i_2 \ldots i_p}_{k j_2 \ldots j_q} u_{i_2} \otimes \cdots \otimes u_{i_p} \otimes u^{j_2} \otimes \cdots \otimes u^{j_q}, \quad (12.3.22)$$

which lies in $\mathcal{T}^{q-1}_{p-1}$. If $p = q$, we can formulate the *total contraction*

$$c_p(T) = \sum_{i_1, \ldots, i_p} \sum_{j_1, \ldots, j_p} T^{i_1 \ldots i_p}_{j_1 \ldots j_p} \langle u_{i_1}, u^{j_1} \rangle \cdots \langle u_{i_p}, u^{j_p} \rangle$$

$$= \sum_{i_1, \ldots, i_p} T^{i_1 \ldots i_p}_{i_1 \ldots i_p}, \quad (12.3.23)$$

which is a scalar and independent of the choice of a basis or coordinates, often referred to as an *invariant*.

In applications, the collection of quantities $\{T^{i_1 \ldots i_p}_{j_1 \ldots j_q}\}$ following the law of transformation (12.3.18) under the changes of bases (12.3.7) and (12.3.9) viewed as the changes of coordinates is often referred to as a (p,q)-type tensor with q and p covariant and contravariant tensor indices.

Of particular interest is when $\{T_{ij}\}$ is a covariant 2-tensor. If we use the matrix notation $T = (T_{ij})$ to denote the tensor in the original coordinates and $\tilde{T} = (\tilde{T}_{ij})$ the same tensor in the transformed coordinates. Then (12.3.8) reads

$$\tilde{T} = ATA^t, \tag{12.3.24}$$

where in $A = (a_j^i)$ we specify the upper and lower indices to denote the column and row indices of the entries in the matrix A, respectively. Thus, in view of (12.3.10), $B = (b_i^j) = A^{-1}$ follows the same index convention. As a consequence, if $S = (S^{ij})$ denotes a contravariant 2-tensor in its matrix representation, then (12.3.14) gives us

$$\tilde{S} = B^t SB = (A^{-1})^t SA^{-1}. \tag{12.3.25}$$

Thus, in view of (12.3.24) and (12.3.25), we see that, if the covariant 2-tensor T is invertible as an $n \times n$ matrix, then its inverse T^{-1} transforms itself following the same property of that of a contravariant 2-tensor, such that T^{-1} is a contravariant 2-tensor. Of course, the opposite statement of this result is true too.

Exercises

12.3.1 Show that $c_1(T)$ defined in (12.3.22) lies in $\mathcal{T}_{p-1}^{q-1}$ by verifying that the contracted coefficients

$$\sum_{k=1}^{n} T_{kj_2\cdots j_q}^{ki_2\cdots i_p}, \quad i_2,\ldots,i_p,j_2,\ldots,j_q = 1,\ldots,n, \tag{12.3.26}$$

satisfy the basis or coordinate transition property (12.3.18).

12.3.2 Show that the total contraction (12.3.23) is an invariant by verifying that the right-hand side of (12.3.23) is independent of the choice of a basis or coordinate system.

12.3.3 Let $\{T^{ij}\}$ $(i,j = 1,\ldots,n)$ be a contravariant tensor of order 2 over a field $\mathbb{F}$ such that the matrix $(T^{ij}) \in \mathbb{F}(n,n)$ is invertible. Show that its inverse is covariant. That is, if $S = (S_{ij}) = (T^{ij})^{-1}$, then $\{S_{ij}\}$ is a covariant tensor of order 2.

12.3.4 Let $\{e_1,\ldots,e_n\}$ be the standard basis of the Euclidean space $\mathbb{R}^n$.

 (a) Find the matrix representation of the 2-form $e_i \otimes e_j$, $i,j = 1,\ldots,n$.
 (b) Determine the 2-form $e_1 \otimes e_1 + \cdots + e_n \otimes e_n$.
 (c) Show by (a) that $\mathcal{M}(\mathbb{R}^n \times \mathbb{R}^n) = \mathbb{R}^n \otimes \mathbb{R}^n = \mathbb{R}(n,n)$.

12.3.5 Let U be a finite-dimensional real vector space with a positive definite scalar product and $f : U \times U \to \mathbb{R}$ a symmetric 2-form so that the induced quadratic form $q(u) = f(u,u)$ $(u \in U)$ is positive definite. Show that U has an orthonormal basis, say $\{u_1,\dots,u_n\}$, such that

$$f = \lambda_1(u_1 \otimes u_1) + \cdots + \lambda_n(u_n \otimes u_n) \qquad (12.3.27)$$

for some positive numbers $\lambda_1,\dots,\lambda_n$.

12.3.6 Let U be a finite-dimensional vector space over a field $\mathbb{F}$ and $f \in U'$ nonzero. Show that the matrix associated with the 2-form $f \otimes f$ over $U \times U$ under any basis of U is of rank 1.

12.4 Symmetric and anti-symmetric tensors

For $T \in \mathcal{T}^m$ (say), $m \geq 2$, on an n-dimensional vector space U over a field $\mathbb{F}$ with $\mathrm{ch}(\mathbb{F}) \neq 2$, we say that T is *symmetric* if

$$T(\dots,u_i,\dots,u_j,\dots) = T(\dots,u_j,\dots,u_i,\dots),$$
$$u_i,u_j \in U, \quad i \neq j, \quad i,j = 1,\dots,n, \qquad (12.4.1)$$

or *anti-symmetric* or *skew-symmetric* if

$$T(\dots,u_i,\dots,u_j,\dots) = -T(\dots,u_j,\dots,u_i,\dots),$$
$$u_i,u_j \in U, \quad i \neq j, \quad i,j = 1,\dots,n. \qquad (12.4.2)$$

Equivalently and more conveniently, the notions of symmetry and anti-symmetry of a tensor or multilinear form may be reformulated using the concepts of *permutations* and the *parity* or *signature* of a permutation.

Consider the set $\{1,\dots,m\}$ with $m \geq 2$. A 1-1 map of $\{1,\dots,m\}$ onto itself is called a permutation over $\{1,\dots,m\}$ and the set of all permutations over $\{1,\dots,m\}$ is a group with the composition of maps as the group product or multiplication operation. This group is called the *permutation group* and denoted by π_m. There are a total of $m!$ group elements because there are $m!$ ways to rearrange the set $\{1,\dots,m\}$. A special kind of permutations are called *transpositions*. As a permutation, a transposition τ flips two elements in $\{1,\dots,m\}$ but leaves the rest of the elements in $\{1,\dots,m\}$ intact:

$$\tau(i) = j, \quad \tau(j) = i, \quad i \neq j, \quad \tau(k) = k, \quad k \neq i,j, \quad i,j,k \in \{1,\dots,m\}. \qquad (12.4.3)$$

Any permutation $\sigma \in \pi_m$ may be expressed as a product of some transpositions $\tau_1,\dots,\tau_l$,

$$\sigma = \tau_1 \cdots \tau_l. \qquad (12.4.4)$$

Although such a representation or factorization is not unique, the evenness or oddness of the integer l in (12.4.4) is unique, which is referred to as the parity of σ, also realized as the signature of σ:

$$\text{sgn}(\sigma) = (-1)^l = \pm 1. \tag{12.4.5}$$

As a direct consequence of (12.4.4) and (12.4.5), we have

$$\text{sgn}(\sigma_1 \sigma_2) = \text{sgn}(\sigma_1)\text{sgn}(\sigma_2), \quad \sigma_1, \sigma_2 \in \pi_m. \tag{12.4.6}$$

For $\sigma \in \pi_m$ and $T \in \mathcal{T}^m$, we define $\sigma T \in \mathcal{T}^m$ by setting

$$\sigma T(u_1, \ldots, u_m) = T\left(u_{\sigma(1)}, \ldots, u_{\sigma(m)}\right), \quad \forall u_1, \ldots, u_m \in U. \tag{12.4.7}$$

Then we have

Definition 12.2 Any $T \in \mathcal{T}^m$ is referred to be symmetric or anti-symmetric if $\sigma T = T$ or $\sigma T = \text{sgn}(\sigma)T$ for all $\sigma \in \pi_m$.

Let $\{u_1, \ldots, u_n\}$ be a basis of U and $\{u^1, \ldots, u^n\}$ the associated dual basis of U'. For $T \in \mathcal{T}^m$, we have

$$T = \sum_{i_1, \ldots, i_m} T_{i_1 \ldots i_m} u^{i_1} \otimes \cdots \otimes u^{i_m}, \quad T_{i_1 \ldots i_m} \in \mathbb{F}, \quad i_1, \ldots, i_m = 1, \ldots, n. \tag{12.4.8}$$

Using the general expression (12.4.7) to $u_{i_1}, \ldots, u_{i_m}$ and consider (12.4.8), we have

$$\sigma T\left(u_{i_1}, \ldots, u_{i_m}\right) = T\left(u_{i_{\sigma(1)}}, \ldots, u_{i_{\sigma(m)}}\right) = T_{i_{\sigma(1)} \ldots i_{\sigma(m)}}, \quad \sigma \in \pi_m. \tag{12.4.9}$$

As a result, we arrive at the following alternative definition of a symmetric or anti-symmetric covariant tensor.

Definition 12.3 Let $T \in \mathcal{T}^m$ be given in its component form by (12.4.8). Then T is symmetric or anti-symmetric following if

$$T_{i_{\sigma(1)} \ldots i_{\sigma(m)}} = T_{i_1 \ldots i_m} \quad \text{or} \quad T_{i_{\sigma(1)} \ldots i_{\sigma(m)}} = \text{sgn}(\sigma) T_{i_1 \ldots i_m} \tag{12.4.10}$$

for all $\sigma \in \pi_m$ and $i_1, \ldots, i_m = 1, \ldots, n$.

Given any $T \in \mathcal{T}^m$, we can *symmetrize* T by setting

$$\text{sym}(T) = \frac{1}{m!} \sum_{\sigma \in \pi_m} \sigma T. \tag{12.4.11}$$

To see how, we pick $\rho \in \pi_m$ and $u_1, \ldots, u_m \in U$ to evaluate

$$\rho \, \text{sym}(T)(u_1, \ldots, u_m) = \text{sym}(T)(u_{\rho(1)}, \ldots, u_{\rho(m)})$$

$$= \frac{1}{m!} \sum_{\sigma \in \pi_m} \sigma T(u_{\rho(1)}, \ldots, u_{\rho(m)}) = \frac{1}{m!} \sum_{\sigma \in \pi_m} (\rho\sigma) T(u_1, \ldots, u_m)$$

$$= \frac{1}{m!} \sum_{(\rho\sigma) \in \pi_m} \rho\sigma T(u_1, \ldots, u_m) = \text{sym}(T)(u_1, \ldots, u_m), \qquad (12.4.12)$$

which establishes the symmetry of $\text{sym}(T)$.

Likewise, for any $T \in \mathcal{T}^m$, we can *anti-symmetrize* T by setting

$$\text{antisym}(T) = \frac{1}{m!} \sum_{\sigma \in \pi_m} \text{sgn}(\sigma) \sigma T. \qquad (12.4.13)$$

In fact, pick again $\rho \in \pi_m$ and $u_1, \ldots, u_m \in U$. We have

$$\rho \, \text{antisym}(T)(u_1, \ldots, u_m) = \text{antisym}(T)(u_{\rho(1)}, \ldots, u_{\rho(m)})$$

$$= \frac{1}{m!} \sum_{\sigma \in \pi_m} \text{sgn}(\sigma) \sigma T(u_{\rho(1)}, \ldots, u_{\rho(m)})$$

$$= \frac{1}{m!} \sum_{\sigma \in \pi_m} \text{sgn}(\sigma)(\rho\sigma) T(u_1, \ldots, u_m)$$

$$= \text{sgn}(\rho) \frac{1}{m!} \sum_{(\rho\sigma) \in \pi_m} \text{sgn}(\rho\sigma) \rho\sigma T(u_1, \ldots, u_m)$$

$$= \text{sgn}(\rho) \, \text{antisym}(T)(u_1, \ldots, u_m), \qquad (12.4.14)$$

which proves the anti-symmetry of the tensor $\text{antisym}(T)$ as desired.

Note that the normalization factor $\frac{1}{m!}$ in (12.4.11) and (12.4.13) is introduced in order to render $\text{sym}(T) = T$ when T is already symmetric and $\text{antisym}(T) = T$ when T is already anti-symmetric. The presence of such a factor prompts the assumption $\text{ch}(\mathbb{F}) = 0$ for convenience. This assumption is otherwise unnecessary and will be observed subsequently for convenience.

For any $f^1, \ldots, f^m \in U'$, we also have

$$\sigma(f^1 \otimes \cdots \otimes f^m) = f^{\sigma^{-1}(1)} \otimes \cdots \otimes f^{\sigma^{-1}(m)}, \quad \sigma \in \pi_m, \qquad (12.4.15)$$

which may readily be verified by applying both sides of this equation to a testing m-tuple, $(u_1, \ldots, u_m)$, with $u_1, \ldots, u_m \in U$. In fact, we have

$$\sigma(f^1 \otimes \cdots \otimes f^m)(u_1, \ldots, u_m) = f^1(u_{\sigma(1)}) \cdots f^m(u_{\sigma(m)})$$

$$= \prod_{i=1}^{m} f^i(u_{\sigma(i)}) = \prod_{i=1}^{m} f^{\sigma^{-1}(i)}(u_{\sigma(\sigma^{-1}(i))}) = \prod_{i=1}^{m} f^{\sigma^{-1}(i)}(u_i)$$

$$= \left(f^{\sigma^{-1}(1)} \otimes \cdots \otimes f^{\sigma^{-1}(m)} \right)(u_1, \ldots, u_m), \qquad (12.4.16)$$

as anticipated.

In view of (12.4.11), (12.4.13), and (12.4.15), we obtain the useful formulas

$$\text{sym}\left(f^1 \otimes \cdots \otimes f^m\right) = \frac{1}{m!} \sum_{\sigma \in \pi_m} f^{\sigma^{-1}(1)} \otimes \cdots \otimes f^{\sigma^{-1}(m)}$$

$$= \frac{1}{m!} \sum_{\sigma \in \pi_m} f^{\sigma(1)} \otimes \cdots \otimes f^{\sigma(m)}, \qquad (12.4.17)$$

$$\text{antisym}\left(f^1 \otimes \cdots \otimes f^m\right) = \frac{1}{m!} \sum_{\sigma \in \pi_m} \text{sgn}(\sigma) f^{\sigma^{-1}(1)} \otimes \cdots \otimes f^{\sigma^{-1}(m)}$$

$$= \frac{1}{m!} \sum_{\sigma \in \pi_m} \text{sgn}(\sigma) f^{\sigma(1)} \otimes \cdots \otimes f^{\sigma(m)} \quad (12.4.18)$$

for any $f^1, \ldots, f^m \in U'$ since $\text{sgn}(\sigma) = \text{sgn}(\sigma^{-1})$. These expressions lead us to define the *symmetric* and *exterior*, also called *wedge*, products of $f^1, \ldots, f^m$, denoted with $\odot$ and $\wedge$, respectively:

$$f^1 \odot \cdots \odot f^m \equiv m! \, \text{sym}(f^1 \otimes \cdots \otimes f^m)$$

$$= \sum_{\sigma \in \pi_m} f^{\sigma(1)} \otimes \cdots \otimes f^{\sigma(m)}, \qquad (12.4.19)$$

$$f^1 \wedge \cdots \wedge f^m \equiv m! \, \text{antisym}(f^1 \otimes \cdots \otimes f^m)$$

$$= \sum_{\sigma \in \pi_m} \text{sgn}(\sigma) f^{\sigma(1)} \otimes \cdots \otimes f^{\sigma(m)}. \qquad (12.4.20)$$

Combining (12.4.15), (12.4.19), and (12.4.20), we have

$$\sigma(f^1 \odot \cdots \odot f^m) = f^1 \odot \cdots \odot f^m, \qquad (12.4.21)$$

$$\sigma(f^1 \wedge \cdots \wedge f^m) = \text{sgn}(\sigma) f^1 \wedge \cdots \wedge f^m, \qquad (12.4.22)$$

which reconfirm the symmetry and anti-symmetry of the tensors (12.4.19) and (12.4.20) so constructed. In particular, we have

$$\cdots \odot f^i \odot \cdots \odot f^j \odot \cdots = \cdots \odot f^j \odot \cdots \odot f^i \odot \cdots, \qquad (12.4.23)$$

$$\cdots \wedge f^i \wedge \cdots \wedge f^j \wedge \cdots = - \cdots \wedge f^j \wedge \cdots \wedge f^i \wedge \cdots \qquad (12.4.24)$$

for $i \neq j, i, j = 1, \ldots, n$. As a consequence of (12.4.24), we have $\cdots \wedge f^i \wedge \cdots \wedge f^j \wedge \cdots = 0$ if it happens that $f^i = f^j$. More generally, $f^1 \wedge \cdots \wedge f^m = 0$ if $f^1, \ldots, f^m$ are linearly dependent.

We use $\mathcal{S}^m(U)$ to denote the subspace of $\mathcal{T}^m$ consisting of symmetric covariant tensors and $\mathcal{A}^m(U)$ or $\bigwedge^m(U)$ the subspace of anti-symmetric ones. As a consequence of Definition 12.3 and (12.4.19), we have, for $T \in \mathcal{S}^m(U)$ given in (12.4.8),

$$T = \sum_{1 \le i_1 \le \cdots \le i_m \le n} T_{i_1 \ldots i_m} \sum_{\sigma \in \pi_m} u^{i_{\sigma(1)}} \otimes \cdots \otimes u^{i_{\sigma(m)}}$$

$$= \sum_{1 \le i_1 \le \cdots \le i_m \le n} T_{i_1 \ldots i_m} u^{i_1} \odot \cdots \odot u^{i_m}. \tag{12.4.25}$$

The set $\{u^{i_1} \odot \cdots \odot u^{i_m} \mid 1 \le i_1 \le \cdots \le i_m \le n\}$ is a basis for $\mathcal{S}^m(U)$. The number of such vectors is the number of combinations of picking m members with replacement from the n-vector set $\{u^1, \ldots, u^n\}$ which is the combinatorial number C_m^{n+m-1}. That is,

$$\dim(\mathcal{S}^m(U)) = C_m^{n+m-1} = \binom{n+m-1}{m} = \frac{(n+m-1)!}{m!(n-1)!}. \tag{12.4.26}$$

Similarly, for $T \in \mathcal{A}^m(U)$ and $m \le n$, we see from Definition 12.3 and (12.4.20) that T given in (12.4.8) assumes the form

$$T = \sum_{1 \le i_1 < \cdots < i_m \le n} T_{i_1 \ldots i_m} \sum_{\sigma \in \pi_m} \mathrm{sgn}(\sigma) u^{i_{\sigma(1)}} \otimes \cdots \otimes u^{i_{\sigma(m)}}$$

$$= \sum_{1 \le i_1 < \cdots < i_m \le n} T_{i_1 \ldots i_m} u^{i_1} \wedge \cdots \wedge u^{i_m}. \tag{12.4.27}$$

The set $\{u^{i_1} \wedge \cdots \wedge u^{i_m} \mid 1 \le i_1 < \cdots < i_m \le n\}$ is a basis for $\mathcal{A}^m(U)$. Thus

$$\dim(\mathcal{A}^m(U)) = \binom{n}{m} = \frac{n!}{m!(n-m)!}, \tag{12.4.28}$$

since this quantity is the number of combinations of picking m members without replacement from the n-vector set $\{u^1, \ldots, u^n\}$.

It is clear that

$$\mathcal{S}^m(U) \cap \mathcal{A}^m(U) = \{0\}. \tag{12.4.29}$$

However, in general, $\mathcal{S}^m(U)$ and $\mathcal{A}^m(U)$ do not sum up to $\mathcal{T}^m$. That is, the set-containment relation

$$\mathcal{S}^m(U) \oplus \mathcal{A}^m(U) \subset \mathcal{T}^m \tag{12.4.30}$$

does not become an equality for general m. The exceptional case is when $m = 2$ because now

$$\dim(\mathcal{S}^2(U)) + \dim(\mathcal{A}^2(U)) = \binom{n+1}{2} + \binom{n}{2} = n^2 = \dim(\mathcal{T}^2(U)). \tag{12.4.31}$$

Thus $\mathcal{S}^2(U) \oplus \mathcal{A}^2(U) = \mathcal{T}^2$. In fact, when $m = 2$, we may rewrite (12.4.8) as

$$
\begin{aligned}
T &= \sum_{i,j=1}^{n} T_{ij} u^i \otimes u^j \\
&= \sum_{i,j=1}^{n} \left(\frac{1}{2}(T_{ij} + T_{ji}) + \frac{1}{2}(T_{ij} - T_{ji}) \right) u^i \otimes u^j \\
&= \sum_{1 \le i \le j \le n} a_{ij} (u^i \otimes u^j + u^j \otimes u^i) + \sum_{1 \le i < j \le n} b_{ij} (u^i \otimes u^j - u^j \otimes u^i) \\
&= \sum_{1 \le i \le j \le n} a_{ij} u^i \odot u^j + \sum_{1 \le i < j \le n} b_{ij} u^i \wedge u^j, \qquad (12.4.32)
\end{aligned}
$$

where we have set

$$
a_{ij} = \frac{1}{2}(T_{ij} + T_{ji}), \quad b_{ij} = \frac{1}{2}(T_{ij} - T_{ji}), \quad i,j = 1,\dots,n, \qquad (12.4.33)
$$

which reestablishes the relation $\mathcal{S}^2(U) \oplus \mathcal{A}^2(U) = \mathcal{T}^2(U)$ constructively as done for square matrices before.

Since $\mathcal{A}^m(U) = \{0\}$ whenever $m > n$, the lowest positive-dimensional situation of $\mathcal{A}^m(U)$ at $m = n$ is of particular interest:

$$
\dim(\mathcal{A}^n(U)) = 1. \qquad (12.4.34)
$$

So $\mathcal{A}^n(U)$ is singly generated by

$$
\Omega = u^1 \wedge \cdots \wedge u^n, \qquad (12.4.35)
$$

say. We have

$$
\Omega(u_1,\dots,u_n) = \sum_{\sigma \in \pi_n} \mathrm{sgn}(\sigma) \left(u^{\sigma(1)} \otimes \cdots \otimes u^{\sigma(n)} \right) (u_1,\dots,u_n) = 1. \quad (12.4.36)
$$

More generally, for any n vectors

$$
v_j = \sum_{i=1}^{n} a_{ij} u_i \in U, \quad a_{ij} \in \mathbb{F}, \quad i,j = 1,\dots,n, \qquad (12.4.37)
$$

we have

$$
\begin{aligned}
\Omega(v_1,\dots,v_n) &= \sum_{\sigma \in \pi_n} \mathrm{sgn}(\sigma) \left(u^{\sigma(1)} \otimes \cdots \otimes u^{\sigma(n)} \right) (v_1,\dots,v_n) \\
&= \sum_{\sigma \in \pi_n} \mathrm{sgn}(\sigma) \left(u^{\sigma(1)} \otimes \cdots \otimes u^{\sigma(n)} \right) \left(\sum_{i=1}^{n} a_{i1} u_i, \dots, \sum_{i=1}^{n} a_{in} u_i \right) \\
&= \sum_{\sigma \in \pi_n} \mathrm{sgn}(\sigma) a_{\sigma(1)1} \cdots a_{\sigma(n)n}. \qquad (12.4.38)
\end{aligned}
$$

In particular, for $U = \mathbb{F}^n$, we consider the standard basis $\{e_1, \ldots, e_n\}$ of $\mathbb{F}^n$ with the associated dual basis $\{e^1, \ldots, e^n\}$. Then for

$$v_j = \sum_{i=1}^n a_{ij} e_i = \begin{pmatrix} a_{1j} \\ \vdots \\ a_{nj} \end{pmatrix} \in \mathbb{F}^n, \quad j = 1, \ldots, n, \tag{12.4.39}$$

we may form the matrix $A \in \mathbb{F}(n,n)$ taking $v_1, \ldots, v_n$ as its respective column vectors,

$$A = (v_1, \ldots, v_n) = \begin{pmatrix} a_{11} & \cdots & a_{1n} \\ \vdots & \ddots & \vdots \\ a_{n1} & \cdots & a_{nn} \end{pmatrix}. \tag{12.4.40}$$

Then the properties of determinant indicates that the function

$$\det(v_1, \ldots, v_n) \equiv \det \begin{pmatrix} a_{11} & \cdots & a_{1n} \\ \vdots & \ddots & \vdots \\ a_{n1} & \cdots & a_{nn} \end{pmatrix} \tag{12.4.41}$$

is n-linear and anti-symmetric in $v_1, \ldots, v_n \in \mathbb{F}^n$. Hence $\det \in \bigwedge^n(\mathbb{F}^n)$. On the other hand, from (12.4.35), we set

$$\Omega = e^1 \wedge \cdots \wedge e^n. \tag{12.4.42}$$

Then (12.4.34) indicates that $\det = a\Omega$ for some $a \in \mathbb{F}$. However, from the fact $\det(e_1, \ldots, e_n) = \det(I_n) = 1$ and $\Omega(e_1, \ldots, e_n) = 1$ by (12.4.36), we obtain $a = 1$. Thus, we arrive at the conclusion

$$\det = \Omega = e^1 \wedge \cdots \wedge e^n. \tag{12.4.43}$$

In other words, determinant is the unique generator Ω of the top nontrivial exterior covector space $\bigwedge^n(\mathbb{F}^n)$ satisfying the normalized condition $\Omega(e_1, \ldots, e_n) = 1$.

Another by-product of this discussion is that there holds the following determinant expansion formula:

$$\det \begin{pmatrix} a_{11} & \cdots & a_{1n} \\ \vdots & \ddots & \vdots \\ a_{n1} & \cdots & a_{nn} \end{pmatrix} = \sum_{\sigma \in \pi_n} \mathrm{sgn}(\sigma) a_{\sigma(1)1} \cdots a_{\sigma(n)n}, \tag{12.4.44}$$

in view of (12.4.38) and (12.4.43), which is an alternative definition of what the determinant of a square matrix is.

Thus, (12.4.38) may be regarded as the definition of the determinant of any n vectors $v_1, \ldots, v_n$ in an n-dimensional vector space U over a field $\mathbb{F}$, or

$$\det(v_1, \ldots, v_n) \equiv \Omega(v_1, \ldots, v_n) = \det(A), \quad v_1, \ldots, v_n \in U, \qquad (12.4.45)$$

where $A = (a_{ij})$ is the coefficient matrix of the vectors $v_1, \ldots, v_n$ with respect to the basis $\{u_1, \ldots, u_n\}$, given as in (12.4.37).

Let $\{\tilde{u}_1, \ldots, \tilde{u}_n\}$ be another basis of U and $\{\tilde{u}^1, \ldots, \tilde{u}^n\}$ the associated dual basis of U'. Then there is some $a \in \mathbb{F}$ such that

$$\tilde{\Omega} = \tilde{u}^1 \wedge \cdots \wedge \tilde{u}^n = a\Omega = a\left(u^1 \wedge \cdots \wedge u^n\right). \qquad (12.4.46)$$

Therefore, in view of the basis transition relation (12.3.7) and (12.4.45), we have

$$1 = \tilde{\Omega}(\tilde{u}_1, \ldots, \tilde{u}_n) = a\Omega(\tilde{u}_1, \ldots, \tilde{u}_n) = a\det(A), \qquad (12.4.47)$$

which gives us the result $a = \frac{1}{\det(A)}$ or

$$\tilde{\Omega} = \frac{1}{\det(A)}\Omega. \qquad (12.4.48)$$

On the other hand, following (12.3.5), pick a covariant 2-tensor

$$T = \sum_{i,j=1}^{n} T_{ij}u^i \otimes u^j. \qquad (12.4.49)$$

For convenience, use $T = (T_{ij})$ and $\tilde{T} = (\tilde{T}_{ij})$ to denote the coefficient matrices of the 2-tensor with respect to the bases $\{u^1, \ldots, u^n\}$ and $\{\tilde{u}^1, \ldots, \tilde{u}^n\}$, respectively. Then T and $\tilde{T}$ are related through (12.3.24). Consequently, we have

$$\det(\tilde{T}) = (\det(A))^2 \det(T). \qquad (12.4.50)$$

Combining (12.4.48) and (12.4.50), we see that

$$\eta = \sqrt{|\det(T)|}\,\Omega = \sqrt{|\det((T_{ij}))|}\,u^1 \wedge \cdots \wedge u^n \qquad (12.4.51)$$

is a *coordinate invariant* in the sense that it is independent of the choice of a basis, although it depends on whether the basis transformation (12.3.7) is *orientation-preserving* or *orientation-reversing*, characterized by the sign of $\det(A)$ being positive or negative, respectively. That is,

$$\tilde{\eta} = \sqrt{|\det((\tilde{T}_{ij}))|}\,\tilde{u}^1 \wedge \cdots \wedge \tilde{u}^n$$
$$= \mathrm{sgn}(\det(A))\sqrt{|\det((T_{ij}))|}\,u^1 \wedge \cdots \wedge u^n = \mathrm{sgn}(\det(A))\eta. \qquad (12.4.52)$$

In applications, if we are only interested in bases with the same orientation, then $\det(A) > 0$ and the quantity (12.4.51) is an invariant,

$$\tilde{\eta} = \eta. \tag{12.4.53}$$

This invariant is of fundamental importance in differential geometry and differential topology.

Exercises

12.4.1   Show that the definitions stated in (12.4.1) and (12.4.2) are equivalent to those given in Definition 12.2.

12.4.2   Formulate the definitions of symmetric and anti-symmetric contravariant tensors in the spirits of Definitions 12.2 and 12.3.

12.4.3   Consider (12.4.15) again. First establish the expression $\tau(f^1 \otimes \cdots \otimes f^m) = f^{\tau(1)} \otimes \cdots \otimes f^{\tau(m)}$ when τ is a transposition and then use the result to reestablish (12.4.15).

12.4.4   Show that (12.4.27) may be rewritten as

$$T = \sum_{\{i_1,\ldots,i_m\} \subset \{1,\ldots,n\}} T_{i_1 \ldots i_m} u^{i_1} \wedge \cdots \wedge u^{i_m}. \tag{12.4.54}$$

12.4.5   A tensor of order m is called *decomposable, separable, simple, elementary*, or *of rank 1*, if it can be expressed as a product of m tensors of order 1. Let $T \in T^2(U)$ be given by (12.4.49). Show that $T \neq 0$ is decomposable if and only if its coefficient matrix (T_{ij}) satisfies $r((T_{ij})) = 1$.

12.4.6   Verify (12.4.29) if $\text{ch}(\mathbb{F}) \neq 2$.

12.4.7   Show that for any $n \geq 3$ $(n = \dim(U))$

$$S^3(U) \oplus A^3(U) \neq T^3(U). \tag{12.4.55}$$

12.4.8   Let U be a finite-dimensional vector field over a field $\mathbb{F}$ with $\text{ch}(\mathbb{F}) \neq 2$. For f and $g \in U'$, use (12.4.11) and (12.4.13) to get

$$\text{sym}(f \otimes g) = \frac{1}{2}(f \otimes g + g \otimes f), \quad \text{antisym}(f \otimes g) = \frac{1}{2}(f \otimes g - g \otimes f), \tag{12.4.56}$$

and study whether the symmetrization or anti-symmetrization of a decomposable tensor of order 2 remains decomposable.

12.4.9   Show that the determinant (12.4.45) is independent of a basis chosen.

12.4.10 Use the relation (12.4.48) to prove directly that, for two non-singular matrices A and $B \in \mathbb{F}(n,n)$, there holds the product formula

$$\det(AB) = \det(A)\det(B). \tag{12.4.57}$$

12.5 Exterior algebra and Hodge duality

As the tensor product $\otimes$, it is clear that the exterior product $\wedge$ may naturally be extended to operate over exterior cotensors, covectors, or forms, such that it is bilinear and associative:

$$(aR+bS)\wedge T = aR\wedge T + bS\wedge T, \quad T\wedge(aR+bS) = aT\wedge R + bT\wedge S, \tag{12.5.1}$$

where $a,b \in \mathbb{F}; R,S \in \bigwedge^k(U); T \in \bigwedge^l(U)$, and

$$R\wedge(S\wedge T) = (R\wedge S)\wedge T, \tag{12.5.2}$$

where $R \in \bigwedge^k(U), S \in \bigwedge^l(U)$, and $T \in \bigwedge^m(U)$. Here we follow the notation $\bigwedge^m(U) = \mathcal{A}^m(U)$ by custom. It is convenient to use the convention $\bigwedge^1(U) = U'$ and $\bigwedge^0(U) = \mathbb{F}$ so that the exterior product of a scalar and a k-form is understood to be the scalar multiplication of the k-form by the scalar. Now consider the direct sum

$$\bigwedge(U) = \bigoplus_{k=0}^{n}\bigwedge^k(U) = \bigwedge^0(U)\oplus\bigwedge^1(U)\oplus\cdots\oplus\bigwedge^n(U). \tag{12.5.3}$$

As a vector space, we have by the binomial formula the result

$$\dim(\bigwedge(U)) = \sum_{k=0}^{n}\binom{n}{k} = (1+1)^n = 2^n. \tag{12.5.4}$$

Any vector $\tau \in \bigwedge(U)$ is of the form

$$\tau = \sum_{k=0}^{n}\tau^k, \quad \tau^k \in \bigwedge^k(U), \quad k = 0,1,\ldots,n. \tag{12.5.5}$$

Thus, within this notation, for ω and $\tau \in \bigwedge(U)$, we define

$$\omega\wedge\tau = \sum_{k,l=0}^{n}\omega^k\wedge\tau^l \tag{12.5.6}$$

such that the bilinearity (12.5.1) and associativity (12.5.2) continue to be valid for vectors in $\bigwedge(U)$. These properties make $\bigwedge(U)$ an associative algebra with the exterior product $\wedge$, or *alternating product*, specifically called the *exterior algebra, alternating algebra*, or *Grassmann algebra*.

Another basic property of the exterior product is the following.

Theorem 12.4 *Let* $S \in \bigwedge^k(U)$ *and* $T \in \bigwedge^l(U)$ *where* k *and* $l = 0, 1, \ldots, n$. *Then*

$$S \wedge T = (-1)^{kl} T \wedge S. \tag{12.5.7}$$

Proof   It suffices to prove (12.5.7) when S and T are decomposable or simple:

$$S = f^1 \wedge \cdots \wedge f^k, \quad T = g^1 \wedge \cdots \wedge g^l, \quad f^1, \ldots, f^k, g^1, \ldots, g^l \in U'. \tag{12.5.8}$$

In fact, in

$$S \wedge T = f^1 \wedge \cdots \wedge f^k \wedge g^1 \wedge \cdots \wedge g^l, \tag{12.5.9}$$

we can switch g^1 with its adjacent left neighbors $f^k, \ldots, f^1$ consecutively k times to place g^1 to the left-most spot. These k switches result in k times of the sign flips. We then conduct the same procedures with $g^2, \ldots, g^l$. As a result, we arrive at

$$g^1 \wedge \cdots \wedge g^l \wedge f^1 \wedge \cdots \wedge f^k = T \wedge S. \tag{12.5.10}$$

It is clear that the total number of such switches is counted to be kl. This proves (12.5.7). □

The subspace sum relation (12.5.3) indicates a stratification structure of the exterior algebra $\bigwedge(U)$ so that the exterior product across subspaces of respective degrees satisfies

$$\bigwedge^k(U) \wedge \bigwedge^l(U) = \bigwedge^{k+l}(U), \quad k, l = 0, 1, \ldots, n. \tag{12.5.11}$$

This refined structure makes $\bigwedge(U)$ also a *graded algebra*.

We now consider the situation where U is a real vector space of $n \geq 1$ dimensions and equipped with a positive definite scalar product, $(\cdot, \cdot)$. Let $\{u_1, \ldots, u_n\}$ be an orthonormal basis of U so that it is also a dual basis of $U' = U$ through the scalar product pairing. So $u^i = u_i$, $i = 1, \ldots, n$, although we still maintain the original covariant and contravariant index notation for clarity and consistency. For $\bigwedge^m(U), m = 0, 1, \ldots, n$, the induced basis formed from $\{u_1, \ldots, u_n\}$ by the wedge products,

$$\{u^{i_1} \wedge \cdots \wedge u^{i_m} \mid 1 \leq i_1 < \cdots < i_m \leq n\}, \tag{12.5.12}$$

is orthonormal with respect to the scalar product defined over decomposable tensors by

$$(v^1 \wedge \cdots \wedge v^m, w^1 \wedge \cdots \wedge w^m) = \det((v^i, w^j)), \quad v^1, \ldots, v^m, w^1, \ldots, w^m \in U',$$
$$\tag{12.5.13}$$

and otherwise following the rules of bilinearity. In fact, it may be seen from (12.5.13) that the quantity

$$(u^{i_1} \wedge \cdots \wedge u^{i_m}, u^{j_1} \wedge \cdots \wedge u^{j_m}) = 0 \text{ or } \pm 1 \qquad (12.5.14)$$

if $\{i_1, \ldots, i_m\} \neq \{j_1, \ldots, j_m\}$ or $\{i_1, \ldots, i_m\} = \{j_1, \ldots, j_m\}$ as sets but as ordered sets differ by a permutation with an even or odd parity. This is the property we use to handle the computation of a scalar product but not the general formula (12.5.13). Thus we shall not make effort to prove this formula here, but only mention that, in $\mathbb{R}^3$, (12.5.13) is known as the *Binet–Cauchy identity* [77]:

$$(\mathbf{a}_1 \times \mathbf{a}_2) \cdot (\mathbf{b}_1 \times \mathbf{b}_2) = \det \begin{pmatrix} \mathbf{a}_1 \cdot \mathbf{b}_1 & \mathbf{a}_1 \cdot \mathbf{b}_2 \\ \mathbf{a}_2 \cdot \mathbf{b}_1 & \mathbf{a}_2 \cdot \mathbf{b}_2 \end{pmatrix}, \quad \mathbf{a}_1, \mathbf{a}_2, \mathbf{b}_1, \mathbf{b}_2 \in \mathbb{R}^3,$$
$$(12.5.15)$$

whose special case when $\mathbf{a}_1 = \mathbf{b}_1 = \mathbf{a}$ and $\mathbf{a}_2 = \mathbf{b}_2 = \mathbf{b}$ is the *Lagrange identity*:

$$|\mathbf{a} \times \mathbf{b}|^2 = |\mathbf{a}|^2 |\mathbf{b}|^2 - (\mathbf{a} \cdot \mathbf{b})^2, \quad \mathbf{a}, \mathbf{b} \in \mathbb{R}^3. \qquad (12.5.16)$$

For $k = 0, 1, \ldots, n$, we now construct a natural isomorphism

$$\bigwedge^k(U) \to \bigwedge^{n-k}(U). \qquad (12.5.17)$$

Recall that, from the orthonormal basis $\{u_1, \ldots, u_n\}$ of U, we have

$$\bigwedge^n(U) = \text{Span}\{\Omega\}, \quad \Omega = u^1 \wedge \cdots \wedge u^n. \qquad (12.5.18)$$

In this case Ω is actually an invariant since we work on an orthonormal basis so that the coefficient matrix of the 2-form that defines the scalar product is the identity matrix.

For $\omega \in \bigwedge^k(U)$ and $\tau \in \bigwedge^{n-k}(U)$, since $\omega \wedge \tau \in \bigwedge^n(U)$, there is a scalar $a \in \mathbb{R}$ depending on ω and τ bilinearly, say $a = a(\omega, \tau)$, such that

$$\omega \wedge \tau = a(\omega, \tau)\Omega. \qquad (12.5.19)$$

For fixed τ, $a(\omega, \tau)$ is a functional in $\omega \in \bigwedge^k(U)$. As a consequence of the Riesz representation theorem, there is some $\gamma \in \bigwedge^k(U)$ such that $a(\omega, \tau) = (\omega, \gamma)$ for any $\omega \in \bigwedge^k(U)$. This γ is clearly uniquely dependent on τ and thus denoted as $h(\tau)$ for the moment. That is, $\gamma = h(\tau)$. We now show that $h : \bigwedge^{n-k}(U) \to \bigwedge^k(U)$ is linear. For τ_1 and $\tau_2 \in \bigwedge^{n-k}(U)$, we substitute $\tau = \tau_1 + \tau_2$ into (12.5.19) to get

$$(\omega, h(\tau_1 + \tau_2)) = a(\omega, \tau_1 + \tau_2)$$
$$= a(\omega, \tau_1) + a(\omega, \tau_2) = (\omega, h(\tau_1) + h(\tau_2)). \qquad (12.5.20)$$

In other words, we arrive at $h(\tau_1 + \tau_2) = h(\tau_1) + h(\tau_2)$. For $c \in \mathbb{R}$ and $\tau \in \bigwedge^{n-k}(U)$, we replace τ by $c\tau$ in (12.5.19) to get $ca(\omega, \tau) = a(\omega, c\tau)$ or

$(\omega, ch(\tau)) = (\omega, h(c\tau))$. Hence $h(c\tau) = ch(\tau)$. This shows that h is linear. It is obvious that the kernel or null-space of h is trivial, $N(h) = \{0\}$. Hence h is an isomorphism.

In summary, we have obtained an isomorphism h satisfying

$$h : \bigwedge^{n-k}(U) \to \bigwedge^{k}(U),$$

$$\omega \wedge \tau = (\omega, h(\tau))\Omega, \quad \omega \in \bigwedge^{k}(U), \quad \tau \in \bigwedge^{n-k}(U). \quad (12.5.21)$$

The *Hodge duality map*, commonly referred to as the *Hodge star operator* and denoted with $\star$, is the inverse of the isomorphism h given in (12.5.21), which satisfies

$$\star : \bigwedge^{k}(U) \to \bigwedge^{n-k}(U), \quad \alpha \wedge \star\beta = (\alpha, \beta)\Omega, \quad \alpha, \beta \in \bigwedge^{k}(U). \quad (12.5.22)$$

A few properties of the Hodge star operator are

$$\star 1 = \Omega, \quad \star\Omega = 1, \tag{12.5.23}$$

$$(\alpha, \beta) = \star(\alpha \wedge \star\beta), \quad \alpha, \beta \in \bigwedge^{k}(U). \tag{12.5.24}$$

$$\alpha \wedge \star\beta = \beta \wedge \star\alpha, \quad \alpha, \beta \in \bigwedge^{k}(U), \tag{12.5.25}$$

$$(\star\alpha, \star\beta) = (\alpha, \beta), \quad \alpha, \beta \in \bigwedge^{k}(U), \tag{12.5.26}$$

$$\star(\star\alpha) = (-1)^{k(n-k)}\alpha, \quad \alpha \in \bigwedge^{k}(U), \tag{12.5.27}$$

$$(\star)^{-1} = (-1)^{k(n-k)}\star : \bigwedge^{n-k}(U) \to \bigwedge^{k}(U). \tag{12.5.28}$$

In fact, (12.5.23) follows from taking $\alpha = \beta = 1, k = 0$ and $\alpha = \beta = \Omega, k = n$ in (12.5.22), respectively. Applying $\star$ to (12.5.22) and using the second property stated in (12.5.23), we get (12.5.24). Interchanging α and β in (12.5.22), we have (12.5.25).

From (12.5.22), we have

$$(u^{i_1} \wedge \cdots \wedge u^{i_k}) \wedge \star(u^{i_1} \wedge \cdots \wedge u^{i_k}) = \Omega = u^1 \wedge \cdots \wedge u^n, \tag{12.5.29}$$

where $1 \le i_1 < \cdots < i_k \le n$, which leads us to read off the result

$$\star(u^{i_1} \wedge \cdots \wedge u^{i_k}) = \pm u^{j_1} \wedge \cdots \wedge u^{j_{n-k}}, \tag{12.5.30}$$

by applying the two sides of (12.5.29) to the testing vector $(u_1, \ldots, u_n)$, where $1 \le j_1 < \cdots < j_{n-k} \le n$ and $\{i_1, \ldots, i_k\} \cup \{j_1, \ldots, j_{n-k}\} = \{1, \ldots, n\}$, and the choice of the plus or minus sign on the right-hand side of (12.5.30) depends on whether the two ordered sets $\{i_1, \ldots, i_k, j_1, \ldots, j_{n-k}\}$ and $\{1, \ldots, n\}$ differ by a permutation of an even or odd parity. In particular, the operator $\star$ maps an orthonormal basis of $\bigwedge^{k}(U)$ onto an orthonormal basis of $\bigwedge^{n-k}(U)$. Hence $\star$ is an isometry.

For convenience, we absorb the sign on the right-hand side of (12.5.30) to write it as

$$\star \omega = u^{l_1} \wedge \cdots \wedge u^{l_{n-k}}, \quad \omega = u^{i_1} \wedge \cdots \wedge u^{i_k}, \tag{12.5.31}$$

where the ordered set $\{l_1,\ldots,l_{n-k}\}$ is a permutation of the ordered set $\{j_1,\ldots,j_{n-k}\}$. By $\omega \wedge \star\omega = \star\omega \wedge \star\star\omega$, we have

$$u^{i_1} \wedge \cdots \wedge u^{i_k} \wedge u^{l_1} \wedge \cdots \wedge u^{l_{n-k}} = u^{l_1} \wedge \cdots \wedge u^{l_{n-k}} \wedge \star\star\omega. \tag{12.5.32}$$

We can now conduct $(n-k)k$ transpositions starting from u^{i_k} to shift $u^{i_1} \wedge \cdots \wedge u^{i_k}$ on the left-hand side of $u^{i_1} \wedge \cdots \wedge u^{i_k} \wedge u^{l_1} \wedge \cdots \wedge u^{l_{n-k}}$ to its right-hand side. As a result, we have

$$u^{i_1} \wedge \cdots \wedge u^{i_k} \wedge u^{l_1} \wedge \cdots \wedge u^{l_{n-k}} = (-1)^{(n-k)k} u^{l_1} \wedge \cdots \wedge u^{l_{n-k}} \wedge u^{i_1} \wedge \cdots \wedge u^{i_k}$$
$$= u^{l_1} \wedge \cdots \wedge u^{l_{n-k}} \wedge (-1)^{(n-k)k}\omega. \tag{12.5.33}$$

Combining (12.5.32) and (12.5.33), we obtain $\star \star \omega = (-1)^{(n-k)k} \omega$, which establishes (12.5.27). Using (12.5.27), we see that (12.5.28) follows immediately.

From its construction, we see that the Hodge duality map (12.5.22) is actually flexible with how the scalar product over $\bigwedge(U)$ is defined. In particular, the scalar product may be indefinite as well, so far as the Riesz representation theorem is applicable.

Besides, in practice, we may directly delve into setting up such a map knowing what is available and what is to be desired. For example, let

$$\sum_{i,j=1}^{n} g_{ij} u^i \otimes u^j, \quad g_{ij} \in \mathbb{R}, \quad g_{ij} = g_{ji}, \quad i,j = 1,\ldots,n, \tag{12.5.34}$$

be a symmetric covariant 2-form that defines a positive definite scalar product over U, where $\{u^1,\ldots,u^n\}$ is a basis of U' whose dual basis for U is denoted by $\{u_1,\ldots,u_n\}$. Then we know that

$$\sum_{i,j=1}^{n} g^{ij} u_i \otimes u_j, \quad (g^{ij}) = (g_{ij})^{-1}, \tag{12.5.35}$$

is contravariant and defines a positive definite scalar product on the dual space U'. It is clear that $\{u_1,\ldots,u_n\}$ is an orthonormal basis of U if and only if $\{u^1,\ldots,u^n\}$ is an orthonormal basis of U'. Then we set

$$\star\left(u^{i_1} \wedge \cdots \wedge u^{i_k}\right)$$
$$= \sum_{j_1,\ldots,j_n} \frac{\sqrt{|\det(g_{ij})|}}{(n-k)!} g^{i_1 j_1} \cdots g^{i_k j_k} \varepsilon_{j_1 \ldots j_k j_{k+1} \ldots j_n} u^{j_{k+1}} \wedge \cdots \wedge u^{j_n}, \tag{12.5.36}$$

$$\star \left(u_{i_1} \wedge \cdots \wedge u_{i_k} \right)$$

$$= \sum_{j_1,\ldots,j_n} \frac{\sqrt{|\det(g^{ij})|}}{(n-k)!} g_{i_1 j_1} \cdots g_{i_k j_k} \varepsilon^{j_1 \cdots j_k j_{k+1} \cdots j_n} u_{j_{k+1}} \wedge \cdots \wedge u_{j_n}, \quad (12.5.37)$$

where

$$\varepsilon_{i_1 \ldots i_n} = \varepsilon^{i_1 \ldots i_n} = \mathrm{sgn}(\sigma), \quad \sigma = \begin{pmatrix} 1 & \cdots & n \\ i_1 & \cdots & i_n \end{pmatrix}, \qquad (12.5.38)$$

are respectively the covariant and contravariant *Levi-Civita permutation tensors*, in which $\sigma \in \pi_n$ is the permutation satisfying $\sigma(1) = i_1, \ldots, \sigma(n) = i_n$, as designated. This defines the dual maps

$$\star : \bigwedge^k (U) \to \bigwedge^{n-k} (U), \quad \star : \bigwedge_k (U) \to \bigwedge_{n-k} (U), \quad k = 0, 1, \ldots, n. \qquad (12.5.39)$$

To see how, we pick α and $\beta \in \bigwedge^k(U)$ with

$$\alpha = \sum_{1 \le i_1 < \cdots < i_k \le n} \alpha_{i_1 \ldots i_k} u^{i_1} \wedge \cdots \wedge u^{i_k},$$

$$\beta = \sum_{1 \le i_1 < \cdots < i_k \le n} \beta_{i_1 \ldots i_k} u^{i_1} \wedge \cdots \wedge u^{i_k}. \qquad (12.5.40)$$

Then

$$\alpha \wedge \star \beta = \sum_{1 \le i_1 < \cdots < i_k \le n} \alpha_{i_1 \ldots i_k} u^{i_1} \wedge \cdots \wedge u^{i_k} \wedge \sum_{1 \le i_{k+1} < \cdots < i_n \le n} \gamma_{i_{k+1} \ldots i_n} u^{i_{k+1}} \wedge \cdots \wedge u^{i_n}, \qquad (12.5.41)$$

where in view of (12.5.36), we have

$$\sum_{1 \le i_{k+1} < \cdots < i_n \le n} \gamma_{i_{k+1} \ldots i_n} u^{i_{k+1}} \wedge \cdots \wedge u^{i_n} = \star \beta$$

$$= \star \left(\sum_{1 \le l_1 < \cdots < l_k \le n} \beta_{l_1 \ldots l_k} u^{l_1} \wedge \cdots \wedge u^{l_k} \right)$$

$$= \sum_{1 \le l_1 < \cdots < l_k \le n} \beta_{l_1 \ldots l_k} \sum_{j_1, \ldots, j_k, i_{k+1}, \ldots, i_n} \frac{\delta^{l_1 j_1} \ldots \delta^{l_k j_k}}{(n-k)!} \varepsilon_{j_1 \ldots j_k i_{k+1} \ldots i_n} u^{i_{k+1}} \wedge \cdots \wedge u^{i_n}, \qquad (12.5.42)$$

over an orthonormal basis. Note that

$$u^{i_1} \wedge \cdots \wedge u^{i_k} \wedge u^{i_{k+1}} \wedge \cdots \wedge u^{i_n} \ne 0 \qquad (12.5.43)$$

only when $\{i_1, \ldots, i_k\} \cup \{i_{k+1}, \ldots, i_n\} = \{1, \ldots, n\}$. Besides, nontrivial terms on the right-hand side of (12.5.42) occur only at $j_1 = l_1, \ldots, j_k = l_k$ and $\{j_1, \ldots, j_k\} = \{i_1, \ldots, i_k\}$ as sets, in view of the properties of the Kronecker δ

and ε tensors. Using $i_1 < \cdots < i_k$ and $l_1 < \cdots < l_k$, we have $l_1 = i_1, \ldots, l_k = i_k$. Applying these results in (12.5.41) and (12.5.42), we see that the quantity $\alpha \wedge \star \beta$ may be expressed as

$$\sum_{i_1 < \cdots < i_k} \alpha_{i_1 \ldots i_k} \beta_{i_1 \ldots i_k} \sum_{i_{k+1} < \cdots < i_n} \varepsilon_{i_1 \ldots i_k i_{k+1} \ldots i_n} u^{i_1} \wedge \cdots \wedge u^{i_k} \wedge u^{i_{k+1}} \wedge \cdots \wedge u^{i_n}$$

$$= (\alpha, \beta) u^1 \wedge \cdots \wedge u^n \qquad (12.5.44)$$

by (12.5.38). That is, we have recovered the identity $\alpha \wedge \star \beta = (\alpha, \beta)\Omega$ under an orthonormal basis.

It is useful to describe the Hodge dual in terms of tensor coefficients. For example, for the k-form α given in (12.5.40), the tensor coefficients of the $(n-k)$-form

$$\omega = \star \alpha = \sum_{1 \le i_1 < \cdots < i_{n-k} \le n} \omega_{i_1 \ldots i_{n-k}} u^{i_1} \wedge \cdots \wedge u^{i_{n-k}} \qquad (12.5.45)$$

are given by

$$\omega_{i_1 \ldots i_{n-k}} = \sqrt{|\det(g_{ij})|} \sum_{j_1, \ldots, j_k, l_1 < \cdots < l_k} g^{j_1 l_1} \cdots g^{j_k l_k} \alpha_{l_1 \ldots l_k} \varepsilon_{j_1 \ldots j_k i_1 \ldots i_{n-k}} \qquad (12.5.46)$$

for $1 \le i_1 < \cdots < i_{n-k} \le n$.

As an application of the Hodge duality correspondence, we now study a fundamental structure in theoretical physics: Electric and magnetic fields are connected through the Hodge dual map.

Consider the Minkowski space $\mathbb{R}^{3,1}$ equipped with the scalar product

$$e^1 \otimes e^1 + e^2 \otimes e^2 + e^3 \otimes e^3 - e^4 \otimes e^4, \qquad (12.5.47)$$

so that $\{e^1, e^2, e^3, e^4\}$ is an orthonormal basis with the coefficient 2-tensor $\{g_{ij}\} = (g_{ij}) = \text{diag}\{1, 1, 1, -1\}$. We have

$$\star(e^1 \wedge e^2) = e^3 \wedge e^4, \quad \star(e^1 \wedge e^3) = -e^2 \wedge e^4, \quad \star(e^1 \wedge e^4) = -e^2 \wedge e^3,$$
$$\star(e^2 \wedge e^3) = e^1 \wedge e^4, \quad \star(e^2 \wedge e^4) = e^1 \wedge e^3, \quad \star(e^3 \wedge e^4) = -e^1 \wedge e^2.$$

$$\qquad (12.5.48)$$

Let

$$\alpha = \frac{1}{2} \sum_{1 \le i, j \le 4} \alpha_{ij} e^i \wedge e^j = \sum_{1 \le i < j \le 4} \alpha_{ij} e^i \wedge e^j \qquad (12.5.49)$$

be a 2-form such that $i = 1, 2, 3$ represent spatial coordinate labels and $i = 4$ temporal one, where $(\alpha_{ij}) \in \mathbb{R}(4,4)$ is skew-symmetric. Then the induced space vectors, $\mathbf{E}$ and $\mathbf{B}$, given by

$$\mathbf{E} = (E_i), \quad E_i = \alpha_{4i}, \quad \mathbf{B} = (B_i), \quad B_i = -\frac{1}{2} \sum_{j,k=1}^{3} \varepsilon_{ijk} \alpha_{jk}, \quad i = 1, 2, 3,$$

(12.5.50)

may represent electric and magnetic fields. More conveniently, we can express such a field designation in the following matrix form:

$$(\alpha_{ij})_{i,j=1,2,3,4} = \begin{pmatrix} 0 & -B_3 & B_2 & -E_1 \\ B_3 & 0 & -B_1 & -E_2 \\ -B_2 & B_1 & 0 & -E_3 \\ E_1 & E_2 & E_3 & 0 \end{pmatrix}.$$

(12.5.51)

On the other hand, using (12.5.48), we have

$$\tilde{\alpha} = \frac{1}{2} \sum_{1 \le i,j \le 4} \tilde{\alpha}_{ij} e^i \wedge e^j = \star \alpha =$$

$$-\alpha_{34} e^1 \wedge e^2 + \alpha_{24} e^1 \wedge e^3 + \alpha_{23} e^1 \wedge e^4 - \alpha_{14} e^2 \wedge e^3 - \alpha_{13} e^2 \wedge e^4 + \alpha_{12} e^3 \wedge e^4.$$

(12.5.52)

Following the designation (12.5.51), we see that the electric and magnetic fields $\tilde{\mathbf{E}}$ and $\tilde{\mathbf{B}}$ defined by $\tilde{\alpha}$ are given by the matrix

$$\begin{pmatrix} 0 & -\tilde{B}_3 & \tilde{B}_2 & -\tilde{E}_1 \\ \tilde{B}_3 & 0 & -\tilde{B}_1 & -\tilde{E}_2 \\ -\tilde{B}_2 & \tilde{B}_1 & 0 & -\tilde{E}_3 \\ \tilde{E}_1 & \tilde{E}_2 & \tilde{E}_3 & 0 \end{pmatrix} = (\tilde{\alpha}_{ij}) = \begin{pmatrix} 0 & E_3 & -E_2 & -B_1 \\ -E_3 & 0 & E_1 & -B_2 \\ E_2 & -E_1 & 0 & -B_3 \\ B_1 & B_2 & B_3 & 0 \end{pmatrix}.$$

(12.5.53)

Consequently, we arrive at the following relation between the electric and magnetic fields $\mathbf{E}$ and $\mathbf{B}$ defined by a 2-form α and those $\tilde{\mathbf{E}}$ and $\tilde{\mathbf{B}}$ defined by the Hodge dual $\tilde{\alpha}$ of α:

$$\tilde{\mathbf{E}} = \mathbf{B}, \quad \tilde{\mathbf{B}} = -\mathbf{E}.$$

(12.5.54)

At this juncture, it is of interest to recall the classical Maxwell equations

$$\nabla \cdot \mathbf{E} = 0, \quad \nabla \times \mathbf{B} - \frac{\partial \mathbf{E}}{\partial t} = \mathbf{0}, \quad \nabla \cdot \mathbf{B} = 0, \quad \nabla \times \mathbf{E} + \frac{\partial \mathbf{B}}{\partial t} = \mathbf{0}, \quad (12.5.55)$$

in vacuum, in normalized units, which are known to be invariant under the transformation

$$\mathbf{E} \mapsto \mathbf{B}, \quad \mathbf{B} \mapsto -\mathbf{E},$$

(12.5.56)

a fact often referred to as the *principle of electromagnetic duality* in theoretical physics. Therefore, the Hodge dual correspondence (12.5.54) is a realization of the electromagnetic duality principle (12.5.56).

See [34, 53, 82] for more studies on multilinear algebra and [15, 62, 75, 78] for some of its applications in various contexts.

Exercises

12.5.1 Establish (12.5.11).

12.5.2 Verify the alternatives stated for (12.5.14).

12.5.3 Use the formulas

$$\mathbf{a} \cdot \mathbf{b} = |\mathbf{a}||\mathbf{b}| \cos \theta, \quad |\mathbf{a} \times \mathbf{b}| = |\mathbf{a}||\mathbf{b}|| \sin \theta|, \quad \mathbf{a}, \mathbf{b} \in \mathbb{R}^3, \quad (12.5.57)$$

where θ is the angle between the vectors $\mathbf{a}$ and $\mathbf{b}$ to establish the Lagrange identity (12.5.16).

12.5.4 Explain why the scalar product over $\bigwedge^m(U)$ defined around (12.5.14) is positive definite so that the basis (12.5.12) is orthonormal.

12.5.5 Let U be an n-dimensional real vector space with a positive-definite scalar product $(\cdot, \cdot)$. For an orthonormal basis $\{u^1, \ldots, u^n\}$ of U, use (12.5.22) to show that

$$\star \left(u^i \right) = (-1)^{i-1} u^1 \wedge \cdots \wedge \widehat{u^i} \wedge \cdots \wedge u^n, \quad i = 1, \ldots, n, \quad (12.5.58)$$

where $\widehat{}$ denotes the factor that is absent in the product. What happens if the scalar product $(\cdot, \cdot)$ is indefinite assuming that (12.5.22) is still valid?

12.5.6 Show that, if no restriction is made on the order of indices in (12.5.46), then there holds the formula

$$\omega_{i_1 \ldots i_{n-k}} = \frac{\sqrt{|\det(g_{ij})|}}{k!} \sum_{j_1, \ldots, j_k, l_1, \ldots, l_k} g^{j_1 l_1} \cdots g^{j_k l_k} \alpha_{l_1 \ldots l_k} \varepsilon_{j_1 \ldots j_k i_1 \ldots i_{n-k}}$$

$$= \frac{\sqrt{|\det(g_{ij})|}}{k!} \sum_{j_1, \ldots, j_k} \alpha^{j_1 \ldots j_k} \varepsilon_{j_1 \ldots j_k i_1 \ldots i_{n-k}},$$

$$i_1, \ldots, i_{n-k} = 1, \ldots, n. \quad (12.5.59)$$

12.5.7 Use (12.5.48) to derive (12.5.52).

12.5.8 Consider $U = \mathbb{R}^2$ as a Euclidean space. For the standard basis $\{e^1, e^2\}$, show that

$$\star e^1 = e^2, \quad \star e^2 = -e^1, \quad (12.5.60)$$

such that $\star : \mathbb{R}^2 \to \mathbb{R}^2$ is represented by the rotation matrix

$$\star = \begin{pmatrix} 0 & -1 \\ 1 & 0 \end{pmatrix}. \qquad (12.5.61)$$

In particular, regard $\mathbb{R}^2$ as the complex plane $\mathbb{C}$. Then $\star$ may be represented by the imaginary unit i since $\star^2 = -1$.

12.5.9   Consider $U = \mathbb{R}^{1,1}$ as a two-dimensional Minkowski plane equipped with the scalar product

$$e^1 \otimes e^1 - e^2 \otimes e^2. \qquad (12.5.62)$$

That is, for any vectors $u = a_1 e^1 + a_2 e^2$ and $v = b_1 e^1 + b_2 e^2$, we have

$$(u, v) = a_1 b_1 - a_2 b_2. \qquad (12.5.63)$$

With this scalar product, show that

$$\star e^1 = e^2, \quad \star e^2 = e^1. \qquad (12.5.64)$$

As a consequence, we now have

$$\star = \begin{pmatrix} 0 & 1 \\ 1 & 0 \end{pmatrix} : \mathbb{R}^{1,1} \to \mathbb{R}^{1,1}, \qquad (12.5.65)$$

replacing (12.5.61). Thus $\star^2 = 1$ which differs from (12.5.27) because the scalar product here is indefinite.

12.5.10 Consider the Euclidean space $U = \mathbb{R}^3$ with the standard basis vectors e^1, e^2, and e^3 usually denoted as $\mathbf{i}, \mathbf{j}$, and $\mathbf{k}$, respectively.

   (a) Show that

$$\star e^1 = e^2 \wedge e^3, \quad \star e^2 = -e^1 \wedge e^3, \quad \star e^3 = e^1 \wedge e^2, \qquad (12.5.66)$$
$$\star (e^1 \wedge e^2) = e^3, \quad \star (e^1 \wedge e^3) = -e^2, \quad \star (e^2 \wedge e^3) = e^1. \qquad (12.5.67)$$

   (b) Use (12.5.67) to show that for $u = a_1 e^1 + a_2 e^2 + a_3 e^3$ and $v = b_1 e^1 + b_2 e^2 + b_3 e^3$, we have

$$\star (u \wedge v) = (a_2 b_3 - a_3 b_2) e^1 - (a_1 b_3 - a_3 b_1) e^2 + (a_1 b_2 - a_2 b_1) e^3. \qquad (12.5.68)$$

   In other words, when u and v are regarded as space vectors, $u = \mathbf{u} = a_1 \mathbf{i} + a_2 \mathbf{j} + a_3 \mathbf{k}$ and $v = \mathbf{v} = b_1 \mathbf{i} + b_2 \mathbf{j} + b_3 \mathbf{k}$, we have the realization

$$\star (u \wedge v) = \mathbf{u} \times \mathbf{v}. \qquad (12.5.69)$$

   Thus, up to the Hodge duality map, the exterior product of two space vectors is the same as the vector cross product of these vectors.

(c) Let $w = c_1 e^1 + c_2 e^2 + c_3 e^3$. Then

$$u \wedge v \wedge w = ([a_2 b_3 - a_3 b_2] c_1 - [a_1 b_3 - a_3 b_1] c_2$$
$$+ [a_1 b_2 - a_2 b_1] c_3) e^1 \wedge e^2 \wedge e^3. \qquad (12.5.70)$$

Prove that the above relation implies

$$\star (u \wedge v \wedge w) = (\mathbf{u} \times \mathbf{v}) \cdot \mathbf{w}. \qquad (12.5.71)$$

Thus, up to the Hodge duality map, the triple exterior product of three space vectors is the same as the mixed triple product of these vectors.

12.5.11 Use the formal association $e^i = \mathrm{d}x^i$ $(i = 1,2,3,4)$ to show that (12.5.49)–(12.5.53) lead to the formulas

$$\alpha = - \left(E_1 \mathrm{d}x^1 + E_2 \mathrm{d}x^2 + E_3 \mathrm{d}x^3 \right) \wedge \mathrm{d}x^4$$
$$- \left(B_3 \mathrm{d}x^1 \wedge \mathrm{d}x^2 + B_1 \mathrm{d}x^2 \wedge \mathrm{d}x^3 + B_2 \mathrm{d}x^3 \wedge \mathrm{d}x^1 \right), \qquad (12.5.72)$$
$$\star \alpha = - \left(B_1 \mathrm{d}x^1 + B_2 \mathrm{d}x^2 + B_3 \mathrm{d}x^3 \right) \wedge \mathrm{d}x^4$$
$$+ \left(E_3 \mathrm{d}x^1 \wedge \mathrm{d}x^2 + E_1 \mathrm{d}x^2 \wedge \mathrm{d}x^3 + E_2 \mathrm{d}x^3 \wedge \mathrm{d}x^1 \right), \qquad (12.5.73)$$

known as *differential forms*. Differentiating the coefficients in (12.5.72) and (12.5.73) following the rules:

$$\mathrm{d}f = \partial_1 f \mathrm{d}x^2 + \partial_2 f \mathrm{d}x^2 + \partial_3 f \mathrm{d}x^3 + \partial_4 f \mathrm{d}x^4, \quad (12.5.74)$$
$$\mathrm{d}(f \mathrm{d}x^i) = \mathrm{d}f \wedge \mathrm{d}x^i, \quad i = 1,2,3,4, \qquad (12.5.75)$$

and linearity, show that the Maxwell equations given in (12.5.55) are compressed into a pair of equations,

$$\mathrm{d}\alpha = 0, \quad \mathrm{d} \star \alpha = 0, \qquad (12.5.76)$$

with, for example,

$$\nabla \cdot \mathbf{E} = \partial_1 E_1 + \partial_2 E_2 + \partial_3 E_3, \quad \partial_4 \mathbf{E} = \frac{\partial \mathbf{E}}{\partial t},$$
$$\nabla \times \mathbf{E} = (\partial_2 E_3 - \partial_3 E_2, \partial_3 E_1 - \partial_1 E_3, \partial_1 E_2 - \partial_2 E_1), \qquad (12.5.77)$$

where we observe the convention that the product between differentials is always the exterior product.

12.5.12 Let $U = \mathbb{R}^n$ and use the differentials $\mathrm{d}x^1, \ldots, \mathrm{d}x^n$ to formally denote an orthonormal basis of U. For a sufficiently differentiable real-valued function f over $\mathbb{R}^n$ of the form $f = f(x^1, \ldots, x^n)$, define its differential

$$\mathrm{d}f = \sum_{i=1}^{n} \partial_i f \mathrm{d}x^i. \qquad (12.5.78)$$

Assume $d^2 = 0$ and the Leibniz rule. Use (12.5.22) and $\Omega = dx^1 \wedge \cdots \wedge dx^n$ to establish the formula

$$\star d \left(\star df \right) = \sum_{i=1}^{n} \partial_i^2 f \equiv \Delta f. \qquad (12.5.79)$$

That is, $\star d \star d = \Delta$ gives rise to the usual *Laplace operator* over $\mathbb{R}^n$.

12.5.13 (Continued from Problem 12.5.12) Explain why $d(\star f) = 0$ for any real-valued differentiable function f so that the Laplace operator may be rewritten as

$$\Delta = d\delta + \delta d, \qquad (12.5.80)$$

where $\delta = \star d \star$ is referred to as the *codifferential operator*.

12.5.14 (Continued from Problem 12.5.13) Explain why $\delta^2 = 0$ on general k-forms as well so that $d + \delta$ is formally a "square root" of the Laplace operator Δ. That is,

$$\Delta = (d + \delta)^2. \qquad (12.5.81)$$

12.5.15 Combine the discussion in Problems 12.5.9 and 12.5.12 to show that for a sufficiently differentiable function f over $\mathbb{R}^{1,1}$, we have

$$\star d(\star df) = -(\partial_1^2 f - \partial_2^2)f \equiv -\Box f. \qquad (12.5.82)$$

That is, $- \star d \star d = \partial_1^2 - \partial_2^2 = \Box$ in this context is the *d'Alembert operator* instead.

12.5.16 Extend the results of Problems 12.5.9, 12.5.12, 12.5.13, and 12.5.15 to the setting of the n-dimensional Minkowski space $\mathbb{R}^{n-1,1}$ with general $n \geq 2$.

12.5.17 Use η defined in (12.4.51) to generate $\bigwedge^n(U)$ and formulate the associated Hodge star map.

Notes on applications of tensors in computational linear algebra

To end the chapter, we mention that tensor methods have been effectively applied to computational linear algebra. This is a vast and very active area. Here we only mention that tensors play a key role in the development and analysis of fast matrix multiplication algorithms, particularly in the context of tensor-based methods that aim to reduce the computational complexity of matrix multiplication. These methods include the following.

Tensor contraction: Fast matrix multiplication algorithms often exploit the concept of tensor contractions, which are generalizations of matrix multiplication to higher-order tensors. In tensor notation, the contraction of two tensors along specific indices corresponds to summing over those indices. This operation can be used to express matrix multiplication as a tensor contraction, which enables the use of tensor decomposition methods to reduce the number of operations required.

Tensor decomposition: Tensor decomposition techniques are used to factorize tensors into a sum of simpler components. These decompositions can be applied to the input matrices to express them as tensors with lower ranks, which can reduce the number of operations required for multiplication.

Strassen-like algorithms: Some fast matrix multiplication algorithms, such as the Strassen-like algorithms, exploit tensor-based formulations to reduce the number of scalar multiplications required for matrix multiplication. By decomposing matrices into tensors and applying tensor contractions, these algorithms achieve better performance by reducing the overall computational complexity.

13

Excursion: Essentials of Quantum Information and Quantum Entanglement

In this chapter, we present an introduction to an important area of contemporary quantum physics: quantum information and quantum entanglement. After a brief introduction regarding why and how linear algebra is so useful in this area, we first consider the concepts of quantum bits and quantum gates in quantum information theory. Next, we explore some geometric features of quantum bits and quantum gates. Then, we study the phenomenon of quantum entanglement. In particular, we shall clarify the notions of untangled and entangled quantum states and establish a necessary and sufficient condition to characterize or divide these two different categories of quantum states. Afterwards, we present the Bell theorem which is of central importance for the mathematical foundation of quantum mechanics implicating that quantum mechanics is nonlocal.

13.1 Why and how linear algebra becomes relevant

Linear algebra plays a fundamental role in the study of quantum information and quantum entanglement problems because it provides a powerful mathematical framework for describing the complex relationships and transformations that occur in quantum systems. Here is a brief account of how linear algebra is used in these and other contexts and why it works so well and is indispensable.

State representation: As seen in Chapter 9, quantum states, which describe the properties of quantum systems, are represented by vectors in a complex vector space. Linear algebra provides the tools for manipulating these quantum state vectors, performing operations such as addition, scalar multiplication, and inner products, enabling the computations of all kinds of quantities of interest of the quantum systems involved.

Quantum gates and operations: Quantum gates are mathematical operations that act on quantum states to perform transformations, such as quantum logic gates used in quantum computing. These gates are represented by unitary matrices, which preserve the norm and inner product of quantum state vectors. Linear algebra enables the composition and analysis of quantum gates, allowing researchers to design and optimize quantum algorithms and protocols. This line of study forms the foundation of quantum computing, a subject that is concerned with quantum information processing.

Entanglement and Bell states: Quantum entanglement, a phenomenon where the properties of particles become correlated in such a way that the state of one particle depends on the state of another, is a central concept in quantum information theory. Entangled states are described using tensor products of state vectors, and linear algebra and multilinear algebra provide the formalism for describing the entanglement structure of such compound multipartite systems. Bell states, which are maximally entangled states of two-qubits or 2-qubits, for instance, are the simplest examples of entangled states constructed from tensor products of vectors in $\mathbb{C}^2$ that play a crucial role in quantum information, quantum communication, as well as quantum cryptography.

Quantum measurements: Quantum measurements, which extract information from quantum systems, are described by operators called measurement operators, or observables. These operators are represented by Hermitian matrices, and linear algebra provides the tools for calculating measurement probabilities and predicting the outcomes of quantum measurements.

Bell's theorem: Bell's theorem is a profound result in quantum mechanics that demonstrates the existence of correlations between distant particles that cannot be explained by classical physics. These correlations violate the principle of local realism, which states that physical properties of particles exist independently of observation and that any influence between particles must be mediated by local interactions. Bell's theorem is typically formulated in the context of the Bell inequalities, which are mathematical inequalities derived based on the assumptions of local realism. Violations of these inequalities imply that the correlations observed in quantum mechanics cannot be reproduced by any theory that satisfies the principles of local realism. In this context, linear algebra is used in particular to analyze the correlations between measurement outcomes of an entangled two particle system. This involves calculating the joint probabilities of obtaining particular outcomes for both particles under different measurement settings. Then Bell's inequality is derived by making certain assumptions about the behavior of local hidden variable theories, which attempt to explain the correlations observed in entangled states without invoking nonlocal effects. These assumptions lead to specific constraints on the correlations between measurement outcomes, which can be expressed as a mathematical inequality involving expectation values, expressed in the form of weighted sums of the eigenvalues of the Hermitian matrices associated with the measurement operators applied to the quantum systems.

Thus, linear algebra provides a necessary and versatile mathematical framework for studying quantum information and entanglement problems and enables researchers to represent quantum states, operations, and measurements using vector spaces and matrices and analyze quantum systems with insight, precision, and efficiency. All these make linear algebra an essential tool in the field of quantum information science and a bridge between classical and quantum physics.

13.2 Qubits and quantum information processing

The simplest form of *information* may be given in the form of either a "No" or "Yes" answer to a specific query, with assigned numerical value 0 or 1, say. Such a form of information, consisting of either 0 or 1, is called a "bit," "classical bit," cbit, or simply 1-bit, which is the building block of the binary number system. Specifically, a 1-bit may be represented as $|a\rangle$ with $a \in \{0,1\}$, for example. Analogously, an n-bit is a sequence of n 1-bits of the form

$$|a_1 \cdots a_n\rangle, \quad a_i \in \{0,1\}, \quad i = 1,\ldots,n, \tag{13.2.1}$$

or conveniently, $a_1 \cdots a_n$, when there is no risk of confusion, which may be interpreted as a binary number or No or Yes answers responding to a sequence of n queries, and referred to an n-bit state or concisely a *state*. The number of all possible states of the form (13.2.1) is

$$N = 2^n, \tag{13.2.2}$$

which may be converted to render

$$n = \log_2 N. \tag{13.2.3}$$

The amount of information contained in the collection of all n-bit states is given by the value n. In other words, n measures the amount of information contained in all n-bit states. In general, the amount of information contained in any collection of N possible states, regardless whether N may be represented as a power of base 2, is measured by (13.2.3), which may not be an integer.

The concept of *quantum bit*, or *qubit* in short, extends that of cbit and employs the ideas of quantum mechanics. Specifically, the state of a qubit, or 1-qubit, say $|\phi\rangle$, is formally expressed as a linear combination of two basis states, represented as two linearly independent vectors in $\mathbb{C}^2$, say

$$|0\rangle = e_1 = \begin{pmatrix} 1 \\ 0 \end{pmatrix}, \quad |1\rangle = e_2 = \begin{pmatrix} 0 \\ 1 \end{pmatrix}, \tag{13.2.4}$$

given as a unit vector in $\mathbb{C}^2$ of the form

$$|\phi\rangle = a_0|0\rangle + a_1|1\rangle, \quad a_0, a_1 \in \mathbb{C}, \quad |a_0|^2 + |a_1|^2 = 1. \tag{13.2.5}$$

In other words, the states of a qubit are comprised of all the points on the unit sphere of $\mathbb{C}^2$. Following the statistical interpretation of quantum mechanics, in the state $|\phi\rangle$ given in (13.2.5), an observation of the state $|\phi\rangle$ has two possible outcomes, $|0\rangle$ and $|1\rangle$, such that $|a_0|^2$ and $|a_1|^2$ are the probabilities of "observing" the states $|0\rangle$ and $|1\rangle$, respectively. The basis states $|0\rangle$ and $|1\rangle$ are sometimes loosely called the eigenstates and the orthonormal basis $\{|0\rangle, |1\rangle\}$ of the state space $\mathbb{C}^2$ is referred to as a *computational basis* of the qubit.

Naturally, growing out of (13.2.1) as well, the basis states or eigenstates of a 2-qubit represent themselves as

$$|00\rangle, \quad |01\rangle, \quad |10\rangle, \quad |11\rangle, \tag{13.2.6}$$

which may also be realized in view of the tensor product by setting

$$|a\rangle|b\rangle \equiv |ab\rangle = |a\rangle \otimes |b\rangle, \quad a, b = 0, 1, \tag{13.2.7}$$

so that the eigenstates (13.2.6) are represented concretely as the standard basis vectors of the state space $\mathbb{C}^4$:

$$|00\rangle = \begin{pmatrix} 1 \\ 0 \\ 0 \\ 0 \end{pmatrix}, \quad |01\rangle = \begin{pmatrix} 0 \\ 1 \\ 0 \\ 0 \end{pmatrix}, \quad |10\rangle = \begin{pmatrix} 0 \\ 0 \\ 1 \\ 0 \end{pmatrix}, \quad |11\rangle = \begin{pmatrix} 0 \\ 0 \\ 0 \\ 1 \end{pmatrix}. \tag{13.2.8}$$

In fact, recall that for the vectors $u = (a_i)$ and $v = (b_i)$ in $\mathbb{F}^n$ where $\mathbb{F}$ is a field, the natural pairing

$$\langle u, v \rangle = \sum_{i=1}^{n} a_i b_i \tag{13.2.9}$$

generates a bilinear form defined by setting

$$F(x, y) = \langle u, x \rangle \langle v, y \rangle, \quad x = (x_i), y = (y_i) \in \mathbb{F}^n, \tag{13.2.10}$$

which is referred to as the tensor product of the vectors u and v:

$$(u \otimes v)(x, y) = \langle u, x \rangle \langle v, y \rangle = \sum_{i,j=1}^{n} x_i (a_i b_j) y_j, \quad x = (x_i), y = (y_i) \in \mathbb{F}^n.$$

(13.2.11)

Thus, identifying $u \otimes v$ with the matrix that defines $u \otimes v$ as a bilinear form, we arrive at the matrix representation of $u \otimes v$:

$$u \otimes v = (a_i b_j), \quad u = (a_i), v = (b_i) \in \mathbb{F}^n. \qquad (13.2.12)$$

Furthermore, there are many ways to realize (13.2.12) as a vector in the column vector space $\mathbb{F}^{n^2}$. For example, two convenient ways are

$$u \otimes v = \begin{pmatrix} a_1 v \\ \vdots \\ a_n v \end{pmatrix}, \quad u = (a_i), v = (b_i) \in \mathbb{F}^n, \qquad (13.2.13)$$

and

$$u \otimes v = \begin{pmatrix} u b_1 \\ \vdots \\ u b_n \end{pmatrix}, \quad u = (a_i), v = (b_i) \in \mathbb{F}^n, \qquad (13.2.14)$$

depending on preference. In our discussion, we choose to follow the convention (13.2.13), which results in the representation (13.2.8).

We now return to qubits.

In view of (13.2.7), the state of a 2-qubit can be generated from the states of two 1-qubits, (13.2.5) and $|\psi\rangle = b_0|0\rangle + b_1|1\rangle, b_0, b_1 \in \mathbb{C}, |b_0|^2 + |b_1|^2 = 1$, by

$$|\phi\rangle|\psi\rangle = |\phi \psi\rangle \equiv |\phi\rangle \otimes |\psi\rangle = c_{00}|00\rangle + c_{01}|01\rangle + c_{10}|10\rangle + c_{11}|11\rangle, \quad (13.2.15)$$

where the complex coefficients c's may be seen to be normalized as well.

Similarly, general n-qubit states can be constructed.

Quantum information is stored in qubit states and processed or measured by "observables" realized as Hermitian matrices applied to qubit states expressed as complex vectors, according to quantum mechanics. Thus, at the fundamental 1-qubit level, these observables are 2×2 Hermitian matrices of the form

$$A = \begin{pmatrix} a & b + ic \\ b - ic & d \end{pmatrix}, \quad a, b, c, d \in \mathbb{R},$$

$$= \alpha_0 \sigma_0 + \alpha_1 \sigma_1 + \alpha_2 \sigma_2 + \alpha_3 \sigma_3, \qquad (13.2.16)$$

where

$$\alpha_0 = \frac{1}{2}(a + d), \quad \alpha_1 = b, \quad \alpha_2 = -c, \quad \alpha_3 = \frac{1}{2}(a - d), \qquad (13.2.17)$$

are updated free real coefficients and the linearly independent 2×2 matrices

$$\sigma_0 = \begin{pmatrix} 1 & 0 \\ 0 & 1 \end{pmatrix}, \; \sigma_1 = \begin{pmatrix} 0 & 1 \\ 1 & 0 \end{pmatrix}, \; \sigma_2 = \begin{pmatrix} 0 & -i \\ i & 0 \end{pmatrix}, \; \sigma_3 = \begin{pmatrix} 1 & 0 \\ 0 & -1 \end{pmatrix},$$

(13.2.18)

are the well-known *Pauli spin matrices* or *Pauli gates* which act or "process" upon qubit states as linear transformations. For example, the gate σ_1 has the property

$$\sigma_1 |0\rangle = |1\rangle, \quad \sigma_1 |1\rangle = |0\rangle, \quad \sigma_1 (a_0 |0\rangle + a_1 |1\rangle) = a_1 |0\rangle + a_0 |1\rangle, \quad (13.2.19)$$

such that σ_1 may be considered as the NOT gate in quantum information processing or "computing" in analogy of the NOT gate in classical information processing or computing, whose action is to flip or "negate" a state. Of course, the gate $\sigma_0 = I$ does not alter any state so that it is called the identity gate. Besides, we also have

$$I = \sigma_0^2 = \sigma_1^2 = \sigma_2^2 = \sigma_3^2 = -i\sigma_1\sigma_2\sigma_3. \tag{13.2.20}$$

In particular, the Pauli gates are *involutory* meaning that they are their own inverse gates, respectively. In quantum computing, the Pauli gates σ_1, σ_2, and σ_3 are also referred to the σ_x, σ_y, and σ_z or simply X, Y, and Z gates, interchangeably.

With the Pauli gates as building blocks, we can construct other useful quantum gates to fulfill various specific purposes. For example, the gate

$$H = \frac{1}{\sqrt{2}}\sigma_1 + \frac{1}{\sqrt{2}}\sigma_3 = \frac{1}{\sqrt{2}} \begin{pmatrix} 1 & 1 \\ 1 & -1 \end{pmatrix} \tag{13.2.21}$$

is called the *Hadamard gate* or *Walsh–Hadamard gate*. Like the Pauli gates σ_1 and σ_3, the Hadamard gate H is also real and unitary, or involutory, $H^2 = I$. Applying the gate H to the qubit basis $\{|0\rangle, |1\rangle\}$, we obtain another orthonormal computational basis $\{|+\rangle, |-\rangle\}$ given by

$$|+\rangle = H|0\rangle = \frac{1}{\sqrt{2}}(|0\rangle + |1\rangle), \quad |-\rangle = H|1\rangle = \frac{1}{\sqrt{2}}(|0\rangle - |1\rangle). \tag{13.2.22}$$

Since in the state $|+\rangle$ or $|-\rangle$, the probabilities of observing the states $|0\rangle$ and $|1\rangle$ are equal, $\frac{1}{2}$, the states $|+\rangle$ and $|-\rangle$ may be viewed as "half ways" between the states $|0\rangle$ and $|1\rangle$.

Sometimes non-Hermitian gates are also useful in quantum computing and quantum information processing. For example, the matrix

$$P(\varphi) = \begin{pmatrix} 1 & 0 \\ 0 & e^{i\varphi} \end{pmatrix}, \quad \varphi \in \mathbb{R}, \tag{13.2.23}$$

defines a quantum gate called the phase-shift gate which does not alter the state $|0\rangle$, $P(\varphi)|0\rangle = |0\rangle$, but introduces a phase shift to the state $|1\rangle$, $P(\varphi)|1\rangle = e^{i\varphi}|1\rangle$. Of course, $P(2\pi) = \sigma_0$ and $P(\pi) = \sigma_3$ are two known gates. Besides, $S = P\left(\frac{\pi}{2}\right) = \sqrt{\sigma_3}$ and $T = P\left(\frac{\pi}{4}\right) = \sqrt{S}$ are referred to as the S gate and the T gate, respectively.

Using single-qubit quantum gates and tensor product operations, multi-qubit quantum gates may be constructed.

Let $\mathbb{F}$ be a field. For $A = (a_{ij})$ and $B = (b_{ij}) \in \mathbb{F}(n, n)$, define the tensor product $A \otimes B$ to be the matrix in $\mathbb{F}(n^2, n^2)$ that represents the linear map in $L\left(\mathbb{F}^{n^2}\right)$ induced from setting

$$(A \otimes B)(x \otimes y) = (Ax) \otimes (By), \quad x = (x_i), y = (y_i) \in \mathbb{F}^n. \tag{13.2.24}$$

It is clear that different representations of the vector $x \otimes y$ will result in different matrix representations of $A \otimes B$. In our context, since we assume the convention (13.2.13), we have

$$A \otimes B = \begin{pmatrix} a_{11}B & \cdots & a_{1n}B \\ \vdots & \ddots & \vdots \\ a_{n1}B & \cdots & a_{nn}B \end{pmatrix}, \quad x \otimes y = \begin{pmatrix} x_1 y \\ \vdots \\ x_n y \end{pmatrix}. \tag{13.2.25}$$

As a comparison, if we assume the convention (13.2.14), then we have

$$A \otimes B = \begin{pmatrix} Ab_{11} & \cdots & Ab_{1n} \\ \vdots & \ddots & \vdots \\ Ab_{n1} & \cdots & Ab_{nn} \end{pmatrix}, \quad x \otimes y = \begin{pmatrix} xy_1 \\ \vdots \\ xy_n \end{pmatrix}. \tag{13.2.26}$$

These tensor products of matrices are also known as the *Kronecker products* which enjoy some rich and useful properties. For example, in view of (13.2.24), we see that A and B are acting on x and y separately or "diagonally." In particular, we have

$$(A \otimes I)(x \otimes y) = (Ax) \otimes y, \quad (I \otimes A)(x \otimes y) = x \otimes Ay, \quad x, y \in \mathbb{F}^n, \tag{13.2.27}$$

which leaves y or x intact, respectively, under tensor products. In the context of quantum information processing, the tensor product $A \otimes B$ is referred to as the *parallel gate* formed from the gates A and B since it "preserves" the tensor product by following the correspondence:

$$x \mapsto Ax, \quad y \mapsto By, \quad x \otimes y \mapsto (Ax) \otimes (By), \quad x, y \in \mathbb{F}^n. \tag{13.2.28}$$

As a consequence, the 2-qubit parallel Hadamard gate, $H_2 = H \otimes H$, formed from the 1-qubit Hadamard H gate, is

$$H_2 = \frac{1}{\sqrt{2}} \begin{pmatrix} H & H \\ H & -H \end{pmatrix} = \frac{1}{2} \begin{pmatrix} 1 & 1 & 1 & 1 \\ 1 & -1 & 1 & -1 \\ 1 & 1 & -1 & -1 \\ 1 & -1 & -1 & 1 \end{pmatrix}, \qquad (13.2.29)$$

in view of (13.2.21) and (13.2.25). To appreciate this quantum gate, we apply (13.2.29) to the 2-qubit state $|00\rangle$ to obtain

$$H_2|00\rangle = \frac{1}{2}(|00\rangle + |01\rangle + |10\rangle + |11\rangle) \qquad (13.2.30)$$

by virtue of (13.2.8). In other words, the H_2 serves to "prepare" an equally distributed state over the four eigenstates $|00\rangle, |01\rangle, |10\rangle$, and $|11\rangle$ from the single eigenstate $|00\rangle$, such that observing any of the four eigenstates has the same probability, $\frac{1}{4}$.

Similarly, using the Pauli gates (13.2.18), the 2-qubit SWAP gate is formulated with the expression

$$\text{SWAP} = \frac{1}{2} \sum_{\mu=0}^{3} \sigma_\mu \otimes \sigma_\mu = \begin{pmatrix} 1 & 0 & 0 & 0 \\ 0 & 0 & 1 & 0 \\ 0 & 1 & 0 & 0 \\ 0 & 0 & 0 & 1 \end{pmatrix}. \qquad (13.2.31)$$

It is clear that SWAP leaves the states $|00\rangle$ and $|11\rangle$ intact but swaps the states $|01\rangle$ and $|10\rangle$:

$$|00\rangle \mapsto |00\rangle, \quad |11\rangle \mapsto |11\rangle, \quad |01\rangle \mapsto |10\rangle, \quad |10\rangle \mapsto |01\rangle. \qquad (13.2.32)$$

Another convenient way to construct 2-qubit quantum gates from 1-qubit ones is to form the gates diagonally directly. For example, the controlled NOT gate, or CNOT, may be given by

$$\text{CNOT} = \text{diag}\{\sigma_0, \sigma_1\} = \text{diag}\{I, X\} = \begin{pmatrix} 1 & 0 & 0 & 0 \\ 0 & 1 & 0 & 0 \\ 0 & 0 & 0 & 1 \\ 0 & 0 & 1 & 0 \end{pmatrix}. \qquad (13.2.33)$$

Thus, it is seen that, under CNOT, the first qubit plays the role of a controller such that, when it assumes the state $|0\rangle$, the state of the second qubit is left intact, but when it assumes the state $|1\rangle$, the state of the second qubit is subject to the NOT gate:

$$|00\rangle \mapsto |00\rangle, \quad |01\rangle \mapsto |01\rangle, \quad |10\rangle \mapsto |11\rangle, \quad |11\rangle \mapsto |10\rangle. \qquad (13.2.34)$$

Since such a CNOT gate is "built" from the Pauli X gate, it is also referred to as the controlled X gate, and denoted simply by CX.

Let $\mathbb{C}^n$ be a given state space. A norm-preserving linear map from $\mathbb{C}^n$ into itself is generally referred to as a quantum gate, and vice versa, since it maps a state into another state in the same state space. In this manner, a quantum gate is simply represented by a unitary matrix. Let $A \in \mathbb{C}(n,n)$ be unitary. Then there is an orthonormal basis of $\mathbb{C}^n$, say $\{u_1, \ldots, u_n\}$, consisting of eigenvectors of A, associated with eigenvalues of unit-modulus, say $e^{i\lambda_1}, \ldots, e^{i\lambda_n}$ where $\lambda_1, \ldots, \lambda_n \in \mathbb{R}$. Thus, for any $r \in \mathbb{R}$, we can define $A^r \in \mathbb{C}(n,n)$ by setting

$$A^r u_i = e^{ir\lambda_i} u_i, \quad i = 1, \ldots, n. \tag{13.2.35}$$

The so-defined matrix A^r is also unitary. In quantum computing, if the matrix A is a quantum gate applied to certain qubit states, then A^r is also a quantum gate applied to the same qubit states. The gate A^r is called a *power gate* built over the base gate A.

If A and $B \in \mathbb{C}(n,n)$ are viewed as two quantum gates applied to certain qubit states, then $AB \equiv C$ is also a quantum gate applied to the same qubit states, because when A and B are both unitary, so is C. Such a gate C is referred to as a *serially wired gate* constructed from the gates A and B. In this context, if A is a quantum gate, then the power gate A^k where k is an integer is a serially wired gate as well.

Other operations on matrices may also be used to build quantum gates mathematically. For example, if $A \in \mathbb{C}(n,n)$ is Hermitian, then $B = e^{iA}$ is unitary, which is referred to as the *exponential gate* generated by A.

It will be interesting to know whether we can copy or clone any qubit state using a suitable quantum gate, referred to as U_c. To do so, we need to reserve a state, say $|\gamma\rangle$, as a spot or holder to place or "overlay" the copied or cloned state, and leave the original state, say $|\phi\rangle$ intact after the process of copying or cloning. Thus, we are to start from a composite state involving $|\gamma\rangle$ and $|\phi\rangle$, say $|\gamma\phi\rangle = |\gamma\rangle|\phi\rangle$, and arrive at the composite state $|\phi\phi\rangle = |\phi\rangle|\phi\rangle$, of the appearance of a "doubled" state:

$$U_c(|\gamma\rangle|\phi\rangle) = |\phi\rangle|\phi\rangle, \quad \forall |\phi\rangle, \tag{13.2.36}$$

through U_c. Of course, here U_c is required to be unitary to preserve the norm unity of states, as described earlier.

To ease our illustration, we consider the cloning of a 1-qubit state and set $|\gamma\rangle = |0\rangle$. Then (13.2.36) becomes

$$U_c(|0\rangle|\phi\rangle) = |\phi\rangle|\phi\rangle, \quad |\phi\rangle = a|0\rangle + b|1\rangle, \quad |a|^2 + |b|^2 = 1, \tag{13.2.37}$$

leading to

$$aU_c(|0\rangle|0\rangle) + bU_c(|0\rangle|1\rangle) = a^2|00\rangle + ab|01\rangle + ab|10\rangle + b^2|11\rangle, \quad (13.2.38)$$

which also reads

$$a|00\rangle + b|11\rangle = a^2|00\rangle + ab|01\rangle + ab|10\rangle + b^2|11\rangle \quad (13.2.39)$$

in view of (13.2.37). Comparing the coefficients in (13.2.39), we have $a = a^2, ab = 0$, and $b = b^2$. Consequently, there are two possibilities, $a = 1, b = 0$ and $a = 0, b = 1$, giving us

$$U_c|0\rangle|0\rangle = |0\rangle|0\rangle, \quad U_c|0\rangle|1\rangle = |1\rangle|1\rangle. \quad (13.2.40)$$

Consequently, the condition (13.2.37) cannot hold in general except for the two cases given in (13.2.40). In other words, for general qubit states, the process of copying or cloning cannot be executed as described. This statement is known as the *no-cloning theorem* in quantum information theory.

However, classically, copying or cloning *can* be executed. In fact, for 2-cbit states, we may expand (13.2.40) to define $U_c : \mathbb{Z}_2^2 \to \mathbb{Z}_2^2$ by

$$U_c|00\rangle = |00\rangle, \quad U_c|01\rangle = |11\rangle, \quad U_c|10\rangle = |10\rangle, \quad U_c|11\rangle = |01\rangle, \quad (13.2.41)$$

for example, as a cloning gate, to fulfill the task (13.2.36). Thus an executable classical process may fail to be realized at quantum level. For cloning, we see that the failure arises simply as a result of the abundance of qubit states.

Exercises

13.2.1 Estimate the amounts of information contained in a six-face die, a collection of seven banners of different colors, and two six-face dice.

13.2.2 With (13.2.7) and the convention (13.2.14), obtain an alternative 4-vector realization of the 2-qubit computational basis $\{|00\rangle, |01\rangle, |10\rangle, |11\rangle\}$ to (13.2.8).

13.2.3 Verify that the 2-qubit state (13.2.15) is indeed a unit vector.

13.2.4 Show that the Pauli matrices (13.2.18) are mutually orthogonal under the usual matrix scalar product that defines the Hilbert–Schmidt norm of a complex matrix, namely,

$$(A, B) = \text{Tr}\,(A^\dagger B), \quad A, B \in \mathbb{C}(n, n). \quad (13.2.42)$$

13.2.5 Show that, with the Dirac bra- and ket-operations, we have the representations

$$\sigma_1 = |0\rangle\langle 1| + |1\rangle\langle 0|, \quad \sigma_2 = -i|0\rangle\langle 1| + i|1\rangle\langle 0| \quad \sigma_3 = |0\rangle\langle 0| - |1\rangle\langle 1|$$

(13.2.43)

for the Pauli gates.

13.2.6 For the Hadamard gate defined by (13.2.21), use the property (13.2.22) to show directly that H enjoys the decomposition

$$H = \frac{1}{\sqrt{2}}(|0\rangle + |1\rangle)\langle 0| + \frac{1}{\sqrt{2}}(|0\rangle - |1\rangle)\langle 1|.$$

(13.2.44)

13.2.7 Use the Hadamard H gate and the Pauli $Z = \sigma_3$ gate to form the other two Pauli gates X and Y. More precisely, establish the expressions

$$X = \sigma_1 = HZH, \quad Y = \sigma_2 = iHZHZ,$$

(13.2.45)

so that X and Y are serially wired from H and Z.

13.2.8 Verify (13.2.25) and (13.2.26).

13.2.9 Show that the application of the 2-qubit Hadamard H_2 gate to any of the 2-qubit eigenstates $|01\rangle, |10\rangle$, and $|11\rangle$ always produces an equally distributed 2-qubit state over the four eigenstates $|00\rangle, |01\rangle, |10\rangle$, and $|11\rangle$.

13.2.10 Define the K gate on 2-qubit states by setting $K = H \otimes I$, where H is the 1-qubit Hadamard gate and I the 1-qubit identity gate. Show that

$$K = \frac{1}{\sqrt{2}} \begin{pmatrix} 1 & 0 & 1 & 0 \\ 0 & 1 & 0 & 1 \\ 1 & 0 & -1 & 0 \\ 0 & 1 & 0 & -1 \end{pmatrix}.$$

(13.2.46)

Then use (13.2.46) to show that the gate K acts on the first qubit by the H gate, while leaving the second qubit intact, with the property

$$K|ab\rangle = (H|a\rangle)|b\rangle, \quad a, b = 0, 1,$$

(13.2.47)

which is a realization of the general property (13.2.27), of course.

13.2.11 Recall that, in quantum mechanics, a state vector $|\psi\rangle$ and its phase shift, $e^{i\theta}|\psi\rangle$ $(\theta \in \mathbb{R})$, are considered to represent the same quantum state. Hence, for a cloning process, the condition (13.2.36) may be replaced by the general expression

$$U_c(|\gamma\rangle|\phi\rangle) = e^{i\theta}|\phi\rangle|\phi\rangle, \quad \forall|\phi\rangle, \quad \theta = \theta(|\gamma\rangle, |\phi\rangle) \in \mathbb{R}, \quad (13.2.48)$$

so that, when cloning a 1-qubit state, (13.2.37) assumes the form

$$U_c(|0\rangle|\phi\rangle) = e^{i\theta}|\phi\rangle|\phi\rangle, \quad |\phi\rangle = a|0\rangle + b|1\rangle, \quad |a|^2 + |b|^2 = 1,$$
$$\theta = \theta(|0\rangle, |\phi\rangle) \in \mathbb{R}. \tag{13.2.49}$$

With (13.2.49), prove the no-cloning theorem.

13.2.12 For a process of cloning a general qubit state defined by the expression (13.2.48), prove the no-cloning theorem.

13.2.13 With (13.2.8), find the matrix representation of the classical cloning gate (13.2.41).

13.2.14 Show that the cbit cloning process (13.2.41) may summarily be defined by the map $f : \mathbb{Z}_2^2 \to \mathbb{Z}_2^2$ with

$$f(x,y) = (x+y,y), \quad x, y \in \mathbb{Z}_2. \tag{13.2.50}$$

13.2.15 How many classical cloning gates may be constructed to satisfy (13.2.40)?

13.3 Geometric properties

In this section, we present a discussion on the geometric properties of the 1-qubit states and some of the quantum gates acting on 1-qubit states.

For convenience of notation, we rewrite the 1-qubit state (13.2.5) as

$$|\phi\rangle = \alpha|0\rangle + \beta|1\rangle, \quad \alpha, \beta \in \mathbb{C}, \quad |\alpha|^2 + |\beta|^2 = 1, \tag{13.3.1}$$

so that the state may be regarded as a point in the unit sphere in $\mathbb{C}^2$. We first show that the unit sphere in $\mathbb{C}^2$ corresponds to the special unitary matrix $SU(2)$ in $\mathbb{C}(2,2)$, where

$$SU(2) = \{U \in \mathbb{C}(2,2) \,|\, U^\dagger U = I, \det(U) = 1\}. \tag{13.3.2}$$

In fact, for any $U \in SU(2)$, write

$$U = \begin{pmatrix} a & b \\ c & d \end{pmatrix}, \quad a,b,c,d \in \mathbb{C}. \tag{13.3.3}$$

Then it is clear that $a, b, c,$ and d satisfy the equations

$$|a|^2 + |b|^2 = 1, \tag{13.3.4}$$
$$|c|^2 + |d|^2 = 1, \tag{13.3.5}$$
$$a\bar{c} + b\bar{d} = 0, \tag{13.3.6}$$
$$ad - bc = 1. \tag{13.3.7}$$

Multiplying (13.3.6) by d and applying (13.3.7), we have

$$\bar{c} + (|c|^2 + |d|^2)b = 0, \tag{13.3.8}$$

leading to $b = -\bar{c}$ in view of (13.3.5). Likewise, multiplying (13.3.6) by c and applying (13.3.7), we get

$$a(|c|^2 + |d|^2) - \bar{d} = 0, \tag{13.3.9}$$

yielding $a = \bar{d}$ in view of (13.3.5). These results characterize $U \in SU(2)$ given in (13.3.3) completely. That is,

$$SU(2) = \left\{ \begin{pmatrix} a & b \\ -\bar{b} & \bar{a} \end{pmatrix} \,\middle|\, |a|^2 + |b|^2 = 1, a, b \in \mathbb{C} \right\}, \tag{13.3.10}$$

which is exactly identified with the unit sphere in $\mathbb{C}^2$.

Consequently, we may represent the 1-qubit state (13.3.1) by an element in $SU(2)$, say U_ϕ, with

$$U_\phi = \begin{pmatrix} \alpha & \beta \\ -\bar{\beta} & \bar{\alpha} \end{pmatrix}, \tag{13.3.11}$$

which accommodates the action of $SU(2)$ on it by the matrix multiplication

$$|\psi\rangle = U|\phi\rangle, \quad U_\psi = UU_\phi, \quad U \in SU(2), \tag{13.3.12}$$

serving the purpose of realizing another sort of unitary quantum gates.

At this juncture, it will be interesting to compare the gate (13.3.12) with the more traditional unitary gate discussed in Section 13.2 whose action on the 1-qubit state (13.3.1) is defined by

$$|\psi\rangle = U|\phi\rangle, \quad |\psi\rangle = \delta|0\rangle + \gamma|1\rangle, \quad \begin{pmatrix} \delta \\ \gamma \end{pmatrix} = U \begin{pmatrix} \alpha \\ \beta \end{pmatrix} = \begin{pmatrix} a\alpha + b\beta \\ -\bar{b}\alpha + \bar{a}\beta \end{pmatrix}, \tag{13.3.13}$$

where U is given by (13.3.3) with the property $c = -\bar{b}$ and $d = \bar{a}$. On the other hand, the action defined by (13.3.12) leads to the result

$$\delta = a\alpha - b\bar{\beta}, \quad \gamma = b\bar{\alpha} + a\beta, \tag{13.3.14}$$

which is rather different from what given in (13.3.13) even when a, b, α, and β are all real.

With the unitary representation (13.3.11) of the 1-qubit state (13.3.1), we may use the adjoint action of $SU(2)$ on $|\phi\rangle$ defined by

$$|\psi\rangle = U|\phi\rangle, \quad |\psi\rangle = \delta|0\rangle + \gamma|1\rangle, \quad U_\psi = U^\dagger U_\phi U, \quad U \in SU(2). \tag{13.3.15}$$

In this situation, we may use (13.3.3) with $c = -\bar{b}$ and $d = \bar{a}$, the representation (13.3.11), and the definition (13.3.15) to obtain the result

$$\delta = |a|^2\alpha + |b|^2\bar{\alpha} - \bar{a}\bar{b}\beta + ab\bar{\beta}, \quad \gamma = \bar{a}b\alpha - \bar{a}b\bar{\alpha} + \bar{a}^2\beta + b^2\bar{\beta}. \quad (13.3.16)$$

This result represents yet another different unitary gate.

To explore the geometric characteristics associated with the correspondences (13.3.13), (13.3.14), and (13.3.16), we may use the amplitude-phase angle or radial-polar representations of complex numbers to recast α, β and a, b given in (13.3.1) and (13.3.10), respectively, into

$$\alpha = \rho_\alpha e^{i\omega_\alpha}, \quad \beta = \rho_\beta e^{i\omega_\beta}, \quad \rho_\alpha, \rho_\beta \geq 0, \quad \rho_\alpha^2 + \rho_\beta^2 = 1, \quad (13.3.17)$$

$$a = \rho_a e^{i\omega_a}, \quad b = \rho_b e^{i\omega_b}, \quad \rho_a, \rho_b \geq 0, \quad \rho_a^2 + \rho_b^2 = 1. \quad (13.3.18)$$

Consequently, the effects of various gates are seen as combined geometric actions comprised of a sequence of amplitude stretches and phase angle translations.

It will be interesting to explore the geometry of the 1-qubit state represented by (13.3.17) further. For this purpose, we resolve the constraints on ρ_α and ρ_β in it with trigonometric functions to arrive at

$$\alpha = \cos\left(\frac{\theta}{2}\right) e^{i\omega_\alpha}, \quad \beta = \sin\left(\frac{\theta}{2}\right) e^{i\omega_\beta}, \quad 0 \leq \omega_\alpha, \omega_\beta \leq 2\pi, \quad 0 \leq \theta \leq \pi.$$

$$(13.3.19)$$

This expression is known as the representation of the unit sphere in $\mathbb{C}^2$ in the *Hopf coordinates*. In quantum mechanics, since the phase angle of a quantum state may be shifted back and forth freely, we may set $\omega_\alpha = 0$ and keep θ and $\omega_\beta = \varphi$ as the remaining coordinate variables. Thus the 1-qubit state (13.3.19) becomes

$$\alpha = \cos\frac{\theta}{2}, \quad \beta = \sin\left(\frac{\theta}{2}\right) e^{i\varphi}, \quad 0 \leq \varphi \leq 2\pi, \quad 0 \leq \theta \leq \pi. \quad (13.3.20)$$

On the other hand, the coordinate variables θ and φ in (13.3.20) allow us to represent the state (13.3.1) by the unit sphere $x_1^2 + x_2^2 + x_3^2 = 1$ in the Euclidean space $\mathbb{R}^3$ with the spherical coordinates

$$x_1 = \cos\varphi \sin\theta, \, x_2 = \sin\varphi \sin\theta, \, x_3 = \cos\theta, \quad 0 \leq \varphi \leq 2\pi, 0 \leq \theta \leq \pi.$$

$$(13.3.21)$$

In this manner, the eigenstate $|0\rangle$ corresponds to the north pole $(0, 0, 1)$ on the sphere and the set of all phase-shifted states from the eigenstate $|1\rangle$, namely, $\{e^{i\varphi}|1\rangle \mid \varphi \in [0, 2\pi]\}$, is represented by the south pole $(0, 0, -1)$ on the sphere. Since any phase shift of a state is identified with the state itself quantum

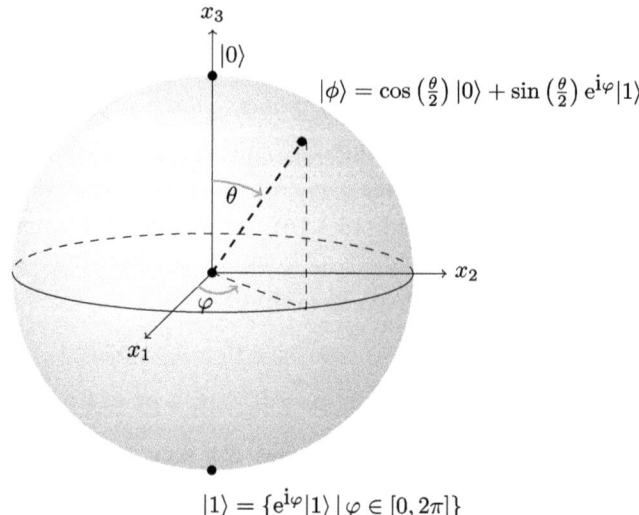

$$|\phi\rangle = \cos\left(\tfrac{\theta}{2}\right)|0\rangle + \sin\left(\tfrac{\theta}{2}\right)e^{i\varphi}|1\rangle$$

$$|1\rangle = \{e^{i\varphi}|1\rangle \mid \varphi \in [0, 2\pi]\}$$

Figure 13.1 A spherical representation of the full set of 1-qubit states known as the Bloch sphere. The two eigenstates $|0\rangle$ and $|1\rangle$ are shown as the north and south poles, respectively, and a general state can be any point on the sphere.

mechanically, we see that it is actually the eigenstate $|1\rangle$ that corresponds to the south pole on the sphere. In general, any state $|\phi\rangle$ given by (13.3.20) is represented as a point on the sphere. This sphere, which houses all possible 1-qubit states, is known as the Bloch sphere in quantum information theory, as drawn in Figure 13.1. Interestingly, any state on the equator $\theta = \frac{\pi}{2}$ is exactly "half way" between the two eigenstates both geometrically and statistically.

There are other geometric methods to represent 1-qubit states and introduce associated quantum gate actions. In the rest of this section, we do so by exploring a classical connection between space rotations and the Pauli spin matrices.

Specifically, consider the standard coordinate system of the Euclidean space $\mathbb{R}^3$ formed with the usual Cartesian x_1-, x_2-, and x_3-axes. Then the rotations about the x_1-, x_2-, and x_3-axes, with the rotation angles θ_1, θ_2, and θ_3, mapping the coordinate vector (x_i) to (y_i), say, are represented by the rotation matrices

$$R_1(\theta_1) = \begin{pmatrix} 1 & 0 & 0 \\ 0 & \cos\theta_1 & -\sin\theta_1 \\ 0 & \sin\theta_1 & \cos\theta_1 \end{pmatrix}, \tag{13.3.22}$$

$$R_2(\theta_2) = \begin{pmatrix} \cos\theta_2 & 0 & \sin\theta_2 \\ 0 & 1 & 0 \\ -\sin\theta_2 & 0 & \cos\theta_2 \end{pmatrix}, \tag{13.3.23}$$

$$R_3(\theta_3) = \begin{pmatrix} \cos\theta_3 & -\sin\theta_3 & 0 \\ \sin\theta_3 & \cos\theta_3 & 0 \\ 0 & 0 & 1 \end{pmatrix}, \tag{13.3.24}$$

respectively. Note that these representations reflect the choice that the angular variables θ_1, θ_2, and θ_3 describe counterclockwise rotations in the (x_2, x_3)-, (x_1, x_3)-, (x_1, x_2)-planes, respectively. The angles θ_1, θ_2, and θ_3 are called the *Euler angles*.

Differentiating (13.3.22)–(13.3.24) with respect to θ_1, θ_2, and θ_3, respectively, we have

$$\dot{R}_1 = J_1 R_1, \quad \dot{R}_2 = J_2 R_2, \quad \dot{R}_3 = J_3 R_3, \tag{13.3.25}$$

where the overdot denotes derivative of R_i with respect to the angular variable θ_i, $i = 1, 2, 3$, and

$$J_1 = \begin{pmatrix} 0 & 0 & 0 \\ 0 & 0 & -1 \\ 0 & 1 & 0 \end{pmatrix}, \quad J_2 = \begin{pmatrix} 0 & 0 & 1 \\ 0 & 0 & 0 \\ -1 & 0 & 0 \end{pmatrix}, \quad J_3 = \begin{pmatrix} 0 & -1 & 0 \\ 1 & 0 & 0 \\ 0 & 0 & 0 \end{pmatrix}, \tag{13.3.26}$$

are given by $J_i = \dot{R}_i(0)$, $i = 1, 2, 3$, respectively. Since $R_i(0) = I$, $i = 1, 2, 3$, we can integrate (13.3.25) to obtain the representations

$$R_i(\theta_i) = e^{\theta_i J_i}, \quad i = 1, 2, 3. \tag{13.3.27}$$

Thus the rotations (13.3.22)–(13.3.24) are generated by the anti-symmetric matrices J_1, J_2, and J_3 as one-parameter groups.

In general, if $\mathbf{a} = (a_1, a_2, a_3)$ is a unit vector in $\mathbb{R}^3$, then the rotation about the vector $\mathbf{a}$ following the right-hand rule is given by

$$R_{\mathbf{a}}(\theta) = e^{\theta(\mathbf{a} \cdot \mathbf{J})}, \quad \mathbf{J} \equiv (J_1, J_2, J_3), \quad \mathbf{a} \cdot \mathbf{J} \equiv \sum_{i=1}^{3} a_i J_i. \tag{13.3.28}$$

Since our study does not rely on (13.3.28), we will not elaborate on it further but only observe that (13.3.22)–(13.3.24) or (13.3.27) are special cases of (13.3.28).

The set of transformations constructed from all possible compositions of the rotations (13.3.22)–(13.3.24) forms a subgroup of the orthogonal group $O(3)$, where the group product is simply the matrix product, called the special orthogonal group $SO(3)$, since it satisfies the condition

$$\det(A) = 1, \quad A \in SO(3). \tag{13.3.29}$$

On the other hand, we may represent any vector $x = (x_i) \in \mathbb{R}^3$ by a traceless Hermitian matrix $h \in \mathbb{C}(2,2)$ with setting

$$h = \frac{x_1\sigma_1 + x_2\sigma_2 + x_3\sigma_3}{\sqrt{2}} = \frac{1}{\sqrt{2}}\begin{pmatrix} x_3 & x_1 - ix_2 \\ x_1 + ix_2 & -x_3 \end{pmatrix}. \qquad (13.3.30)$$

With (13.3.30), the Hilbert–Schmidt norm of h is given by

$$\|h\|^2 = \mathrm{Tr}\,(h^\dagger h) = x_1^2 + x_2^2 + x_3^2 = \|x\|^2. \qquad (13.3.31)$$

Hence, rotations on vectors in $\mathbb{R}^3$, which preserves the Euclidean norm, correspond to unitary maps acting on traceless Hermitian matrices in $\mathbb{C}(2,2)$, which we now uncover.

In fact, for the rotation (13.3.24), with the notation $y = R_3(\theta_3)x$ and the correspondence $\mathbb{R}^3 \to \mathbb{C}(2,2)$, $y \mapsto g$, following (13.3.30), we have

$$\begin{aligned}
g &= \frac{y_1\sigma_1 + y_2\sigma_2 + y_3\sigma_3}{\sqrt{2}} = \frac{1}{\sqrt{2}}\begin{pmatrix} y_3 & y_1 - iy_2 \\ y_1 + iy_2 & -y_3 \end{pmatrix} \\
&= \frac{1}{\sqrt{2}}\begin{pmatrix} x_3 & e^{-i\theta_3}(x_1 - ix_2) \\ e^{i\theta_3}(x_1 + ix_2) & -x_3 \end{pmatrix} \\
&= U_3^\dagger h U_3, \quad U_3 = U_3(\theta_3) = \begin{pmatrix} e^{i\frac{\theta_3}{2}} & 0 \\ 0 & e^{-i\frac{\theta_3}{2}} \end{pmatrix} = e^{i\frac{\theta_3}{2}\sigma_3}. \qquad (13.3.32)
\end{aligned}$$

In general, if $U \in \mathbb{C}(2,2)$ is unitary, or $U \in U(2)$, then the map

$$h \mapsto U^\dagger h U = \mathrm{ad}_U(h), \qquad (13.3.33)$$

referred to as the *adjoint representation* of the matrix group $U(2)$ as before, is seen to enjoy the property that it preserves the Hilbert–Schmidt norm (13.3.31), Hermiticity, and trace.

For the rotation (13.3.23), with $y = R_2(\theta_2)x$, the matrix $g \in \mathbb{C}(2,2)$ corresponding to y becomes

$$\begin{aligned}
g &= \frac{1}{\sqrt{2}}\begin{pmatrix} -\sin\theta_2 x_1 + \cos\theta_2 x_3 & \cos\theta_2 x_1 + \sin\theta_2 x_3 - ix_2 \\ \cos\theta_2 x_1 + \sin\theta_2 x_3 + ix_2 & \sin\theta_2 x_1 - \cos\theta_2 x_3 \end{pmatrix} \\
&= U_2^\dagger h U_2, \quad U_2 = U_2(\theta_2) = \begin{pmatrix} \cos\frac{\theta_2}{2} & \sin\frac{\theta_2}{2} \\ -\sin\frac{\theta_2}{2} & \cos\frac{\theta_2}{2} \end{pmatrix} = e^{i\frac{\theta_2}{2}\sigma_2}, \qquad (13.3.34)
\end{aligned}$$

after a somewhat lengthy calculation.

For the rotation (13.3.22), with $y = R_1(\theta_1)x$, the corresponding matrix g in $\mathbb{C}(2,2)$ reads

$$g = \frac{1}{\sqrt{2}} \begin{pmatrix} \sin\theta_1 x_2 + \cos\theta_1 x_3 & x_1 - i(\cos\theta_1 x_2 - \sin\theta_1 x_3) \\ x_1 + i(\cos\theta_1 x_2 - \sin\theta_1 x_3) & -(\sin\theta_1 x_2 + \cos\theta_1 x_3) \end{pmatrix}$$

$$= U_1^\dagger h U_1, \quad U_1 = U_1(\theta_1) = \begin{pmatrix} \cos\frac{\theta_1}{2} & i\sin\frac{\theta_1}{2} \\ i\sin\frac{\theta_1}{2} & \cos\frac{\theta_1}{2} \end{pmatrix} = e^{i\frac{\theta_1}{2}\sigma_1}. \quad (13.3.35)$$

Note that the right-hand sides of (13.3.32), (13.3.34), and (13.3.35) are special cases of the general exponential identity

$$e^{i\theta(\mathbf{a}\cdot\boldsymbol{\sigma})} = \cos\theta\, I + i\sin\theta\,(\mathbf{a}\cdot\boldsymbol{\sigma}), \quad \theta \in \mathbb{R}, \quad (13.3.36)$$

for the Pauli matrices, which is the $U(2)$ extension of the Euler exponential formula for complex numbers of unit modulus, or $U(1)$, where

$$U(1) = \{e^{i\theta} = \cos\theta + i\sin\theta \,|\, \theta \in \mathbb{R}\}. \quad (13.3.37)$$

Here, in (13.3.36), $\mathbf{a} = (a_1, a_2, a_3)$ is a unit vector in $\mathbb{R}^3$, and

$$\boldsymbol{\sigma} \equiv (\sigma_1, \sigma_2, \sigma_3), \quad \mathbf{a}\cdot\boldsymbol{\sigma} = \sum_{i=1}^{3} a_i\sigma_i, \quad (13.3.38)$$

as in (13.3.28). An immediate consequence of (13.3.36) is that the matrix exponential on the left-hand side of (13.3.36) is 2π-periodic and

$$e^{i(\theta+\pi)(\mathbf{a}\cdot\boldsymbol{\sigma})} = -e^{i\theta(\mathbf{a}\cdot\boldsymbol{\sigma})}. \quad (13.3.39)$$

Using the formula (13.3.36), we may proceed as in the derivations of (13.3.32), (13.3.34), and (13.3.35) to arrive at the conclusion that the rotation (13.3.28) corresponds to the unitary matrix

$$U_\mathbf{a}(\theta) = e^{i\frac{\theta}{2}(\mathbf{a}\cdot\boldsymbol{\sigma})}, \quad \theta \in \mathbb{R}. \quad (13.3.40)$$

It is clear that $\det(U_\mathbf{a}(\theta)) = 1$ since all the Pauli matrices σ_1, σ_2, and σ_3 are traceless. Recall that, like the special orthogonal group $SO(3)$, the subgroup of the matrix group $U(2)$ comprised of elements of determinant 1 is referred to as the special unitary group, $SU(2)$. As a result of the correspondence between the rotation (13.3.28) and the unitary matrix (13.3.40), we see that the map

$$U_\mathbf{a}(\theta) = e^{i\frac{\theta}{2}(\mathbf{a}\cdot\boldsymbol{\sigma})} \mapsto R_\mathbf{a}(\theta) = e^{\theta(\mathbf{a}\cdot\mathbf{J})}, \quad \theta \in \mathbb{R}, \quad (13.3.41)$$

gives rise to a natural correspondence between $SU(2)$ and $SO(3)$. This map is onto and two-to-one since the property (13.3.39) and (13.3.41) lead to

$$U_{\mathbf{a}}(\theta) \mapsto R_{\mathbf{a}}(\theta), \quad -U_{\mathbf{a}}(\theta) = U_{\mathbf{a}}(\theta + 2\pi) \mapsto R_{\mathbf{a}}(\theta + 2\pi) = R_{\mathbf{a}}(\theta), \quad \theta \in \mathbb{R}. \tag{13.3.42}$$

In other words, we may regard $SU(2)$ as a *two-fold covering* of $SO(3)$.

Therefore, for any (x_1, x_2, x_3) on the unit sphere in $\mathbb{R}^3$, we may construct a 1-qubit state (13.3.1) by setting

$$\alpha = x_3, \quad \beta = x_1 + ix_2, \tag{13.3.43}$$

which is also represented by the traceless Hermitian matrix (13.3.30) or

$$h = \frac{1}{\sqrt{2}} \begin{pmatrix} \alpha & \overline{\beta} \\ \beta & -\alpha \end{pmatrix}. \tag{13.3.44}$$

Then, with the unitary matrix (13.3.40), we can build the *adjoint gate*:

$$h \mapsto \mathrm{ad}_{U_{\mathbf{a}}(\theta)}(h) = (U_{\mathbf{a}}(\theta))^{\dagger} h U_{\mathbf{a}}(\theta) = e^{-i\frac{\theta}{2}(\mathbf{a}\cdot\boldsymbol{\sigma})} h e^{i\frac{\theta}{2}(\mathbf{a}\cdot\boldsymbol{\sigma})}, \tag{13.3.45}$$

corresponding to applying the rotation (13.3.28) to the associated vector (x_1, x_2, x_3) in the Euclidean space $\mathbb{R}^3$ with

$$x_1 = \Re\beta, \quad x_2 = \Im\beta, \quad x_3 = \alpha, \quad \alpha \in \mathbb{R}, \quad \beta \in \mathbb{C}, \quad \alpha^2 + |\beta|^2 = 1. \tag{13.3.46}$$

Multiple qubit gates associated with the 1-qubit ones discussed here may be constructed as in Section 13.2.

Exercises

13.3.1 Derive (13.3.14).

13.3.2 Derive (13.3.16).

13.3.3 Derive the equations stated in (13.3.25).

13.3.4 Establish (13.3.27).

13.3.5 For $R_{\mathbf{a}}(\theta)$ defined by (13.3.28), verify that $\det(R_{\mathbf{a}}(\theta)) = 1$.

13.3.6 Establish (13.3.34).

13.3.7 Prove (13.3.36) by going over the following steps:

    (a) Show that the Pauli matrices σ_1, σ_2, and σ_3 satisfy the following anti-commutation relations:

$$\{\sigma_i, \sigma_j\} = \sigma_i\sigma_j + \sigma_j\sigma_i = 2\delta_{ij}I, \quad i, j = 1, 2, 3. \tag{13.3.47}$$

(b) Establish the identity

$$(\mathbf{a} \cdot \boldsymbol{\sigma})^2 = \sum_{i,j=1}^{3} a_i a_j \sigma_i \sigma_j = I \qquad (13.3.48)$$

for the matrix $\mathbf{a} \cdot \boldsymbol{\sigma}$ given in (13.3.38) by applying (13.3.47).

(c) Use U and V to denote the left- and right-hand sides of (13.3.36), respectively. Show that U and V satisfy the same differential equation and initial condition:

$$\dot{U} = i(\mathbf{a} \cdot \boldsymbol{\sigma})U, \quad U(0) = I; \quad \dot{V} = i(\mathbf{a} \cdot \boldsymbol{\sigma})V, \quad V(0) = I, \quad (13.3.49)$$

by applying (13.3.48), where the overdot again denotes derivative with respect to the variable θ.

(d) Apply the uniqueness theorem for the solutions to the initial value problems of ordinary differential equations to establish the conclusion that $U(\theta)$ and $V(\theta)$ agree for all θ, which is (13.3.36).

13.4 Quantum entanglement

In order to understand what an entangled quantum state is, we start from a discussion of an untangled quantum state.

Let A and B be two observers who are measuring two 1-qubit states, $|\phi\rangle$ and $|\psi\rangle$, independently, where

$$|\phi\rangle = a_0|0\rangle + a_1|1\rangle, \quad |\psi\rangle = b_0|0\rangle + b_1|1\rangle. \qquad (13.4.1)$$

For convenience and clarity, use the same letters in mathematical italic fonts, A and B, to denote their measurement readings which are two binary valued random variables with the probability distributions:

$$P(A = 0) = |a_0|^2, \, P(A = 1) = |a_1|^2; \, P(B = 0) = |b_0|^2, \, P(B = 1) = |b_1|^2. \qquad (13.4.2)$$

Now compose the 2-qubit state from (13.4.1) by using the tensor product as described in Section 13.2:

$$|\phi\psi\rangle = a_0 b_0|00\rangle + a_0 b_1|01\rangle + a_1 b_0|10\rangle + a_1 b_1|11\rangle. \qquad (13.4.3)$$

Consequently, we have

$$P(A = i, B = j) = |a_i|^2 |b_j|^2 = P(A = i)P(B = j), \quad i, j = 0, 1. \qquad (13.4.4)$$

Thus the events $\{A = i\}$ and $\{B = j\}$ $(i, j = 0, 1)$ are independent so that the knowledge of the measurement of one observer does not affect that of the other observer. Statistically, the correlation of the two independent random variables A and B is zero. In this situation, we refer to the 2-qubit state as an *untangled* state. Therefore, in view of multilinear algebra, we conclude that all separable or decomposable 2-qubit states are untangled.

However, it should be noted that a 2-qubit state being decomposable is a sufficient but *not* a necessary condition for the state to be untangled. For example, consider the 2-qubit state:

$$|\sigma\rangle = \frac{1}{2}(|00\rangle - |01\rangle - |10\rangle - |11\rangle). \tag{13.4.5}$$

It can be verified that $|\sigma\rangle$ is untangled. Assume that $|\sigma\rangle$ is decomposable such that $|\sigma\rangle$ is given by (13.4.3). Then we have

$$a_0 b_0 = \frac{1}{2}, \quad a_0 b_1 = -\frac{1}{2}, \quad a_1 b_0 = -\frac{1}{2}, \quad a_1 b_1 = -\frac{1}{2}. \tag{13.4.6}$$

The first two equations in (13.4.6) indicate that $b_0 + b_1 = 0$. Hence we have

$$0 = (a_0 + a_1)(b_0 + b_1) = a_0 b_0 + a_0 b_1 + a_1 b_0 + a_1 b_1, \tag{13.4.7}$$

which contradicts (13.4.6). Therefore we see that the state (13.4.5) is not separable or decomposable.

In contrast, if A and B denote the random variables of the two measurement readings of the two observers A and B where A and B measure the first and second qubits, respectively, a 2-qubit state is called *entangled* if A and B are not independent random variables.

With this definition of entangled 2-qubit states, we now derive a necessary and sufficient condition to characterize such states. To this end, consider the following 2-qubit state:

$$|\sigma\rangle = c_{00}|00\rangle + c_{01}|01\rangle + c_{10}|10\rangle + c_{11}|11\rangle. \tag{13.4.8}$$

Then we have

$$P(A = i, B = j) = |c_{ij}|^2, \quad i, j = 0, 1, \tag{13.4.9}$$

$$P(A = i) = |c_{i0}|^2 + |c_{i1}|^2, \ P(B = i) = |c_{0i}|^2 + |c_{1i}|^2, \ i = 0, 1. \tag{13.4.10}$$

In view of (13.4.4), (13.4.9), and (13.4.10), we conclude that the state (13.4.8) is untangled if and only if the amplitude coefficients of the state satisfy the condition:

$$|c_{ij}|^2 = \left(|c_{i0}|^2 + |c_{i1}|^2\right)\left(|c_{0j}|^2 + |c_{1j}|^2\right), \quad i, j = 0, 1. \tag{13.4.11}$$

In other words, the state (13.4.8) is entangled if and only if the condition (13.4.11) is violated for some $i, j = 0, 1$.

To evaluate the strength of entanglement, we can compute the correlation of the random variables A and B or compare the probability $P(A = i)$ with the conditional probability $P(A = i|B = j)$, say, for $i, j = 0, 1$, where the latter is given by

$$P(A = i|B = j) = \frac{P(A = i, B = j)}{P(B = j)} = \frac{|c_{ij}|^2}{|c_{0j}|^2 + |c_{1j}|^2}, \quad i, j = 0, 1. \quad (13.4.12)$$

For example, we study the 2-qubit state:

$$|\gamma\rangle = \frac{1}{\sqrt{2}}|00\rangle + \frac{1}{2\sqrt{2}}|01\rangle + \frac{1}{2\sqrt{2}}|10\rangle + \frac{1}{2}|11\rangle. \quad (13.4.13)$$

It is straightforward to check that the condition (13.4.11) is violated for all $i, j = 0, 1$. Besides, in view of (13.4.10), (13.4.12), and (13.4.13), we have

$$P(A = 0) = \frac{5}{8}, \quad P(A = 1) = \frac{3}{8}, \quad (13.4.14)$$

$$P(A = 0|B = 0) = \frac{4}{5}, \quad P(A = 0|B = 1) = \frac{1}{3}, \quad (13.4.15)$$

$$P(A = 1|B = 0) = \frac{1}{5}, \quad P(A = 1|B = 1) = \frac{2}{3}. \quad (13.4.16)$$

These results show clearly how the measurement of B affects the measurement of A in both directions in terms of likelihood or probability estimates, indicating onset of quantum entanglement. For example, the knowledge on the occurrence of the event $B = 0$ increases the probability of the event $A = 0$ from $\frac{5}{8}$ to $\frac{4}{5}$ and decreases the probability of the event $A = 1$ from $\frac{3}{8}$ to $\frac{1}{5}$, in comparison with the situations that no knowledge about B is given.

With (13.4.12), we can consider the question under what condition the 2-qubit state exhibits the strongest entanglement for which the occurrence $B = 0$ implies the occurrence $A = 0$ with certainty and the occurrence $B = 1$ implies the occurrence $A = 1$ with certainty, as well, for example. In other words, we have

$$P(A = 0|B = 0) = 1, \quad P(A = 1|B = 1) = 1. \quad (13.4.17)$$

Inserting (13.4.17) into (13.4.12), we get $c_{10} = c_{01} = 0$ so that (13.4.8) becomes

$$|\sigma\rangle = c_{00}|00\rangle + c_{11}|11\rangle, \quad |c_{00}|^2 + |c_{11}|^2 = 1, \quad c_{00}c_{11} \neq 0. \quad (13.4.18)$$

Similarly, we may also request that the occurrence $B = 0$ implies the occurrence $A = 1$ with certainty and the occurrence $B = 1$ implies the occurrence $A = 0$ with certainty, such that

$$P(A = 1|B = 0) = 1, \quad P(A = 0|B = 1) = 1. \quad (13.4.19)$$

Using (13.4.19) in (13.4.12), we find $c_{00} = c_{11} = 0$ so that (13.4.8) assumes the form

$$|\sigma\rangle = c_{01}|01\rangle + c_{10}|10\rangle, \quad |c_{01}|^2 + |c_{10}|^2 = 1, \quad c_{01}c_{10} \neq 0. \qquad (13.4.20)$$

As concrete examples of such 2-qubit states exhibiting the strongest entanglement, we have the *Bell* or *EPR states*:

$$\frac{1}{\sqrt{2}}(|00\rangle \pm |11\rangle), \quad \frac{1}{\sqrt{2}}(|01\rangle \pm |10\rangle), \qquad (13.4.21)$$

after Bell [10] and Einstein, Podolsky, and Rosen [25] who used these states to study the quantum entanglement phenomenon and its implications to quantum mechanics. Since these states possess the property that probability is equally distributed for the occurrence of either eigenstate, they are also referred to as *states of equal superposition*.

Exercises

13.4.1 Show that the Hadamard H_2 gate prepared state

$$H_2|00\rangle = \frac{1}{2}(|00\rangle + |01\rangle + |10\rangle + |11\rangle) \qquad (13.4.22)$$

is untangled by checking that this state is separable or decomposable.

13.4.2 Consider the 2-qubit state (13.4.5) and let A and B be the random variables of the measurement readings of the first and second qubits of the state. Show that the state (13.4.5) is untangled by verifying the property (13.4.4).

13.4.3 Show that if the condition (13.4.17) or (13.4.19) is replaced by

$$P(A = 0|B = 0) = 1, \quad P(A = 0|B = 1) = 1, \qquad (13.4.23)$$

say, then the state (13.4.8) is untangled.

13.4.4 Show that the condition $c_{00}c_{11} \neq 0$ or $c_{01}c_{10} \neq 0$ in (13.4.18) or (13.4.20), respectively, is a necessary and sufficient condition for the 2-qubit state shown to be entangled.

13.4.5 Consider the 2-qubit state (13.4.8) and set the conditional probabilities

$$p_{0i} = P(A = 0|B = i), \quad p_{1i} = P(A = 1|B = i), \quad i = 0, 1. \quad (13.4.24)$$

It is clear that $p_{0i} + p_{1i} = 1$ for $i = 0, 1$. Show that, to achieve largest possible values for p_{0i} and p_{1i} simultaneously, or equivalently to achieve the largest *Shannon entropy* E_i associated defined by

$$E_i = -p_{0i}\ln p_{0i} - p_{1i}\ln p_{1i}, \quad i = 0, 1, \qquad (13.4.25)$$

the amplitude coefficients in the state (13.4.8) must satisfy

$$|c_{0i}|^2 = |c_{1i}|^2, \quad i = 0, 1. \tag{13.4.26}$$

In particular, the state (13.4.22) is such a state. The condition (13.4.26) spells out a state of partially equal superpositions between the eigenstates $|00\rangle, |10\rangle$ and $|01\rangle, |11\rangle$.

Similarly, set the conditional probabilities:

$$p_{i0} = P(B = 0|A = i), \quad p_{i1} = P(B = 1|A = i), \quad i = 0, 1. \tag{13.4.27}$$

Then, to achieve largest possible values for p_{i0} and p_{i1} simultaneously, we are led to the condition:

$$|c_{i0}|^2 = |c_{i1}|^2, \quad i = 0, 1. \tag{13.4.28}$$

13.4.6 Show that any 2-qubit state (13.4.8) satisfying the two sets of the conditions found in Problem 13.4.5, namely, (13.4.26) and (13.4.28), must be an equal superposition state, satisfying

$$|c_{00}|^2 = |c_{01}|^2 = |c_{10}|^2 = |c_{11}|^2 = \frac{1}{4}, \tag{13.4.29}$$

and thus untangled, in particular.

13.4.7 Show that the 2-qubit state:

$$|\sigma\rangle = \frac{1}{2}|00\rangle + a|01\rangle + \frac{1}{2}|10\rangle + b|11\rangle, \quad a, b \in \mathbb{C}, \quad |a|^2 + |b|^2 = \frac{1}{2}, \tag{13.4.30}$$

is entangled if and only if $|a|^2 \neq |b|^2$.

13.5 Bell theorem

To prepare for a discussion of the Bell theorem, we begin with a general study on the measurement of a system consisting of two identical particles, having the same space $\mathbb{C}^n$ ($n \geq 2$), say, as their respective state spaces, which are separated far apart so that they can be measured individually, by Observer A and Observer B. So, for convenience, we can label these two particles as particle A and particle B, observed at the stations occupied by the two so-named observers, respectively. See Figure 13.2 for an illustration.

Use O_A and O_B to denote two observables to be measured on particles A and B, respectively, which are realized by two Hermitian matrices in $\mathbb{C}(n, n)$. Let

$$\{|u_1\rangle, \lambda_1, \ldots, |u_n\rangle, \lambda_n\}, \quad \{|v_1\rangle, \mu_1, \ldots, |v_n\rangle, \mu_n\}, \tag{13.5.1}$$

Observer A $\longleftrightarrow$ • - - - - - - - - ◉ - - - - - - - - • $\longrightarrow$ Observer B

Figure 13.2 Two identical particles are set to travel in opposite directions and to be measured by two observers, A and B, being separated and stationed far apart, so that the two particles measured by the observers are also labeled correspondingly without causing confusion.

be two full sets of eigenpairs of O_A and O_B comprised of orthonormal eigenvectors $\{|u_i\rangle\}$ and $\{|v_i\rangle\}$, which are two bases of $\mathbb{C}^n$, associated with the real eigenvalues $\{\lambda_i\}$ and $\{\mu_i\}$, respectively. Thus measurement of O_A and O_B yields the following list of possible readings:

$$O_A : \lambda_1,\ldots,\lambda_n, \quad O_B : \mu_1,\ldots,\mu_n, \tag{13.5.2}$$

respectively.

On the other hand, the state space of the two-particle system is the tensor product space $\mathbb{C}^n \otimes \mathbb{C}^n = \mathbb{C}^{n^2}$ with the induced basis

$$\{|u_i\rangle \otimes |v_j\rangle \,|\, i,j = 1,\ldots,n\}. \tag{13.5.3}$$

Since the Hermitian scalar product $(\cdot,\cdot)$ over the space $\mathbb{C}^n \otimes \mathbb{C}^n$ satisfies or is defined by setting

$$(|u\rangle \otimes |v\rangle, |x\rangle \otimes |y\rangle) = \langle u|x\rangle \langle v|y\rangle, \quad |u\rangle, |v\rangle, |x\rangle, |y\rangle \in \mathbb{C}^n, \tag{13.5.4}$$

we have

$$(|u_i\rangle \otimes |v_j\rangle, |u_k\rangle \otimes |v_l\rangle) = \delta_{ik}\delta_{jl}, \quad i,j,k,l = 1,\ldots,n. \tag{13.5.5}$$

As a consequence, we see that the set of vectors given in (13.5.3) form an orthonormal basis of the state space $\mathbb{C}^n \otimes \mathbb{C}^n$.

For the two-particle system with the state space $\mathbb{C}^n \otimes \mathbb{C}^n$, we construct the *parallel observable* $O_A \otimes O_B$ from the observables O_A and O_B following (13.2.24) or

$$(O_A \otimes O_B)(|u\rangle \otimes |v\rangle) = (O_A|u\rangle) \otimes (O_B|v\rangle), \quad |u\rangle, |v\rangle \in \mathbb{C}^n. \tag{13.5.6}$$

Hence we have

$$(O_A \otimes O_B)(|u_i\rangle \otimes |v_j\rangle) = \lambda_i \mu_j (|u_i\rangle \otimes |v_j\rangle), \quad i,j = 1,\ldots,n. \tag{13.5.7}$$

Thus, the orthonormal basis (13.5.3) of $\mathbb{C}^n \otimes \mathbb{C}^n$ is a set of eigenvectors of $O_A \otimes O_B$ associated with the corresponding eigenvalues

$$\lambda_i \mu_j, \quad i,j = 1,\ldots,n, \tag{13.5.8}$$

which are all real. Consequently, $O_A \otimes O_B$ is also Hermitian, and an observable indeed. The quantities listed in (13.5.8) are the products of all possible readings of the observables O_A and O_B measured on the particles A and B, respectively. In other words, a measurement of $O_A \otimes O_B$ yields a reading which is the product of a reading of the observable O_A and a reading of the observable O_B. For this reason, we may refer $O_A \otimes O_B$ as the *product observable* of the observables O_A and O_B.

Let $|\sigma\rangle$ be a two-particle state which is a unit vector in $\mathbb{C}^n \otimes \mathbb{C}^n$. Then we have the expansion

$$|\sigma\rangle = \sum_{i,j=1}^{n} c_{ij}(|u_i\rangle \otimes |v_j\rangle), \quad \sum_{i,j=1}^{n} |c_{ij}|^2 = 1. \qquad (13.5.9)$$

Therefore, in the state $|\sigma\rangle$, the expected value of the product observable $O_A \otimes O_B$ reads

$$\langle\sigma|O_A \otimes O_B|\sigma\rangle = \sum_{i,j=1}^{n} \lambda_i\mu_j|c_{ij}|^2. \qquad (13.5.10)$$

We next work out some calculation.
We choose

$$O_A^1 = \sigma_3 = \begin{pmatrix} 1 & 0 \\ 0 & -1 \end{pmatrix}, \qquad (13.5.11)$$

$$O_A^2 = \sigma_1 = \begin{pmatrix} 0 & 1 \\ 1 & 0 \end{pmatrix}, \qquad (13.5.12)$$

$$O_B^1 = -\frac{1}{\sqrt{2}}(\sigma_1 + \sigma_3) = \frac{1}{\sqrt{2}}\begin{pmatrix} -1 & -1 \\ -1 & 1 \end{pmatrix}, \qquad (13.5.13)$$

$$O_B^2 = \frac{1}{\sqrt{2}}(\sigma_1 - \sigma_3) = \frac{1}{\sqrt{2}}\begin{pmatrix} -1 & 1 \\ 1 & 1 \end{pmatrix}, \qquad (13.5.14)$$

to be two sets of observables to be measured simultaneously on particles A and B, respectively, as denoted. Since the eigenvalues of these observables are 1 and -1, we know that the possible measurement readings of these observables are 1 and -1 only.

With (13.5.11)–(13.5.14), we build all four possible product observables as follows:

$$O_A^1 \otimes O_B^1 = \frac{1}{\sqrt{2}}\begin{pmatrix} -1 & -1 & 0 & 0 \\ -1 & 1 & 0 & 0 \\ 0 & 0 & 1 & 1 \\ 0 & 0 & 1 & -1 \end{pmatrix}, \qquad (13.5.15)$$

$$O_A^1 \otimes O_B^2 = \frac{1}{\sqrt{2}} \begin{pmatrix} -1 & 1 & 0 & 0 \\ 1 & 1 & 0 & 0 \\ 0 & 0 & 1 & -1 \\ 0 & 0 & -1 & -1 \end{pmatrix}, \tag{13.5.16}$$

$$O_A^2 \otimes O_B^1 = \frac{1}{\sqrt{2}} \begin{pmatrix} 0 & 0 & -1 & -1 \\ 0 & 0 & -1 & 1 \\ -1 & -1 & 0 & 0 \\ -1 & 1 & 0 & 0 \end{pmatrix}, \tag{13.5.17}$$

$$O_A^2 \otimes O_B^2 = \frac{1}{\sqrt{2}} \begin{pmatrix} 0 & 0 & -1 & 1 \\ 0 & 0 & 1 & 1 \\ -1 & 1 & 0 & 0 \\ 1 & 1 & 0 & 0 \end{pmatrix}. \tag{13.5.18}$$

Consider, for example, the Bell state

$$|\sigma\rangle = \frac{1}{\sqrt{2}}(|01\rangle - |10\rangle) = \frac{1}{\sqrt{2}} \begin{pmatrix} 0 \\ 1 \\ -1 \\ 0 \end{pmatrix}, \tag{13.5.19}$$

which is known to possess the strongest entanglement. In view of (13.5.15)–(13.5.18) and (13.5.19), we obtain the expected values of the product observables (13.5.15)–(13.5.18) in the 2-qubit state (13.5.19) to be

$$\langle \sigma | O_A^1 \otimes O_B^1 | \sigma \rangle = \langle \sigma | O_A^1 \otimes O_B^2 | \sigma \rangle = \langle \sigma | O_A^2 \otimes O_B^1 | \sigma \rangle = \frac{1}{\sqrt{2}}, \tag{13.5.20}$$

$$\langle \sigma | O_A^2 \otimes O_B^2 | \sigma \rangle = -\frac{1}{\sqrt{2}}. \tag{13.5.21}$$

Historically, entangled quantum states were formulated by Einstein, Podolsky, and Rosen in [25] as a paradox against the statistical interpretation of quantum mechanics, which we now study.

Consider the entangled 2-qubit state (13.5.19) as an example governing a system of two widely separated identical particles, A and B, for which the first and second qubits describe the states of these particles, respectively. When a measurement is made of the system, the probability of observing either of the two eigenstates, $|01\rangle$ or $|10\rangle$, is $\frac{1}{2}$. However, if a measurement is made on particle A and renders the observation of the state $|0\rangle$ or $|1\rangle$ for the particle, a *simultaneous* measurement on particle B will yield the state $|1\rangle$ or $|0\rangle$ with certainty since the 2-qubit state observed must either be $|01\rangle$ or $|10\rangle$, respectively. In particular, we see that no time is needed for particle A to communicate its measurement result to particle B so that the latter is able to respond correctly to exhibit its state in correspondence to the observation made on the former. On the other hand, special relativity forbids faster-than-light motion of any kind of signal, the demonstrated dependence of the measurement of particle B on that of particle A raises a paradox to the entanglement phenomenon based on the statistical interpretation of the 2-qubit state. Thus, to maintain the entanglement, there must be some hidden mechanism that serves to predetermine the two particles when they start to fly apart and this hidden mechanism must continue to function during the entire separation process of the two particles if we hold the viewpoint that physical interaction should be local rather than nonlocal, meaning that such interaction always exerts its influence first to its immediate vicinity. Consequently, in order to uphold the predicted observations about the entangled state as well as special relativity and the principle that physical interaction must be local, the statistical interpretation of quantum mechanics should be discarded, and, instead, the entanglement properties owe their presence to a hidden mechanism, coined "hidden variables," having remained unveiled, in quantum mechanics. In other words, the paradox indicates that quantum mechanics is incomplete as a physical theory, since it cannot explain the entanglement phenomenon it predicts, in consistency with special relativity and locality principle. A would-be complete theory of quantum mechanics is thus called a local hidden-variable theory.

In [10], Bell derived some constraints to certain measurable quantities of an entangled quantum state if these quantities demonstrate themselves in predetermined ways following locality principle rather than statistical interpretation of quantum mechanics. Constraints of this nature are commonly referred to in literature as the *Bell inequalities*. Therefore, by experimentally testing these inequalities, we would be able to examine the correctness of a hypothesized theory of hidden-variable quantum mechanics against the statistical interpretation of quantum mechanics developed.

We now proceed to derive a Bell inequality for the two-particle system we have discussed earlier. Recall the measurement readings of the observables O_A^1, O_A^2, O_B^1, and O_B^2 are either 1 or -1. For convenience, we continue to use these same letters to denote the random variables of these measurement readings. Thus

$$O_A^1, O_A^2, O_B^1, O_B^2 = \pm 1. \qquad (13.5.22)$$

With (13.5.22), we form the quantity [19, 63], a random variable,

$$Q = O_A^1 O_B^1 + O_A^1 O_B^2 + O_A^2 O_B^1 - O_A^2 O_B^2, \qquad (13.5.23)$$

which is a sum of some combined products of the random variables O_A^1, O_A^2, O_B^1, and O_B^2. To estimate (13.5.23), we rewrite it as

$$Q = (O_A^1 + O_A^2)O_B^1 + (O_A^1 - O_A^2)O_B^2. \tag{13.5.24}$$

In view of (13.5.22), we see that, if $O_A^1 = O_A^2$, then $O_A^1 + O_A^2 = \pm 2$, and if $O_A^1 = -O_A^2$, then $O_A^1 - O_A^2 = \pm 2$. In either case, we have $Q = \pm 2$ in view of (13.5.22) again. Hence we conclude that the expected value, $\langle Q \rangle$, of the random variable Q satisfies the inequality

$$|\langle Q \rangle| \leq 2. \tag{13.5.25}$$

This is a Bell inequality. It is also called the *CHSH inequality* [63] after the initials of the names of the people who found it [19]. It is important to note that this is a universal inequality independent of the probability distribution of the random variable Q.

On the other hand, with the product observables (13.5.15)–(13.5.18), and in analogy of (13.5.23), we may form the observable matrix

$$Q = O_A^1 \otimes O_B^1 + O_A^1 \otimes O_B^2 + O_A^2 \otimes O_B^1 - O_A^2 \otimes O_B^2. \tag{13.5.26}$$

Thus, in view of (13.5.20), (13.5.21), and (13.5.26), we have

$$\langle \sigma | Q | \sigma \rangle = 2\sqrt{2}, \tag{13.5.27}$$

over the entangled state (13.5.19), which is seen to violate the Bell inequality (13.5.25). In other words, quantum mechanics, with its statistical interpretation, is not reconcilable in a local hidden-variable theory, and any theory in agreement with quantum mechanics cannot be a local theory [11]. Such a statement is known as the *Bell theorem*.

Note that if we replace O_B^1 given in (13.5.13) by

$$O_B^1 = \frac{1}{\sqrt{2}}(\sigma_1 + \sigma_3), \tag{13.5.28}$$

then the quantities in (13.5.20) and (13.5.21) become

$$\langle \sigma | O_A^1 \otimes O_B^1 | \sigma \rangle = \langle \sigma | O_A^2 \otimes O_B^1 | \sigma \rangle = \langle \sigma | O_A^2 \otimes O_B^2 | \sigma \rangle = -\frac{1}{\sqrt{2}}, \tag{13.5.29}$$

$$\langle \sigma | O_A^1 \otimes O_B^2 | \sigma \rangle = \frac{1}{\sqrt{2}}. \tag{13.5.30}$$

In view of (13.5.29) and (13.5.30), we see that the expected value of the observable (13.5.26) over the entangled state $|\sigma\rangle$ given by (13.5.19) is

$$\langle \sigma | Q | \sigma \rangle = 0. \tag{13.5.31}$$

This result is compatible with the Bell inequality (13.5.25), of course.

Thus we may draw the conclusion that violation of a Bell inequality in quantum mechanics is *occasional*.

See [22, 52] for some other highly accessible discussions on the Bell inequalities and their consequences.

The subsequent experimental contributions of Clauser [27], Aspect [6], and Zeilinger [64], along with their collaborators, on the phenomenon of quantum entanglement confirming violations of the Bell inequalities and verifying the Bell theorem and hence the statistical interpretation of quantum mechanics were jointly awarded the Nobel Physics Prize in 2022 [29].

Exercises

13.5.1 Show that, with (13.2.8) and (13.5.5), the Hermitian scalar product over $\mathbb{C}^4$ agrees with the scalar product over the tensor product space $\mathbb{C}^2 \otimes \mathbb{C}^2$ as defined by (13.5.4).

13.5.2 Verify (13.5.18), for example.

13.5.3 For $O_A = \sigma_1$ and $O_B = \sigma_2$, find the Hermitian matrix representation of the product observable $O_A \otimes O_B$ and compute the expected value of $O_A \otimes O_B$ in the Bell state $|\sigma\rangle = \frac{1}{\sqrt{2}}(|00\rangle + |11\rangle)$.

13.5.4 Show that if we replace (13.5.23) by

$$Q = O_A^1 O_B^1 + O_A^1 O_B^2 + O_A^2 O_B^1 + O_A^2 O_B^2, \qquad (13.5.32)$$

then the Bell inequality (13.5.25) should be revised into

$$|Q| \leq 4. \qquad (13.5.33)$$

In this case, study whether there is violation of the inequality (13.5.33) by the Hermitian matrix version of (13.5.32) evaluated over the entangled state (13.5.19).

Solutions to Selected Problems

Section 1.1

1.1.2 Let n satisfy $1 \leq n < p$ and write $2n = kp + l$ for some integers k and l where $0 \leq l < p$. Then $n + (n - l) = kp$. Thus $[n - l] = -[n]$.

If $[n] \neq [0]$, then n is not a multiple of p. Since p is a prime, so the greatest common divisor of n and p is 1. Thus there are integers k and l such that

$$kp + ln = 1. \tag{S1}$$

Consequently $[l]$ is the multiplicative inverse of $[n]$. That is, $[l] = [n]^{-1}$.

We may also prove the existence of $[n]^{-1}$ without using (S1) but by using an interesting statement called *Fermat's Little Theorem*: For any non-negative integer m and positive integer p, the integer $m^p - m$ is divisible by p. We prove this theorem by induction. When $m = 0$, there is nothing to show. Assume the statement is true at $m \geq 0$. At $m + 1$ we have

$$(m + 1)^p - (m + 1) = \sum_{k=1}^{p-1} \binom{p}{k} m^k + (m^p - m), \tag{S2}$$

which is clearly divisible by p in view of the inductive assumption and the definition of binomial coefficients. So the theorem follows.

Now we come back to the construction of $[n]^{-1}$. Since p is a prime and divides $n^p - n = n(n^{p-1} - 1)$ but not n, p divides $n^{p-1} - 1$. So there is an integer k such that $n^{p-1} - 1 = kp$. Therefore $nn^{p-2} = 1$ modulo p. That is, $[n^{p-2}] = [n]^{-1}$.

1.1.9 Multiplying the relation $AB = I$ by C from the left we have $C(AB) = C$, which gives us $B = IB = (CA)B = C(AB) = C$.

Section 1.2

1.2.7 We have

$$u^t v = \begin{pmatrix} a_1 \\ \vdots \\ a_n \end{pmatrix} (b_1,\ldots,b_n) = \begin{pmatrix} a_1 b_1 & a_1 b_2 & \cdots & a_1 b_n \\ a_2 b_1 & a_2 b_2 & \cdots & a_2 b_n \\ \cdots & \cdots & \cdots & \cdots \\ a_n b_1 & a_n b_2 & \cdots & a_n b_n \end{pmatrix},$$

whose *i*th and *j*th rows are

$$a_i(b_1,b_2,\ldots,b_n) \quad \text{and} \quad a_j(b_1,b_2,\ldots,b_n),$$

which are clearly linearly dependent with each other.

For column vectors we proceed similarly.

1.2.8 If either $U_1 \subset U_2$ or $U_2 \subset U_1$, the statement is trivially true. Suppose otherwise and pick $u_1 \in U_1$ but $u_1 \notin U_2$ and $u_2 \in U_2$ but $u_2 \notin U_1$. We assert that $u = u_1 + u_2 \notin U_1 \cup U_2$. If $u \in U_1 \cup U_2$ then $u \in U_1$ or $u \in U_2$. So, respectively, $u_2 = u + (-u_1) \in U_1$ or $u_1 = u + (-u_2) \in U_2$, which is false.

1.2.9 Without loss of generality we may assume that there are no such i and j, $i, j = 1,\ldots,k$, $i \neq j$, that $U_i \subset U_j$. Thus, in view of the previous problem, we know that for $k = 2$ we can find $u_1 \in U_1$ and $u_2 \in U_2$ such that $u = u_1 + u_2 \notin U_1 \cup U_2$. Assume that the statement of the problem is true at $k = m \geq 2$. We proceed to prove the statement at $k = m + 1$.

Assume otherwise that $U = U_1 \cup \cdots \cup U_{m+1}$. By the inductive assumption there is some $u \in U$ such that

$$u \notin U_1 \cup \cdots \cup U_m. \tag{S3}$$

Thus $u \in U_{m+1}$. Pick $v \notin U_{m+1}$ and consider the following $m + 1$ vectors:

$$w_1 = u + v, \quad w_2 = 2u + v, \quad \ldots, \quad w_{m+1} = (m+1)u + v.$$

There is some $i = 1,\ldots,m + 1$ such that $w_i \notin U_1 \cup \cdots \cup U_m$. Otherwise, if $w_i \in U_1 \cup \cdots \cup U_m$ for all $i = 1,\ldots,m + 1$, then there are $i, j = 1,\ldots,m + 1$, $i \neq j$, such that w_i and w_j lie in one of the subspaces $U_1,\ldots,U_m$, say U_l ($1 \leq l \leq m$), which leads to $w_i - w_j = (i - j)u \in U_l$ or $u \in U_l$, a contradiction to (S3).

Now assume $w_i \notin U_1 \cup \cdots \cup U_m$ for some $i = 1,\ldots,m + 1$. Thus $w_i \in U_{m+1}$. So $v = w_i + (-iu) \in U_{m+1}$ since $u \in U_{m+1}$, which is again false.

Section 1.4

1.4.3 Without loss of generality assume $f \neq 0$. Then there is some $u \in U$ such that $f(u) \neq 0$. For any $v \in U$ consider

$$w = v - \frac{f(v)}{f(u)}u.$$

We have $f(w) = 0$. Hence $g(w) = 0$ as well which gives us

$$g(v) = \frac{g(u)}{f(u)}f(v) \equiv af(v), \quad v \in U.$$

That is, $g = af$.

1.4.4 Let $\{v_1, \ldots, v_{n-1}\}$ be a basis of V. Extend it to get a basis of $\mathbb{F}^n$, say $\{v_1, \ldots, v_{n-1}, v_n\}$. Let $\{f_1, \ldots, f_{n-1}, f_n\}$ be a basis of $(\mathbb{F}^n)'$ dual to $\{v_1, \ldots, v_{n-1}, v_n\}$. It is clear that $V^0 = \text{Span}\{f_n\}$. On the other hand, for any $(x_1, \ldots, x_n) \in \mathbb{F}^n$ we have

$$v = (x_1, x_2, \ldots, x_n) - (x_1 + x_2 + \cdots + x_n, 0, \ldots, 0) \in V.$$

So

$$0 = f_n(v) = f_n(x_1, \ldots, x_n) - (x_1 + \cdots + x_n)f_n(e_1), \quad (x_1, \ldots, x_n) \in \mathbb{F}^n. \tag{S4}$$

For any $f \in V^0$, there is some $a \in \mathbb{F}$ such that $f = af_n$. Hence in view of (S4) we obtain

$$f(x_1, \ldots, x_n) = af_n(e_1) \sum_{i=1}^{n} x_i, \quad (x_1, \ldots, x_n) \in \mathbb{F}^n.$$

Section 1.7

1.7.2 Assume the nontrivial situation $u \neq 0$. It is clear that $\|u\|_p \leq \|u\|_\infty$ for any $p \geq 1$. Thus $\limsup_{p \to \infty} \|u\|_p \leq \|u\|_\infty$.

Let $t_0 \in [a, b]$ be such that $|u(t_0)| = \|u\|_\infty > 0$. For any $\varepsilon \in (0, \|u\|_\infty)$ we can find an interval around t_0, say $I_\varepsilon \subset [a, b]$, such that $|u(t)| > \|u\|_\infty - \varepsilon$ when $t \in I_\varepsilon$. Thus

$$\|u\|_p \geq \left(\int_{I_\varepsilon} |u(t)|^p \, dt \right)^{\frac{1}{p}} \geq (\|u\|_\infty - \varepsilon) |I_\varepsilon|^{\frac{1}{p}}, \tag{S5}$$

where $|I_\varepsilon|$ denotes the length of the interval I_ε. Letting $p \to \infty$ in (S5) we find $\liminf_{p \to \infty} \|u\|_p \geq \|u\|_\infty - \varepsilon$. Since $\varepsilon > 0$ may be chosen arbitrarily small, we arrive at $\liminf_{p \to \infty} \|u\|_p \geq \|u\|_\infty$.

Thus the limit $\|u\|_p \to \|u\|_\infty$ as $p \to \infty$ follows.

1.7.3 (a) Positivity: Of course $\|u'\|' \geq 0$ for any $u' \in U'$. If $\|u'\|' = 0$, then $|u'(u)| = 0$ for any $u \in U$ satisfying $\|u\| = 1$. Thus, for any $v \in U, v \neq 0$, we have $u'(v) = \|v\| u'(\frac{1}{\|v\|}v) = 0$, which shows $u'(v) = 0$ for any $v \in U$. So $u' = 0$.

(b) Homogeneity: For any $a \in \mathbb{F}$ and $u' \in U'$, we have

$$\|au'\|' = \sup\{|au'(u)| \,|\, u \in U, \|u\| = 1\}$$
$$= |a| \sup\{|u'(u)| \,|\, u \in U, \|u\| = 1\} = |a| \|u'\|'.$$

(c) Triangle inequality: For any $u', v' \in U'$, we have

$$\|u' + v'\|' = \sup\{|u'(u) + v'(u)| \,|\, u \in U, \|u\| = 1\}$$
$$\leq \sup\{|u'(u)| + |v'(u)| \,|\, u \in U, \|u\| = 1\}$$
$$\leq \sup\{|u'(u)| \,|\, u \in U, \|u\| = 1\}$$
$$+ \sup\{|v'(v)| \,|\, v \in U, \|v\| = 1\}$$
$$= \|u'\|' + \|v'\|'.$$

1.7.4 (a) Positivity: Of course $\|u\| \geq 0$. If $\|u\| = 0$, then $|u'(u)| = 0$ for all $u' \in U'$ satisfying $\|u'\|' = 1$. Then for any $v' \in U'$, $v' \neq 0$, we have $v'(u) = \|v'\|' \left(\frac{1}{\|v'\|'}v'\right)(u) = 0$, which shows $v'(u) = 0$ for any $v' \in U'$. Hence $u = 0$.

(b) Homogeneity: Let $a \in \mathbb{F}$ and $u \in U$. We have

$$\|au\| = \sup\{|u'(au)| \,|\, u' \in U', \|u'\|' = 1\}$$
$$= |a| \sup\{|u'(u)| \,|\, u' \in U', \|u'\|' = 1\} = |a| \|u\|.$$

(c) Triangle inequality: For u and $v \in U$, we have

$$\|u + v\| = \sup\{|u'(u + v)| \,|\, u' \in U', \|u'\|' = 1\}$$
$$\leq \sup\{|u'(u)| + |u'(v)| \,|\, u' \in U', \|u'\|' = 1\}$$
$$\leq \sup\{|u'(u)| \,|\, u' \in U', \|u'\|' = 1\}$$
$$+ \sup\{|v'(v)| \,|\, v' \in U', \|v'\|' = 1\}$$
$$= \|u\| + \|v\|.$$

Section 2.1

2.1.9 (a) Since f and $g \neq 0$ we have $\dim(f^0) = \dim(g^0) = n - 1$. Let $\{u_1, \ldots, u_{n-1}\}$ be a basis of f^0 and take $v \in g^0$ but $v \notin f^0$. Then $u_1, \ldots, u_{n-1}$, and v are linearly independent and, hence, form a basis of U. In particular, $U = f^0 + g^0$.

(b) From the dimensionality equation

$$n = \dim(f^0) + \dim(g^0) - \dim(f^0 \cap g^0),$$

the answer follows.

2.1.10 We have $N(T) \subset N(T^2)$ and $R(T^2) \subset R(T)$. On the other hand, if $n = \dim(U)$, the rank equation gives us

$$n(T) + r(T) = n = n(T^2) + r(T^2).$$

Thus $n(T) = n(T^2)$ if and only if $r(T) = r(T^2)$. So $N(T^2) = N(T)$ if and only if $R(T^2) = R(T)$.

2.1.12 Let $w_1, \ldots, w_k \in W$ be a basis of $R(S \circ T)$. Then there are $u_1, \ldots, u_k \in U$ such that $S(T(u_i)) = w_i, i = 1, \ldots, k$.

Let $v_i = T(u_i), i = 1, \ldots, k$. Then $\mathrm{Span}\{v_1, \ldots, v_k\} \subset R(T)$. Hence $k \leq r(T)$. Of course, $R(S \circ T) \subset R(S)$. So $k \leq r(S)$ as well.

Next, let $u_1, \ldots, u_k \in U$ form a basis of $N(T) \subset N(S \circ T)$. Let $z_1, \ldots, z_l \in N(S \circ T)$ so that $u_1, \ldots, u_k, z_1, \ldots, z_l$ form a basis of $N(S \circ T)$. Of course $T(z_1), \ldots, T(z_l) \in N(S)$. We assert that these are linearly independent. In fact, if there are scalars $a_1, \ldots, a_l$ such that $a_1 T(z_1) + \cdots + a_l T(z_l) = 0$, then $T(a_1 z_1 + \cdots + a_l z_l) = 0$, which indicates that $a_1 z_1 + \cdots + a_l z_l \in N(T)$. Hence $a_1 z_1 + \cdots + a_l z_l = b_1 u_1 + \cdots + b_k u_k$ for some scalars $b_1, \ldots, b_k$. However, we know that $u_1, \ldots, u_k, z_1, \ldots, z_l$ are linearly independent. So a's and b's are all zero. In particular, $T(z_1), \ldots, T(z_l)$ are linearly independent. Hence $l \leq n(S)$, which proves $k + l \leq n(T) + n(S)$.

2.1.14 Define $T \in L(\mathbb{F}^m, \mathbb{F}^n)$ by $T(x) = Bx, x \in \mathbb{F}^m$. Then, since $R(T) \subset \mathbb{R}^n$, we have $r(T) \leq n$. By the rank equation $n(T) + r(T) = m$ and the condition $m > n$, we see that $n(T) > 0$. Hence there is a nonzero vector $x \in \mathbb{F}^m$ such that $T(x) = 0$ or $Bx = 0$. Thus $(AB)x = A(Bx) = 0$ which proves that the $m \times m$ matrix AB cannot be invertible.

2.1.16 Use $\mathcal{N}$ and $\mathcal{R}$ to denote the null-space and range of a mapping. Let

$$\mathcal{N}(R) \cap \mathcal{R}(S \circ T) = \mathrm{Span}\{w_1, \ldots, w_k\} \subset W,$$

where $w_1, \ldots, w_k$ are independent vectors.

Expand $\{w_1, \ldots, w_k\}$ to get a basis for $\mathcal{R}(S \circ T)$ so that

$$\mathcal{R}(S \circ T) = \mathrm{Span}\{w_1, \ldots, w_k, y_1, \ldots, y_l\}.$$

Then $\{R(y_1), \ldots, R(y_l)\}$ is a basis for $\mathcal{R}(R \circ S \circ T)$ since

$$R(w_1) = \ldots = R(w_k) = 0$$

and $R(y_1), \ldots, R(y_l)$ are independent. In particular, $r(R \circ S \circ T) = l$.

Since $\mathcal{R}(S \circ T) \subset \mathcal{R}(S)$, we can expand $\{w_1, ..., w_k, y_1, ..., y_l\}$ to get a basis for $\mathcal{R}(S)$ so that

$$\mathcal{R}(S) = \text{Span}\{w_1, ..., w_k, y_1, ..., y_l, z_1, ..., z_m\}.$$

Now we can count the numbers as follows:

$$
\begin{aligned}
r(R \circ S) &= \dim(\text{Span}\{R(w_1), ..., R(w_k), R(y_1), ..., R(y_l), R(z_1), ..., R(z_m)\}) \\
&= \dim(\text{Span}\{R(y_1), ..., R(y_l), R(z_1), ..., R(z_m)\}) \\
&\leq l + m,
\end{aligned}
$$

$$r(S \circ T) = k + l.$$

So

$$
\begin{aligned}
r(R \circ S) + r(S \circ T) &\leq l + m + k + l \\
&= (k + l + m) + l = r(S) + r(R \circ S \circ T).
\end{aligned}
$$

2.1.21 Define a mapping $T : \mathbb{F}^{k+l} \to V + W$ by

$$T(y_1, ..., y_k, z_1, ..., z_l) = \sum_{i=1}^{k} y_i v_i + \sum_{j=1}^{l} z_j w_j, \quad (y_1, ..., y_k, z_1, ..., z_l) \in \mathbb{F}^{k+l}.$$

Then $N(T) = S$. From the rank equation we have $n(T) + r(T) = k + l$ or $\dim(S) + r(T) = k + l$. On the other hand, it is clear that $R(T) = V + W$. So $r(T) = \dim(V + W)$. From the dimensionality equation (1.5.6), we have $\dim(V + W) = \dim(V) + \dim(W) - \dim(V \cap W)$, or $r(T) + \dim(V \cap W) = k + l$. Therefore $\dim(S) = \dim(V \cap W)$.

Section 2.2

2.2.4 Define $T \in L(\mathbb{R}^2)$ by setting

$$T(x) = \begin{pmatrix} a & b \\ c & d \end{pmatrix} x, \quad x = \begin{pmatrix} x_1 \\ x_2 \end{pmatrix} \in \mathbb{R}^2.$$

Then we have

$$T(e_1) = a e_1 + c e_2,$$
$$T(e_2) = b e_1 + d e_2,$$

and

$$T(u_1) = \frac{1}{2}(a + b + c + d)u_1 + \frac{1}{2}(a + b - c - d)u_2,$$
$$T(u_2) = \frac{1}{2}(a - b + c - d)u_1 + \frac{1}{2}(a - b - c + d)u_2,$$

so that the problem follows.

2.2.5 Let $A = (a_{ij})$ and define $T \in L(\mathbb{F}^n)$ by setting $T(x) = Ax$ where $x \in \mathbb{F}^n$ is taken to be a column vector. Then

$$T(e_j) = \sum_{i=1}^{n} a_{ij} e_i, \quad j = 1,\ldots,n.$$

Consider a new basis of $\mathbb{F}^n$ given by

$$f_1 = e_n, \quad f_2 = e_{n-1}, \quad \ldots, \quad f_n = e_1,$$

or $f_i = e_{n-i+1}$ for $i = 1,\ldots,n$. Then we obtain

$$T(f_j) = \sum_{i=1}^{n} a_{n-i+1,n-j+1} f_i \equiv \sum_{i=1}^{n} b_{ij} f_i, \quad j = 1,\ldots,n.$$

Since A and $B = (b_{ij})$ are the matrix representations of the mapping T under the bases $\{e_1,\ldots,e_n\}$ and $\{f_1,\ldots,f_n\}$, respectively, we conclude that $A \sim B$.

Section 2.3

2.3.3 That the equation $T(u) = v$ has a solution for some $u \in U$ is equivalent to $v \in R(T)$ which is equivalent to $v \in N(T')^0$ (by Theorem 2.8) which is equivalent to $\langle v, v' \rangle = 0$ for all $v' \in N(T')$ or $T'(v') = 0$.

Section 2.4

2.4.3 Let $k = r(\tilde{T}) = \dim(R(\tilde{T}))$. Let $\{[v_1]_Y,\ldots,[v_k]_Y\}$ be a basis of $R(\tilde{T})$ in V/Y. Then there are $[u_1]_X,\ldots,[u_k]_X \in U/X$ such that

$$\tilde{T}([u_1]_X) = [v_1]_Y, \quad \ldots, \quad \tilde{T}([u_k]) = [v_k]_Y.$$

On the other hand, from the definition of $\tilde{T}$, we have $\tilde{T}([u_i]_X) = [T(u_i)]_Y$ for $i = 1,\ldots,k$. We claim that $T(u_1),\ldots,T(u_k)$ are linearly independent. In fact, let $a_1,\ldots,a_k$ be scalars such that

$$a_1 T(u_1) + \cdots + a_k T(u_k) = 0.$$

Taking cosets, we get $a_1 [T(u_1)]_Y + \cdots + a_k [T(u_k)]_Y = [0]_Y$, which leads to $a_1 [v_1]_Y + \cdots + a_k [v_k]_Y = [0]_Y$. Thus $a_1 = \cdots = a_k = 0$. This proves $r(T) \geq k$.

Section 2.5

2.5.2 Let the right-hand side of (2.5.54) be denoted by R. Then we have

$$(I + T \circ S) \circ R$$
$$= I + T \circ S - T \circ (I + S \circ T)^{-1} \circ S - T \circ S \circ T \circ (I + S \circ T)^{-1} \circ S$$
$$= I + T \circ S - T \circ \left((I + S \circ T)^{-1} + S \circ T \circ (I + S \circ T)^{-1} \right) \circ S$$
$$= I + T \circ S - T \circ \left((I + S \circ T) \circ (I + S \circ T)^{-1} \right) \circ S = I.$$

2.5.12 (a) Assume $R(S) = R(T)$. Since S projects U onto $R(S) = R(T)$ along $N(S)$, we have for any $u \in U$ the result $S(T(u)) = T(u)$. Thus $S \circ T = T$. Similarly, $T \circ S = S$.

Furthermore, from $S \circ T = T$, then $T(U) = (S \circ T)(U)$ implies $R(T) \subset R(S)$. Likewise, $T \circ S = S$ implies $R(S) \subset R(T)$. So $R(S) = R(T)$.

(b) Assume $N(S) = N(T)$. For any $u \in U$, rewrite u as $u = v + w$ with $v \in R(T)$ and $w \in N(T)$. Then $T(v) = v$, $T(w) = 0$, and $S(w) = 0$ give us

$$(S \circ T)(u) = S(T(v) + T(w)) = S(v) = S(v + w) = S(u).$$

Hence $S \circ T = S$. Similarly, $T \circ S = T$.

Assume $S \circ T = S, T \circ S = T$. Let $u \in N(T)$. Then $S(u) = S(T(u)) = 0$. So $u \in N(S)$. So $N(T) \subset N(S)$. Interchange S and T. We get $N(S) \subset N(T)$.

2.5.14 (a) We have

$$T^2(u) = T(\langle u, u_1' \rangle u_1 + \langle u, u_2' \rangle u_2)$$
$$= \langle u, u_1' \rangle T(u_1) + \langle u, u_2' \rangle T(u_2)$$
$$= \langle u, u_1' \rangle (\langle u_1, u_1' \rangle u_1 + \langle u_1, u_2' \rangle u_2)$$
$$\quad + \langle u, u_2' \rangle (\langle u_2, u_1' \rangle u_1 + \langle u_2, u_2' \rangle u_2)$$
$$= (\langle u, u_1' \rangle \langle u_1, u_1' \rangle + \langle u, u_2' \rangle \langle u_2, u_1' \rangle) u_1$$
$$\quad + (\langle u, u_1' \rangle \langle u_1, u_2' \rangle + \langle u, u_2' \rangle \langle u_2, u_2' \rangle) u_2$$
$$= T(u) = \langle u, u_1' \rangle u_1 + \langle u, u_2' \rangle u_2. \tag{S6}$$

Since u_1 and u_2 are independent, we have

$$\langle u, u_1' \rangle \langle u_1, u_1' \rangle + \langle u, u_2' \rangle \langle u_2, u_1' \rangle = \langle u, u_1' \rangle,$$
$$\langle u, u_1' \rangle \langle u_1, u_2' \rangle + \langle u, u_2' \rangle \langle u_2, u_2' \rangle = \langle u, u_2' \rangle;$$

namely,

$$\langle u, \langle u_1, u_1' \rangle u_1' + \langle u_2, u_1' \rangle u_2' \rangle = \langle u, u_1' \rangle,$$
$$\langle u, \langle u_1, u_2' \rangle u_1' + \langle u_2, u_2' \rangle u_2' \rangle = \langle u, u_2' \rangle.$$

Since $u \in U$ is arbitrary, we have

$$\langle u_1, u_1' \rangle u_1' + \langle u_2, u_1' \rangle u_2' = u_1', \quad \langle u_1, u_2' \rangle u_1' + \langle u_2, u_2' \rangle u_2' = u_2'.$$

Since u_1' and u_2' are independent, we have

$$\langle u_1, u_1' \rangle = 1, \quad \langle u_2, u_1' \rangle = 0, \quad \langle u_1, u_2' \rangle = 0, \quad \langle u_2, u_2' \rangle = 1. \quad \text{(S7)}$$

Conversely, if (S7) holds, then we can use (S6) to get

$$T^2(u) = \langle u, u_1' \rangle u_1 + \langle u, u_2' \rangle u_2 = T(u), \quad u \in U.$$

In conclusion, (S7) is a necessary and sufficient condition to ensure that T is a projection.

(b) Assume T is a projection. Then (S7) holds. From $V = \{u \in U \mid T(u) = u\}$ and the definition of T, we see that $V \subset \text{Span}\{u_1, u_2\}$. Using (S7), we also have u_1 and $u_2 \in V$. So $V = \text{Span}\{u_1, u_2\}$.

If $T(u) = 0$, then $\langle u, u_1' \rangle u_1 + \langle u, u_2' \rangle u_2 = 0$. Since u_1 and u_2 are independent, we have $\langle u, u_1' \rangle = 0$ and $\langle u, u_2' \rangle = 0$. So $W = (\text{Span}\{u_1', u_2'\})^0$.

2.5.15　We have $T(T - aI) = 0$ or $(aI - T)T = T(aI - T) = 0$. Besides,

$$I = \left(I - \frac{1}{a}T\right) + \frac{1}{a}T = \frac{1}{a}(aI - T) + \frac{1}{a}T.$$

Therefore, we may rewrite any $u \in U$ as

$$u = I(u) = \frac{1}{a}(aI - T)(u) + \frac{1}{a}T(u) \equiv v + w. \quad \text{(S8)}$$

Thus

$$T(v) = T\left(\frac{1}{a}(aI - T)(u)\right) = 0,$$

$$(aI - T)(w) = (aI - T)\left(\frac{1}{a}T(u)\right) = 0.$$

That is, $v \in N(T)$ and $w \in N(aI - T)$. Hence $U = N(T) + N(aI - T)$.

Let $u \in N(aI - T) \cap N(T)$. Inserting this into (S8), we get $u = 0$. So $N(aI - T) \cap N(T) = \{0\}$ and $U = N(aI - T) \oplus N(T)$.

Since T commutes with $aI - T$ and T, it is clear that $N(aI - T)$ and $N(T)$ are invariant under T.

2.5.24　(a) Since T is nilpotent of degree k, there is an element $u \in U$ such that $T^{k-1}(u) \neq 0$ but $T^k(u) = 0$. It is clear that

$$V = \text{Span}\{u, T(u), \ldots, T^{k-1}(u)\}$$

is a k-dimensional subspace of U invariant under T. Let W be the complement of $\text{Span}\{T^{k-1}(u)\}$ in $N(T)$. Then $\dim(W) = l-1$. Consider the subspace $X = V + W$ of U. Pick $v \in V \cap W$ and write v as

$$v = a_0 u + a_1 T(u) + \cdots + a_{k-1} T^{k-1}(u), \quad a_0, a_1, \ldots, a_{k-1} \in \mathbb{F}.$$

Then $T(v) = 0$ gives us $a_0 T(u) + \cdots + a_{k-2} T^{k-1}(u) = 0$, which leads to $a_0 = a_1 = \cdots = a_{k-2} = 0$. So $v \in \text{Span}\{T^{k-1}(u)\}$ and $v \in W$ which indicates $v = 0$. Hence $X = V \oplus W$. However, since $\dim(X) = \dim(V) + \dim(W) = k + (l-1) = n$, we have $X = U$ so that T is reducible over V, W.

(b) It is clear that $R(T) = \text{Span}\{T(u), \ldots, T^{k-1}(u)\}$ and $r(T) = k-1$.

2.5.28 (a) Write $S - T$ as $S \circ (I - S^{-1} \circ T)$ and set $P = S^{-1} \circ T$. Then P is nilpotent as well. Let the degree of P be k. Assume $P \neq 0$. Then $k \geq 2$. It may be checked that $I - P$ is invertible since $I + P + \cdots + P^{k-1}$ is the inverse of $I - P$.

(b) For $A \in \mathbb{R}(2,2)$ define $T_A \in L(\mathbb{R}^2)$ by $T_A(x) = Ax$, where $x \in \mathbb{R}^2$ is a column vector. Now set

$$A = \begin{pmatrix} 0 & 1 \\ 1 & 0 \end{pmatrix}, \quad B = \begin{pmatrix} 0 & 1 \\ 0 & 0 \end{pmatrix}.$$

Then T_A is invertible, T_B is nilpotent, but $T_A - T_B$ is not invertible. It is direct to check that T_A and T_B do not commute.

2.5.29 Let $v \in V$. Then $T(v) \in R(T) \cap V$ since V is invariant under T. Since $R(T) \cap V = \{0\}$ we have $T(v) = 0$. So $v \in N(T)$. That is, $V \subset N(T)$. On the other hand, the rank equation indicates that $\dim(V) = n(T) = \dim(N(T))$. So $V = N(T)$.

Section 2.6

2.6.1 Write $S_n = T_n - T$. Then

$$T = T_n - S_n = T_n \circ (I - R_n), \tag{S9}$$

where $R_n = T_n^{-1} \circ S_n$. If $\|T_n^{-1}\| \not\to \infty$ as $n \to \infty$, we may assume that $\{\|T_n^{-1}\|\}$ is bounded without loss of generality. Since $\|S_n\| \to 0$ as $n \to \infty$ we have $\|R_n\| \leq \|T_n^{-1}\| \|S_n\| \to 0$ as $n \to \infty$ as well. So in view of Theorem 2.25 we see that $I - R_n$ is invertible when n is sufficiently large which leads to the false conclusion that T is invertible in view of (S9).

2.6.5 Let $P \in \mathcal{N}$ and consider $T_\lambda = \lambda I + P$ where λ is a scalar. It is clear that T_λ is invertible for all $\lambda \neq 0$ and $\|P - T_\lambda\| \to 0$ as $\lambda \to 0$. Since T_λ is invertible so it can never be nilpotent.

Section 3.1

3.1.4 (a) Let C be a closed curve around but away from the origin of $\mathbb{R}^2$ and consider

$$u : C \to S^1, \quad u(x,y) = \frac{f(x,y)}{\|f(x,y)\|}, \quad (x,y) \in C.$$

Take $R > 0$ to be a constant and let C be parametrized by θ: $x = \frac{R}{a}\cos\theta, y = \frac{R}{b}\sin\theta, 0 \le \theta \le 2\pi$. Then $u = (u_1, u_2) = (\cos\theta, \sin\theta)$ (on C). So $\text{ind}(f|_C) = \deg(u) = 1$. On the other hand, on C, we have

$$\|g(x,y)\|^2 = (a^2x^2 - b^2y^2)^2 + 4a^2b^2x^2y^2 = (a^2x^2 + b^2y^2)^2 = R^4.$$

Now we can construct $v : C \to S^1$ by setting

$$v(x,y) = \frac{g(x,y)}{\|g(x,y)\|}$$
$$= \frac{1}{R^2}(a^2x^2 - b^2y^2, 2abxy) = (\cos^2\theta - \sin^2\theta, 2\cos\theta\sin\theta).$$

Therefore, $\text{ind}(g|_C) = \deg(v) = 2$.

(b) The origin of $\mathbb{R}^2$ is the only zero of f and g which is a simple zero of f and a double zero of g.

(c) Since f_ε and g_ε are small perturbations of f and g, by stability we deduce $\text{ind}(f_\varepsilon|_C) = 1, \text{ind}(g_\varepsilon|_C) = 2$. However, f_ε still has exactly one simple zero, $x = \varepsilon, y = 0$, but g_ε has two simple zeros, $x = \pm\sqrt{\varepsilon}, y = 0$. Thus, when going from $\varepsilon = 0$ to $\varepsilon > 0$, the double zero of the latter splits into two simple zeros. Note that, algebraically, a double zero and two simple zeros are both counted as two zeros, which is, loosely speaking, indicative of the result $\text{ind}(g_\varepsilon|_C) = 2$.

3.1.5 Consider the vector field

$$v = \left(x^3 - 3xy^2 - 5\cos^2(x+y), 3x^2y - y^3 + 2e^{-x^2y^2}\right).$$

We are to show that v has a zero somewhere. We use the deformation method again to simplify the problem. For this purpose, we set

$$v^t = \left(x^3 - 3xy^2 - 5t\cos^2(x+y), 3x^2y - y^3 + 2te^{-x^2y^2}\right), \quad 0 \le t \le 1.$$

It is not hard to show that there is some $R > 0$ such that $\|v^t\| \ge 1$ for $\sqrt{x^2 + y^2} = r \ge R, t \in [0,1]$. Thus, we only need to show that over a suitably large circle given by $C_R = \{r = R\}$, we have $\text{ind}(v^0|_{C_R}) \neq 0$. Over C_R

we have $x = R\cos\theta, y = R\sin\theta, \theta \in [0, 2\pi]$. On the other hand, for $z = x + iy$, we have $z^3 = (x^3 - 3xy^2) + i(-y^3 + 3x^2y)$. Hence, with $z = Re^{i\theta}$, we get $v^0 = R^3(\cos 3\theta, \sin 3\theta)$, so that $u = \frac{1}{\|v^0\|}v^0 = (\cos 3\theta, \sin 3\theta)$. Thus $\deg(u) = 3$. So $\mathrm{ind}(v|_{C_R}) = \mathrm{ind}(v^1|_{C_R}) = \mathrm{ind}(v^0|_{C_R}) = 3 \neq 0$ as expected and v must vanish at some point inside C_R.

3.1.7 Inserting the expression for the stereographic projection given and going through a tedious calculation, we obtain

$$\deg(u) = \frac{1}{\pi}\int_{\mathbb{R}^2}\frac{1}{(1+x^2+y^2)^2}\,dxdy = \frac{1}{\pi}\int_0^{2\pi}d\theta\int_0^\infty\frac{r}{(1+r^2)^2}\,dr = 1.$$

3.1.8 Inserting the hedgehog expression and integrating, we obtain

$$\deg(u) = \frac{1}{4\pi}\int_0^\infty\int_0^{2\pi}u\cdot\left(\frac{\partial u}{\partial r}\times\frac{\partial u}{\partial\theta}\right)d\theta dr = -\frac{n}{2}\int_\pi^0\sin f\,df = n,$$

which indicates that the map covers the 2-sphere, while preserving the orientation, n times.

Section 3.2

3.2.6 Consider the matrix $C = aA + bB$. The entries of C not in the jth column are the corresponding entries of A multiplied by $(a+b)$ but the jth column of C is equal to the sum of the a-multiple of the jth column of A and b-multiple of the jth column of B. So by the properties of determinants, we have

$$\det(C) = (a+b)^{n-1}\det\begin{pmatrix} a_{11} & \cdots & aa_{1j} + bb_{1j} & \cdots & a_{1n} \\ \vdots & \cdots & \cdots & \cdots & \vdots \\ a_{n1} & \cdots & aa_{nj} + bb_{nj} & \cdots & a_{nn} \end{pmatrix}$$

$$= (a+b)^{n-1}(a\det(A) + b\det(B)).$$

3.2.7 Let the column vectors in $A(t)$ be denoted by $A_1(t),\dots,A_n(t)$. Then

$$\det(A(t+h)) - \det(A(t))$$
$$= |A_1(t+h), A_2(t+h),\dots,A_n(t+h)| - |A_1(t), A_2(t),\dots,A_n(t)|$$
$$= |A_1(t+h), A_2(t+h),\dots,A_n(t+h)| - |A_1(t), A_2(t+h),\dots,A_n(t+h)|$$
$$+ |A_1(t), A_2(t+h), A_3(t+h)\dots,A_n(t+h)|$$
$$- |A_1(t), A_2(t), A_3(t+h),\dots,A_n(t+h)|$$

$$+ |A_1(t), A_2(t), A_3(t+h), \ldots, A_n(t+h)|$$
$$+ \cdots + |A_1(t), A_2(t), \ldots, A_n(t+h)| - |A_1(t), A_2(t), \ldots, A_n(t)|$$
$$= |A_1(t+h) - A_1(t), A_2(t+h), \ldots, A_n(t+h)|$$
$$+ |A_1(t), A_2(t+h) - A_2(t), A_3(t+h), \ldots, A_n(t+h)|$$
$$+ \cdots + |A_1(t), A_2(t), \ldots, A_n(t+h) - A_n(t)|.$$

Dividing the above by $h \neq 0$, we get

$$\frac{1}{h}(\det(A(t+h)) - \det(A(t)))$$

$$= \left| \frac{1}{h}(A_1(t+h) - A_1(t)), A_2(t+h), \ldots, A_n(t+h) \right|$$

$$+ \left| A_1(t), \frac{1}{h}(A_2(t+h) - A_2(t)), \ldots, A_n(t+h) \right|$$

$$+ \cdots + \left| A_1(t), A_2(t), \ldots, \frac{1}{h}(A_n(t+h) - A_n(t)) \right|.$$

Now taking the $h \to 0$ limits on both sides of the above, we arrive at

$$\frac{d}{dt} \det(A(t)) = |A_1'(t), A_2(t), \ldots, A_n(t)| + |A_1(t), A_2'(t), \ldots, A_n(t)|$$

$$+ \cdots + |A_1(t), A_2(t), \ldots, A_n'(t)|.$$

Finally, expanding the jth determinant along the jth column on the right-hand side of the above by the cofactors, $j = 1, 2, \ldots, n$, we have

$$\frac{d}{dt} \det(A(t)) = \sum_{i=1}^{n} a_{i1}'(t) C_{i1}(t) + \sum_{i=1}^{n} a_{i2}'(t) C_{i2}(t) + \cdots + \sum_{i=1}^{n} a_{in}'(t) C_{in}(t).$$

3.2.9 Adding all columns to the first column of the matrix, we get

$$
\begin{vmatrix}
x + \sum_{i=1}^{n} a_i & a_1 & a_2 & \cdots & a_n \\
x + \sum_{i=1}^{n} a_i & x & a_2 & \cdots & a_n \\
x + \sum_{i=1}^{n} a_i & a_2 & x & \cdots & a_n \\
\vdots & \vdots & \vdots & \ddots & \vdots \\
x + \sum_{i=1}^{n} a_i & a_2 & a_3 & \cdots & x
\end{vmatrix}
= \left(x + \sum_{i=1}^{n} a_i \right)
\begin{vmatrix}
1 & a_1 & a_2 & \cdots & a_n \\
1 & x & a_2 & \cdots & a_n \\
1 & a_2 & x & \cdots & a_n \\
\vdots & \vdots & \vdots & \ddots & \vdots \\
1 & a_2 & a_3 & \cdots & x
\end{vmatrix}.
$$

Consider the $(n+1) \times (n+1)$ determinant on the right-hand side of the above. We now subtract row n from row $n+1$, row $n-1$ from row n,..., row 2 from row 3, and row 1 from row 2. Then that determinant becomes

$$
a \equiv \begin{vmatrix}
1 & a_1 & a_2 & \cdots & a_n \\
0 & x-a_1 & 0 & \cdots & 0 \\
0 & \cdots & x-a_2 & \cdots & 0 \\
\vdots & \vdots & \vdots & \ddots & \vdots \\
0 & \cdots & \cdots & \cdots & x-a_n
\end{vmatrix}.
$$

Since the minor of the entry at the position $(1,1)$ is the determinant of a lower triangular matrix, we get $a = \prod_{i=1}^{n}(x-a_i)$.

3.2.10 If $c_1,\ldots,c_{n+2}$ are not all distinct then it is clear that the determinant is zero since there are two identical columns. Assume now $c_1,\ldots,c_{n+2}$ are distinct and consider the function of x given by

$$
p(x) = \det \begin{pmatrix}
p_1(x) & p_1(c_2) & \cdots & p_1(c_{n+2}) \\
p_2(x) & p_2(c_2) & \cdots & p_2(c_{n+2}) \\
\vdots & \vdots & \ddots & \vdots \\
p_{n+2}(x) & p_{n+2}(c_2) & \cdots & p_{n+2}(c_{n+2})
\end{pmatrix}.
$$

By the cofactor expansion along the first column, we see that $p(x)$ is a polynomial of degree at most n which vanishes at $n+1$ points: $c_2,\ldots,c_{n+2}$. Hence $p(x) = 0$ for all x. In particular, $p(c_1) = 0$ as well.

3.2.11 We use D_{n+1} to denote the determinant and implement induction to do the computation. When $n = 1$, we have $D_2 = a_1x+a_0$. At $n-1$, we assume $D_n = a_{n-1}x^{n-1}+\cdots+a_1x+a_0$. At n, by the cofactor expansion according to the first column, we get

$$
D_{n+1} = xD_n + (-1)^{n+2}a_0(-1)^n = x(a_nx^{n-1}+\cdots+a_2x+a_1)+a_0.
$$

3.2.12 Use $D(\lambda)$ to denote the determinant on the left-hand side of (3.2.45). It is clear that $D(0) = 0$. So (3.2.45) is true at $\lambda = 0$.

Now assume $\lambda \neq 0$ and rewrite $D(\lambda)$ as

$$
D(\lambda) = \det \begin{pmatrix}
1 & 0 & 0 & \cdots & 0 \\
1 & a_1-\lambda & a_2 & \cdots & a_n \\
1 & a_1 & a_2-\lambda & \cdots & a_n \\
\vdots & \vdots & \vdots & \ddots & \vdots \\
1 & a_1 & a_2 & \cdots & a_n-\lambda
\end{pmatrix}.
$$

Adding the $(-a_1)$ multiple of column 1 to column 2, the $(-a_2)$ multiple of column 1 to column 3, $\ldots$, and the $(-a_n)$ multiple of column 1 to column $n+1$, we see that $D(\lambda)$ becomes

$$D(\lambda) = \det \begin{pmatrix} 1 & -a_1 & -a_2 & \cdots & -a_n \\ 1 & -\lambda & 0 & \cdots & 0 \\ 1 & 0 & -\lambda & \cdots & 0 \\ \vdots & \vdots & \vdots & \ddots & \vdots \\ 1 & 0 & 0 & \cdots & -\lambda \end{pmatrix}$$

$$= \lambda^n \det \begin{pmatrix} 1 & -\frac{a_1}{\lambda} & -\frac{a_2}{\lambda} & \cdots & -\frac{a_n}{\lambda} \\ 1 & -1 & 0 & \cdots & 0 \\ 1 & 0 & -1 & \cdots & 0 \\ \vdots & \vdots & \vdots & \ddots & \vdots \\ 1 & 0 & 0 & \cdots & -1 \end{pmatrix}.$$

On the right-hand side of the above, adding column 2, column 3, ..., and column $n+1$ to column 1, we get

$$D(\lambda) = \lambda^n \det \begin{pmatrix} 1 - \frac{1}{\lambda}\sum_{i=1}^{n} a_i & -\frac{a_1}{\lambda} & -\frac{a_2}{\lambda} & \cdots & -\frac{a_n}{\lambda} \\ 0 & -1 & 0 & \cdots & 0 \\ 0 & 0 & -1 & \cdots & 0 \\ \vdots & \vdots & \vdots & \ddots & \vdots \\ 0 & 0 & 0 & \cdots & -1 \end{pmatrix}$$

$$= \lambda^n \left(1 - \frac{1}{\lambda}\sum_{i=1}^{n} a_i \right)(-1)^n = (-1)^n \lambda^{n-1}\left(\lambda - \sum_{i=1}^{n} a_i \right).$$

3.2.13 From $\det(AA^t) = \det(I_n) = 1$ and $\det(A)^2 = \det(AA^t)$ we get $\det(A)^2 = 1$. Using $\det(A) < 0$, we get $\det(A) = -1$. Hence $\det(A + I_n) = \det(A + AA^t) = \det(A)\det(I_n + A^t) = -\det(I_n + A)$ which results in $\det(I_n + A) = 0$.

3.2.14 Since $a_{11} = 1$ or -1, we may add or subtract the first row from the other rows of A so that we reduce A into a matrix $B = (b_{ij})$ satisfying $b_{11} = 1$ or -1, $b_{i1} = 0$ for $i \geq 2$, and all entries in the submatrix of B with the first row and column of B deleted are even numbers. By the cofactor expansion along the first column of B we see that $\det(B) =$ an even integer. Since $\det(A) = \det(B)$, the proof follows.

3.2.15 To see that $\det(A) \neq 0$, it suffices to show that A is invertible, or equivalently, $N(A) = \{x \in \mathbb{R}^n \mid Ax = 0\} = \{0\}$. In fact, take $x \in N(A)$ and assume otherwise $x = (x_1, \ldots, x_n)^t \neq 0$. Let $i = 1, \ldots, n$ be such that

$$|x_i| = \max_{1 \leq j \leq n} \{|x_j|\}. \tag{S10}$$

Then $|x_i| > 0$. On the other hand, the ith component of the equation $Ax = 0$ reads

$$a_{i1}x_1 + \cdots + a_{ii}x_i + \cdots + a_{in}x_n = 0. \qquad \text{(S11)}$$

Combining (S10) and (S11) we arrive at

$$0 = |a_{i1}x_1 + \cdots + a_{ii}x_i + \cdots + a_{in}x_n| \geq |a_{ii}||x_i| - \sum_{1 \leq j \leq n, j \neq i} |a_{ij}||x_j|$$

$$\geq |x_i| \left(|a_{ii}| - \sum_{1 \leq j \leq n, j \neq i} |a_{ij}| \right) > 0,$$

which is a contradiction.

3.2.16 Consider the modified matrix:

$$A(t) = D + t(A - D), \quad 0 \leq t \leq 1,$$

where $D = \text{diag}\{a_{11}, \ldots, a_{nn}\}$. Then $A(t)$ satisfies the condition stated in the previous problem. So $\det(A(t)) \neq 0$. Furthermore, $\det(A(0)) = \det(D) > 0$. So $\det(A(1)) > 0$ as well otherwise there is a point $t_0 \in (0, 1)$ such that $\det(A(t_0)) = 0$ which is false.

3.2.18 Let $\alpha = (a_1, \ldots, b_n)$ and $\beta = (b_1, \ldots, b_n)$.

To compute

$$\det(I_n - \alpha^t \beta) = \begin{vmatrix} 1 - a_1 b_1 & -a_1 b_2 & \cdots & -a_1 b_n \\ -a_2 b_1 & 1 - a_2 b_2 & \cdots & -a_2 b_n \\ \vdots & \vdots & \ddots & \vdots \\ -a_n b_1 & -a_n b_2 & \cdots & 1 - a_n b_n \end{vmatrix},$$

we artificially enlarge it into an $(n+1) \times (n+1)$ determinant of the form

$$\det(I_n - \alpha^t \beta) = \begin{vmatrix} 1 & b_1 & b_2 & \cdots & b_n \\ 0 & 1 - a_1 b_1 & -a_1 b_2 & \cdots & -a_1 b_n \\ 0 & -a_2 b_1 & 1 - a_2 b_2 & \cdots & -a_2 b_n \\ \vdots & \vdots & \vdots & \ddots & \vdots \\ 0 & -a_n b_1 & -a_n b_2 & \cdots & 1 - a_n b_n \end{vmatrix}.$$

Now adding the a_1 multiple of row 1 to row 2, a_2 multiple of row 1 to row 3, ..., and a_n multiple of row 1 to the last row, of the above determinant, we get

$$\det(I_n - \alpha^t \beta) = \begin{vmatrix} 1 & b_1 & b_2 & \cdots & b_n \\ a_1 & 1 & 0 & \cdots & 0 \\ a_2 & 0 & 1 & \cdots & 0 \\ \vdots & \vdots & \vdots & \ddots & \vdots \\ a_n & 0 & 0 & \cdots & 1 \end{vmatrix}.$$

Next, subtracting the b_1 multiple of row 2 from row 1, b_2 multiple of row 3 from row 1, ..., and b_n multiple of the last row from row 1, we obtain

$$
\det(I_n - \alpha^t \beta) =
\begin{vmatrix}
1 - \sum_{i=1}^{n} a_i b_i & 0 & 0 & \cdots & 0 \\
a_1 & 1 & 0 & \cdots & 0 \\
a_2 & 0 & 1 & \cdots & 0 \\
\vdots & \vdots & \vdots & \ddots & \vdots \\
a_n & 0 & 0 & \cdots & 1
\end{vmatrix}
= 1 - \sum_{i=1}^{n} a_i b_i.
$$

3.2.19 The method is contained in the solution of Problem 18. In fact, adding the $(-x_1)$ multiple of row 2 to row 1, $(-x_2)$ multiple of row 3 to row 1,..., and $(-x_n)$ multiple of the last row to row 1, we get

$$
f(x_1,\ldots,x_n) = \det
\begin{pmatrix}
100 - \sum_{i=1}^{n} x_i^2 & 0 & 0 & \cdots & 0 \\
x_1 & 1 & 0 & \cdots & 0 \\
x_2 & 0 & 1 & \cdots & 0 \\
\vdots & \vdots & \vdots & \ddots & \vdots \\
x_n & 0 & 0 & \cdots & 1
\end{pmatrix}
= 100 - \sum_{i=1}^{n} x_i^2.
$$

So $f(x_1,\ldots,x_n) = 0$ represents the sphere in $\mathbb{R}^n$ centered at the origin and of radius 10.

Section 3.3

3.3.3 First, we observe that for $A \in \mathbb{F}(m,n)$ and $B \in \mathbb{F}(n,l)$ written as $B = (B_1,\ldots,B_l)$, where $B_1,\ldots,B_l \in \mathbb{F}^n$ are l column vectors, we have

$$
AB = (AB_1,\ldots,AB_l), \tag{S12}
$$

where $AB_1,\ldots,AB_l$ are l column vectors in $\mathbb{F}^m$.

Now, for $A \in \mathbb{F}(n,n)$, we recall the relation (3.3.5). If $\det(A) \neq 0$, we can evaluate determinants on both sides of (3.3.5) to arrive at (3.3.15). If $\det(A) = 0$, then in view of (3.3.5) and (S12), we have

$$
\mathrm{adj}(A)A_1 = 0, \quad \ldots, \quad \mathrm{adj}(A)A_n = 0, \tag{S13}
$$

where $A_1,\ldots,A_n$ are the n column vectors of A. If $A \neq 0$, then at least one of $A_1,\ldots,A_n$ is nonzero. In other words, $n(\mathrm{adj}(A)) \geq 1$ so that $r(\mathrm{adj}(A)) \leq n-1$. Hence $\det(\mathrm{adj}(A)) = 0$ and (3.3.15) is valid. If $A = 0$, then $\mathrm{adj}(A) = 0$ and (3.3.15) trivially holds.

3.3.4 If $r(A) = n$, then both A and $\text{adj}(A)$ are invertible by (3.3.5). So we obtain $r(\text{adj}(A)) = n$. If $r(A) = n-1$, then A contains an $(n-1) \times (n-1)$ submatrix whose determinant is nonzero. So $\text{adj}(A) \neq 0$ which leads to $r(\text{adj}(A)) \geq 1$. On the other hand, from (S13), we see that $n(\text{adj}(A)) \geq n-1$. So $r(\text{adj}(A)) \leq 1$. This proves $r(\text{adj}(A)) = 1$. If $r(A) \leq n-2$, then any $n-1$ row vectors of A are linearly dependent, which indicates that all row vectors of any $(n-1) \times (n-1)$ submatrix of A are linearly dependent. In particular, all cofactors of A are zero. Hence $\text{adj}(A) = 0$ and $r(\text{adj}(A)) = 0$.

3.3.6 Let $n \geq 3$ and assume A is invertible. Using (3.3.5) and (3.3.15) we have

$$\text{adj}(A)\text{adj}(\text{adj}(A)) = \det(\text{adj}(A))I_n = (\det(A))^{n-1}I_n.$$

Comparing this with (3.3.5) again, we obtain $\text{adj}(\text{adj}(A)) = (\det(A))^{n-2}A$. If A is not invertible, then $\det(A) = 0$ and $r(\text{adj}(A)) \leq 1$. Since $n \geq 3$, we have $r(\text{adj}(A)) \leq 1 \leq n-2$. So in view of Problem 3.3.4, we have $r(\text{adj}(\text{adj}(A))) = 0$ or $\text{adj}(\text{adj}(A)) = 0$. So $\text{adj}(\text{adj}(A)) = (\det(A))^{n-2}A$ is trivially true.

If $n = 2$, it is direct to verify the relation $\text{adj}(\text{adj}(A)) = A$.

3.3.7 For $A = (a_{ij})$, we use M_{ij}^A and C_{ij}^A to denote the minor and cofactor of the entry a_{ij}, respectively. Then we have

$$M_{ij}^A = M_{ji}^{A^t}, \tag{S14}$$

which leads to $C_{ij}^A = C_{ji}^{A^t}, i, j = 1, \ldots, n$. Hence

$$(\text{adj}(A))^t = (C_{ij}^A) = (C_{ij}^{A^t})^t = \text{adj}(A^t).$$

(If A is invertible, we may use (3.3.5) to write $\text{adj}(A) = \det(A)A^{-1}$. Therefore $(\text{adj}(A))^t = \det(A)(A^{-1})^t = \det(A^t)(A^t)^{-1} = \text{adj}(A^t)$.)

3.3.8 From (3.3.5), we have $AA^t = \det(A)I_n$. For $B = (b_{ij}) = AA^t$, we get

$$b_{ii} = \sum_{j=1}^n a_{ij}^2, \quad i = 1, \ldots, n.$$

So $\det(A) = 0$ if and only if $b_{ii} = 0$ for all $i = 1, \ldots, n$, or $A = 0$.

3.3.9 First, assume that A and B are both invertible. Then, in view of (3.3.5), we have

$$\text{adj}(AB) = \det(AB)(AB)^{-1} = \det(A)\det(B)B^{-1}A^{-1}$$
$$= (\det(B)B^{-1})(\det(A)A^{-1}) = \text{adj}(B)\text{adj}(A).$$

Next, we prove the conclusion without assuming that A and B are invertible. Since $\det(A - \lambda I)$ and $\det(B - \lambda I)$ are polynomials of the variable

λ, of degree n, they have at most n roots. Let $\{\lambda_k\}$ be a sequence in $\mathbb{R}$ so that $\lambda_k \to 0$ as $k \to \infty$ and $\det(A - \lambda_k I) \neq 0, \det(B - \lambda_k I) \neq 0$ for $k = 1, 2, \ldots$. Hence we have

$$\mathrm{adj}((A - \lambda_k I)(B - \lambda_k I)) = \mathrm{adj}(B - \lambda_k I)\mathrm{adj}(A - \lambda_k I), \quad k = 1, 2, \ldots.$$

Letting $k \to \infty$ in the above, we arrive at $\mathrm{adj}(AB) = \mathrm{adj}(B)\mathrm{adj}(A)$ again.

Section 3.4

3.4.1 Using C_{ij}^A to denote the cofactor of the entry a_{ij} of the matrix $A = (a_{ij})$ and applying (3.2.40), we have

$$a_1 = \frac{\mathrm{d}}{\mathrm{d}\lambda} p_A(\lambda)\bigg|_{\lambda=0} = \frac{\mathrm{d}}{\mathrm{d}\lambda}(\det(\lambda I_n - A))\bigg|_{\lambda=0}$$
$$= \sum_{i=1}^n C_{ii}^{(-A)} = (-1)^{n-1} \sum_{i=1}^n C_{ii}^A.$$

3.4.2 It suffices to show that $\mathcal{M} = \mathbb{C}(n,n) \setminus \mathcal{D}$ is closed in $\mathbb{C}(n,n)$. Let $\{A_k\}$ be a sequence in $\mathcal{M}$ which converges to some $A \in \mathbb{C}(n,n)$. We need to prove that $A \in \mathcal{M}$. To this end, consider $p_{A_k}(\lambda)$ and let $\{\lambda_k\}$ be multiple roots of $p_{A_k}(\lambda)$. Then we have

$$p_{A_k}(\lambda_k) = 0, \quad p'_{A_k}(\lambda_k) = 0, \quad k = 1, 2, \ldots. \tag{S15}$$

On the other hand, we know that the coefficients of $p_A(\lambda)$ are continuously dependent on the entries of A. So the coefficients of $p_{A_k}(\lambda)$ converge to those of $p_A(\lambda)$, respectively. In particular, the coefficients of $p_{A_k}(\lambda)$, say $\{a_{n-1}^k\}, \ldots, \{a_1^k\}, \{a_0^k\}$, are bounded sequences. Thus

$$|\lambda_k|^n \leq |\lambda_k^n - p_{A_k}(\lambda_k)| + |p_{A_k}(\lambda_k)| \leq \sum_{i=0}^{n-1} |a_i^k||\lambda_k|^i,$$

which indicates that $\{\lambda_k\}$ is a bounded sequence in $\mathbb{C}$. Passing to a subsequence if necessary, we may assume without loss of generality that $\lambda_k \to$ some $\lambda_0 \in \mathbb{C}$ as $k \to \infty$. Letting $k \to \infty$ in (S15), we obtain $p_A(\lambda_0) = 0, p'_A(\lambda_0) = 0$. Hence λ_0 is a multiple root of $p_A(\lambda)$ which proves $A \in \mathcal{M}$.

3.4.4 Using (3.4.31), we have

$$\mathrm{adj}(\lambda I_n - A) = A_{n-1}\lambda^{n-1} + \cdots + A_1\lambda + A_0,$$

where the matrices $A_{n-1}, A_{n-2}, \ldots, A_1, A_0$ are determined through (3.4.32). Hence, setting $\lambda = 0$ in the above, we get

$$\text{adj}(-A) = A_0 = a_1 I_n + a_2 A + \cdots + a_{n-1} A^{n-2} + A^{n-1}.$$

That is,

$$\text{adj}(A) = (-1)^{n-1}(a_1 I_n + a_2 A + \cdots + a_{n-1} A^{n-2} + A^{n-1}). \qquad \text{(S16)}$$

Note that (S16) may be used to prove some known facts more easily. For example, the relation $\text{adj}(A^t) = (\text{adj}(A))^t$ (see Problem 3.3.7) follows immediately.

3.4.9 First, assume that A is invertible. Then

$$\begin{aligned} p_{AB}(\lambda) &= \det(\lambda I_n - AB) = \det(A[\lambda A^{-1} - B]) \\ &= \det(A)\det(\lambda A^{-1} - B) = \det([\lambda A^{-1} - B]A) \\ &= \det(\lambda I_n - BA) = p_{BA}(\lambda). \end{aligned}$$

Next, assume A is arbitrary. If $\mathbb{F} = \mathbb{R}$ or $\mathbb{C}$, we may proceed as follows. Let $\{a_k\}$ be a sequence in $\mathbb{F}$ such that $A - a_k I_n$ is invertible and $a_k \to 0$ as $k \to \infty$. Then we have

$$p_{(A-a_k I_n)B}(\lambda) = p_{B(A-a_k I_n)}(\lambda), \quad k = 1, 2, \ldots.$$

Letting $k \to \infty$ in the above, we arrive at $p_{AB}(\lambda) = p_{BA}(\lambda)$ again.

If A is not invertible or $\mathbb{F}$ is not $\mathbb{R}$ or $\mathbb{C}$, the above methods fail and a different method needs to be used. To this end, we recognize the following matrix relations in $\mathbb{F}(2n, 2n)$:

$$\begin{pmatrix} I_n & -A \\ 0 & \lambda I_n \end{pmatrix} \begin{pmatrix} A & \lambda I_n \\ I_n & B \end{pmatrix} = \begin{pmatrix} 0 & \lambda I_n - AB \\ \lambda I_n & \lambda B \end{pmatrix},$$

$$\begin{pmatrix} I_n & 0 \\ -B & \lambda I_n \end{pmatrix} \begin{pmatrix} A & \lambda I_n \\ I_n & B \end{pmatrix} = \begin{pmatrix} A & \lambda I_n \\ \lambda I_n - BA & 0 \end{pmatrix}.$$

The determinants of the left-hand sides of the above are the same. The determinants of the right-hand sides of the above are

$$(-1)^n \lambda^n \det(\lambda I_n - AB), \quad (-1)^n \lambda^n \det(\lambda I_n - BA).$$

Hence the conclusion $p_{AB} = (\lambda) = p_{BA}(\lambda)$ follows.

3.4.10 We consider the nontrivial case $n \geq 2$. Using the notation and method in the solution to Problem 3.2.18, we have

$$\det(\lambda I_n - \alpha^t \beta) = \begin{vmatrix} 1 & b_1 & b_2 & \cdots & b_n \\ a_1 & \lambda & 0 & \cdots & 0 \\ a_2 & 0 & \lambda & \cdots & 0 \\ \vdots & \vdots & \vdots & \ddots & \vdots \\ a_n & 0 & 0 & \cdots & \lambda \end{vmatrix}.$$

Assume $\lambda \neq 0$. Subtracting the b_1/λ multiple of row 2 from row 1, b_2/λ multiple of row 3 from row 1, ..., and b_n/λ multiple of the last row from row 1, we obtain

$$\det(\lambda I_n - \alpha^t \beta) = \begin{vmatrix} 1 - \frac{1}{\lambda}\sum_{i=1}^{n} a_i b_i & 0 & 0 & \cdots & 0 \\ a_1 & \lambda & 0 & \cdots & 0 \\ a_2 & 0 & \lambda & \cdots & 0 \\ \vdots & \vdots & \vdots & \ddots & \vdots \\ a_n & 0 & 0 & \cdots & \lambda \end{vmatrix}$$

$$= \left(1 - \frac{1}{\lambda}\sum_{i=1}^{n} a_i b_i\right)\lambda^n = \lambda^{n-1}(\lambda - \alpha\beta^t).$$

Since both sides of the above also agree at $\lambda = 0$, they are identical for all λ.

3.4.11 (a) We have $p_A(\lambda) = \lambda^3 - 2\lambda^2 - \lambda + 2$. From $p_A(A) = 0$, we have $A(A^2 - 2A - I_n) = -2I_n$. So

$$A^{-1} = -\frac{1}{2}(A^2 - 2A - I_n) = \begin{pmatrix} 1 & -1 & 0 \\ 0 & \frac{1}{2} & 0 \\ -2 & \frac{3}{2} & -1 \end{pmatrix}.$$

(b) We divide λ^{10} by $p_A(\lambda)$ to find

$$\lambda^{10} = p_A(\lambda)(\lambda^7 + 2\lambda^7 + 5\lambda^5 + 10\lambda^4 + 21\lambda^3 + 42\lambda^2 + 85\lambda + 170)$$
$$+ 341\lambda^2 - 340.$$

Consequently, inserting A in the above, we have

$$A^{10} = 341A^2 - 340I_n = \begin{pmatrix} 1 & 2046 & 0 \\ 0 & 1024 & 0 \\ 0 & -1705 & 1 \end{pmatrix}.$$

Section 4.1

4.1.8 Let u_1 and u_2 be linearly independent vectors in U and assume $(u_1, u_1) \neq 0$ and $(u_2, u_2) \neq 0$, otherwise there is nothing to show. Set $u_3 = a_1 u_1 + u_2$, where $a_1 = -(u_1, u_2)/(u_1, u_1)$. Then $u_3 \neq 0$ and is perpendicular to u_1. So u_1 and u_3 are linearly independent as well. Consider $u = cu_1 + u_3$

$(c \in \mathbb{C})$. Of course $u \neq 0$ for any c. Since $(u_1, u_3) = 0$, we have $(u,u) = c^2(u_1,u_1) + (u_3,u_3)$. Thus, in order to have $(u,u) = 0$, we may choose

$$c = \sqrt{-\frac{(u_3,u_3)}{(u_1,u_1)}}.$$

Section 4.2

4.2.7 We first show that the definition has no ambiguity. For this purpose, assume $[u_1] = [u]$ and $[v_1] = [v]$. We need to show that $(u_1,v_1) = (u,v)$. In fact, since $u_1 \in [u]$ and $v_1 \in [v]$, we know that $u_1 - u \in U_0$ and $v_1 - v \in U_0$. Hence there are x and $y \in U_0$ such that $u_1 - u = x$ and $v_1 - v = y$. So $(u_1,v_1) = (u+x,v+y) = (u,v) + (u,y) + (x,v+y) = (u,v)$.

It is obvious that $([u],[v])$ is bilinear and symmetric.

To show that the scalar product is nondegenerate, we assume $[u] \in U/U_0$ satisfies $([u],[v]) = 0$ for any $[v] \in U/U_0$. Thus $(u,v) = 0$ for all $v \in U$ which implies $u \in U_0$. In other words, $[u] = [0]$.

4.2.9 Let $w \in V^\perp$. Then $(v,w) = 0$ for any $v \in V$. Hence $0 = (v,w) = \langle v, \rho(w) \rangle, \forall v \in V$, which implies $\rho(w) \in V^0$. Hence $\rho(V^\perp) \subset V^0$. On the other hand, take $w' \in V^0$. Then $\langle v, w' \rangle = 0$ for all $v \in V$. Since $\rho : U \to U'$ is an isomorphism, there is a unique $w \in U$ such that $\rho(w) = w'$. Thus $(v,w) = \langle v, \rho(w) \rangle = \langle v, w' \rangle = 0, v \in V$. That is, $w \in V^\perp$. Thus $w' \in \rho(V^\perp)$ which proves $V^0 \subset \rho(V^\perp)$.

4.2.11 Recall (4.1.26). Replace V and W in (4.1.26) by $V^\perp$ and $W^\perp$ and use $(V^\perp)^\perp = V$ and $(W^\perp)^\perp = W$. We get $(V^\perp + W^\perp)^\perp = V \cap W$. Taking $\perp$ on both sides of this equation, we see that the conclusion follows.

4.2.12 From Problem 4.1.8, we know that there is a nonzero vector $u \in U$ such that $(u,u) = 0$. Since $(\cdot,\cdot)$ is nondegenerate, there is some $w \in U$ such that $(u,w) \neq 0$. It is clear that u and w are linearly independent. If w satisfies $(w,w) = 0$, then we take $v = w/(u,w)$. Hence $(v,v) = 0$ and $(u,v) = 1$. Assume now $(w,w) \neq 0$ and consider $x = u + cw$ ($c \in \mathbb{C}$). Then x can never be zero for any $c \in \mathbb{C}$. Set $(x,x) = 0$. We get $2c(u,w) + c^2(w,w) = 0$. So we may choose $c = -2(u,w)/(w,w)$. Thus, with $v = x/(u,x)$ where

$$(u,x) = -\frac{2(u,w)^2}{(w,w)} \neq 0,$$

we have $(v,v) = 0$ and $(u,v) = 1$ again.

Section 4.3

4.3.1 We consider the real case for simplicity. The complex case is similar.

Suppose M is non-singular and consider $a_1 u_1 + \cdots + a_k u_k = 0$, where $a_1, \ldots, a_k$ are scalars. Taking scalar products of this equation with the vectors $u_1, \ldots, u_k$ consecutively, we get

$$a_1(u_1, u_i) + \cdots + a_k(u_k, u_i) = 0, \quad i = 1, \ldots, k.$$

Since M^t is non-singular, we get $a_1 = \cdots = a_k = 0$.

Suppose M is singular. Then so is M^t and the above system of equations will have a solution $(a_1, \ldots, a_k) \neq (0, \ldots, 0)$. We rewrite the above system as:

$$\left(\sum_{i=1}^{k} a_i u_i, u_1 \right) = 0, \quad \ldots, \quad \left(\sum_{i=1}^{k} a_i u_i, u_k \right) = 0.$$

Then modify the above into

$$a_1 \left(\sum_{i=1}^{k} a_i u_i, u_1 \right) = 0, \quad \ldots, \quad a_k \left(\sum_{i=1}^{k} a_i u_i, u_k \right) = 0$$

and sum up the k equations. We get

$$\left(\sum_{i=1}^{k} a_i u_i, \sum_{j=1}^{k} a_j u_j \right) = 0.$$

That is, we have $(v, v) = 0$ where $v = \sum_{i=1}^{k} a_i u_i$. Since the scalar product is positive definite, we arrive at $v = 0$. Thus $u_1, \ldots, u_k$ are linearly dependent.

4.3.2 Suppose $u \in \text{Span}\{u_1, \ldots, u_k\}$. Then there are scalars $a_1, \ldots, a_k$ such that $u = a_1 u_1 + \cdots + a_k u_k$. Taking scalar products of this equation with $u_1, \ldots, u_k$, we have

$$(u_i, u) = a_1(u_i, u_1) + \cdots + a_k(u_i, u_k), \quad i = 1, \ldots, k,$$

which establishes

$$\begin{pmatrix} (u_1, u) \\ \vdots \\ (u_k, u) \end{pmatrix} \in \text{Span} \left\{ \begin{pmatrix} (u_1, u_1) \\ \vdots \\ (u_k, u_1) \end{pmatrix}, \ldots, \begin{pmatrix} (u_1, u_k) \\ \vdots \\ (u_k, u_k) \end{pmatrix} \right\}.$$

On the other hand, let $V = \text{Span}\{u_1, \ldots, u_k\}$. If $k < \dim(U)$, then $V^{\perp} \neq \{0\}$. Take $u \in V^{\perp}$, $u \neq 0$. Then

$$\begin{pmatrix} (u_1, u) \\ \vdots \\ (u_k, u) \end{pmatrix} = \begin{pmatrix} 0 \\ \vdots \\ 0 \end{pmatrix} \in \text{Span} \left\{ \begin{pmatrix} (u_1, u_1) \\ \vdots \\ (u_k, u_1) \end{pmatrix}, \dots, \begin{pmatrix} (u_1, u_k) \\ \vdots \\ (u_k, u_k) \end{pmatrix} \right\},$$

but $u \notin \text{Span}\{u_1, \dots, u_k\}$.

4.3.4 For any $u \in U$, we have

$$\begin{aligned} \|(I \pm S)u\|^2 &= ((I \pm S)u, (I \pm S)u) \\ &= \|u\|^2 + \|Su\|^2 \pm (u, Su) \pm (Su, u) \\ &= \|u\|^2 + \|Su\|^2 \pm (S'u, u) \pm (Su, u) \\ &= \|u\|^2 + \|Su\|^2 \mp (Su, u) \pm (Su, u) \\ &= \|u\|^2 + \|Su\|^2. \end{aligned}$$

So $(I \pm S)u \neq 0$ whenever $u \neq 0$. Thus $n(I \pm S) = 0$ and $I \pm S$ must be invertible.

4.3.6 Set $R(A) = \{u \in \mathbb{C}^m \mid u = Ax \text{ for some } x \in \mathbb{C}^n\}$. Rewrite $\mathbb{C}^m$ as $\mathbb{C}^m = R(A) \oplus (R(A))^\perp$ and assert $(R(A))^\perp = \{y \in \mathbb{C}^m \mid A^\dagger y = 0\} = N(A^\dagger)$.

In fact, if $A^\dagger y = 0$, then $(y, Ax) = y^\dagger Ax = (x^\dagger A^\dagger y)^\dagger = 0$ for any $x \in \mathbb{C}^n$. That is, $N(A^\dagger) \subset R(A)^\perp$. On the other hand, take $z \in R(A)^\perp$. Then for any $x \in \mathbb{C}^n$ we have $0 = (z, Ax) = z^\dagger Ax = (A^\dagger z)^\dagger x = (A^\dagger z, x)$ which indicates $A^\dagger z = 0$. So $z \in N(A^\dagger)$. Thus $R(A)^\perp \subset N(A^\dagger)$.

Therefore the equation $Ax = b$ has a solution if and only if $b \in R(A)$ or, if and only if b is perpendicular to $R(A)^\perp = N(A^\dagger)$.

Section 4.4

4.4.2 It is clear that $P \in L(U)$. Let $\{w_1, \dots, w_l\}$ be an orthogonal basis of $W = V^\perp$. Then $\{v_1, \dots, v_k, w_1, \dots, w_l\}$ is an orthogonal basis of U. Now it is direct to check that $P(v_j) = v_j, j = 1, \dots, k, P(w_s) = 0, s = 1, \dots, l$. So $P^2(v_j) = P(v_j), j = 1, \dots, k, P^2(w_s) = P(w_s), s = 1, \dots, l$. Hence $P^2 = P$, $R(P) = V$, and $N(P) = W$ as claimed. In other words, $P : U \to U$ is the projection of U onto V along W.

4.4.4 (a) We know that we have the direct sum $U = V \oplus V^\perp$. For any x and $y \in [u]$ we have $x = v_1 + w_1$ and $y = v_2 + w_2, v_i \in V, w_i \in V^\perp, i = 1, 2$. So $x - y = (v_1 - v_2) + (w_1 - w_2)$. However, since $x - y \in V$, we have $w_1 = w_2$. In other words, for any $x \in [u]$, there is a unique $w \in V^\perp$ so that $x = v + w$ for some $v \in V$. Of course, $[x] = [w] = [u]$ and $\|x\|^2 = \|v\|^2 + \|w\|^2$ whose minimum is attained at $v = 0$. This proves $\|[u]\| = \|w\|$.

(b) So we need to find the unique $w \in V^\perp$ such that $[w] = [u]$. First find an orthogonal basis $\{v_1, \ldots, v_k\}$ of V. Then for u we know that the projection of u onto V along $V^\perp$ is given by the Fourier expansion (see Problem 4.4.2)

$$v = \sum_{i=1}^{k} \frac{(v_i, u)}{(v_i, v_i)} v_i.$$

Thus, from $u = v + w$ where $w \in V^\perp$, we get

$$w = u - \sum_{i=1}^{k} \frac{(v_i, u)}{(v_i, v_i)} v_i.$$

Section 4.5

4.5.5 (a) It is direct to check that $\|T(x)\| = \|x\|$ for $x \in \mathbb{R}^n$.

(b) Since $T(T(x)) = x$ for $x \in \mathbb{R}^n$, we have $T^2 = I$. Consequently, $T = T'$. So if λ is an eigenvalue of T then $\lambda^2 = 1$. Thus the eigenvalues of T may only be ± 1. In fact, for $u = (1, 0, \ldots, 0, 1)^t$ and $v = (1, 0, \ldots, 0, -1)^t$, we have $T(u) = u$ and $T(v) = -v$. So ± 1 are both eigenvalues of T.

(c) Let x and $y \in \mathbb{R}^n$ satisfy $T(x) = x$ and $T(y) = -y$. Then $(x, y) = (T(x), T(y)) = (x, -y) = -(x, y)$. So $(x, y) = 0$.

(d) Since $T^2 - I = 0$ and $T \neq \pm I$, we see that the minimal polynomial of T is $m_T(\lambda) = \lambda^2 - 1$.

Section 5.2

5.2.9 We can write A as $A = P^t DP$ where P is orthogonal and

$$D = \text{diag}\{d_1, \ldots, d_n\}, \quad d_i = \pm 1, \quad i = 1, \ldots, n.$$

It is easily seen that the column vectors of D are mutually orthogonal and of length 1. So D is orthogonal as well. Hence A must be orthogonal as a product of three orthogonal matrices.

5.2.18 Use $\| \cdot \|$ to denote the standard norm of $\mathbb{R}^n$ and let $u_1 = \frac{1}{\|u\|} u$. Expand $\{u_1\}$ to get an orthonormal basis $\{u_1, u_2, \ldots, u_n\}$ of $\mathbb{R}^n$ (with the usual scalar product), where $u_1, \ldots, u_n$ are all column vectors. Let Q be the $n \times n$ matrix whose 1st, 2nd,$\ldots$, nth column vectors are $u_1, u_2, \ldots, u_n$. Then

$$u^t Q = u^t Q = u^t (u_1, u_2, \ldots, u_n)$$
$$= (u^t u_1, u^t u_2, \ldots, u^t u_n) = (\|u\|, 0, \ldots, 0).$$

Therefore

$$Q^t(uu^t)Q = (u^tQ)^t(u^tQ)$$

$$= \begin{pmatrix} \|u\| \\ 0 \\ \vdots \\ 0 \end{pmatrix} (\|u\|, 0, \ldots, 0) = \text{diag}\{\|u\|^2, 0, \ldots, 0\}.$$

Section 5.3

5.3.9 For definiteness, assume $m < n$. So $r(A) = m$. Use $\|\cdot\|$ to denote the standard norm of $\mathbb{R}^l$. Now check

$$q(x) = x^t(A^tA)x = (Ax)^t(Ax) = \|Ax\|^2 \geq 0, \quad x \in \mathbb{R}^n,$$

$$Q(y) = y^t(AA^t)y = (A^ty)^t(A^ty) = \|A^ty\|^2 \geq 0, \quad y \in \mathbb{R}^m.$$

Hence $A^tA \in \mathbb{R}(n,n)$ and $AA^t \in \mathbb{R}(m,m)$ are both semi-positive definite.

If $q(x) = 0$, we have $Ax = 0$. Since $r(A) = m < n$, we know that $n(A) = n - r(A) > 0$. Thus there is some $x \neq 0$ such that $Ax = 0$. Hence q cannot be positive definite.

If $Q(y) = 0$, we have $A^ty = 0$. Since $r(A^t) = r(A) = m$, we know that $n(A^t) = m - r(A^t) = 0$. Thus $y = 0$. Hence $Q(y) > 0$ whenever $y \neq 0$ and Q is positive definite.

5.3.12 Assume A is semi-positive definite. Then there is a semi-positive definite matrix B such that $A = B^2$. If $x \in S$, then $0 = x^tAx = x^tB^2x = (Bx)^t(Bx) = \|Bx\|^2$. That is, $Bx = 0$, which implies $Ax = B^2x = 0$. Conversely, $Ax = 0$ clearly implies $x \in S$.

If A is indefinite, $N(A)$ may not be equal to S. For example, take $A = \text{diag}\{1, -1\}$. Then $N(A) = \{0\}$ and

$$S = \left\{ x = \begin{pmatrix} x_1 \\ x_2 \end{pmatrix} \in \mathbb{R}^2 \,\middle|\, x_1^2 - x_2^2 = 0 \right\},$$

which is the union of the two lines $x_1 = x_2$ and $x_1 = -x_2$, which cannot even be a subspace of $\mathbb{R}^2$.

5.3.13 Since A is positive definite, there is an invertible matrix $P \in \mathbb{R}(n,n)$ such that $P^tAP = I_n$. Since P^tBP is symmetric, there is an orthogonal matrix $Q \in \mathbb{R}(n,n)$ such that $Q^t(P^tBP)Q$ is a real diagonal matrix. Set $C = PQ$. Then both C^tAC and C^tBC are diagonal.

5.3.14 First, we observe that if C is symmetric then the eigenvalues of C are greater than or equal to c if and only if $x^tCx \geq cx^tx = c\|x\|^2, x \in \mathbb{R}^n$. In fact, if $\lambda_1, \ldots, \lambda_n$ are eigenvalues of C, then there is an orthogonal matrix P such that $C = P^t DP$, where $D = \text{diag}\{\lambda_1, \ldots, \lambda_n\}$. So with $y = Px$ we have

$$x^t Cx = (Px)^t D(Px) = y^t Dy = \sum_{i=1}^{n} \lambda_i y_i^2 \geq \lambda_0 \|y\|^2,$$

where λ_0 is the smallest eigenvalue of C. If $\lambda_0 \geq c$ then the above gives us $x^t Cx \geq c \|y\|^2 = c \|x\|^2$. Conversely, we see that

$$x^t Cx - c\|x\|^2 \geq 0 \quad \text{implies} \quad \sum_{i=1}^{n} \lambda_i y_i - c\|y\|^2 = \sum_{i=1}^{n} (\lambda_i - c) y_i^2 \geq 0.$$

Since x and hence y can be arbitrarily chosen, we have $\lambda_i \geq c$ for all i.

Now, from $x^t Ax \geq a\|x\|^2$ and $x^t Bx \geq b\|x\|^2$, we infer $x^t (A + B)x \geq (a + b)\|x\|^2$. Thus the eigenvalues of $A + B$ are greater than or equal to $a + b$.

5.3.15 Since A is positive definite, there is another positive definite matrix $P \in \mathbb{R}(n, n)$, such that $A = P^2$. Let $C = AB$. Then we have $C = P^2 B$. Thus $P^{-1} CP = PBP$. Since the right-hand side of this relation is positive definite, its eigenvalues are all positive. Hence all the eigenvalues of C are positive.

5.3.17 Let P be a non-singular matrix such that $A = P^t P$. Then $\det(A) = \det(P^t P) = \det(P)^2$. Now

$$\det(A + B) = \det(P^t P + B)$$
$$= \det(P^t)\det(I + [P^{-1}]^t B[P^{-1}])\det(P).$$

Consider $C = [P^{-1}]^t B[P^{-1}]$ which is clearly semi-positive definite. Let the eigenvalues of C be $\lambda_1, \ldots, \lambda_n \geq 0$. Then there is an orthogonal matrix Q such that $Q^t CQ = D$ where $D = \text{diag}\{\lambda_1, \ldots, \lambda_n\}$. Thus $\det(C) = \det(Q^t DQ) = \det(D) = \lambda_1 \cdots \lambda_n$. Combining the above results and using $\det(Q)^2 = 1$, we obtain

$$\det(A + B) = \det(Q^t)\det(P^t)\det(I + C)\det(P)\det(Q)$$
$$= \det(A)\det(I + Q^t CQ) = \det(A)\det(I + D)$$
$$= \det(A) \prod_{i=1}^{n} (1 + \lambda_i)$$
$$\geq \det(A)(1 + \lambda_1 \cdots \lambda_n)$$
$$= \det(A) + \det(A)\det(C).$$

However, recall the relation between B and C we get $B = P^t CP$ which gives us $\det(B) = \det(P)^2 \det(C) = \det(A)\det(C)$ so that the desired inequality is established.

If the inequality becomes an equality, then all $\lambda_1, \ldots, \lambda_n$ vanish. So $C = 0$ which indicates that $B = 0$.

5.3.18 Since A is positive definite, there is an invertible matrix $P \in \mathbb{R}(n,n)$ such that $P^t A P = I_n$. Hence we have

$$(\det(P))^2 \det(\lambda A - B) = \det(P^t((\lambda - 1)A - (B - A))P)$$
$$= \det((\lambda - 1)I_n - P^t(B - A)P).$$

Since $P^t(B - A)P$ is positive semi-definite whose eigenvalues are all non-negative, we see that the roots of the equation $\det(\lambda A - B) = 0$ must satisfy $\lambda - 1 \geq 0$.

5.3.19 (a) Let $x \in U$ be a minimum point of f. Then for any $y \in U$ we have $g(\varepsilon) \geq g(0)$, where $g(\varepsilon) = f(x + \varepsilon y)$ ($\varepsilon \in \mathbb{R}$). Thus

$$0 = \left(\frac{dg}{d\varepsilon}\right)_{\varepsilon=0} = \frac{1}{2}(y, T(x)) + \frac{1}{2}(x, T(y)) - (y, b) = (y, T(x) - b).$$

Since $y \in U$ is arbitrary, we get $T(x) - b = 0$ and (5.3.25) follows. Conversely, if $x \in U$ satisfies (5.3.25), then

$$f(u) - f(x) = \frac{1}{2}(u, T(u)) - (u, T(x)) - \frac{1}{2}(x, T(x)) + (x, T(x))$$
$$= \frac{1}{2}(u - x, T(u - x)) \geq \lambda_0 \|u - x\|^2, \quad u \in U, \quad \text{(S17)}$$

for some constant $\lambda_0 > 0$. That is, $f(u) \geq f(x)$ for all $u \in U$.

(b) If x and $y \in U$ solve (5.3.25), then (a) indicates that x and y are the minimum points of f. Hence $f(x) = f(y)$. Replacing u by y in (S17) we arrive at $x = y$.

5.3.20 Let $S \in L(U)$ be positive semi-definite so that $S^2 = T$. Then $q(u) = (S(u), S(u)) = \|S(u)\|^2$ for any $u \in U$. Hence the Schwarz inequality (4.3.10) gives us

$$q(\alpha u + \beta v) = (S(\alpha u + \beta v), S(\alpha u + \beta v))$$
$$= \alpha^2 \|S(u)\|^2 + 2\alpha\beta(S(u), S(v)) + \beta^2 \|S(v)\|^2$$
$$\leq \alpha^2 \|S(u)\|^2 + 2\alpha\beta \|S(u)\| \|S(v)\| + \beta^2 \|S(v)\|^2$$
$$\leq \alpha^2 \|S(u)\|^2 + \alpha\beta(\|S(u)\|^2 + \|S(v)\|^2) + \beta^2 \|S(v)\|^2$$
$$= \alpha(\alpha + \beta)\|S(u)\|^2 + \beta(\alpha + \beta)\|S(v)\|^2$$
$$= \alpha q(u) + \beta q(v), \quad u, v \in U,$$

where we have also used the inequality $2ab \leq a^2 + b^2$ for $a, b \in \mathbb{R}$.

Section 5.4

5.4.1 Suppose otherwise that A is positive definite. Using Theorem 5.11 we have

$$a_1, a_2, a_3 > 0, \quad \begin{vmatrix} a_1 & a_2 \\ a_2 & a_3 \end{vmatrix}, \begin{vmatrix} a_1 & a_3 \\ a_3 & a_2 \end{vmatrix}, \begin{vmatrix} a_3 & a_1 \\ a_1 & a_2 \end{vmatrix} > 0.$$

Hence we arrive at

$$a_1 a_3 > a_2^2, \quad a_1 a_2 > a_3^2, \quad a_2 a_3 > a_1^2.$$

Adding these results, we get

$$\begin{aligned} a_1^2 + a_2^2 + a_3^2 &< a_1 a_3 + a_1 a_2 + a_2 a_3 \\ &\leq \frac{1}{2}(a_1^2 + a_3^2) + \frac{1}{2}(a_1^2 + a_2^2) + \frac{1}{2}(a_2^2 + a_3^2) \\ &= a_1^2 + a_2^2 + a_3^2, \end{aligned} \tag{S18}$$

which is false.

5.4.2 In the general case, $n \geq 3$, we have

$$q(x) = \sum_{i=1}^{n-1} (x_i + a_i x_{i+1})^2 + (x_n + a_n x_1)^2, \quad x = \begin{pmatrix} x_1 \\ \vdots \\ x_n \end{pmatrix} \in \mathbb{R}^n.$$

We can rewrite $q(x)$ as $\sum_{i=1}^{n} y_i^2$ with

$$y_i = x_i + a_i x_{i+1}, \quad i = 1, \ldots, n-1, \quad y_n = x_n + a_n x_1. \tag{S19}$$

It is seen that $q(x)$ is positive definite if and only if the change of variables given in (S19) is invertible, which is equivalent to the condition $1 + (-1)^{n+1} a_1 a_2 \cdots a_n \neq 0$.

5.4.3 If A is positive semi-definite, then $A + \lambda I_n$ is positive definite for all $\lambda > 0$. Hence all the leading principal minors of $A + \lambda I_n$ are positive when $\lambda > 0$. Taking $\lambda \to 0^+$ in these positive minors we arrive at the conclusion. The converse is not true, however. For example, take $A = \mathrm{diag}\{1, 0, -1\}$. Then all the leading principal minors of A are non-negative but A is indefinite.

5.4.5 (a) The left-hand side of the equation (5.4.24) reads

$$\begin{pmatrix} A_{n-1} & A_{n-1}\beta + \alpha \\ \beta^t A_{n-1} + \alpha^t & \beta^t A_{n-1}\beta + \alpha^t\beta + \beta^t\alpha + a_{nn} \end{pmatrix}.$$

Comparing the above with the right-hand side of (5.4.24), we arrive at the equation $A_{n-1}\beta = -\alpha$, which has a unique solution since A_{n-1} is invertible. Inserting this result into the entry at the position (n,n) in the above matrix and comparing with the right-hand side of (5.4.24) again, we get $a = -\alpha^t(A_{n-1}^{-1})\alpha \le 0$ since A_{n-1}^{-1} is positive definite.

(b) Taking determinants on both sides of (5.4.24) and using the facts $\det(A_{n-1}) > 0$ and $a \le 0$, we obtain

$$\det(A) = \det(A_{n-1})(a_{nn} + a) \le \det(A_{n-1})a_{nn}.$$

(c) This follows directly.

(d) If A is positive semi-definite, then $A + \lambda I_n$ is positive definite for $\lambda > 0$. Hence $\det(A + \lambda I_n) \le (a_{11} + \lambda) \cdots (a_{nn} + \lambda), \lambda > 0$. Letting $\lambda \to 0^+$ in the above inequality we see that (5.4.26) holds again.

5.4.6 (a) It is clear that the entry at the position (i, j) of the matrix $A^t A$ is (u_i, u_j).

(b) Thus, applying (5.4.26), we have $\det(A^t A) \le (u_1, u_1) \cdots (u_n, u_n)$ or $(\det(A))^2 \le \|u_1\|^2 \cdots \|u_n\|^2$.

5.4.7 Using the definition of the standard Euclidean scalar product we have $\|u_i\| \le an^{\frac{1}{2}}, i = 1, \ldots, n$. Thus (5.4.29) follows.

Section 5.5

5.5.1 Assume the nontrivial case $\dim(U) \ge 2$. First suppose that T satisfies $T(u) \in \mathrm{Span}\{u\}$ for any $u \in U$. If $T \ne aI$ for some $a \in \mathbb{F}$, then there are some linearly independent vectors u and $v \in U$ such that $T(u) = au$ and $T(v) = bv$ for some a and $b \in \mathbb{F}$ with $a \ne b$. Since $T(u+v) \in \mathrm{Span}\{u+v\}$, we have $T(u+v) = c(u+v)$ for some $c \in \mathbb{F}$. Thus $c(u+v) = au + bv$ which implies $a = c$ and $b = c$ since u and v are linearly independent. This is false. Next we assume that there is some $u \in U$ such that $v = T(u) \notin \mathrm{Span}\{u\}$. Thus u and v are linearly independent. Let $S \in L(U)$ be such that $S(u) = v$ and $S(v) = u$. Then $T(v) = T(S(u)) = S(T(u)) = S(v) = u$. Let $R \in L(U)$ be such that $R(u) = v$ and $R(v) = 0$. Then $u = T(v) = T(R(u)) = R(T(u)) = R(v) = 0$, which is again false.

5.5.3 Let $\lambda_1, \ldots, \lambda_n \in \mathbb{F}$ be the n distinct eigenvalues of T and $u_1, \ldots, u_n$ the associated eigenvectors, respectively, which form a basis of U. We have $T(S(u_i)) = S(T(u_i)) = \lambda_i S(u_i)$. So $S(u_i) \in E_{\lambda_i}$. In other words, there is some $b_i \in \mathbb{F}$ such that $S(u_i) = b_i u_i$ for $i = 1, \ldots, n$. It is clear that the scalars $b_1, \ldots, b_n$ determine S uniquely. On the other hand, consider a mapping $R \in L(U)$ defined by

$$R = a_0 I + a_1 T + \cdots + a_{n-1} T^{n-1}, \quad a_0, a_1, \ldots, a_{n-1} \in \mathbb{F}. \tag{S20}$$

Then

$$R(u_i) = (a_0 + a_1 \lambda_i + \cdots + a_{n-1} \lambda_i^{n-1}) u_i, \quad i = 1, \ldots, n.$$

Therefore, if $a_0, a_1, \ldots, a_{n-1}$ are so chosen that

$$a_0 + a_1 \lambda_i + \cdots + a_{n-1}\lambda_i^{n-1} = b_i, \quad i = 1, \ldots, n, \qquad (S21)$$

then $R = S$. However, using the Vandermonde determinant, we know that the non-homogeneous system of equations (S21) has a unique solution in $a_0, a_1, \ldots, a_{n-1}$. Thus R may be constructed by (S20) to yield $R = S$.

5.5.4 From Problem 5.5.3, we have $C_T = \mathrm{Span}\{I, T, \ldots, T^{n-1}\}$. It remains to show that, as vectors in $L(U)$, the mappings $I, T, \ldots, T^{n-1}$ are linearly independent. To see this, we consider

$$c_0 I + c_1 T + \cdots + c_{n-1} T^{n-1} = 0, \quad c_0, c_1, \ldots, c_{n-1} \in \mathbb{F}.$$

Applying the above to u_i, we obtain

$$c_0 + c_1 \lambda_i + \cdots + c_{n-1}\lambda_i^{n-1} = 0, \quad i = 1, \ldots, n,$$

which lead to $c_0 = c_1 = \cdots = c_{n-1} = 0$, in view of the Vandermonde determinant. So $\dim(C_T) = n$.

5.5.5 (a) We have

$$\|T(u)\|^2 = (T(u), T(u)) = (u, T' \circ T(u)) = (u, T \circ T'(u))$$
$$= (T'(u), T'(u)) = \|T'(u)\|^2, \quad \forall u \in U.$$

(b) Given $u \in U$, we have

$$u = (I - T)(u) + T(u).$$

However, since $(I - T)(u) \in N(T) = N(T')$, we have $T'(u) = T' \circ T(u)$. Besides, we also have

$$u = (I - T')(u) + T'(u)$$

and $(I - T')(u) \in N(T') = N(T)$ (since T' is idempotent too), we get $T(u) = T \circ T'(u)$. Thus we arrive at

$$T'(u) = T(u), \quad \forall u \in U.$$

Section 5.6

5.6.3 (a) Let $v \in R(T)^\perp$. Then for any $u \in U$, we have

$$0 = (T(u), v)_V = (u, T'(v))_U.$$

Hence $T'(v) = 0$ or $v \in N(T')$. Conversely, if $v \in N(T')$, then for any $u \in U$, we have

$$0 = (u, T'(v))_U = (T(u), v)_V,$$

which proves $v \in R(T)^{\perp}$. So we have established the fact $R(T)^{\perp} = N(T')$. Taking $\perp$ in this equation, we obtain $R(T) = R(T)^{\perp\perp} = N(T')^{\perp}$.

(b) From (4.2.38), we have

$$\dim(N(T)) + \dim(N(T)^{\perp}) = \dim(U).$$

Using (a), we have $R(T') = N(T)^{\perp}$. Inserting this into the above equation, we have $n(T) + r(T') = \dim(U)$. However, the nullity-rank equation says $n(T) + r(T) = \dim(U)$. So we arrive at $r(T) = r(T')$.

(c) It is clear that $N(T) \subset N(T' \circ T)$. On the other hand, we have $N(T' \circ T) \subset N(T)$, since $u \in N(T' \circ T)$ implies that $\|T(u)\|_V^2 = (T(u), T(u))_V = (u, N(T' \circ T))_U = 0$, or $u \in N(T)$. So $N(T) = N(T' \circ T)$ such that the equation

$$n(T) + r(T) = \dim(U) = n(T' \circ T) + r(T' \circ T)$$

leads to $r(T' \circ T) = r(T)$. As a consequence, we can apply this result to T' to get $r(T \circ T') = r(T') = r(T)$. So the problem follows.

Section 6.2

6.2.5 Since $P^2 = P$, we know that P projects U onto $R(P)$ along $N(P)$. If $N(P) = R(P)^{\perp}$, we show that $P' = P$. To this end, for any u_1 and $u_2 \in U$ we rewrite them as $u_1 = v_1 + w_1$ and $u_2 = v_2 + w_2$, where v_1 and $v_2 \in R(P), w_1$ and $w_2 \in R(P)^{\perp}$. Then we have $P(v_1) = v_1, P(v_2) = v_2$, and $P(w_1) = P(w_2) = 0$. Hence $(u_1, P(u_2)) = (v_1 + w_1, v_2) = (v_1, v_2)$, and $(P(u_1), u_2) = (v_1, v_2 + w_2) = (v_1, v_2)$. This proves $(u_1, P(u_2)) = (P(u_1), u_2)$ which indicates $P = P'$. Conversely, we note that for any $T \in L(U)$ there holds $R(T)^{\perp} = N(T')$ (cf. Problem 5.6.3). If $P = P'$, we have $R(P)^{\perp} = N(P)$.

6.2.8 (a) If $T^k = 0$ for some $k \geq 1$ then eigenvalues of T vanish. Let $\{u_1, \ldots, u_n\}$ be a basis of U consisting of eigenvectors of T. Then $T(u_i) = 0$ for all $i = 1, \ldots, n$. Hence $T = 0$.

(b) If $k = 1$, then there is nothing to show. Assume that the statement is true up to $k = l \geq 1$. We show that the statement is true at $k = l + 1$. If $l = $ odd, then $k = 2m$ for some integer $m \geq 1$. Hence $T^{2m}(u) = 0$ gives us $(u, T^{2m}(u)) = (T^m(u), T^m u) = 0$. So $T^m(u) = 0$. Since $m \leq l$, we deduce $T(u) = 0$. If $l = $ even, then $l = 2m$ for some integer $m \geq 1$ and $T^{2m+1}(u) = 0$ gives us $T^{2m}(v) = 0$ where $v = T(u)$. Hence $T(v) = 0$. That is, $T^2(u) = 0$. So it follows again that $T(u) = 0$.

Section 6.3

6.3.11 Since $A^\dagger A$ is positive definite, there is a positive definite Hermitian matrix $B \in \mathbb{C}(n,n)$ such that $A^\dagger A = B^2$. Thus we can rewrite A as $A = PB$ with $P = (A^\dagger)^{-1}B$ which may be checked to satisfy

$$PP^\dagger = (A^\dagger)^{-1}BB^\dagger A^{-1} = (A^\dagger)^{-1}B^2 A^{-1} = (A^\dagger)^{-1}A^\dagger A A^{-1} = I_n.$$

6.3.12 From Problem 6.3.11, we may rewrite A as $A = CB$, where C and B are in $\mathbb{C}(n,n)$ such that C is unitary and B positive definite. Hence $A^\dagger A = B^2$. So if $\lambda_1,\ldots,\lambda_n > 0$ are the eigenvalues of $A^\dagger A$ then $\sqrt{\lambda_1},\ldots,\sqrt{\lambda_n}$ are those of B. Let $Q \in \mathbb{C}(n,n)$ be unitary such that $B = Q^\dagger DQ$ where D is given in (6.3.26). Then $A = CQ^\dagger DQ$. Let $P = CQ^\dagger$. We see that P is unitary and the problem follows.

6.3.13 Since T is positive definite, $(u,v)_T \equiv (u, T(v))$, $u, v \in U$, also defines a positive definite scalar product over U. So the problem follows from the Schwarz inequality stated in terms of this new scalar product.

Section 6.4

6.4.4 Assume that $A = (a_{ij})$ is upper triangular. Then we see that the diagonal entries of $A^\dagger A$ and $AA^\dagger$ are

$$|a_{11}|^2,\ldots, \sum_{j=1}^{i} |a_{ji}|^2,\ldots, \sum_{j=1}^{n} |a_{jn}|^2,$$

$$\sum_{j=1}^{n} |a_{1j}|^2,\ldots, \sum_{j=i}^{n} |a_{ij}|^2,\ldots, |a_{nn}|^2,$$

respectively. Comparing these entries in the equation $A^\dagger A = AA^\dagger$, we get $a_{ij} = 0$ for $i < j$. So A is diagonal.

6.4.6 (a) The conclusion $T = 0$ follows from the fact that all eigenvalues of T vanish and U has a basis consisting of eigenvectors of T.

(b) We have $(T' \circ T)^k(u) = 0$ since T' and T commute. Using Problem 6.2.8, we have $(T' \circ T)(u) = 0$. Hence $0 = (u, (T' \circ T)(u)) = (T(u), T(u))$ which gives us $T(u) = 0$.

6.4.10 Assume that there are positive semi-definite R and unitary S, in $L(U)$, such that

$$T = R \circ S = S \circ R. \tag{S22}$$

Then $T' = S' \circ R = R \circ S'$. So $T' \circ T = T \circ T'$ and T is normal. Now assume that T is normal. Then U has an orthonormal basis, say $\{u_1, \ldots, u_n\}$, consisting of eigenvectors of T. Set $T(u_i) = \lambda_i u_i$ ($i = 1, \ldots, n$). Make polar decompositions of these eigenvalues, $\lambda_i = |\lambda_i| e^{i\theta_i}$, $i = 1, \ldots, n$, with the convention that $\theta_i = 0$ if $\lambda_i = 0$. Define R and $S \in L(U)$ by setting

$$R(u_i) = |\lambda_i| u_i, \quad S(u_i) = e^{i\theta_i} u_i, \quad i = 1, \ldots, n.$$

Then it is clear that R is positive semi-definite, S unitary, and (S22) holds.

6.4.12 Assume that T is normal. Let $\lambda_1, \ldots, \lambda_k \in \mathbb{C}$ be all the distinct eigenvalues of T and $E_{\lambda_1}, \ldots, E_{\lambda_k}$ the corresponding eigenspaces. Then $U = E_{\lambda_1} \oplus \cdots \oplus E_{\lambda_k}$. Let $\{u_{i,1}, \ldots, u_{i,m_i}\}$ be an orthonormal basis of E_{λ_i} ($i = 1, \ldots, k$). Then $T'(u) = \overline{\lambda}_i u$ for any $u \in E_{\lambda_i}$ ($i = 1, \ldots, k$). On the other hand, using the Vandermonde determinant, we know that we can find $a_0, a_1, \ldots, a_{k-1} \in \mathbb{C}$ such that

$$a_0 + a_1 \lambda_i + \cdots + a_{k-1} \lambda_i^{k-1} = \overline{\lambda}_i, \quad i = 1, \ldots, k.$$

Now set $p(t) = a_0 + a_1 t + \cdots + a_{k-1} t^{k-1}$. Then $p(T)(u) = \overline{\lambda}_i u$ for any $u \in E_{\lambda_i}$ ($i = 1, \ldots, k$). Thus $T' = p(T)$.

6.4.13 Assume that T is normal. If $\dim(U) = n = 1$, there is nothing to do. Assume $n \geq 2$. Then the previous problem establishes that there is a polynomial $p(t)$ such that $T' = p(T)$. Hence any subspace invariant under T is also invariant under T'. However, $T' = p(T)$ implies $T = q(T')$ where q is the polynomial obtained from p by replacing the coefficients of p by the complex conjugates of the corresponding coefficients of p. Thus any invariant subspace of T' is also invariant under T.

Now assume that T and T' have the same invariant subspaces. We show that T is normal. We proceed inductively on $\dim(U) = n$. If $n = 1$, there is nothing to do. Assume that the conclusion is true at $n = k \geq 1$. We consider $n = k + 1$. Let $\lambda \in \mathbb{C}$ be an eigenvalue of T and $u \in U$ an associated eigenvector of T. Then $V = \text{Span}\{u\}$ is invariant under T and T'. We claim that $V^\perp$ is invariant under T and T' as well. In fact, take $w \in V^\perp$. Since $T'(u) = au$ for some $a \in \mathbb{C}$, we have $(u, T(w)) = (T'(u), w) = a(u, w) = 0$. So $T(w) \in V^\perp$. Hence $V^\perp$ is invariant under T and T' as well. Since $\dim(V^\perp) = n - 1 = k$, we see that $T \circ T' = T' \circ T$ on $V^\perp$. Besides, we have $T(T'(u)) = T(au) = a\lambda u$ and $T'(T(u)) = T'(\lambda u) = \lambda a u$. So $T \circ T' = T' \circ T$ on V. This proves that T is normal on U since $U = V \oplus V^\perp$.

Section 6.5

6.5.4 Since T is normal, we have $(T^2)' \circ T^2 = (T' \circ T)^2$ so that the eigenvalues of $(T^2)' \circ T^2$ are those of $T' \circ T$ squared. This proves $\|T^2\| = \|T\|^2$ in particular. Likewise, $(T^m)' \circ T^m = (T' \circ T)^m$ so that the eigenvalues of $(T^m)' \circ T^m$ are those of $T' \circ T$ raised to the mth power. So $\|T^m\| = \|T\|^m$ is true in general.

6.5.8 Note that some special forms of this problem have appeared as Problems 6.3.11 and 6.3.12 By the singular value decomposition for A, we may rewrite A as $A = P\Lambda Q$, where Q and $P \in \mathbb{C}(n,n)$ are unitary and $\Lambda \in \mathbb{R}(n,n)$ is diagonal whose diagonal entries are all non-negative. Alternatively, we also have $A = (PQ)(Q^\dagger \Lambda Q) = (P\Lambda P^\dagger)(PQ)$, as products of a unitary and positive semi-definite matrices, expressed in two different orders.

Section 7.1

7.1.4 Since $g \neq 0$, we also have $f \neq 0$. Assume $f^n | g^n$. We show that $f | g$. If $n = 1$, there is nothing to do. Assume $n \geq 2$. Let $h = \gcd(f,g)$. Then we have $f = hp$, $g = hq$, $p,q \in \mathcal{P}$, and $\gcd(p,q) = 1$. If p is a scalar, then $f | g$. Suppose otherwise that p is not a scalar. We rewrite f^n and g^n as $f^n = h^n p^n$ and $g^n = h^n q^n$. Since $f^n | g^n$, we have $g^n = f^n r$, where $r \in \mathcal{P}$. Hence $h^n q^n = h^n p^n r$. Therefore $q^n = p^n r$. In particular, $p | q^n$. However, since $\gcd(p,q) = 1$, we have $p | q^{n-1}$. Arguing repeatedly we arrive at $p | q$ which is a contradiction.

7.1.5 Let $h = \gcd(f,g)$. Then $f = hp$, $g = hq$, and $\gcd(p,q) = 1$. Thus $f^n = h^n p^n$, $g^n = h^n q^n$, and $\gcd(p^n, q^n) = 1$, which implies $\gcd(f^n, g^n) = h^n$.

Section 7.2

7.2.1 If $p_S(\lambda)$ are $p_T(\lambda)$ are relatively prime, then there are polynomials f and g such that $f(\lambda)p_S(\lambda) + g(\lambda)p_T(\lambda) = 1$. Thus, $I = f(T)p_S(T) + g(T)p_T(T) = f(T)p_S(T)$ which implies $p_S(T)$ is invertible and $p_S(T)^{-1} = f(T)$. Similarly we see that $p_T(S)$ is also invertible.

7.2.2 Using the notation of the previous problem, we have $I = f(T)p_S(T)$. Thus, applying $R \circ S = T \circ R$, we get

$$R = f(T)p_S(T) \circ R = R \circ f(S)p_S(S) = 0.$$

7.2.3 It is clear that $N(T) \subset R(I - T)$. It is also clear that $R(I - T) \subset N(T)$ when $T^2 = T$. So $N(T) = R(I-T)$ if $T^2 = T$. Thus (7.2.14) follows from the rank equation $r(T) + n(T) = \dim(U)$. Conversely, assume (7.2.14) holds. From this and the rank equation again, we get $r(I - T) = n(T)$. So $N(T) = R(I - T)$, which establishes $T^2 = T$.

7.2.4 We have $f_1 g_1 + f_2 g_2 = 1$ for some polynomials f_1 and f_2. So

$$I = f_1(T)g_1(T) + f_2(T)g_2(T). \qquad (S23)$$

As a consequence, for any $u \in U$, we have $u = v + w$, where $v = f_1(T)g_1(T)u$ and $w = f_2(T)g_2(T)u$. Hence $g_2(T)v = f_1(T)p_T(T)u = 0$ and $g_1(T)w = f_2(T)p_T(T)u = 0$. This shows $v \in N(g_2(T))$ and $w \in N(g_1(T))$. Therefore $U = N(g_1(T)) + N(g_2(T))$. Pick $u \in N(g_1(T)) \cap N(g_2(T))$. Applying (S23) to u, we see that $u = 0$. Hence the problem follows.

7.2.6 We proceed by induction on k. If $k = 1$, (7.2.16) follows from Problem 7.2.3. Assume that, at $k - 1 \geq 1$, the relation (7.2.16) holds. That is,

$$U = R(T_1) \oplus \cdots \oplus R(T_{k-1}) \oplus W, \quad W = N(T_1) \cap \cdots \cap N(T_{k-1}). \quad (S24)$$

Now we have $U = R(T_k) \oplus N(T_k)$. We assert $N(T_k) = R(T_1) \oplus \cdots \oplus R(T_{k-1}) \oplus V$. In fact, pick any $u \in N(T_k)$. Then (S24) indicates that $u = u_1 + \cdots + u_{k-1} + w$ for $u_1 \in R(T_1), \ldots, u_{k-1} \in R(T_{k-1}), w \in W$. Since $T_k \circ T_i = 0$ for $i = 1, \ldots, k - 1$, we see that $u_1, \ldots, u_{k-1} \in N(T_k)$. Hence $w \in N(T_k)$. So $w \in V$. This establishes the assertion and the problem follows.

Section 7.3

7.3.2 (b) From (7.3.41), we see that, if we set

$$p(\lambda) = \lambda^n - a_{n-1}\lambda^{n-1} - \cdots - a_1\lambda - a_0,$$

then $p(T)(T^k(u)) = T^k(p(T)(u)) = 0$ for $k = 0, 1, \ldots, n - 1$, where $T^0 = I$. This establishes $p(T) = 0$ since $\{T^{n-1}(u), \ldots, T(u), u\}$ is a basis of U. So $p_T(\lambda) = p(\lambda)$. It is clear that $m_T(\lambda) = p_T(\lambda)$.

7.3.3 Let u be a cyclic vector of T. Assume that S and T commute. Let $a_0, a_1, \ldots, a_{n-1} \in \mathbb{F}$ be such that $S(u) = a_{n-1}T^{n-1}(u) + \cdots + a_1 T(u) + a_0 u$. Set $p(t) = a_{n-1}t^{n-1} + \cdots + a_1 t + a_0$. Hence $S(T^k(u)) = T^k(S(u)) = (T^k p(T))(u) = p(T)(T^k(u))$ for $k = 1, \ldots, n - 1$. This proves $S = p(T)$ since $u, T(u), \ldots, T^{n-1}(u)$ form a basis of U.

7.3.4 (a) Let u be a cyclic vector of T. Since T is normal, U has a basis consisting of eigenvectors, say $u_1, \ldots, u_n$, of T, associated with the corresponding eigenvalues, $\lambda_1, \ldots, \lambda_n$. Express u as $u = a_1 u_1 + \cdots + a_n u_n$, for some $a_1, \ldots, a_n \in \mathbb{C}$. Hence

$$T(u) = a_1\lambda_1 u_1 + \cdots + a_n\lambda_n u_n,$$

$$\cdots\cdots\cdots\cdots\cdots\cdots$$

$$T^{n-1}(u) = a_1\lambda_1^{n-1}u_1 + \cdots + a_n\lambda_n^{n-1}u_n.$$

Inserting the above relations into the equation

$$x_1 u + x_2 T(u) + \cdots + x_n T^{n-1}(u) = 0, \qquad (S25)$$

we obtain

$$\begin{cases} a_1(x_1 + \lambda_1 x_2 + \cdots + \lambda_1^{n-1}x_n) &= 0, \\ a_2(x_1 + \lambda_2 x_2 + \cdots + \lambda_2^{n-1}x_n) &= 0, \\ \cdots\cdots\cdots\cdots\cdots\cdots\cdots & \cdots \quad \cdots \\ a_n(x_1 + \lambda_n x_2 + \cdots + \lambda_n^{n-1}x_n) &= 0. \end{cases} \qquad (S26)$$

If $\lambda_1,\ldots,\lambda_n$ are not all distinct, then (S26) has a solution $(x_1,\ldots,x_n) \in \mathbb{C}^n$, which is not the zero vector for any given $a_1,\ldots,a_n$, contradicting the linear independence of the vectors $u, T(u),\ldots,T^{n-1}(u)$.

(b) With (7.3.42), we consider the linear dependence of the vectors

$$u, T(u), \quad \ldots, \quad T^{n-1}(u),$$

as in part (a), and come up with (S25) and (S26). Since $\lambda_1,\ldots,\lambda_n$ are distinct, in view of the Vandermonde determinant, we see that the system (S26) has only the zero solution $x_1 = 0,\ldots,x_n = 0$, if and only if $a_i \neq 0$ for any $i = 1,\ldots,n$.

7.3.5 Assume there is such an S. Then S is nilpotent of degree m where m satisfies $2(n-1) < m \leq 2n$, since $S^{2(n-1)} = T^{n-1} \neq 0$ and $S^{2n} = T^n = 0$. By Theorem 2.22, we arrive at $2(n-1) < m \leq n$, which is false.

Section 7.4

7.4.5 Let the invertible matrix $C \in \mathbb{C}(n,n)$ be decomposed as $C = P + iQ$, where P and $Q \in \mathbb{R}(n,n)$. If $Q = 0$, there is nothing to show. Assume $Q \neq 0$. Then, from $CA = BC$, we have $PA = BP$ and $QA = BQ$. Thus, for any real number λ, we have $(P + \lambda Q)A = B(P + \lambda Q)$. It is clear that there is some $\lambda \in \mathbb{R}$ such that $\det(P + \lambda Q) \neq 0$. For such λ, set $K = P + \lambda Q$. Then $A = K^{-1}BK$.

7.4.9 (a) Let $u \in \mathbb{C}^n \setminus \{0\}$ be an eigenvector associated to the eigenvalue λ. Then $Au = \lambda u$. Thus $A^k u = \lambda^k u$. Since there is an invertible matrix

$B \in \mathbb{C}(n,n)$ such that $A^k = B^{-1}AB$, we obtain $(B^{-1}ABu) = \lambda^k u$ or $A(Bu) = \lambda^k(Bu)$, so that the problem follows.

(b) From (a), we see that if $\lambda \in \mathbb{C}$ is an eigenvalue, then $\lambda^k, \lambda^{k^2}, \ldots, \lambda^{k^l}$ are all eigenvalues of A, which cannot be all distinct when l is large enough. So there are some integers $1 \leq l < m$ such that $\lambda^{k^l} = \lambda^{k^m}$. Since A is non-singular, then $\lambda \neq 0$. Hence λ satisfies $\lambda^{k^{m-l}} = 1$ as asserted.

7.4.11 Use (3.4.37) without assuming $\det(A) \neq 0$ or (S16).

7.4.12 Take $u = (1,\ldots,1)^t \in \mathbb{R}^n$. It is clear that $Au = nu$. It is also clear that $n(A) = n - 1$, since $r(A) = 1$. So $n(A - nI_n) = 1$ and $A \sim$ diag$\{n,0,\ldots,0\}$. Take $v = (1, \frac{b_2}{n}, \ldots, \frac{b_n}{n})^t \in \mathbb{R}^n$. Then $Bv = nv$. Since $r(B) = 1$, we get $n(B) = n - 1$. So $n(B - nI_n) = 1$. Consequently, $B \sim$ diag$\{n,0,\ldots,0\}$. Thus $A \sim B$.

7.4.17 Suppose otherwise that there is an $A \in \mathbb{R}(3,3)$, such that $m(\lambda) = \lambda^2 + 3\lambda + 4$ is the minimal polynomial of A. Let $p_A(\lambda)$ be the characteristic polynomial of A. Since $p_A(\lambda) \in \mathcal{P}_3$ and the coefficients of $p_A(\lambda)$ are all real, so $p_A(\lambda)$ has a real root. On the other hand, recall that the roots of $m(\lambda)$ are all the roots of $p_A(\lambda)$, but the former has no real root. So we arrive at a contradiction.

7.4.18 (a) Let $m_A(\lambda)$ be the minimal polynomial of A. Then $m_A(\lambda) | \lambda^2 + 1$. However, $\lambda^2 + 1$ is prime over $\mathbb{R}$, so $m_A(\lambda) = \lambda^2 + 1$.

(b) Let $p_A(\lambda)$ be the characteristic polynomial of A. Then the degree of $p_A(\lambda)$ is n. Since $m_A(\lambda) = \lambda^2 + 1$ contains all the roots of $p_A(\lambda)$ in $\mathbb{C}$, which are $\pm i$, which must appear in conjugate pairs because $p_A(\lambda)$ is of real coefficients, so $p_A(\lambda) = (\lambda^2 + 1)^m$ for some integer $m \geq 1$. Hence $n = 2m$.

(c) Since $p_A(\lambda) = (\lambda - i)^m(\lambda + i)^m$ and $m_A(\lambda) = (\lambda - i)(\lambda + i)$ (i.e., $\pm i$ are single roots of $m_A(\lambda)$), we know that

$$N(A - iI_n) = \{x \in \mathbb{C}^n \mid Ax = ix\},$$
$$N(A + iI_n) = \{y \in \mathbb{C}^n \mid Ay = -iy\}$$

are both of dimension m in $\mathbb{C}^n$ and $\mathbb{C}^n = N(A - iI_n) \oplus N(A + iI_n)$. Since A is real, we see that if $x \in N(A - iI_n)$, then $\bar{x} \in N(A + iI_n)$, and vice versa. Moreover, if $\{w_1, \ldots, w_m\}$ is a basis of $N(A - iI_n)$, then $\{\bar{w}_1, \ldots, \bar{w}_m\}$ is a basis of $N(A + iI_n)$, and vice versa. Thus $\{w_1, \ldots, w_m, \bar{w}_1, \ldots, \bar{w}_m\}$ is a basis of $\mathbb{C}^n$. We now make the decomposition

$$w_i = u_i + iv_i, \quad u_i, v_i \in \mathbb{R}^n, \quad i = 1, \ldots, m. \tag{S27}$$

Then u_i, v_i ($i = 1,\ldots,m$) satisfy (7.4.23). It remains to show that these vectors are linearly independent in $\mathbb{R}^n$. In fact, consider

$$\sum_{i=1}^{m} a_i u_i + \sum_{i=1}^{m} b_i v_i = 0, \quad a_1,\ldots,a_m,b_1,\ldots,b_m \in \mathbb{R}. \quad (S28)$$

From (S27) we have

$$u_i = \frac{1}{2}(w_i + \overline{w}_i), \quad v_i = \frac{1}{2i}(w_i - \overline{w}_i), \quad i = 1,\ldots,m. \quad (S29)$$

Inserting (S29) into (S28), we obtain

$$\sum_{i=1}^{m}(a_i - ib_i)w_i + \sum_{i=1}^{m}(a_i + ib_i)\overline{w}_i = 0,$$

which leads to $a_i = b_i = 0$ for all $i = 1,\ldots,m$.

(d) Take the ordered basis $\mathcal{B} = \{u_1,\ldots,u_m,v_1,\ldots,v_m\}$. Then it is seen that, with respect to $\mathcal{B}$, the matrix representation of the mapping $T_A \in L(\mathbb{R}^n)$ defined by $T_A(u) = Au, u \in \mathbb{R}^n$ is simply

$$C = \begin{pmatrix} 0 & I_m \\ -I_m & 0 \end{pmatrix}.$$

More precisely, if a matrix called B is formed by using the vectors in the ordered basis $\mathcal{B}$ as its first, second, $\ldots$, and the nth column vectors, then $AB = BC$, which establishes (7.4.24).

Section 8.1

8.1.5 If A is normal, there is a unitary matrix $P \in \mathbb{C}(n,n)$ such that $A = P^\dagger DP$, where D is a diagonal matrix of the form diag$\{\lambda_1,\ldots,\lambda_n\}$ with $\lambda_1,\ldots,\lambda_n$ the eigenvalues of A which are assumed to be real. Thus $A^\dagger = A$. However, since A is real, we have $A = A^t$.

8.1.6 We proceed inductively on dim(U). If dim$(U) = 1$, there is nothing to show. Assume that the problem is true at dim$(U) = n - 1 \geq 1$. We prove the conclusion at dim$(U) = n \geq 2$. Let $\lambda \in \mathbb{C}$ be an eigenvalue of T and E_λ the associated eigenspace of T. Then, for $u \in E_\lambda$, we have $T(S(u)) = S(T(u)) = \lambda S(u)$. Hence $S(u) \in E_\lambda$. So E_λ is invariant under S. As an element in $L(E_\lambda)$, S has an eigenvalue $\mu \in \mathbb{C}$. Let $u \in E_\lambda$ be an eigenvector of S associated with μ. Then u is a common eigenvector of S and T. Applying this observation to S' and T', since S' and T' commute as well, we know that S' and T' also have a common eigenvector, say

w, satisfying $S'(w) = \sigma w$ and $T'(w) = \gamma w$, for some σ and $\gamma \in \mathbb{C}$. Let $V = (\text{Span}\{w\})^{\perp}$. Then V is invariant under S and T, as can be seen from

$$(w, S(v)) = (S'(w), v) = (\sigma w, v) = \overline{\sigma}(w, v) = 0,$$
$$(w, T(v)) = (T'(w), v) = (\gamma w, v) = \overline{\gamma}(w, v) = 0,$$

for $v \in V$. Since $\dim(V) = n - 1$, we may find an orthonormal basis of V, say $\{u_1, \dots, u_{n-1}\}$, under which the matrix representations of S and T are upper triangular. Let $u_n = w / \|w\|$. Then $\{u_1, \dots, u_{n-1}, u_n\}$ is an orthonormal basis of U, under which the matrix representations of S and T are upper triangular.

8.1.9 Let $\lambda \in \mathbb{C}$ be any eigenvalue of T and $v \in U$ an associated eigenvector. Then we have

$$((T' - \overline{\lambda} I)(u), v) = (u, (T - \lambda I)(v)) = 0, \quad u \in U, \qquad \text{(S30)}$$

which implies that $R(T' - \overline{\lambda} I) \subset (\text{Span}\{v\})^{\perp}$. Therefore $r(T' - \overline{\lambda} I) \le n - 1$. Thus, in view of the rank equation, we obtain $n(T' - \overline{\lambda} I) \ge 1$. In other words, this shows that $\overline{\lambda}$ must be an eigenvalue of T'.

Section 8.2

8.2.3 Let $\mathcal{B} = \{u_1, \dots, u_n\}$ be a basis of U and x and $y \in \mathbb{F}^n$ the coordinate vectors of u and $v \in U$, with respect to $\mathcal{B}$. With $A = (a_{ij}) = (f(u_i, u_j))$ and (8.2.2), we see that $u \in U_0$, if and only if $x^t A y = 0$ for all $y \in \mathbb{F}^n$ or $Ax = 0$. In other words, $u \in U_0$, if and only if $x \in N(A)$. So $\dim(U_0) = n(A) = n - r(A) = \dim(U) - r(A)$.

8.2.4 (a) As in Problem 8.2.3, we use $\mathcal{B} = \{u_1, \dots, u_n\}$ to denote a basis of U and x and $y \in \mathbb{F}^n$ the coordinate vectors of any vectors u and $v \in U$, with respect to $\mathcal{B}$. With $A = (a_{ij}) = (f(u_i, u_j))$ and (8.2.2), we see that $u \in V^{\perp}$, if and only if

$$(Ax)^t y = 0, \quad v \in V. \qquad \text{(S31)}$$

Let $\dim(V) = m$. We can find m linearly independent vectors $y^{(1)}, \dots, y^{(m)}$ in $\mathbb{F}^n$ to replace the condition (S31) by

$$(y^{(1)})^t (Ax) = 0, \dots, (y^{(m)})^t (Ax) = 0.$$

These equations indicate that, if we use B to denote the matrix formed by taking $(y^{(1)})^t, \dots, (y^{(m)})^t$ as its first, ..., and mth row vectors, then $Ax \in N(B) = \{z \in \mathbb{F}^n \mid Bz = 0\}$. In other words, the subspace of $\mathbb{F}^n$ consisting of the coordinate vectors of the vectors in $V^{\perp}$ is given by $X = \{x \in \mathbb{F}^n \,|\, Ax \in N(B)\}$. Since A is invertible, we have $\dim(X) = \dim(N(B)) = n(B) = n - r(B) = n - m$. This establishes $\dim(V^\perp) = \dim(X) = \dim(U) - \dim(V)$.

(b) For $v \in V$, we have $f(u, v) = 0$ for any $u \in V^\perp$. So $V \subset (V^\perp)^\perp$. On the other hand, from (i), we get

$$\dim(V) + \dim(V^\perp) = \dim(V^\perp) + \dim((V^\perp)^\perp).$$

So, $\dim(V) = \dim((V^\perp)^\perp)$, which implies $V = (V^\perp)^\perp$.

8.2.7 If $V = V^\perp$, then V is isotropic and $\dim(V) = \frac{1}{2}\dim(U)$, in view of Problem 8.2.4. Thus V is Lagrangian. Conversely, if V is Lagrangian, then V is isotropic such that $V \subset V^\perp$ and $\dim(V) = \frac{1}{2}\dim(U)$. From Problem 8.2.4, we have $\dim(V^\perp) = \frac{1}{2}\dim(U) = \dim(V)$. So $V = V^\perp$.

Section 8.3

8.3.1 Let $u = (a_i) \in \mathbb{R}^n$ be a positive eigenvector associated to r. Then the ith component of the relation $Au = ru$ reads

$$ra_i = \sum_{j=1}^{n} a_{ij}a_j, \quad i = 1, \ldots, n. \tag{S32}$$

Choose $k, l = 1, \ldots, n$ such that

$$a_k = \min\{a_i \,|\, i = 1, \ldots, n\}, \quad a_l = \max\{a_i \,|\, i = 1, \ldots, n\}.$$

Inserting these into (S32), we find

$$ra_k \geq a_k \sum_{j=1}^{n} a_{kj}, \quad ra_l \leq a_l \sum_{j=1}^{n} a_{lj}.$$

From these, we see that the bounds stated in (8.3.27) follow.

8.3.3 Use the notation $\Lambda_A = \{\lambda \in \mathbb{R} \,|\, \lambda \geq 0, Ax \geq \lambda x \text{ for some } x \in S\}$, where S is defined by (8.3.3). Recall the construction (8.3.7). We see that $r_A = \sup\{\lambda \in \Lambda_A\}$. Since $A \leq B$ implies $\Lambda_A \subset \Lambda_B$, we deduce $r_A \leq r_B$.

Section 8.4

8.4.4 From $\lim_{m \to \infty} A^m = K$, we obtain $\lim_{m \to \infty} (A^t)^m = \lim_{m \to \infty} (A^m)^t = K^t$. Since all the row vectors of K and K^t are identical, we see that all entries of K are identical. By the condition (8.4.13), we deduce (8.4.24).

8.4.5 It is clear that all the entries of A and A^t are non-negative. It remains to show that 1 and $u = (1,\dots,1)^t \in \mathbb{R}^n$ are a pair of eigenvalue and eigenvector of both A and A^t. In fact, applying $A_i u = u$ and $A_i^t u = u$ $(i = 1,\dots,k)$ consecutively, we obtain $Au = A_1 \cdots A_k u = u$ and $A^t u = A_k^t \cdots A_1^t u = u$, respectively.

Section 9.4

9.4.2 (a) In the uniform state (9.4.30), we have

$$\langle A \rangle = \frac{1}{n} \sum_{i=1}^{n} \lambda_i, \quad \langle A^2 \rangle = \frac{1}{n} \sum_{i=1}^{n} \lambda_i^2.$$

Hence, the uncertainty σ_A of the observable A in the state (9.4.30) is given by the formula

$$\sigma_A^2 = \langle A^2 \rangle - \langle A \rangle^2 = \frac{1}{n} \sum_{i=1}^{n} \lambda_i^2 - \left(\frac{1}{n} \sum_{i=1}^{n} \lambda_i \right)^2. \tag{S33}$$

(b) From (9.4.25) and (S33), we obtain the comparison

$$\frac{1}{n} \sum_{i=1}^{n} \lambda_i^2 - \left(\frac{1}{n} \sum_{i=1}^{n} \lambda_i \right)^2 \leq \frac{1}{4} (\lambda_{\max} - \lambda_{\min})^2.$$

Section 10.1

10.1.6 For notational convenience, let $y = x'$. Then, with

$$K(x,y) = (x^t y + a)^2, \quad x = (x_i), y = (y_i) \in \mathbb{R}^2,$$

we have

$$\begin{aligned}
K(x,y) &= (x_1 y_1 + x_2 y_2 + a)^2 \\
&= x_1^2 y_1^2 + x_2^2 y_2^2 + 2x_1 x_2 y_1 y_2 + 2ax_1 y_1 + 2ax_2 y_2 + a^2 \\
&= x_1^2 y_1^2 + x_2^2 y_2^2 + \sqrt{2} x_1 x_2 \sqrt{2} y_1 y_2 + \sqrt{2a} x_1 \sqrt{2a} y_1 \\
&\quad + \sqrt{2a} x_2 \sqrt{2a} y_2 + a \cdot a \\
&= \phi^t(x) \phi(y),
\end{aligned}$$

where

$$\phi(x) = (x_1^2, x_2^2, \sqrt{2} x_1 x_2, \sqrt{2a} x_1, \sqrt{2a} x_2, a)^t.$$

So, it is seen that, when $a > 0$, we have $m = 6$; when $a = 0$, which is the homogeneous situation, the map ϕ is simplified into the form

$$\phi(x) = (x_1^2, x_2^2, \sqrt{2}x_1 x_2)^t,$$

such that we have $m = 3$.

Section 10.3

10.3.3 To solve this problem, we first state and prove the *Carathéodory theorem* that any $x \in H_A$ may be viewed as a convex combination of $m + 1$ elements in A.

In fact, given $x \in H_A$, we have

$$x = \sum_{i=0}^{k} a_i u_i, \quad a_0, a_1, \ldots, a_k \geq 0, \quad \sum_{i=0}^{k} a_i = 1,$$

for some $u_0, u_1, \ldots, u_k \in A$. Of course, we may assume that this sum is in a nontrivial form such that each a_i in the sum is positive. If $k \leq m$, there is nothing to do. Assume $k > m$. Then the k vectors

$$u_1 - u_0, \quad \ldots, \quad u_k - u_0,$$

are linearly dependent. Thus there is a nonzero vector $(b_1, \ldots, b_k) \in \mathbb{R}^k$ such that

$$\sum_{i=1}^{k} b_i (u_i - u_0) = 0,$$

which may be rewritten as

$$\sum_{i=0}^{k} b_i u_i = 0, \quad b_0 = \sum_{i=1}^{k} b_i.$$

Making a sign flipping if necessary, we may assume there is some $i \in \{0, 1, \ldots, k\}$ such that $b_i > 0$. With this fact in mind, consider

$$x = \sum_{i=0}^{k} a_i u_i - r \sum_{i=0}^{k} b_i u_i = 0$$

$$= \sum_{i=0}^{k} (a_i - r b_i) u_i = 0,$$

where $r > 0$ is to be chosen suitably. It is clear that

$$a_i - r b_i \geq 0, \quad i = 0, 1, \ldots, k,$$

if $r > 0$ is sufficiently small. Now make $r > 0$ the largest possible quantity such that the above holds true. Then there is some $j \in \{0, 1, \ldots, k\}$ such that

$$a_j - rb_j = 0; \quad a_i - rb_i \geq 0, \quad i \in \{0, 1, \ldots, k\}.$$

In other words, the convex combination of x is shortened into the form

$$x = \sum_{i=0}^{k} a_i u_i, \quad a_0, a_1, \ldots, a_k \geq 0, \quad \sum_{i=0}^{k} a_i = 1, \quad a_j = 0.$$

We can repeat this procedure within a few steps to reach $k \leq m$. So the theorem is proved.

We now prove that H_A is compact, if A is so.

To this end, we show that any sequence in H_A, say $\{x_n\}$, has a convergent subsequence with limit in H_A. To see this, we use the notation

$$x_n = \sum_{i=0}^{m} a_{n,i} u_{n,i}, \quad a_{n,i} \geq 0, i = 0, 1, \ldots, m, \quad \sum_{i=0}^{m} a_{n,i} = 1,$$
$$u_{n,i} \in A, i = 0, 1, \ldots, m; \quad n = 1, 2, \ldots.$$

By extracting a subsequence if necessary, we may assume that each of the sequences

$$\{a_{n,i}\}, \quad i = 0, 1, \ldots, m,$$

is convergent. Thus, setting

$$a_{0,i} = \lim_{n \to \infty} a_{n,i}, \quad i = 0, 1, \ldots, m,$$

we have

$$a_{0,i} \geq 0, \quad i = 0, 1, \ldots, m, \quad \sum_{i=0}^{m} a_{0,i} = 1.$$

Furthermore, by extracting a suitable subsequence if necessary, we may assume that each of the sequences

$$\{u_{n,i}\}, \quad i = 0, 1, \ldots, m,$$

is convergent so that there are $u_{0,i}, u_{1,i}, \ldots, u_{m,i} \in A$ such that

$$u_{0,i} = \lim_{n \to \infty} u_{n,i}, \quad i = 0, 1, \ldots, m.$$

These establish the convergence of the sequence $\{x_n\}$ in H_A, since its limit

$$u_0 = \sum_{i=0}^{m} a_{0,i} u_{0,i}$$

has been shown to lie in H_A. Hence the compactness of H_A follows.

Section 11.2

11.2.7 Assume that $A = (a_{ij})$ is doubly substochastic, but not doubly stochastic. Then, there is some $i = 1, \ldots, n$, such that

$$\sum_{k=1}^{n} a_{ik} < 1,$$

or some $j = 1, \ldots, n$ such that

$$\sum_{k=1}^{n} a_{kj} < 1.$$

That is, the ith row or jth column of A does not sum up to 1. If the former is true, then the sum of all rows of A is less than n, which implies that the sum of all columns of A is also less than n. Consequently, there is a column of A for which the sum of its entries is less than 1, so that the latter in the above is also true. In other words, there is a position (i, j) in the matrix A, whose ith row and jth column do not sum up to 1. Now set

$$\varepsilon = \min \left\{ 1 - \sum_{k=1}^{n} a_{ik}, 1 - \sum_{k=1}^{n} a_{kj} \right\},$$

and define $\tilde{A} \in \mathbb{R}(n, n)$, such that the (i, j)th entry of $\tilde{A}$ is

$$\tilde{a}_{ij} = a_{ij} + \varepsilon,$$

and all other entries of $\tilde{A}$ are the same as those of A. Then $A \leq \tilde{A}$, $\tilde{A}$ is substochastic, and $\tilde{A}$ has one more row or column, whose entries sum up to 1. Thus, after a few such steps, we make A into a doubly stochastic matrix.

11.2.8. We rewrite the left-hand side of (11.2.27) as

$$c_{11} \sum_{i=1}^{k} a_i \sum_{j=1}^{k} b_j + c_{12} \sum_{i=1}^{k} a_i \sum_{j=2}^{k} b_j + \cdots + c_{1k} \left(\sum_{i=1}^{k} a_i \right) b_k$$

$$+ c_{21} \sum_{i=2}^{k} a_i \sum_{j=1}^{k} b_j + c_{22} \sum_{i=2}^{k} a_i \sum_{j=2}^{k} b_j + \cdots + c_{2k} \left(\sum_{i=2}^{k} a_i \right) b_k$$

$$+ \cdots \cdots$$

$$+ c_{k1} a_k \sum_{j=1}^{k} b_j + c_{k2} a_k \sum_{j=2}^{k} b_j + \cdots + c_{kk} a_k b_k$$

$$= a_1 b_1 c_{11} + a_1 b_2 (c_{11} + c_{12}) + \cdots + a_1 b_k (c_{11} + c_{12} + \cdots + c_{1k})$$
$$+ a_2 b_1 (c_{11} + c_{21}) + a_2 b_2 (c_{11} + c_{12} + c_{21} + c_{22}) + \cdots$$
$$+ a_2 b_k (c_{11} + c_{12} + \cdots + c_{1k} + c_{21} + c_{22} + \cdots + c_{2k})$$
$$+ \cdots \cdots$$
$$+ a_k b_1 (c_{11} + c_{21} + \cdots + c_{k1})$$
$$+ a_k b_2 (c_{11} + c_{21} + \cdots + c_{k1} + c_{12} + c_{22} + \cdots + c_{k2})$$
$$+ \cdots \cdots$$
$$+ a_k b_k (c_{11} + c_{12} + \cdots + c_{1k} + c_{21} + c_{22} + \cdots + c_{2k}$$
$$+ \cdots + c_{k1} + c_{k2} + \cdots + c_{kk})$$
$$= \sum_{1 \le i, j \le k} a_i b_j \sum_{1 \le r \le i, 1 \le s \le j} c_{rs},$$

which agrees with the right-hand side of (11.2.27).

Section 12.4

12.4.8 Let $\{u_1, \dots, u_n\}$ be a basis of U and set $T = \mathrm{sym}(f \otimes g)$. Then, by (12.4.11), we have

$$T(u_i, u_j) = \frac{1}{2}((f \otimes g)(u_i, u_j) + (f \otimes g)(u_j, u_i))$$
$$= \frac{1}{2}(f(u_i)g(u_j) + f(u_j)g(u_i))$$
$$= \frac{1}{2}(f \otimes g + g \otimes f)(u_i, u_j), \quad i, j = 1, \dots, n.$$

That is, $T = \frac{1}{2}(f \otimes g + g \otimes f)$. Besides, let $T = \mathrm{antisym}(f \otimes g)$. Then, by (12.4.13), we have

$$T(u_i, u_j) = \frac{1}{2}((f \otimes g)(u_i, u_j) - (f \otimes g)(u_j, u_i))$$
$$= \frac{1}{2}(f(u_i)g(u_j) - f(u_j)g(u_i))$$
$$= \frac{1}{2}(f \otimes g - g \otimes f)(u_i, u_j), \quad i, j = 1, \dots, n.$$

That is, $T = \frac{1}{2}(f \otimes g - g \otimes f)$.

Let T be an order 2 tensor over a two-dimensional space whose coefficient matrix with respect to a given basis reads

$$A = \begin{pmatrix} 1 & 1 \\ 0 & 0 \end{pmatrix}.$$

Since $r(A) = 1$, we know by Problem 12.4.5 that T is decomposable. The coefficient matrices of $\text{sym}(T)$ and $\text{antisym}(T)$, with respect to the same basis, are now given by

$$B = \text{sym}(A) = \frac{1}{2}(A + A^t) = \begin{pmatrix} 1 & 1 \\ 0 & 0 \end{pmatrix} = \frac{1}{2}\begin{pmatrix} 2 & 1 \\ 1 & 0 \end{pmatrix},$$

$$C = \text{antisym}(A) = \frac{1}{2}(A - A^t) = \begin{pmatrix} 1 & 1 \\ 0 & 0 \end{pmatrix} = \frac{1}{2}\begin{pmatrix} 0 & 1 \\ -1 & 0 \end{pmatrix}.$$

Since $r(B) = r(C) = 2$, we see that $\text{sym}(T)$ and $\text{antisym}(T)$ are not decomposable.

Section 12.5

12.5.15 We use the 2-form

$$h = e^1 \otimes e^1 - e^2 \otimes e^2 - \cdots - e^n \otimes e^n$$

to equip $\mathbb{R}^{n-1,1}$ with a scalar product $(\cdot, \cdot)$ such that $\{e^1, e^2, \ldots, e^n\}$ is an orthonormal basis satisfying

$$(e^1, e^1) = 1; \quad (e^i, e^i) = -1, \quad i = 2, \ldots, n.$$

With this, we derive from (12.5.22) the property

$$\star e^1 = e^2 \wedge \cdots \wedge e^n; \quad \star e^i = (-1)^i e^1 \wedge \cdots \wedge \widehat{e^i} \wedge \cdots \wedge e^n, \quad i = 2, \ldots, n.$$

Besides, with $\Omega = e^1 \wedge e^2 \wedge \cdots \wedge e^n$ and (12.5.13), we have

$$(\Omega, \Omega) = (-1)^{n-1}.$$

Therefore, from (12.5.22), we arrive at

$$\Omega \wedge \star \Omega = (\Omega, \Omega)\Omega,$$

which gives us the result $\star \Omega = (-1)^{n-1}$.

Now we consider differential forms. Let f be a differentiable function. Then we have

$$\star df = \star \left(\sum_{i=1}^{n} \partial_i f dx^i \right)$$

$$= \partial_1 f dx^2 \wedge \cdots \wedge dx^n + \sum_{i=2}^{n} \partial_i f (-1)^i dx^1 \wedge \cdots \wedge \widehat{dx^i} \wedge \cdots \wedge dx^n,$$

$$d \star df = \partial_1^2 f dx^1 \wedge dx^2 \wedge \cdots \wedge dx^n$$

$$+ \sum_{i=2}^{n} \partial_i^2 f (-1)^i dx^i \wedge dx^1 \wedge \cdots \wedge \widehat{dx^i} \wedge \cdots \wedge dx^n$$

$$= \left(\partial_1^2 f - \sum_{i=2}^{n} \partial_i^2 f \right) dx^1 \wedge dx^2 \wedge \cdots \wedge dx^n.$$

Thus, using $\star dx^1 \wedge dx^2 \wedge \cdots \wedge dx^n = (-1)^{n-1}$, we obtain

$$\star d \star df = (-1)^{n-1} \left(\partial_1^2 f - \sum_{i=2}^{n} \partial_i^2 f \right) = (-1)^{n-1} \square f.$$

That is, we have the following representation for the d'Alembert operator:

$$\square = \partial_1^2 - \sum_{i=2}^{n} \partial_i^2 = (-1)^{n-1} \star d \star d,$$

over the n-dimensional Minkowski space $\mathbb{R}^{n-1,1}$.

With $\delta = \star d \star$, we still have $d \star f = 0$ for any scalar function f, since $\star 1 = dx^1 \wedge \cdots \wedge dx^n$. Hence

$$\square = \partial_1^2 - \sum_{i=2}^{n} \partial_i^2 = (-1)^{n-1} (\delta d + d \delta),$$

over functions.

Section 13.2

13.2.8 Let $u = (a_i) = Ax$ and $v = (b_i) = By$ be given in (13.2.24). Following (13.2.13), we have

$$u \otimes v = \begin{pmatrix} a_1 v \\ \vdots \\ a_n v \end{pmatrix} = \begin{pmatrix} (a_{11}x_1 + \cdots + a_{1n}x_n)By \\ \vdots \\ (a_{n1}x_1 + \cdots + a_{nn}x_n)By \end{pmatrix}$$

$$= \begin{pmatrix} a_{11}Bx_1y + \cdots + a_{1n}Bx_ny \\ \vdots \\ a_{n1}Bx_1y + \cdots + a_{nn}Bx_ny \end{pmatrix}$$

$$= \begin{pmatrix} a_{11}B & \cdots & a_{1n}B \\ \vdots & \ddots & \vdots \\ a_{n1}B & \cdots & a_{nn}B \end{pmatrix} \begin{pmatrix} x_1y \\ \vdots \\ x_ny \end{pmatrix}$$

$$= (A \otimes B)(x \otimes y),$$

as given in (13.2.24) and (13.2.25).

Besides, following (13.2.14), we have

$$u \otimes v = \begin{pmatrix} ub_1 \\ \vdots \\ ub_n \end{pmatrix} = \begin{pmatrix} Ax(b_{11}y_1 + \cdots + b_{1n}y_n) \\ \vdots \\ Ax(b_{n1}y_1 + \cdots + b_{nn}y_n) \end{pmatrix}$$

$$= \begin{pmatrix} Ab_{11}xy_1 + \cdots + Ab_{1n}xy_n \\ \vdots \\ Ab_{n1}xy_1 + \cdots + Ab_{nn}xy_n \end{pmatrix}$$

$$= \begin{pmatrix} Ab_{11} & \cdots & Ab_{1n} \\ \vdots & \ddots & \vdots \\ Ab_{n1} & \cdots & Ab_{nn} \end{pmatrix} \begin{pmatrix} xy_1 \\ \vdots \\ xy_n \end{pmatrix}$$

$$= (A \otimes B)(x \otimes y),$$

as given in (13.2.24) and (13.2.26).

Bibliographic Notes

We end the book by mentioning a few important but more specialized subjects that are not touched in this book. We only point out some relevant references for the interested.

Convex sets. In Lang [46] basic properties and characterizations of convex sets in $\mathbb{R}^n$ are presented. For a deeper study of convex sets using advanced tools such as the Hahn–Banach theorem, see Lax [49].

Tensor products and alternating forms. These topics are covered elegantly in Halmos [36]. In particular, there, the determinant is seen to arise as the unique scalar, associated with each linear mapping, defined by the one-dimensional space of top-degree alternating forms, over a finite-dimensional vector space.

Minmax principle for computing the eigenvalues of self-adjoint mappings. This is a classical variational resolution of the eigenvalue problem known as the method of the Rayleigh–Ritz quotients. For a thorough treatment, see Bellman [12], Lancaster and Tismenetsky [45], and Lax [49].

Calculus of matrix-valued functions. These techniques are useful and powerful in applications. For an introduction, see Lax [49].

Irreducible matrices. Such a notion is crucial for extending the Perron–Frobenius theorem and for exploring the Markov matrices further under more relaxed conditions. See Berman and Plemmons [13], Horn and Johnson [42], Lancaster and Tismenetsky [45], Meyer [57], and Xu [81] for related studies.

Transformation groups and bilinear forms. Given a nondegenerate bilinear form over a finite-dimensional vector space, the set of all linear mappings on the space which preserve the bilinear form is a group under the operation of composition. With a specific choice of the bilinear form, a particular such transformation group may thus be constructed and investigated. For a concise introduction to this subject in the context of linear algebra, see Hoffman and Kunze [40].

Computational methods for solving linear systems. Practical methods for solving systems of linear equations are well investigated and documented. See Lax [49], Golub and Ortega [30], and Stoer and Bulirsch [70] for a description of some of the methods.

Computing the eigenvalues of symmetric matrices. This is a much developed subject and many nice methods are available. See Lax [49] for methods based on the QR factorization and differential flows. See also Stoer and Bulirsch [70].

Computing the eigenvalues of general matrices. Some nicely formulated iterative methods may be employed to approximate the eigenvalues of a general matrix under certain conditions. These methods include the QR convergence algorithm and the power method. See Golub and Ortega [30] and Stoer and Bulirsch [70].

Random matrices. The study of random matrices, the matrices whose entries are random variables, was pioneered by Wigner [79, 80] to model the spectra of large atoms and has recently become the focus of active mathematical research. See Akemann, Baik, and Di Francesco [3], Anderson, Guionnet, and Zeitouni [5], Mehta [56], and Tao [72], for textbooks, and Beenakker [9], Diaconis [23], and Guhr, Müller-Groeling, and Weidenmüller [35], for survey articles.

Besides, for a rich variety of applications of linear algebra and its related studies, see Bai, Fang, and Liang [7], Bapat [8], Bellman [12], Berman and Plemmons [13], Berry, Dumais, and O'Brien [14], Brualdi and Ryser [17], Datta [20], Davis [21], Gomide *et al* [31], Graham [32], Graybill [33], Horadam [41], Latouche and Vaidyanathan [48], Leontief [50], Lyubich, Akin, Vulis, and Karpov [51], Meyn and Tweedie [58], Stinson [69], Taubes [73], Van Dooren and Wyman [74], and references therein.

Notes added for the second edition

The subjects of the four chapters added for the second edition are vast with rich literature. In addition to those already referenced in the text, here are some more of interest. For machine learning, see [2, 28]; for data science, see [37, 71]; see also [26] for a new book on both subjects and signal processing; for multilinear algebra, see [1, 18, 47, 65]; and for linear algebra and quantum information and quantum computation, see [38, 43, 44, 61, 76, 84].

Bibliography

[1] R. Abraham, *Linear and Multilinear Algebra*, W. A. Benjamin Publishing, New York, 1966.

[2] C. C. Aggarwal, *Linear Algebra and Optimization for Machine Learning: A Textbook*, Springer-Verlag, New York, 2020.

[3] G. Akemann, J. Baik, and P. Di Francesco, *The Oxford Handbook of Random Matrix Theory*, Oxford University Press, Oxford, 2011.

[4] E. Alpaydin, *Introduction to Machine Learning*, MIT Press, Cambridge, MA, 2004.

[5] G. W. Anderson, A. Guionnet, and O. Zeitouni, *An Introduction to Random Matrices*, Cambridge University Press, Cambridge, 2010.

[6] A. Aspect, J. Dalibard, and G. Roger, Experimental test of Bell's inequalities using time-varying analyzers, *Phys. Rev. Lett.* **49** (1982) 1804–1987.

[7] Z. Bai, Z. Fang, and Y.-C. Liang, *Spectral Theory of Large Dimensional Random Matrices and Its Applications to Wireless Communications and Finance Statistics*, World Scientific, Singapore, 2014.

[8] R. B. Bapat, *Linear Algebra and Linear Models*, 3rd ed., Universitext, Springer-Verlag and Hindustan Book Agency, New Delhi, 2012.

[9] C. W. J. Beenakker, Random-matrix theory of quantum transport, *Rev. Mod. Phys.* **69** (1997) 731–808.

[10] J. S. Bell, On the Einstein–Podolsky–Rosen paradox, *Physics* **1** (1964) 195–200.

[11] J. S. Bell, *Speakable and Unspeakable in Quantum Mechanics: Collected Papers on Quantum Philosophy*, 2nd ed., Cambridge University Press, Cambridge, 2004.

[12] R. Bellman, *Introduction to Matrix Analysis*, 2nd ed., Society of Industrial and Applied Mathematics, Philadelphia, 1997.

[13] A. Berman and R. J. Plemmons, *Non-negative Matrices in the Mathematical Sciences*, Society of Industrial and Applied Mathematics, Philadelphia, 1994.

[14] M. Berry, S. Dumais, and G. O'Brien, Using linear algebra for intelligent information retrieval, *SIAM Rev.* **37** (1995) 573–595.

[15] R. Bott and L. W. Tu, *Differential Forms in Algebraic Topology*, Springer-Verlag, Berlin and New York, 1982.

[16] S. Boyd and L. Vandenberghe, *Convex Optimization*, Cambridge University Press, Cambridge, 2004.

[17] R. A. Brualdi and H. J. Ryser, *Combinatorial Matrix Theory*, Encyclopedia of Mathematics and Its Applications **39**, Cambridge University Press, Cambridge, 1991.

[18] E. Castillo and J. R. Ruiz-Tolosa, *From Vectors to Tensors* (Universitext), Springer-Verlag, New York, 2004.

[19] J. F. Clauser, M. A. Horne, A. Shimony, and R. A. Holt, Proposed experiment to test local hidden-variable theories, *Phys. Rev. Lett.* **23** (1969) 880–884. Erratum, *ibid* **24** (1970) 547.

[20] B. N. Datta, *Numerical Linear Algebra and Applications*, 2nd ed., Society of Industrial and Applied Mathematics, Philadelphia, 2010.

[21] E. Davis, *Linear Algebra and Probability for Computer Science Applications*, A. K. Peters/CRC Press, Boca Raton, FL, 2012.

[22] D. Dehlinger and M. W. Mitchell, Entangled photons, nonlocality and Bell inequalities in the undergraduate laboratory, *Am. J. Phys.* **70** (2002) 903–910.

[23] P. Diaconis, Patterns in eigenvalues: the 70th Josiah Willard Gibbs lecture, *Bull. Am. Math. Soc.* (New Series) **40** (2003) 155–178.

[24] C. Eckart and G. Young, The approximation of one matrix by another of lower rank, *Psychometrika* **1** (1936) 211–218.

[25] A. Einstein, B. Podolsky, and N. Rosen, Can quantum-mechanical description of physical reality be considered complete? *Phys. Rev.* **47** (1935) 777–780.

[26] J. A. Fessler and R. R. Nadakuditi, *Linear Algebra for Data Science, Machine Learning, and Signal Processing*, Cambridge University Press, Cambridge, 2024.

[27] S. J. Freedman and J. F. Clauser, Experimental test of local hidden-variable theories, *Phys. Rev. Lett.* **28** (1972) 938–941.

[28] J. H. Gallier and J. Quaintance, *Linear Algebra and Optimization with Applications to Machine Learning.* Volume I: Linear Algebra for Computer Vision, Robotics, and Machine Learning, Volume II: Fundamentals of Optimization Theory with Applications to Machine Learning, World Scientific, Singapore, 2020.

[29] D. Garisto, The universe is not locally real, and the Physics Nobel Prize winners proved it, *Scientific American*, October 6, 2022.

[30] G. H. Golub and J. M. Ortega, *Scientific Computing and Differential Equations*, Academic Press, Boston and New York, 1992.

[31] J. Gomide, R. Melo-Minardi, M. A, dos Santos, G. Neshich, W. Meira, Jr., J. C. Lopes, and M. Santoro, Using linear algebra for protein structural comparison and classification, *Genet. Mol. Biol.* **32** (2009) 645–651.

[32] A. Graham, *Non-negative Matrices and Applicable Topics in Linear Algebra*, John Wiley & Sons, New York, 1987.

[33] F. A. Graybill, *Introduction to Matrices with Applications in Statistics*, Wadsworth Publishing Company, Belmont, California, 1969.

[34] W. H. Greub, *Multilinear Algebra*, Springer-Verlag, New York, 1967.

[35] T. Guhr, A. Müller-Groeling, and H. A. Weidenmüller, Random-matrix theories in quantum physics: common concepts, *Phys. Rep.* **299** (1998) 189–425.

[36] P. R. Halmos, *Finite-Dimensional Vector Spaces*, 2nd ed., Springer-Verlag, New York, 1987.

[37] M. Haviv, *Linear Algebra for Data Science*, World Scientific, Singapore, 2023.

[38] J. D. Hidary, *Quantum Computing: An Applied Approach*, 2nd ed., Springer-Verlag, New York, 2021.

[39] N. J. Higham, Matrix nearness problems and applications, in *Applications of Matrix Theory*, edited by M. J. C. Gover and S. Barnett, Inst. Math. Appl. Conf. Ser. New Ser. **22**, Oxford University Press, New York, 1989.

[40] K. Hoffman and R. Kunze, *Linear Algebra*, Prentice-Hall, Eaglewood Cliffs, New Jersey, 1965.

[41] K. J. Horadam, *Hadamard Matrices and Their Applications*, Princeton University Press, Princeton, NJ, 2007.

[42] R. A. Horn and C. R. Johnson, *Matrix Analysis*, Cambridge University Press, Cambridge, New York, and Melbourne, 1985.

[43] R. Horodecki, P. Horodecki, M. Horodecki, and K. Horodecki, Quantum entanglement, *Rev. Mod. Phys.* **81** (2009) 865–942.

[44] P. Kaye, R. Laflamme, and M. Mosca, *An Introduction to Quantum Computing*, Oxford University Press, Oxford, 2007.

[45] P. Lancaster and M. Tismenetsky, *The Theory of Matrices*, 2nd ed., Academic Press, San Diego, New York, London, Sydney, and Tokyo, 1985.

[46] S. Lang, *Linear Algebra*, 3rd ed., Springer-Verlag, New York, 1987.

[47] S. Lang, *Algebra* (Graduate Texts in Mathematics, Volume 211), 3rd ed., Springer-Verlag, New York and Berlin, 2002.

[48] G. Latouche and R. Vaidyanathan, *Introduction to Matrix Analytic Methods in Stochastic Modeling*, Society of Industrial and Applied Mathematics, Philadelphia, 1999.

[49] P. D. Lax, *Linear Algebra and Its Applications*, John Wiley & Sons, Hoboken, NJ, 2007.

[50] W. Leontief, *Input-Output Economics*, Oxford University Press, New York, 1986.

[51] Y. I. Lyubich, E. Akin, D. Vulis, and A. Karpov, *Mathematical Structures in Population Genetics*, Springer-Verlag, New York, 2011.

[52] L. Maccone, A simple proof of Bell's inequality, *Am. J. Phys.* **81** (2013) 854–859.

[53] M. Marcus, *Finite Dimensional Multilinear Algebra*, Part II (Monographs and Textbooks in Pure and Applied Mathematics, Volume 23), Dekker, New York, 1975.

[54] I. Markovsky, *Low-Rank Approximation: Algorithms, Implementation, Applications*, Springer-Verlag, Berlin and New York, 2012.

[55] A. W. Marshall, I. Olkin, and B. C. Arnold, *Inequalities: Theory of Majorization and Its Applications*, Springer-Verlag, New York, 2010.

[56] M. L. Mehta, *Random Matrices*, Elsevier Academic Press, Amsterdam, 2004.

[57] C. Meyer, *Matrix Analysis and Applied Linear Algebra*, Society of Industrial and Applied Mathematics, Philadelphia, 2000.

[58] S. P. Meyn and R. L. Tweedie, *Markov Chains and Stochastic Stability*, Springer-Verlag, London, 1993; 2nd ed., Cambridge University Press, Cambridge, 2009.

[59] L. Mirsky, Symmetric gauge functions and unitarily invariant norms, *Q. J. Math.* **11** (1960) 50–59.

[60] L. Mirsky, A trace inequality of John von Neumann, *Monatshefte Math.* **79** (1975) 303–306.

[61] M. Nakahara and T. Ohmi, *Quantum Computing – From Linear Algebra to Physical Realizations*, CRC Press, Boca Raton, 2008.

[62] C. Nash and S. Sen, *Topology and Geometry for Physicists*, Academic Press, London and New York, 1983.

[63] M. A. Nielsen and I. L. Chuang, *Quantum Computation and Quantum Information*, Cambridge University Press, Cambridge, 2000.

[64] J.-W. Pan, D. Bouwmeester, M. Daniell, H. Weinfurter, and A. Zeilinger, (2000). Experimental test of quantum nonlocality in three-photon GHZ entanglement, *Nature* **403** (2000) 515–519.

[65] S. Roman, *Advanced Linear Algebra* (Graduate Texts in Mathematics, Volume 135), 3rd ed., Springer-Verlag, New York and Berlin, 2007.

[66] S. J. Russell and P. Norvig, *Artificial Intelligence: A Modern Approach*, 4th ed., Pearson, Hoboken, 2021.

[67] M. Schechter, *Principles of Functional Analysis*, Academic Press, New York and London, 1971.

[68] E. Schmidt, Zur Theorie der linearen und nichtlinearen Integralgleichungen, *Math. Ann.* **63** (1907) 433–476.

[69] D. R. Stinson, *Cryptography, Discrete Mathematics and Its Applications*, Chapman & Hall/CRC Press, Boca Raton, Florida, 2005.

[70] J. Stoer and R. Bulirsch, *Introduction to Numerical Analysis*, Springer-Verlag, New York, Heidelberg, and Berlin, 1980.

[71] G. Strang, *Linear Algebra and Learning from Data*, Wellesley-Cambridge Press, Wellesley, 2019.

[72] T. Tao, *Topics in Random Matrix Theory*, American Mathematical Society, Providence, 2012.

[73] C. H. Taubes, *Lecture Notes on Probability, Statistics, and Linear Algebra*, Department of Mathematics, Harvard University, Cambridge, Spring, 2010.

[74] P. Van Dooren and B. Wyman, *Linear Algebra for Control Theory*, IMA Volumes in Mathematics and Its Applications, Springer-Verlag, New York, 2011.

[75] C. von Westenholz, *Differential Forms in Mathematical Physics*, North-Holland, Amsterdam, New York, and Oxford, 1978.

[76] J. Watrous, *The Theory of Quantum Information*, Cambridge University Press, Cambridge, 2018.

[77] E. W. Weisstein, *CRC Concise Encyclopedia of Mathematics*, 3rd ed., CRC Press, Boca Raton, Florida, 2009.

[78] R. O. Wells, *Differential Analysis on Complex Manifolds*, Springer-Verlag, New York and Berlin, 1986.

[79] E. Wigner, Characteristic vectors of bordered matrices with infinite dimensions, *Ann. Math.* **62** (1955) 548–564.

[80] E. Wigner, On the distribution of the roots of certain symmetric matrices, *Ann. Math.* **67** (1958) 325–327.

[81] Y. Xu, *Linear Algebra and Matrix Theory* (in Chinese), 2nd ed., Higher Education Press, Beijing, 2008.

[82] T. Yokonuma, *Tensor Spaces and Exterior Algebra*, American Mathematical Society, Providence, 1992.

[83] K. Yosida, *Functional Analysis*, 6th ed., Springer-Verlag, Berlin and New York, 1980.

[84] L. W. Yu, N. L. Wang, and S. Kanemitsu, From linear algebra to quantum information, *Ann. Math. Phys.* **4** (2021) 032–047.

Index

For EU product safety concerns, contact us at Calle de José Abascal, 56–1°,
28003 Madrid, Spain or eugpsr@cambridge.org.

www.ingramcontent.com/pod-product-compliance
Ingram Content Group UK Ltd.
Pitfield, Milton Keynes, MK11 3LW, UK
UKHW020615191225
466164UK00017B/323